IUTAM Bookseries 41

The IUTAM Bookseries publishes the refereed proceedings of symposia organised by the International Union of Theoretical and Applied Mechanics (IUTAM).

Every two years the IUTAM General Assembly decides on the list of IUTAM Symposia. The Assembly calls upon the advice of the Symposia panels. Proposals for Symposia are made through the Assembly members, the Adhering Organizations, and the Affiliated Organizations, and are submitted online when a call is launched on the IUTAM website.

The IUTAM Symposia are reserved to invited participants. Those wishing to participate in an IUTAM Symposium are therefore advised to contact the Chairman of the Scientific Committee in due time in advance of the meeting. From 1996 to 2010, Kluwer Academic Publishers, now Springer, was the preferred publisher of the refereed proceedings of the IUTAM Symposia. Proceedings have also been published as special issues of appropriate journals. From 2018, this bookseries is again recommended by IUTAM for publication of Symposia proceedings.

Indexed in Ei Compendex and Scopus.

Xiaojing Zheng · S. Balachandar
Editors

Proceedings of the IUTAM Symposium on Turbulent Structure and Particles-Turbulence Interaction

Springer

Editors
Xiaojing Zheng
Center for Particle-laden Turbulence
Lanzhou University
Lanzhou, China

S. Balachandar
Mechanical and Aerospace Engineering
University of Florida
Gainesville, FL, USA

ISSN 1875-3507 ISSN 1875-3493 (electronic)
IUTAM Bookseries
ISBN 978-3-031-47257-2 ISBN 978-3-031-47258-9 (eBook)
https://doi.org/10.1007/978-3-031-47258-9

This Springer imprint is published by the registered company Springer Nature Switzerland AG
The registered company address is: Gewerbestrasse 11, 6330 Cham, Switzerland

Paper in this product is recyclable.

Preface

Particle-turbulence interactions are common in natural environments, such as sand/dust storms and pollutant dispersion, and in industrial applications such as fluidized beds and combustion. In recent years, more and more fluid dynamicists, as well as scholars in geophysics, energy and power engineering, chemical engineering and environmental science have contributed their efforts in advancing our knowledge on particle-turbulence interactions from various aspects. However, owing to the stochastic nature of particles, the strong nonlinearity of wall turbulence, and the complex coupling among them, the mechanisms and laws of the interactions among the carrier and disperse phases are not fully understood up to now. Particularly, when focusing on the multiphase turbulent flows in natural environments, it is more urgent than ever to carry out full-scale modeling and numerical simulations with improved accuracy, high-resolution experiments and comprehensive field measurements, so as to unravel the underlying flow mechanisms and to validate the reliability of the theoretical models and numerical simulations.

For this purpose, the IUTAM Symposium on Turbulent Structure and Particles-Turbulence Interaction was approved in 2018 by the IUTAM fluid mechanics panel to be held in Lanzhou, China. We, Prof. Xiaojing Zheng on behalf of the Center for Particle-laden Turbulence, Lanzhou University and Prof. Sivaramakrishnan Balachandar from the University of Florida, were assigned as the co-chairs of the symposium. One noteworthy point is that it is the second time we both are hosting an IUTAM Symposium. My (XZ's) group at Lanzhou University held the first one themed 'the Dynamics of Extreme Events Influenced by Climate Change' in 2012. Similarly, SB organized an IUTAM Symposium with Prof. Prosperetti in 2004 titled 'Recent developments in computational approaches to multiphase flow.' It is a rare thing in the history of IUTAM Symposium for the same group hosting a symposium with similar topics twice, which demonstrates the hydrodynamic mechanism behind extreme weather events and disasters typically exemplified by desertification and dust storms has become one of the hotspots and frontier of fluid dynamics. Also, it is a recognition from IUTAM to the advances of my (XZ's) team achieved on the high-Reynolds-number gas-solid turbulence with the background of the windblown sand movements in the last 20 years.

After overcoming the difficulties posed by the three-year COVID-19, the symposium was finally successfully held in-person at Lanzhou University, from July 7 to 11, 2023. Twenty-five invited lectures were presented at the symposium. The invited lecturers are Profs. *S. Balachandar*, *Rodney O. Fox*, *Lian Shen*, *Jesse Capecelatro*, *David Richter*, *Jiarong Hong*, *Eckart Meiburg*, and *Rui Ni* from the USA, Profs. *Detlef Lohse*, *Devaraj van der Meer* and *Yali Tang* from the Netherland, Prof. *Gaetano Sardina* from Sweden; Prof. *Alfredo Soldati* from Austria, Prof. *René van Hout* from Israel, Profs. *Cristian Marchioli* and *Francesco Picano* from Italy, and Prof. *Susumu Goto* from Japan, as well as Profs. *Xiaojing Zheng*, *Ke-Qing Xia*, *Lian-Ping Wang*, *Chunxiao Xu*, *Chao Sun*, *Zhaosheng Yu*, *Kun Luo* and *Wei Ge* from China. In order to give young scholars an opportunity to present their research progress in this field, an 'Early-career forum' with oral presentations by 24 young scholars and PhDs was organized on the last day of the conference, and additional 15 contributions were presented and exchanged in the form of posters. In total, this symposium brought together more than 100 participants attending the symposium.

During the symposium, all participants fully exchanged their latest research result in theoretical modeling, numerical computational methods and experimental techniques of multiphase turbulent flows. On this basis, the participating scholars also fully discussed and explored how to more widely apply the recent advances of multiphase turbulent flow in environment and industrial scenes, and how to promote more cooperation and sharing in the future, etc. Several consensuses on the further research were achieved: Firstly, it should be encouraged to pay more attention to real-world multiphase turbulence occurring in nature and industry, and reflect the role of multiphase turbulence research in solving practical problems; secondly, we should encourage the development and standardization of multiphase turbulence observation and experimental analysis methods and tools, so as to form basic standards that can be referred to by newcomers; thirdly, a unified paradigm of multiphase turbulence research, and the formation of consistent numerical methods should be encouraged; fourthly, we should encourage the development of systematic research including experimental observations, mechanism analysis and prediction of multiphase turbulence under extreme conditions; and finally more young scholars should be encouraged and attracted into the field of turbulent multiphase flow.

A general outcome was that the scientific communities learned much from one another, and this meeting brought new research ideas and new concepts on the characteristics of turbulence structure, mechanisms of particle-turbulence interactions, numerical calculations and experimental methods in turbulent multiphase flow. We received more than 60 contributions concerning the principal objectives of the symposium. After peer review, 22 full papers are accepted for the publication in the IUTAM Proceedings. On behalf of the Scientific Committee of the workshop, we would like to express our gratitude to all the invited lecturers and participants of the symposium for their contributions.

Xiaojing Zheng
S. Balachandar
Co-chairs

Contents

Two-Phase Structures in High-Reynolds-Number Sand-Laden Wall-Bounded Turbulence

Xiaojing Zheng(✉), Yanxiong Shi, and Hongyou Liu

Center for Particle-laden Turbulence, Lanzhou University, Lanzhou 730000, People's Republic of China
xjzheng@lzu.edu.cn

Abstract. Sandstorms, a high Reynolds number gas-solid two-phase wall turbulence, occur on the erodible sandy surfaces. The quantitative analysis of the nonstationary stages of sandstorms can help us better understand their occurrence and development, and provide scientific basis for optimally designing the prevention measures. It also contributes to quantitative research on the nonstationary sandstorms process, which is a open issue with multiphase modulation, multiphysical coupling, multiscale and highly nonlinear interlacing. In this study, a nonstationary signal processing method is proposed based on the empirical mode decomposition and adaptive segmentation. Then it is used to analyze the sandstorm data obtained from the occurrence and evolution process of sandstorms. Results revealed that the competitive evolutionary process of the main driving factors in the early, middle and late stages of sandstorms, as shear turbulence becomes dominant and is then suppressed by enhanced thermal stability. This results in some different characteristics and rules of turbulent motions between the nonstationary and stationary stages during the sandstorms: Even in the steady stage of sandstorms, the turbulence exhibits significant differences with the sand-free flow due to the present of dust/sand particles. Additionally, studies based on sandstorm data originally revealed that in addition to flow filed, the dust field is also found to exist coherent structures with streamwise scales larger than 5 times the surface layer thickness. The difference and connection between the dust concentration and the flow field structure is also revealed.

Keywords: nonstaionary · sandstorm · coherent structure

1 Introduction

Sandstorms are a typical strong wind-blown sand. Its formation and development is influenced by the interaction of atmospheric circulation, local atmosphere, near-surface airflow and erodible surface materials [1–4]. On the on hand, the sandstorms cause the loss of fine particulate matter and nutrient elements in

X. Zheng and S. Balachandar (Eds.): IUTAM 2023, IUTAM Bookseries 41, pp. 1–15, 2024.
https://doi.org/10.1007/978-3-031-47258-9_1

surface soil through wind erosion, which accelerates the expansion of desertification and bring various impacts and hazards to human health and life [5–7]. On the other hand, sandstorms carry a lot of minerals, nutrients and organic and inorganic matter to oceans from the world's deserts and semi-deserts, which is a principal driver of oceanic primary productivity and has a range of effects on marine biodiversity [8,9]. Studies have shown that sandstorms have hysteretic, continuous and long-term effects on the environment, biology, climate and human activities and health [10–17]. Analyzing the temporal and spatial characteristics and the causes of sandstorms can help us better understand their occurrence and evolution, as well as the sand/dust transport characteristics, which provide great help for dust forecasting, early warning and ecological environmental impact assessment [4,5].

The analysis of a large number of sandstorms data showed that there are obvious rising, steady and declining stages [18]. Its rising and declining stages are typical nonstationary process. The primary techniques commonly performed for nonstationary data analysis are the time-frequency analysis, including the short-time Fourier transform [19], Wigner-Ville distribution [20], the wavelet transform [21], the evolutionary spectra [22] and Hilbert-Huang transform [23], etc. However, it is noted that the analysis and processing of nonstationary signals requires not only time-frequency spectrum analysis, but also the statistical features, spatial structure and interactions. Additionally, statistical analysis needs suitable mean intervals. The averaging interval for the statistical analysis is usually selected as 1 h [24,25], 10 min [26,27]. However, a fixed length may defeat the purpose of segmentation [28,29] and these experienced intervals are often used in some specific situations, and can ignore the nonstationary characteristics of the signals. Therefore, for analysing nonstationary signals, appropriate analytical tool is necessary.

Sandstorms are high-Reynolds-number particle-laden wall-bounded turbulence with multiphase modulation, multiphysical coupling, multiscale and highly nonlinear interlacing. However, the existing studies on turbulence characteristics in sandstorms are mostly focused on the steady stage where the wind velocity remains constant, these flows are dominated by shear-driven turbulence [30–32]. The observation in atmospheric surfuce layer suggestes that not only the large scale coherent structure exists in the steady stage of sandstorms, but also the structures enhance the mass transport in streamwise [33,34]. Additionally, Zhao et al. [35] studied the turbulence coherent-structure sand emission mechanism and found that turbulence coherent-structure contribution to sand/dust emissions has a stable value which is more than 50%. However, the evolution of turbulence properties in the nonstationary stage of sandstorms is problematic due to the lack of powerful tools for analyzing and processing the nonstationary signals.

The atmospheric inhalable particles (PM10) is one of the main hazards affecting atmospheric environmental quality and human health [36,37]. Sandstorm processes are the main sources of PM10 emissions [38,39]. Therefore, studying the characteristics of PM10 during sandstorms can help to better understand the mechanism of dust transport. It also provide a scientific basis for reducing dust emissions [40,41]. It has been found that sand-laden flow has not only the particle streaks appearing on the ground, but also the very-large-scale sand structure similar to a very large-scale motion(VLSM) in the sandstorm flow field [34], which has been confirmed by LES results [42]. The presence of practical would also have an influence on the flow field. The addition of a large number of sand particles reduces the wall-normal length scale by approximately 40% and also reduces the velocity gradient in the flow field, which leads to an increase in the inclination angle [34]. Additionally, the addition of particles changes the distribution of energy between multi-scale turbulent motions, that is, sand particles increase the energy of VLSMs and the total streamwise turbulent kinetic energy in sand-laden flows [33]. Moreover, the particles produce a large damping in the degree of the inter-layer and multiscale amplitude modulation [43].

2 Observation of Sandstorm and Data Processing

The sandstorm observation data employed in this work were obtained from a high-tech observatory called the Qingtu Lake observation array (QLOA), which is located in a flat dry bed of Qingtu Lake between the Tengger Desert and the Badain Jaran Desert(as shown in Fig. 1(a)). The QLOA is composed of a 32 m high main tower and 22 lower towers that are 5 m in height and the arrangement of these towers is shown in Fig. 1(b). To capture the wind velocity and dust concentration signal, eleven sonic anemometers (Campbell CSAT3B, with a sampling frequency of 50 Hz) and aerosol monitors (TSI, Dustrak II-8530-EP, with a sampling frequency of 1 Hz) are installed on the main tower from 0.9 to 30 m in a logarithmic gradient. Thus, the QLOA can perform synchronous multipoint measurements of the wind velocity, temperature and sand concentration during sandstorms. More details of the QLOA and experimental apparatus can be found in [18,34].

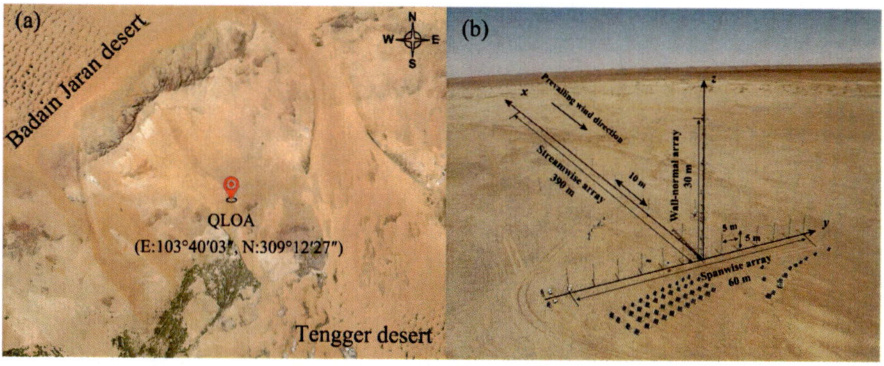

Fig. 1. The field observational site and measurement array: (a) satellite photographs of the QLOA site, (b) panoramic view of the QLOA.

Preliminary studies shown that the nonstationary wind speed series can be considered as an ergodic stochastic process composed of a time-varying mean wind speed and three fluctuation components. The variation of the mean wind velocity is the main factor causing the nonstationarity in the nonstationary stages of the sandstorms, and the turbulence information is contained in the fluctuation part. Therefore, a key issue of nonstationary wind velocity processing is how to extract an appropriate time-varying wind velocity from a wind time history recorded. In this work, the empirical modal decomposition (EMD) based on the local feature time scale of the data [23] was used to extract time-varying wind velocity, and the residual term and IMFs with average fluctuation periods greater than 1 h are taken as the time-varying mean flow.

However, the nonstationarity is not only reflected in the time-varying mean, but also its fluctuation shows time-varying. Let $u(t)$ be a nonstationary random process, if there are a set of instants points t_k such that $u(t_k+1), \cdots, u(t_{k+1})$ is stationary, then $u(t)$ is piecewise stationary [44,45]. Therefore, this work proposes an adaptive segmented stationary method based on the concept of piecewise stationary, which divides the nonstationary fluctuation into a number of stationary signals. It is achieved by performing a sequential procedure on the data, where two time lengths need to be defined. First, the period of the minimum segment is estimated considering the characteristic time of typical energetic eddies in the outer region of the wall-bounded turbulence, i.e., $\Delta t = 10\delta/U_c$(where U_c is the convection velocity taken as the local mean). Then, the adaptive segmented length of interval $\Delta t_i = (N_i - 1)\Delta t(N_i = 1, 2, \cdots)$ is determined by checking the stationarity and statistical convergence of the divided segment of data. The flow chart of the adaptive segmented stationary method is shown in Fig. 2. After applying the adaptive segmentation procedure to the turbulence signals extracted by removing the time-varying mean, a nonstationary signal can be divided into several segments of stationary signals,

$$u(t) = \{u_1(\Delta t_1), u_2(\Delta t_2), \cdots, u_m(\Delta t_m)\} \tag{1}$$

where u_i is stationary and $t = \Delta t_1 + \Delta t_2 + \cdots + \Delta t_m$.

3 Characteristics of Flow Structures in Sandstorms

The large coherent structures that are formed by two or more hairpin vortices aligned in flow direction are responsible for the generation and dissipation of turbulence, which is important for understanding the turbulence dynamics [46]. Many studies have suggested that coherent structures are widely present in stationary shear turbulence [31,32,47–49] and thermal turbulence [50,51]. In Fig. 3(a-c), the instantaneous velocity fluctuations of the rising, steady and declining stages of the sandstorm are shown, and the presence of high- and low-speed regions implies the existence of large-scale coherent structure. Additionally, two-point correlation contours of the streamwise velocity fluctuations in three stages are given in Fig. 3(d-f). The results indicates that the VLSMs exist in both the rising stage and the steady stage of the sandstorm, while the VLSMs

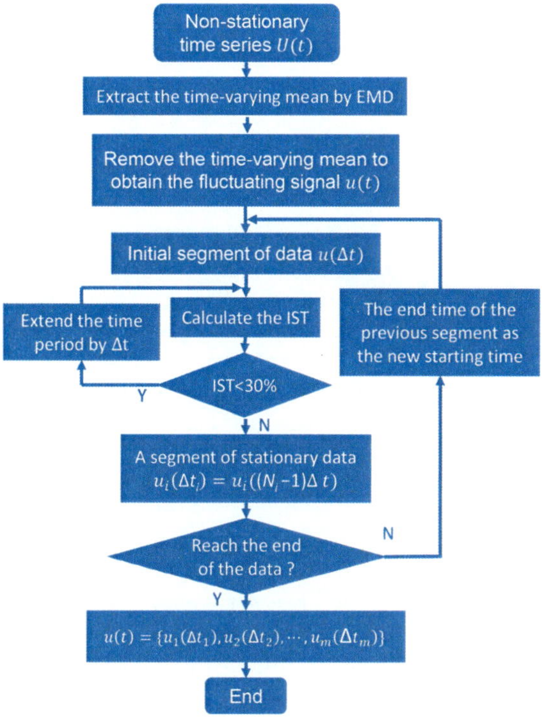

Fig. 2. Flow chart of the adaptive segmented stationary method.

gradually disappear in the declining stage. Moreover, there are significant morphological differences in the structure of the different stages, suggesting that the origination of this structure may be distinct.

The variation in the wavelength $(\lambda_x)_{max}/\delta$ of the most energetic scale with height in the different stages shown in Fig. 4(a) indicates that there are significant differences at three stages. In the early stage of sandstorms, the large coherent structures are generated by the breaking process of synoptic-scale structures, which indicates the variation of structures scale with height is insignificant. As the development of the sandstorms, the shearing breaking is enhanced, resulting in the scale varies significantly with height. When the sandstorms are attenuated, the large coherent structures cannot maintain their coherence due to the decreasing wind velocity.

To evaluate the energy contributions of the VLSMs at different stages of sandstorms, the variations in the energy fraction of the VLSMs with height at different stages are shown in Fig. 4(b). Albeit the VLSMs energy fraction increases with height in an approximately log-linear trend at different stages, it is observed that the energy fraction of the VLSMs in the rising stage is larger than that of the other two stages. In the rising stage, the energy fraction is 54% at the near ground and can reach up to 75% at the highest position($z/\delta = 0.2$, where

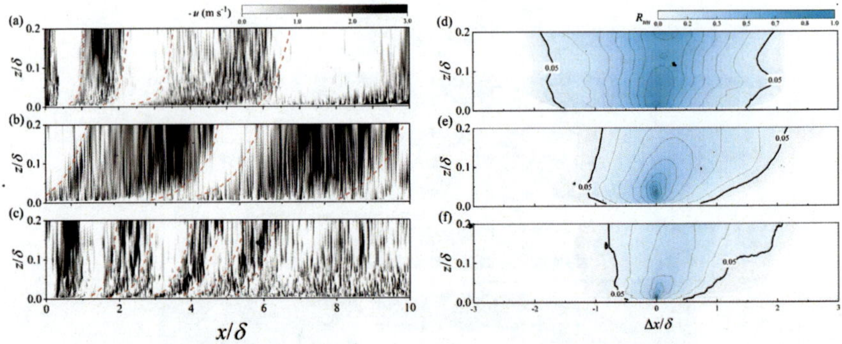

Fig. 3. The large coherent structures in the different stage of sandstorms. Instantaneous streamwise velocity fluctuations at (a) the rising stage, (b) the steady stage, and (c) the declining stage. Two-dimensional correlation contours of Ruu in the $x-y$ plane at (d) the rising stage, (e) the steady stage, and (f) the declining stage.

corresponds approximately to the top of the logarithmic region) in the present ASL observation. The value is significantly larger than the energy fraction of the VLSMs that were previously documented. The variation of energy fraction is qualitatively consistent with the results in the laboratory turbulent boundary layer [48,52] in the steady stage. In the declining stage, the energy fraction of VLSMs is reduced to 40%.

To clarify the evolution of the driving factors in the sandstorm, Fig. 5(a) show the ambient temperature θ and the Monin-Obukhov thermal stability parameter z/L during the entire sandstorm process. The z/L is given as

$$z/L = -\frac{\kappa z g \overline{w\theta'}}{\bar{\theta} u_\tau^3}, \tag{2}$$

where $\kappa = 0.41$ is the Kármán constant, z is the local height, $g = 9.8m \cdot s^{-2}$ is the gravitational acceleration, w and θ' is the fluctuating vertical velocity and temperature after removing the time-varying mean, $\overline{w\theta'}$ is the average vertical heat flux obtained by the covariance between w and θ', $\bar{\theta}$ the average temperature, and u_τ is the fraction velocity. It is shown that the thermal stability evolves from unstably stratification to neutral stratification and then to stably stratification as the sandstorm evolves. This indicates that the thermal turbulence is dominant in the rising stage of the sandstorm due to the impact of the cold front transit. As the wind velocity increases, the shearing gradually enhances and becomes the main driving factors. Finally, in the declining stage, the exhausted energy brought by the cold air mass leads to the decreased wind velocity and the atmosphere is transformed into one of stable stratification.

Figures 5(b) shows the quadrant analysis of the streamwise and vertical velocity fluctuations. It is shown that the frequency of all quadrant event is approximately equal in the rising and declining stages, but the Q2 and Q4 which are related to gradient-type motions are larger than the Q1 and Q3 which are related

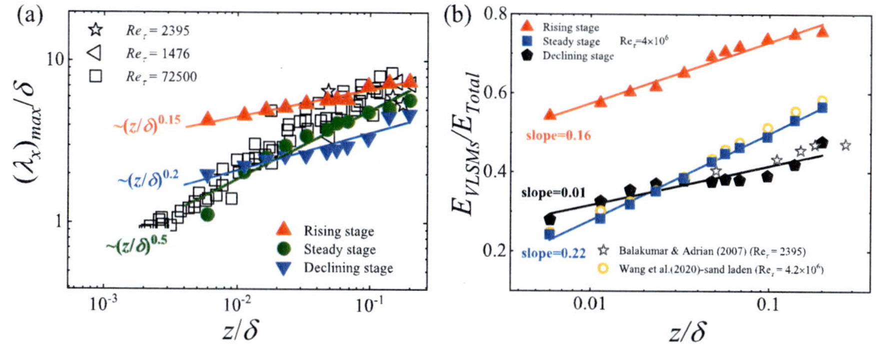

Fig. 4. (a) Variation in the wavelength $(\lambda_x)_{max}/\delta$ of the most energetic scale with height in the different stages. Hollow symbols are the turbulent boundary layer (TBL) results in Balakumar and Adrian [48] (Re_τ = 1476, 2395) and Vallikivi et al. [52] (Re_τ = 72500). (b) Variations in the energy fraction contributed by VLSMs to the total streamwise turbulent kinetic energy with height at different stages of sandstorms. The gray star and the yellow cycle are the results of the laboratory [48] and the atmosphere surface layer [33].

to countergradient-type motions in the steady stage [53](Wallace 2016). In other words, the Q1 and Q3 are increasing and Q2 and Q4 are decreasing in the non-stationary stage. The hairpin vortex generated by shear, which causes the lower velocity fluids to throw up and the higher velocity fluids to sweep down (i.e. "ejection" and "sweep" events). Therefore, the relatively equal ejection/sweep events and outward/inward interactions in the rising stage implies weak shear-driven turbulence. The enhancing ejection and sweep events represents strong shear-driven turbulence is in the steady stage. And the frequency of every quadrant event back to uniform in the declining stage indicates attenuated shear turbulence.

In summary, the evolution of the large coherent structures can be illustrated. At the beginning of a sandstorm, the cold air sinks towards the ground [54,55], and the synoptic-scale structures are gradually broken into large turbulent structures by the nonlinear interaction [18]. The energetic structures are dominated by the VLSMs which arises primarily from the top-down mechanism. With the continuous development and evolution of the sandstorm, the steady stage is mainly affected by the downward momentum transport [56,57] and the shear turbulence is dominant. The VLSMs are created by the bottom-up mechanism and the top-down mechanism, those structures have characteristics consistent with laboratory results. When the sandstorm fades, the effects of strong synoptic events are weakened and the exhausted energy brought by the cold air mass lead to the reduced wind velocity. The VLSMs cannot retain their coherence, and their scales became smaller and smaller.

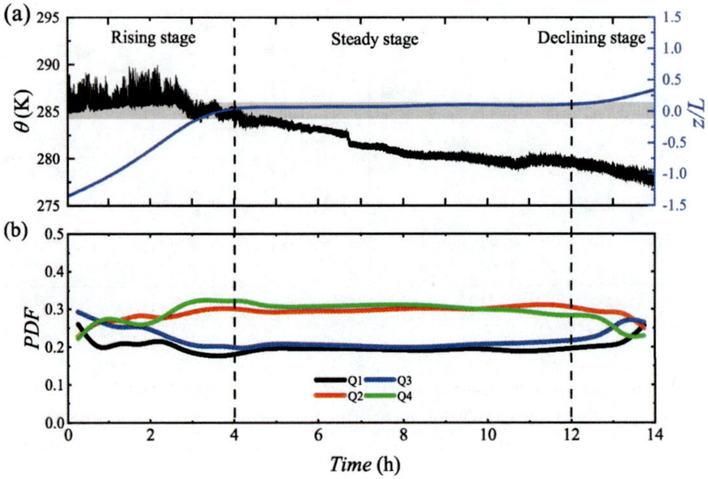

Fig. 5. (a) The ambient temperature(black line) and the Monin-Obukhov thermal stability parameter (blue line); (b) Evolution of the quadrant events at the height of $z = 5\,\mathrm{m}$ over the entire sandstorm process.

4 Characteristics of Dust Structures in Sandstorms

As the Reynolds number increases, the VLSMs becomes increasingly important in the transport of scalar material [58,59]. Observations in the atmospheric surface layer show that VLSMs are present in the sand-laden flow field [60]. However, similar VLSMs also exist in the sand concentration scalar field in wind-blown sand flows/sandstorms [34,40]. With the development of sandstorms, the structural characteristics of the scalar material in the flow field evolve with the change of wind velocity. Therefore, it is necessary to further consider the evolution of the structural features of dust concentration in the scalar field during the sandstorms.

Figure 6(a–c) and Fig. 6(d–f) show the instantaneous PM10 concentration and the PM10 concentration correlation contours at the rising, steady and declining stages. It is can found that the PM10 concentration field also appear in the high- and low- concentration regions which are similar to the high- and low-speed regions in the flow field, implying that there is also a large-scale coherent structure in the PM10 concentration field. However, the PM10 concentration field does not show a coherent structure at higher positions in the rising and declining stages. This is beacuse the low wind velocity is insufficient to transport dust to higher locations in the nonstationary stage. In rising stage, the turbulence VLSMs are mainly derived from the synoptic-scale fluctuations by the top-down mechanism, which cannot transport dust particles upward. When the sandstorm weakens, the system energy is gradually depleted and the large-scale turbulence coherent structures are no longer maintained, which weakens the flow field's ability to transport dust particles. Meanwhile, under the action of gravity, dust particles begins to settle toward the ground. Therefore, there is only coherent structures of dust near the ground.

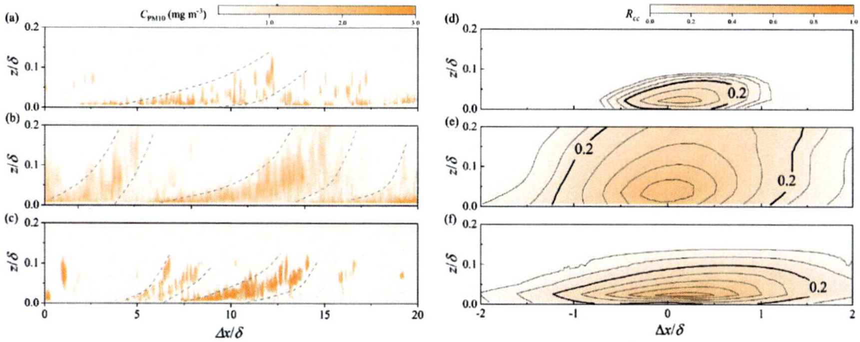

Fig. 6. The large coherent structures in the different stages of sandstorms. Instantaneous PM10 concentration contours at (a) the rising stage, (b) the steady stage, and (c) the declining stage. Two-dimensional correlation maps of Rcc in the $x-y$ plane at (d) the rising stage, (e) the steady stage, and (f) the declining stage.

The VLSMs are associated with the wavelengths of the lower wavenumber peaks in the pre-multiplied spectra [52]. Figure 7 (a)-(c) show the pre-multiplied spectra of the dust concentration $k_x\Phi_{cc}/\sigma_{cc}$ versus the streamwise wavenumber $k_x\delta$ at different heights in different stages of the sandstorm. The distinct peaks of the pre-multiplied spectra appear in the lower wavenumber region (on the left side of the vertical line) in the steady stage, but not exist in both the rising stage and declining, indicating that the VLSMs exist in the steady stage rather than rising and declining stage. Combining instantaneous concentration fluctuations and two-point correlation contours of the dust concentration, it is shown that the VLSMs of dust concentration have an evolution process that the VLSMs are absent in the early stage to gradually appear with the development of sandstorm and finally disappear gradually with the extinction of the sandstorm.

To gain insight into the evolution of the large coherent structure of the dust concentration and the difference with the flow structure, Fig. 8(a) shows the evolution of the lower wavenumber peak in the pre-multiplied spectra of dust concentration(yellow solid line) and streamwise velocity fluctuations(blue solid line) at a height of 3.49 m with time during the sandstorms. It is seen from Fig. 8(a) that there are significant difference between the evolution of the VLSMs, which implies their origin may be different. With the increase of wind velocity, the dust structures gradually appear and its scale becomes progressively larger with the development of the sandstorm. This is contrary to the results that the VLSMs are initially present in the flow field at the beginning and then gradually become smaller. When the sandstorm evolves to steady stage dominated by dynamical turbulence, the scales of the dust structure and turbulence structure remain constant with time, but the scale of dust structure is significantly larger than the scale of flow field structure, which can be explained by the reduced coherence of the turbulence structures due to the small-scale turbulent motions near the ground. In the declining stage, both the turbulence structure and the

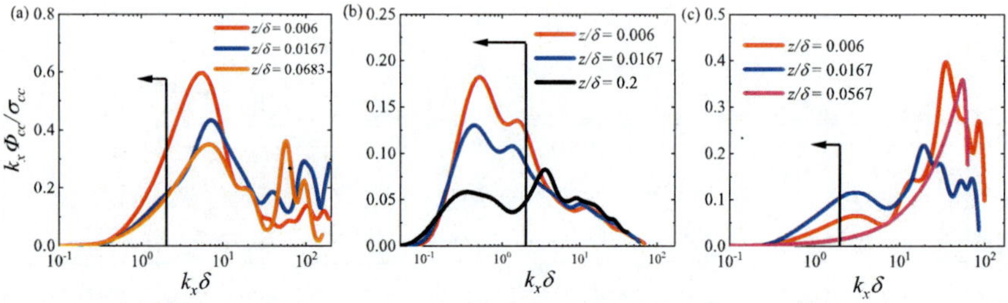

Fig. 7. Pre-multiplied spectra of dust concentration $k_x\Phi_{cc}/\sigma_{cc}$(where $k_x = 2\pi/\lambda_x$ denotes the streamwise wavenumber, λ_x is the streamwise wavelength, Φ_{cc} is the spectral density of the dust concentration, and σ_{cc} is the variance of the dust concentration) versus streamwise wavenumber $k_x\delta$, where (a) is the rising stage, (b) is the steady stage, and (c) is the declining stage.

dust structure gradually disappear as the wind velocity decreases. The energy brought by the cold air mass is gradually exhausted, which makes the large-scale structure cannot retain its coherence [18].

Figure 8(b) shows the variations in the energy fraction contributed by the VLSMs of the dust concentration and streamwise velocity fluctuations. During the rising stage, the energy fraction of VLSMs in dust concentration field is comparable to that of streamwise velocity fluctuations near the ground. However, as the height increases, the energy fraction of VLSMs in dust concentration decreases and becomes smaller than that of the flow VLSMs. The energy fraction of the dust VLSMs is only 10% at the highest measurement point, while it reaches 75% at the same position in the flow field. This provided the evidence for the different origin of the two structures. In the steady stage, the energy fraction of the VLSMs in the dust concentration at the near ground is much larger than that in the streamwise velocity fluctuations, and the energy fraction of both is gradually equal as the height increases. In the declining stage, the varying trends are similar to those of the steady stage, but the energy fraction of the VLSMs in the dust concentration is smaller. Additionally, the energy fraction of VLSMs in the dust concentration at different stages gradually decreases with the height, which is exactly different from variation pattern of the streamwise velocity fluctuations. The difference between dust and flow field structure is related to the release of dust and the turbulence mixing at the atmospheric surface layer. During the sandstorms, when large scale structures sweep across the sandy surface, the strong shearing effect of the large scale structures drives the surface to release a large amounts of sand and dust particles, which are mixed and transported in turbulence motions. Near the ground, the small-scale turbulence motions enhanced the dust mixing, which decreases the small-scale fluctuations and increases the large-scale fluctuations of dust concentration. This leads to the spatial scale of dust is large than that of the turbulence. As the height increases, the dust mixing is almost uniform, which leads to the large scales of the dust fluctuations are approximately equal to that of the streamwise velocity fluctuations, and the streamwise scale of the dust gradually also equal to that of the streamwise velocity.

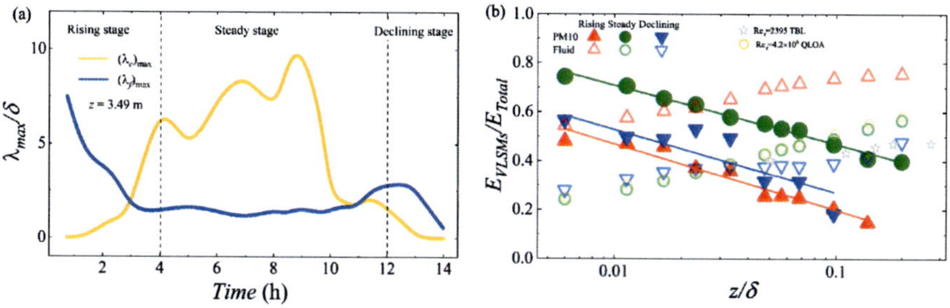

Fig. 8. (a) The evolution of the lower wavenumber peak in pre-multiplied spectra during the sandstorms, where the blue and yellow lines are the results of the steamwise velocity and dust concentration, respectively. (b) Variations in the energy fraction contributed by VLSMs with height at different stages of sandstorms, where the solid and hollow markers are the results of the steamwise velocity and dust concentration, respectively.

5 Conclusions

In this work, a nonstationary data processing method based on EMD and adaptive segmentation was proposed, which can effectively extract the time-varying mean values of nonstationary signals and segment the data according to the stationary index and statistical convergence. The multiphysics data of the entire sandstorms process is analyzed by this method. The results show that the large-scale coherent structures of turbulence and dust exist in the sandstorm, but a competitive evolutionary process of dynamic turbulence and thermal turbulence makes these structures have different spatial structure morphology and evolution patterns in different stages. The rising stage is dominated by thermal turbulence. The VLSMs of streamwise velocity fluctuations are apparent at the beginning, with an energy fraction of 75%, which is larger than the previously documented results. In contrast, the VLSMs of dust concentration are evolved gradually and their energy fraction is only 10%. With the development of sandstorm, the thermal turbulence is weaken and the steady stage is dominated by shear turbulence. There are the VLSMs with $L_x > 3\delta$ in turbulence and dust concentration field. However, the dust structures is significantly larger than the turbulence structures at the near ground and the energy fraction of dust VLSMs decreases with height, which is contrary to the results of turbulence. A plausible explanation is that the small-scale turbulence motions enhanced the dust maxing at the near ground, resulting in a decrease of small-scale and an increase of large-scale in the dust concentration fluctuations. In the declining stage, the stable thermal condition dampens the turbulence, resulting in attenuating the dust and turbulence structure scales and decreasing the energy fraction of the VLSMs. Additionally, the differences in turbulence and dust structures suggest that their origins are different. At the beginning of the sandstorm, the VLSMs in flow field are mainly dominated by the top-down mechanism, i.e., formed by breaking of synoptic-scale turbulent structures. As the sandstorm develops, the shear enhances and the dynamic turbulence gradually becomes dominant, which indicated that the VLSMs is dominated by both the bottom-up mechanism and the top-down mechanism. At the end of the sandstorm, the wind velocity decreases and the shear weakens, implying that the bottom-up mechanism also weakens. However, the VLSMs of dust concentration are dominated by the bottom-up mechanism throughout the sandstorm, which is different from the origination of the flow VLSMs that are dominated by both mechanisms with a competitive evolutionary process.

References

1. Knippertz, P.: Meteorological aspects of dust storms. In: Knippertz, P., Stuut, J.-B.W. (eds.) Mineral Dust, pp. 121–147. Springer, Dordrecht (2014). https://doi.org/10.1007/978-94-017-8978-3_6
2. Hamzeh, N., Karami, S., Kaskaoutis, D.G., Tegen, I., Moradi, M., Opp, C.: Atmospheric dynamics and numerical simulations of six frontal dust storms in the middle east region. Atmosphere **12**, 125 (2021). https://doi.org/10.3390/atmos12010125
3. Gui, K., et al.: Record-breaking dust loading during two mega dust storm events over northern China in March 2021: aerosol optical and radiative properties and meteorological drivers. Atmos. Chem. Phys. **22**, 7905–7932 (2022). https://doi.org/10.5194/acp-22-7905-2022
4. Miri, A., Maleki, S., Middleton, N.: An investigation into climatic and terrestrial drivers of dust storms in the Sistan region of Iran in the early twenty-first century. Sci. Total Environ. **757**, 143–952 (2021). https://doi.org/10.1016/j.scitotenv.2020.143952
5. Gasch, P., et al.: Revealing the meteorological drivers of the September 2015 severe dust event in the Eastern Mediterranean. Atmos. Chem. Phys. **17**, 13 573-13 604 (2017). , acp-17-13573-2017
6. Xu, C., et al.: Spatiotemporal variations and driving factors of dust storm events in northern China based on high-temporal-resolution analysis of meteorological data (1960–2007). Environ. Pollut. **260**, 114 084 (2020). https://doi.org/10.1016/j.envpol.2020.114084
7. Hamzeh, N., Karami, S., Opp, C., Fattahi, E., Francois, V.: Spatial and temporal variability in dust storms in the Middle East, 2002–2018: three case studies in July 2009. Arab. J. Geosci. **14**, 538 (2021). https://doi.org/10.1007/s12517-021-06859-0
8. Middleton, N., Kappelle, M.: Impacts of sand and dust storms on oceans: a scientific environmental assessment for policy makers, Technical report, United Nations Environment Programme, Nairobi, Kenya (2020). https://www.unep.org/resources/report/impacts-sand-and-dust-storms-oceans
9. Behzad, H., Ohyanagi, H., Alharbi, B., Ibarra, M., Alarawi, M., Saito, Y., Duarte, C.M., Bajic, V., Mineta, K., Gojobori, T.: A cautionary signal from the Red Sea on the impact of increased dust activity on marine microbiota. BMC Genomics **23**, 277 (2022). https://doi.org/10.1186/s12864-022-08485-w
10. Li, J., Wang, Z., Zhuang, G., Luo, G., Sun, Y., Wang, Q.: Mixing of Asian mineral dust with anthropogenic pollutants over East Asia: a model case study of a super-duststorm in March 2010. Atmos. Chem. Phys. **12**, 7591–7607 (2012). https://doi.org/10.5194/acp-12-7591-2012

11. Formenti, P., et al.: Mapping the physico-chemical properties of mineral dust in western Africa: mineralogical composition. Atmos. Chem. Phys. **14**, 10 663-10 686 (2014). https://doi.org/10.5194/acp-14-10663-2014
12. Lovett, C., Sowlat, M.H., Saliba, N.A., Shihadeh, A.L., Sioutas, C.: Oxidative potential of ambient particulate matter in Beirut during Saharan and Arabian dust events. Atmos. Environ. **188**, 34–42 (2018). https://doi.org/10.1016/j.atmosenv.2018.06.016
13. Oduber, F., et al.: Unusual winter Saharan dust intrusions at Northwest Spain: air quality, radiative and health impacts. Sci. Total Environ. **669**, 213–228 (2019). https://doi.org/10.1016/j.scitotenv.2019.02.305
14. Shao, Y., Klose, M., Wyrwoll, K.-H.: Recent global dust trend and connections to climate forcing. JJ. Geophys. Res. Atmos. **118**, 11 107-11 118 (2013). https://doi.org/10.1002/jgrd.50836
15. Kosmopoulos, P.G., et al.: Dust impact on surface solar irradiance assessed with model simulations, satellite observations and ground-based measurements. Atmos. Meas. Tech. **10**, 2435–2453 (2017). https://doi.org/10.5194/amt-10-2435-2017
16. Mahowald, N.M., et al.: Atmospheric global dust cycle and iron inputs to the ocean. Global Biogeochem. Cy. **19**, GB4025 (2005). https://doi.org/10.1029/2004GB002402
17. Lawrence, C.R., Neff, J.C.: The contemporary physical and chemical flux of aeolian dust: a synthesis of direct measurements of dust deposition. Chem. Geol. **267**, 46–63 (2009). https://doi.org/10.1016/j.chemgeo.2009.02.005
18. Liu, H., Shi, Y., Zheng, X.: Evolution of turbulent kinetic energy during the entire sandstorm process. Atmos. Chem. Phys. **22**, 8787–8803 (2022). https://doi.org/10.5194/acp-22-8787-2022
19. Potter, R.K.: Visible speech. Am. Ann. Deaf **91**, 447–452 (1947)
20. Ville, J.: Théorie et applications de la notion de signal analytique. Cables et transmission **2**, 61–47 (1948)
21. Morlet, J., Arens, G., Fourgeau, E., Giard, D.: Wave propagation and sampling theory-Part II: sampling theory and complex waves. Geophysics **47**, 222–236 (1982). https://doi.org/10.1190/1.1441329
22. Priestley, M.B.: Evolutionary spectra and non-stationary processes. J. R. Stat. Soc. B **27**, 204–229 (1965). https://doi.org/10.1111/j.2517-6161.1965.tb01488.x
23. Huang, N.E., et al.: The empirical mode decomposition and the Hilbert spectrum for nonlinear and non-stationary time series analysis. P. Roy. Soc. A-Math. Phy. **454**, 903–995 (1998). https://doi.org/10.1098/rspa.1998.0193
24. Chen, J., Xu, Y.: On modelling of typhoon-induced non-stationary wind speed for tall buildings. Design Tall Spec. Build. **13**, 145–163 (2004). https://doi.org/10.1002/tal.247
25. Wang, S., Zhao, L., Cao, S., Zhang, Y., Yin, F.: A comparison study on gust factor considering non-stationary effects under typhoon and monsoon conditions. Adv. Struct. Eng. **21**, 1853–1864 (2018). https://doi.org/10.1177/1369433218758474
26. Tamura, Y., Kawai, H., Uematsu, Y., Marukawa, H., Fujii, K., Taniike, Y.: Wind load and wind- induced response estimations in the recommendations for loads on buildings, AIJ 1993. Eng. Struct. **18**, 399–411 (1996). https://doi.org/10.1016/0141-0296(95)00121-2
27. Wang, G.H., Zheng, X.J.: Very large scale motions in the atmospheric surface layer: a field investigation. J. Fluid Mech. **802**, 464–489 (2016). https://doi.org/10.1017/jfm.2016.439

28. Hassanpour, H., Shahiri, M., and IEEE: Adaptive segmentation using wavelet transform. In: 2007 International Conference on Electrical Engineering, pp. 365–369, IEEE, IEEE, New York (2007). <Go to ISI>://WOS:000254289200068
29. Azami, H., Sanei, S., Mohammadi, K., Hassanpour, H.: A hybrid evolutionary approach to segmentation of non-stationary signals. Digit. Signal Process. **23**, 1103–1114 (2013). https://doi.org/10.1016/j.dsp.2013.02.019
30. Smits, A.J., McKeon, B.J., Marusic, I.: High-Reynolds number wall turbulence. Annu. Rev. Fluid Mech. **43**, 353–375 (2011). https://doi.org/10.1146/annurev-fluid-122109-160753
31. Jiménez, J.: Coherent structures in wall-bounded turbulence. J. Fluid Mech. **842**, P1 (2018). https://doi.org/10.1017/jfm.2018.144
32. Marusic, I., Monty, J.P.: Attached eddy model of wall turbulence. Annu. Rev. Fluid Mech. **51**, 49–74 (2019). https://doi.org/10.1146/annurev-fluid-010518-040427
33. Wang, G.H., Gu, H.H., Zheng, X.J.: Large scale structures of turbulent flows in the atmospheric surface layer with and without sand. Phy. Fluids **32**, 106–604 (2020). https://doi.org/10.1063/5.0023126
34. Liu, H., Zheng, X.: Large-scale structures of wall-bounded turbulence in single- and two-phase flows: advancing understanding of the atmospheric surface layer during sandstorms. Flow **1**, E5 (2021). https://doi.org/10.1017/flo.2021.6
35. Zhao, J.H., Long, X., Zhang, F., Yang, Y.L., Liu, S.X., Yun, L.: The role of turbulent coherent structure in sand-dust emissions in a sanddust storm of the middle China-Mongolia regime, Chinese. J. Geophys.-Ch. **63**, 3967–3980 (2020). https://doi.org/10.6038/cjg2020O0179
36. Kampa, M., Castanas, E.: Human health effects of air pollution. Environ. Pollut. **151**, 362–367 (2008). https://doi.org/10.1016/j.envpol.2007.06.012
37. Kim, K., Kabir, E., Kabir, S.: A review on the human health impact of airborne particulate matter. Environ. Int. **74**, 136–143 (2015). https://doi.org/10.1016/j.envint.2014.10.005
38. Choi, H., Zhang, Y., Kim, K.: Sudden high concentration of TSP affected by atmospheric boundary layer in Seoul metropolitan area during duststorm period. Environ. Int. **34**, 635–647 (2008). https://doi.org/10.1016/j.envint.2007.12.023
39. Thalib, L., Al-Taiar, A.: Dust storms and the risk of asthma admissions to hospitals in Kuwait. Sci. Total Environ. **433**, 347–351 (2012). https://doi.org/10.1016/j.scitotenv.2012.06.082
40. Zheng, X.J., Zhang, J.H., Wang, G.H., Liu, H.Y., Zhu, W.: Investigation on very large scale motions (VLSMs) and their influence in a dust storm. Sci. China-Phys. Mech. Astron. **56**, 306–314 (2013). https://doi.org/10.1007/s11433-012-4985-1
41. Zhang, F., et al.: Seasonal variations and chemical characteristics of PM2.5 in Wuhan, central China. Sci. Total Environ. **518–519**, 97–105 (2015). https://doi.org/10.1016/j.scitotenv.2015.02.054
42. Zhang, Y.Y., Hu, R.F., Zheng, X.J.: Large-scale coherent structures of suspended dust concentration in the neutral atmospheric surface layer: a large-eddy simulation study. Phy. Fluids **30**, 046 601 (2018). https://doi.org/10.1063/1.5022089
43. Liu, H., He, X., Zheng, X.: Amplitude modulation in particle-laden atmospheric surface layers. J. Fluid Mech. **957**, A14 (2023). https://doi.org/10.1017/jfm.2022.1092
44. Lavielle, M.: Optimal segmentation of random processes. IEEE T. Signal Proces. **46**, 1365–1373 (1998). https://doi.org/10.1109/78.668798
45. Djuric, P.M., Kay, S.M., Boudreaux-Bartels, G.F.: Segmentation of nonstationary signals. In: [Proceedings] ICASSP-92: 1992 IEEE International Conference on

Acoustics. Speech, and Signal Processing, San Francisco, vol. 5, pp. 161–164. IEEE Computer Society (1992)
46. Robinson, S.K.: Coherent motions in the turbulent boundary layer. Annu. Rev. Fluid Mech. **23**, 601–639 (1991). https://doi.org/10.1146/annurev.fl.23.010191.003125
47. Guala, M., Hommema, S.E., Adrian, R.J.: Large-scale and very-large-scale motions in turbulent pipe flow. J. Fluid Mech. **554**, 521–542 (2006). https://doi.org/10.1017/S0022112006008871
48. Balakumar, B., Adrian, R.: Large- and very-large-scale motions in channel and boundary-layer flows. Philos. T. R. Soc. A **365**, 665–681 (2007). https://doi.org/10.1098/rsta.2006.1940
49. Adrian, R.J.: Hairpin vortex organization in wall turbulencea. Physics of Fluids **19**, 041 301 (2007). https://doi.org/10.1063/1.2717527
50. Sun, C., Xi, H., Xia, K.: Azimuthal symmetry, flow dynamics, and heat transport in turbulent thermal convection in a cylinder with an aspect ratio of 0.5. Phys. Rev. Lett. **95**, 074 502 (2005). https://doi.org/10.1103/PhysRevLett.95.074502
51. Xie, Y., Ding, G., Xia, K.: Flow topology transition via global bifurcation in thermally driven turbulence. Phys. Rev. Lett. **120**, 214 501,c (2018). https://doi.org/10.1103/PhysRevLett.120.214501
52. Vallikivi, M., Ganapathisubramani, B., Smits, A.: Spectral scaling in boundary layers and pipes at very high Reynolds numbers. J. Fluid Mech. **771**, 303–326 (2017). https://doi.org/10.1017/jfm.2015.181
53. Wallace, J.M.: Quadrant analysis in turbulence research: history and evolution. Annu. Rev. Fluid Mech. **48**, 131–158 (2016). https://doi.org/10.1146/annurev-fluid-122414-034550
54. Dragani, W.C.: A feature model of surface pressure and wind fields associated with the passage of atmospheric cold fronts. Comput. Geosci. **25**, 1149–1157 (1999). https://doi.org/10.1016/S0098-3004(99)00074-6
55. Helfer, K.C., Nuijens, L.: The morphology of simulated trade-wind convection and cold pools under wind shear. J. Geophys. Res.- Atmos. **126**, e2021JD035 148 (2021). https://doi.org/10.1029/2021JD035148
56. Todd, M.C., Washington, R., Raghavan, S., Lizcano, G., Knippertz, P.: Regional model simulations of the Bodélé Low-Level Jet of Northern Chad during the Bodélé Dust Experiment (BoDEx 2005). J. Climate **21**, 995–1012 (2008). https://doi.org/10.1175/2007JCLI1766.1
57. Kaskaoutis, D., et al.: Meteorological aspects associated with dust storms in the Sistan region, southeastern Iran. Clim. Dyn. **45**, 407–424 (2015). https://doi.org/10.1007/s00382-014-2208-3
58. Marusic, I., McKeon, B.J., Monkewitz, P.A., Nagib, H.M., Smits, A.J., Sreenivasan, K.R.: Wall-bounded turbulent flows at high Reynolds numbers: recent advances and key issues. Phy. Fluids, **22** (2010). https://doi.org/10.1063/1.3453711
59. Jacob, C., Anderson, W.: Conditionally averaged large-scale motions in the neutral atmospheric boundary layer: insights for aeolian processes. Boundary-Layer Meteorol. **162**, 21–41 (2017). https://doi.org/10.1007/s10546-016-0183-4
60. Marusic, I.and Hutchins, N.: Study of the log-layer structure in wall turbulence over a very large range of reynolds number. Flow Turbul. Combust **81** (2008). https://doi.org/10.1007/s10494-007-9116-0

Wake of a Finite-Size Particle in Wall Turbulence Over a Rough Bed

Xing Li[1,2], S. Balachandar[2(✉)], Hyungoo Lee[3], and Bofeng Bai[1(✉)]

[1] State Key Laboratory of Multiphase Flow in Power Engineering, School of Chemical Engineering and Technology, Xi'an Jiaotong University, Xi'an 710049, China
bfbai@mail.xjtu.edu.cn

[2] Mechanical & Aerospace Engineering, University of Florida, Gainesville, FL 32608, USA
bala1s@ufl.edu

[3] Power Reactor Development Division, Korea Atomic Energy Research Institute, Daedeok-Daero 989, Yuseong-Gu, Daejeon 305–353, Republic of Korea

Abstract. For finite-size particles in turbulence, the particle's wake causes an additional force on the particle, alters the ambient turbulence and thus changes the properties of the two-phase flow. The mechanism of the interaction between the wake and turbulence, is still ambiguous. Here, we performed fully resolved direct numerical simulations with an immersed boundary method and investigated the wake of a stationary finite-size particle in wall turbulence over a rough bed. We focused on the time-dependent features of the wake, and studied the effect of the ambient turbulence on the wake and the turbulence modulation by the wake. Results show that there exists a strong correlation between the wake and the oncoming turbulent flow, which is beneficial for the precise prediction of the wake-induced force on the particle and the turbulence modulation by the wake. Turbulence augmentation is observed far downstream, while turbulence attenuation region lies near the sphere and the wall where viscous effect is dominant. We also found the effect of roughness is prominent in the near-wall region and is weakened far from the wall. In addition, the vortex shedding happens on only one side of the sphere, and the Strouhal number based on the particle center velocity is around 0.262, about twice that in an unbounded flow. The mechanisms we revealed here provide a basis for building the model of the wake-induced force on a finite-size particle and the turbulence modulation by the particle's wake.

Keywords: Particle Wake · Wall Turbulence · Turbulence Modulation

1 Introduction

Finite-size particles, which are comparable to or larger than the Kolmogorov microscale of ambient turbulence, are usually encountered in natural and industrial applications, such as the seabed, aero-engines, fluidized beds, etc. The particle Reynolds number Re_p of a finite-size particle is usually larger than unity and even of the order of $Re_p \sim 100$ or above. In this case, the wake oscillation or vortex shedding of the finite-size particle can induce a non-negligible force on the particle as well as modulate the downstream

X. Zheng and S. Balachandar (Eds.): IUTAM 2023, IUTAM Bookseries 41, pp. 16–39, 2024.
https://doi.org/10.1007/978-3-031-47258-9_2

turbulence. The wake is an important aspect of dispersed multiphase flows and can be the main factor in deciding the qualitative and quantitative behavior of multiphase systems [1]. For example, in case of fuel droplets in engines and large particles or bubbles in chemical equipment, their performance can be profoundly affected by the particle wakes.

The problem of the wake of a particle in laminar flows has been widely investigated and has formed the basis of analyzing the more complex situation of particle-laden turbulence [2–9]. Generally, the wake evolves from the axisymmetric stationary state to a planar-symmetric double-threaded stationary state and then to a periodic vortex shedding state as the particle Reynolds number increases, and finally towards chaos. Specifically, for a particle in a uniform flow, the wake maintains the stationary axisymmetric state up to $Re_p \approx 212$. Then, the first bifurcation occurs and leads to a stationary state that is no longer axisymmetric but planar symmetric. A second bifurcation, which is a Hopf bifurcation, occurs at $Re_p \approx 273$ and leads to a periodic oscillatory state, and the wake still preserves planar symmetry. The Strouhal number of the periodic oscillation is about 0.127. As the particle Reynolds number grows to $Re_p \approx 285$, the wake oscillation evolves into the vortex shedding. A tertiary Hopf bifurcation occurs at $Re_p \approx 320$, leading to a secondary shedding frequency that is about three times smaller than the primary one. Beyond $Re_p \approx 355$, the wake becomes fully three-dimensional and chaotic [5]. Recently, 3D vortex structures of the sphere wake have been measured via the tomo-PIV method, and different structures were observed in the cases of a stationary and moving sphere [10, 11]. A numerical research was also conducted to investigate the stability and sensitivity of the secondary bifurcation in the sphere wake [12]. Particle wake dynamics at higher Reynolds numbers have also been widely investigated [13–17]. Recently, David et al. investigated the wake of a sphere with a rough surface, and found that the roughness lowers the critical Re_p for the clear, temporally averaged recirculating wake to disappear [18].

In wall-bounded laminar flows, the evolution of the particle wake follows a similar path. The differences lie in the loss of the axisymmetry, which influences the critical Reynolds numbers of the Hopf bifurcations and the structure of the wake. Zeng et al. [19] reported an asymmetric double-threaded wake at a lower critical Reynolds number than that in the unbounded case due to the presence of the nearby wall. Zeng et al. [20] extended this research to examine gap ratio down to 0.005 sphere diameter, where the particle is practically sitting on the bottom wall. Stewart et al. [21] examined the wake of a rolling sphere near a wall, and they found that the critical Reynolds number for transition to the unsteady state depends on the rotation rate, and it is much lower than the critical Reynolds number in the unbounded case. Besides, the Strouhal number of vortex shedding is around 0.12, slightly lower than that in the unbounded case. Rao et al. [22] found the wake of a rolling sphere near a wall to become unsteady at $Re_p \approx 139$, and a lateral oscillation occurs at $Re_p \approx 192$, whose frequency is (3/7)th of the primary shedding frequency.

In turbulence, the wake is prominently influenced by the oncoming turbulent disturbance and is chaotic, unlike its counterpart in a laminar flow which is regular and symmetric. In a turbulent flow, the interaction between the particle wake and the ambient flow is much more complicated. First, because the oncoming flow is turbulent, the onset of wake unsteadiness occurs earlier [23, 24]. Second, the turbulence influences the wake-induced force on a particle. The oscillating wake or the vortex shedding induces an additional fluctuating force on the particle and changes the particle trajectory. The wake has been shown to enhance the hydrodynamic force fluctuation on a stationary finite-size particle in a uniform turbulent flow [24] and in a wall-bounded turbulent flow [25]. Third, the particle wake modulates the ambient turbulence. Turbulence augmentation usually occurs due to the enhanced velocity fluctuation induced by the unsteady wake [26]. Wu and Faeth [27] experimentally investigated the interaction between the particle wake and the turbulent pipe flow, suggesting that the turbulent wakes behave like self-preserving laminar wakes. Zeng et al. [28] studied the wake response of a stationary finite-size particle in a turbulent channel flow, revealing that the total kinetic energy is reduce in the wake, but the turbulent kinetic energy (TKE) is enhanced due to vortex shedding and wake oscillation. In addition, an enhancement of the wall shear stress was also noticed below the particle. Naso and Properetti [29] investigated the interaction between a fixed sphere and isotropic turbulence and observed that the particle has a profound influence on the turbulence, affecting a region that extends more than 10 times farther downstream than the viscous layer. Cisse et al. [30] investigated the interaction between a large particle and a homogeneous turbulent flow, finding that the particle calms turbulent fluctuations in its wake up to distances of the order of its diameter, probably due to the lower particle Reynolds number ($Re_p = 27 \sim 126$). Recently, van Hout et al. [31] experimentally studied the wake of a stationary sphere in a turbulent boundary layer using time-resolved tomo-PIV, and 3D hairpin-like structures were observed to shed with their "heads" tilted away from the wall, which was believed to cause a negative lift force. Besides, statistics of the wall turbulence modulated by the sphere were also analyzed.

Unlike its apparent counterpart in laminar flows, the natural vortex shedding frequency in a turbulent flow is hidden in turbulent fluctuations [24, 30]. Numerical [25] and experimental [31] results show that the Strouhal number of vortex shedding behind a finite-size particle in wall turbulence is lower than that in a uniform flow, except when the particle is very close to the wall [31].

For the case of many particles, the interaction between particle wakes and ambient turbulence is even more complex. The research of Xu & Bodenschatz [32] indicates that preferential accumulation of finite-size particles can possibly be affected by their wake interaction. Bellani [33] compared the turbulence modulation of spherical and ellipsoid particles and conjectured that the lower reduction in TKE by ellipsoids than the spheres is due to the more energetic wakes produced by ellipsoids. It was also noticed that the wake behind falling elliptical particles due to gravity enhances the TKE production [34]. Yu et al. [35] investigated the statistics of finite-size particles in turbulent channel flows and found the velocity fluctuations in the vicinity of the particle are dissipated more than generated by the presence of the wake. In general, the mechanism of interaction between the particle wakes and ambient turbulence is far from clear.

To our knowledge, the interaction between the wake of a finite-size particle and wall turbulence over a rough bed has not been investigated. The interaction mechanism between a finite-size particle and the ambient turbulence is still ambiguous. How the wake reacts to the oncoming turbulent disturbance is still unclear. In our previously published paper (Li et al., Phys. Rev. Fluids, 2019), we conducted an investigation on the finite-size particle over a rough bed. The previous work is focused on the drag and lift force on the particle, while the current research is focused on the wake characteristics. The objective of the current research is to investigate the interaction between the wake of a stationary finite-size particle and the wall turbulence over a rough bed at $Re_p = 99 \sim 310$, with special emphasis on the time-dependent features of the wake. In Sect. 2, the direct numerical simulation method is introduced. In Sect. 3, the vortex structures behind the particle are analyzed. In Sect. 4, we analyze the turbulence modulation by the wake by examining the time-averaged statistics of the wake. In Sect. 5, time-dependent features of the turbulence\wake interaction are examined, and the correlation between the wake and the oncoming flow is thoroughly discussed. In Sect. 6, the correlation between the wake and the oncoming flow, the turbulence modulation and the Strouhal number of the natural vortex shedding are discussed.

2 Problem Description and Numerical Method

2.1 Problem Description

The set-up of the flow is similar to the one considered in Li et al. [36]. We consider a fully developed open channel flow. The flow is characterized by the friction Reynolds number $Re_\tau = H_* u_{*\tau}/\nu_*$, in which H_* is the channel height and $u_{*\tau}$ is the friction velocity, which are also the length and velocity scales. The time and pressure scales are $H_*/u_{*\tau}$ and $\rho_* u_{*\tau}^2$, respectively, in which ρ_* is the density of the fluid. The friction Reynolds number is $Re_\tau = 180$. The rough bed consists of hemispheres of diameter $d_* = 0.1H_*$, which is shown in Fig. 1. So, in the non-dimensional form, the diameter of the hemispheres is $d = 0.1$. In wall units, the diameter of the particle and the distance from the wall are defined as $d^+ = d_* u_{*\tau}/\nu_*$ and $y_p^+ = y_{p*} u_{*\tau}/\nu_*$, respectively. In this research, we investigate an idealized rough bed which is made up of a single layer of monodispersed spheres. The hemispheres are located on the bottom wall of the channel and their centers are at the "base level", as shown in Fig. 1. The hemispheres are closely packed. To save on computational cost, the bottom wall of the computational domain is not totally covered with hemispheres, and there are only seven rows and ten columns (C10 × R7) of hemispheres (see Fig. 2). It was shown in [36] that this configuration provides a good representation of the rough bed. The flow is fully adjusted to the rough bed after five rows.

As shown in Fig. 1, the "reference particle" we study is located above the rough bed and is stationary. The gap (δ) between the particle and the rough bed is crucial. Several definitions are proposed by Lee and Balachandar [37], and we choose the definition where the "bed level" is the lowest possible position for a particle sitting on the rough bed (i.e. corresponds to $\delta = 0$). In this case, the location of bed level is at $0.3165d$, i.e. $h_{BL} = 0.3165d$. The particle above the bed has the same diameter as the hemispheres. In the research, three vertical locations of the particle center, which are $\delta = 0.0d$, $\delta = d/3$ and $\delta = 1.0d$ ($y_p^+ = 14.7$, $y_p^+ = 20.7$, and $y_p^+ = 32.7$ in wall units) are investigated.

The instantaneous particle Reynolds number $Re_p = u_*(y_{*p})d_*/\nu_* = u^+\left(y_p^+\right)d^+$. As the oncoming flow seen by the particle is fluctuating with time, Re_p also varies. For the particle sitting on the bed, the Re_p ranges from 99 to 218. For the other two cases,

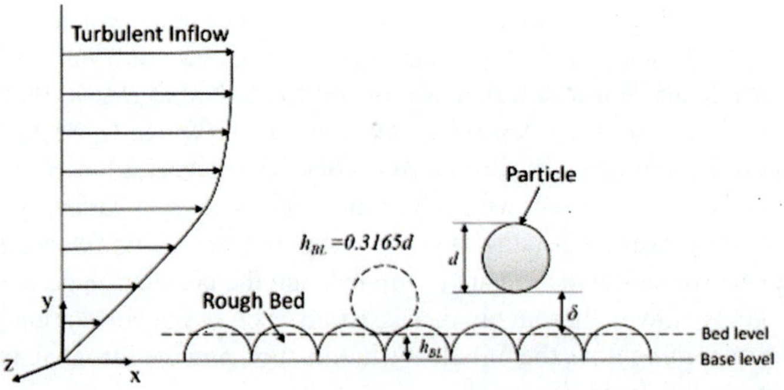

Fig. 1. Schematic of a particle over a rough bed made of hemispheres in a wall-bounded turbulent flow.

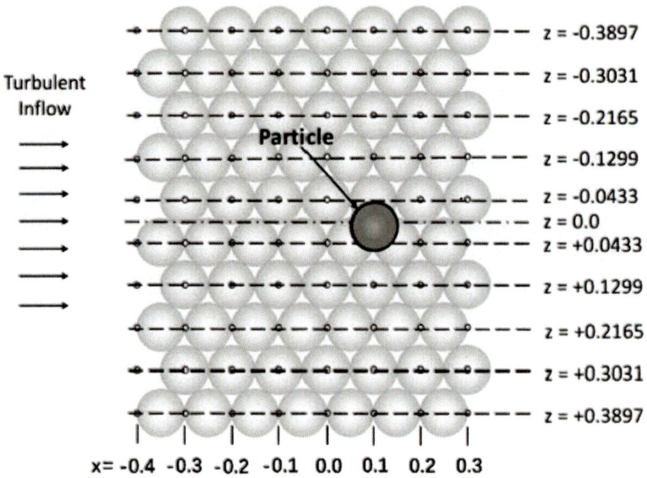

Fig. 2. Top view of the arrangement of the rough bed and the particle.

the oncoming velocity is larger, and the ranges of Re_p are 157 ~ 280 and 174 ~ 310, respectively.

On the bottom boundary of the computational domain, no-slip condition is employed. On the top boundary we enforce free-slip for the streamwise and spanwise velocities and no-penetration for the vertical velocity. In the spanwise direction, periodic boundary conditions are employed. A fully developed turbulent open channel flow is introduced from the inlet of the computational domain in the same way used by Li et al. [36]. In this method, two simulations are performed simultaneously. A typical fully-developed turbulent open channel flow simulation at $Re_\tau = 180$ with periodic boundary conditions employed along the streamwise direction with the bottom wall being smooth is conducted first. The second simulation is the inflow-outflow open channel flow simulation with a rough bed at the bottom and a particle placed above the rough bed, which is described above. The two simulation setups are such that the computational domain of the first simulation has the same wall-normal and spanwise lengths as the second one, and the outflow velocity of the first simulation is exactly applied as the inlet velocity of the second simulation. In this way, a huge amount of computational cost is saved due to the far less grid points in the first simulation.

2.2 Numerical Methodology

The dimensionless continuity and Navier-Stokes equations for incompressible flow can be written as:

$$\nabla \cdot \mathbf{u} = 0, \tag{1}$$

$$\frac{\partial \mathbf{u}}{\partial t} + \mathbf{u} \cdot \nabla \mathbf{u} = -\nabla p + \frac{1}{Re_\tau} \nabla^2 \mathbf{u} + \mathbf{f}, \tag{2}$$

where Re_τ is the frictional Reynolds number. A finite volume method on a collocated Cartesian grid system is used. For the space discretization, the second-order central difference scheme is employed. For the time advancement, a fractional-step method is used. The nonlinear advection term is discretized explicitly with the second-order Adams-Bashforth scheme, and the diffusion term is treated implicitly with the Crank-Nicolson scheme.

An immersed boundary method is used to resolve the flow around the particle and to capture flow details of all scales [38]. Multilayers of Lagrangian markers are placed within the particle and the hemispheres in order to enhance the accuracy. 20 spherical layers are used and the total number of Lagrangian markers in the particle is 36077. No-slip and no-penetration conditions are employed on the surface of the particle and the hemispheres.

In the first simulation, the extent of the computational domain in non-dimensional terms is $[0, 4\pi] \times [0, 1.0] \times [-4\pi/3, 4\pi/3]$ along the streamwise, wall-normal and spanwise directions, respectively, and the resolution is $257 \times 129 \times 129$. The grid is uniform in the streamwise and spanwise directions and is non-uniform in the wall-normal direction. In the second simulation, the extent is $[-1.0, 1.0] \times [0, 1.0] \times [-4\pi/3, 4\pi/3]$ along the three directions, and the resolution is $481 \times 161 \times 481$. We employed a finer uniform grid in the extent of $[-0.45, 0.4] \times [0, 0.25] \times [-0.44, 0.44]$ along the streamwise, wall-normal and spanwise directions, respectively in order to resolve the flow around the particle and the rough bed more precisely. In this region, the grid sizes along three directions are the same and it is $d/40$. The validity of the domain size and grid resolution are stated in [36, 39, 40].

3 Vortex Structure of the Wake

The three-dimensional vortex structure around the particle in the three cases studied are shown in Fig. 3. It is noticed that the fluid rolls up above each hemisphere composing the rough bed. The sphere above the rough bed, which has the same radius as the hemispheres, is covered by the vortex structure. In Fig. 3(a), the particle Reynolds number is relatively low, ranging from 99 to 218. So no obvious threaded wake is observed behind the sphere.

However, in Fig. 3(b), a double-threaded wake starts forming behind the particle as the particle Reynolds number increases to the range of 157 to 280, which is due to the increase of the particle-to-wall distance. Though the double-threaded wake interferes with the turbulent structures of the oncoming flow, it is still clearly observed. In Fig. 3(c) the threaded wake becomes much more obvious as the particle-to-wall distance increases further. The particle Reynolds number in this case rises to the range of 174 to 310.

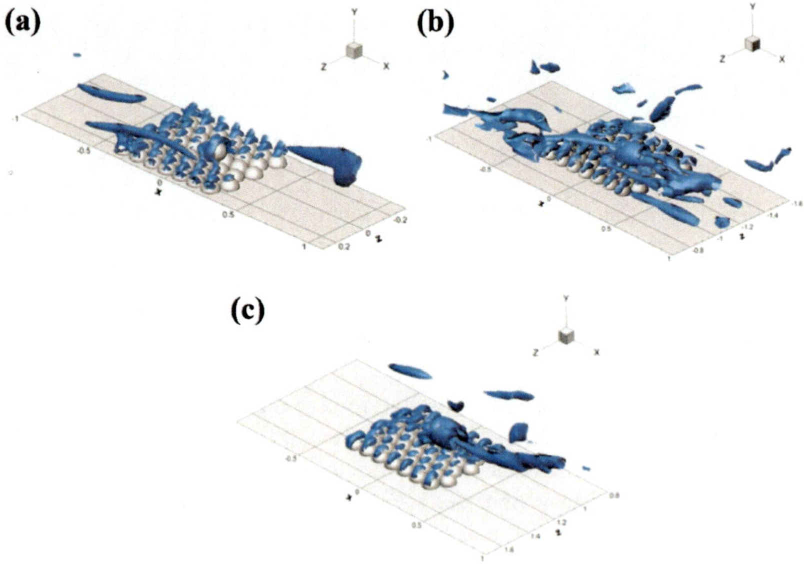

Fig. 3. Three-dimensional nature of the turbulent flow over the rough bed and around the sphere. (a) $\delta^+/d^+ = 0.0$; ; (b) $\delta^+/d^+ = 1/3$; ; (c) $\delta^+/d^+ = 1.0$. The vortex structures are shown by the iso-surface of the swirling strength $\lambda_{ci} = 30$.

In order to examine the unsteady features of the wake, streamlines around each sphere at different time instants are shown in Fig. 4, 5, and 6, respectively. Figure 4 shows the evolution of a vortex behind the sphere sitting on the bed. The most notable phenomenon is the asymmetry of the flow field around the sphere due to the presence of the wall, which resembles the flow features of the particle in a laminar shear layer investigated by Zeng et al. [20]. Another notable feature is the unsteady evolution of the vortices. A vortex emerges near the particle centered around $x = 0.21$ (see Fig. 4a). Then, it grows larger, moves downstream at the same time and is centered at $x = 0.41$ in Fig. 4b. Finally, it migrates downstream to the position $x = 0.8$ and towards the wall, which is shown in Fig. 4c. At the same time, the size of the vortex is shrinking. Figure 4 documents the process of vortex shedding in the present case. It is noticed in Fig. 4c that another vortex forms centered at around $x = 0.4$, which implies that successive vortices form and shed behind the sphere.

For the sphere of $\delta^+/d^+ = 1/3$, the flow is still highly asymmetric and the vortex behind the sphere changes over time, though no obvious shedding process is observed. This does not mean the shedding does not exist, but it is affected by the faster moving

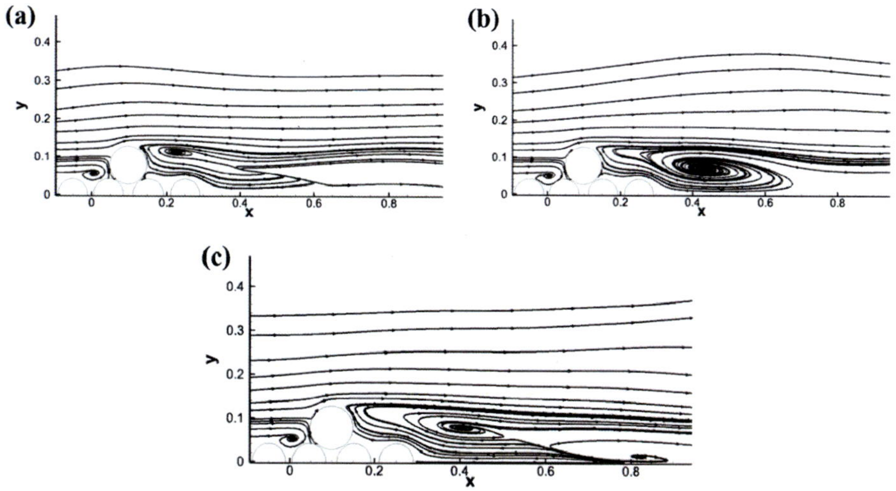

Fig. 4. Streamlines around the sphere of $\delta^+/d^+ = 0.0$. Shown is the x-y plane (side view) through the particle center. (a) $t^+ = 60.66$; (b) $t^+ = 68.22$; (c) $t^+ = 99.54$.

fluid of larger velocity compared with the case of $\delta^+/d^+ = 0.0$. This phenomenon is more noticeable in the case of $\delta^+/d^+ = 1.0$, which is shown in Fig. 6. It can be observed in Fig. 6 that the flow is much less asymmetric than the other two cases due to the enhancement of the particle-to-wall distance. A standing vortex exists at the lower part of the wake near the sphere throughout the whole period. The vortices at the higher part of the wake are shedding vortices that form and migrate downstream successively. It should be noticed that Fig. 6 shows that the shed vortices disappear quickly. The reason is that the turbulent fluctuation deforms or breaks the vortices during the shedding process, making the shedding happen much earlier.

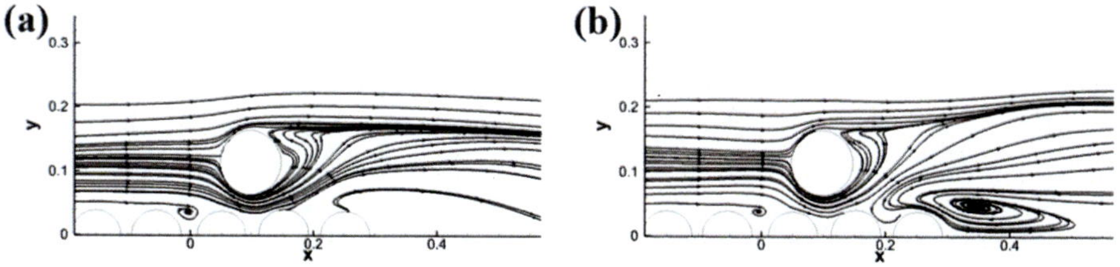

Fig. 5. Streamlines around the sphere of $\delta^+/d^+ = 1/3$. Shown is the x-y plane (side view) through the particle center. (a) $t^+ = 15.318$; (b) $t^+ = 31.518$.

In laminar flows, the threaded wake oscillates or sheds regularly. However, in turbulent flows, the oscillation or shedding process is profoundly affected by the oncoming turbulence. Figure 7 shows the contours of the vorticity magnitude on the x-z plane through the particle center at $y_p^+ = 32.7$ (i.e. the third case). Two oncoming fluctuations are analyzed here. It is noticed when the fluctuation marked by the black circle is passing the sphere, the wake oscillates downward. While the fluctuation marked by the red circle

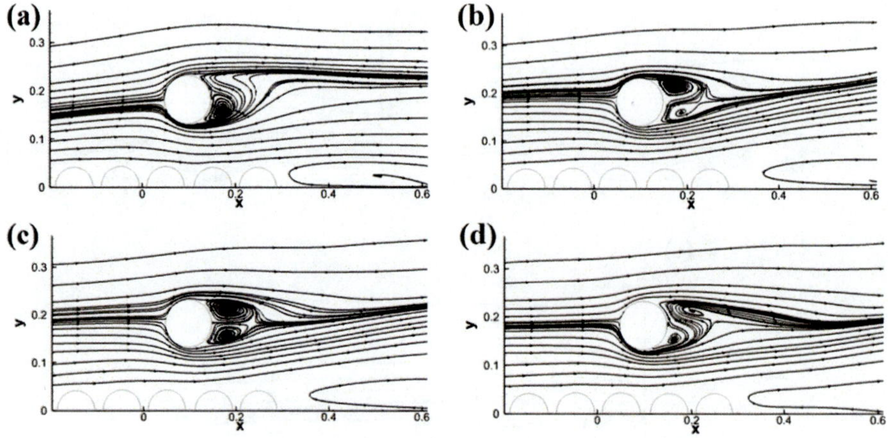

Fig. 6. Streamlines around the sphere of $\delta^+/d^+ = 1.0$. Shown is the x-y plane (side view) through the particle center. (a) $t^+ = 18.558$; (b) $t^+ = 29.358$; (c) $t^+ = 30.438$; (d) $t^+ = 32.598$.

acts on the sphere, the wake oscillates to the opposite direction. Therefore, in this case, the direction of oscillation of the sphere wake is dominated by the oncoming fluctuation, and the self-induced natural shedding frequency is overwhelmed.

4 Turbulence Modulation by the Wake

In this section, time-averaged statistics in the streamwise, wall-normal and spanwise directions of three cases are analyzed. The modulation of wall turbulence by the wake is of the most importance, thus the root mean square velocities and turbulent kinetic energy are investigated.

4.1 Root Mean Square Velocities and Turbulent Kinetic Energy

The r.m.s. velocities and the turbulent kinetic energy for the case of $\delta^+/d^+ = 0.0$ is shown in Fig. 8. It is noticed that behind the sphere, as shown Fig. 8a, the streamwise r.m.s. velocity near it is quite low at $x = 0.2$ (1.0d behind the sphere). The low streamwise r.m.s. velocity region extends to the wall. Above the sphere, the streamwise r.m.s. velocity is high, but it is limited in a thin region. This region expands as the flow moves downstream, which is caused by the vortex shedding behind the sphere. The r.m.s. velocity becomes much higher far downstream, which does not even reach its highest value at the outlet of the computational domain. It indicates that in the moderate particle Reynolds range, the turbulence modulation by the particle wake is definitely not restricted in the area near it. We can notice the obvious turbulence augmentation extends well more than 10.0*d* behind the particle. The contour of the turbulent kinetic energy in Fig. 8(d) indicates the same.

For the wall-normal and spanwise directions, the fluctuation is inhibited near the sphere, but unlike the streamwise fluctuation, the other two components are not augmented far downstream. Their variation is primarily in the wall-normal direction, and

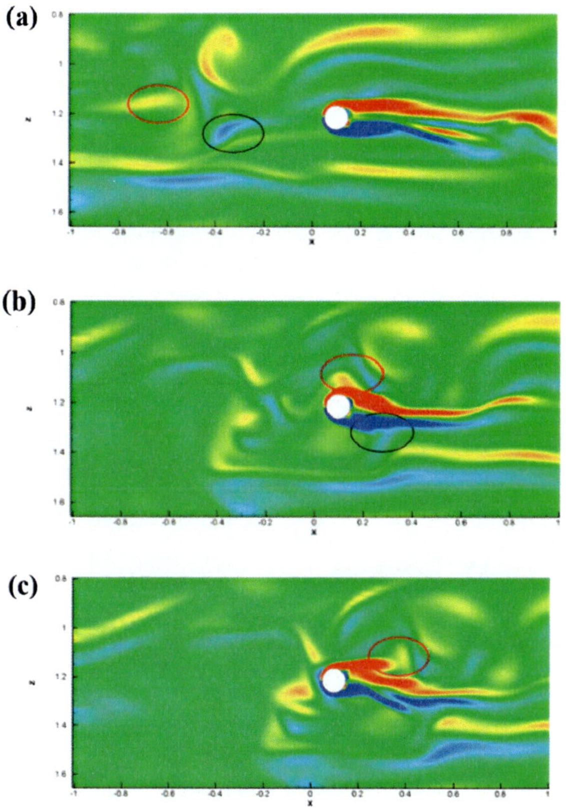

Fig. 7. Contours of vorticity magnitude on the x-z plane through the particle center ($y_p^+ = 32.7$). $\delta^+/d^+ = 1.0$. (a), (b) and (c) show contours of different time instants. Red and black circles mark two different oncoming fluctuations acting on the sphere and its wake.

the similar behavior for both upstream and downstream of the particle suggests a weaker influence of the particle. Only at $x = 0.9$, the wall-normal turbulent kinetic energy starts being augmented, but only by a little. The spanwise turbulent kinetic energy is attenuated near the sphere, but it is not augmented downstream.

As the distance between the sphere and the wall increases to $\delta^+/d^+ = 1/3$, the streamwise turbulent kinetic energy (or r.m.s. velocity) is no longer augmented, as shown in Fig. 9a. The fluctuation is inhibited near the sphere. However, the intensity of the fluctuation does not rise higher than that in the oncoming flow. The highest r.m.s. velocity in the wake is still a little lower than that in the oncoming flow. Turbulence augmentation in the wall-normal and spanwise directions is observed, as shown in Fig. 9b and 9c. The augmentation starts at about $x = 0.7$ and $x = 0.5$ for the wall-normal and spanwise direction, respectively.

For $\delta^+/d^+ = 1.0$, the turbulence augmentation region is much closer to the sphere due to the higher oncoming velocity. In Fig. 10a and 10d, we can see that the turbulent attenuation region still exists near the sphere. Because of the higher particle Reynolds

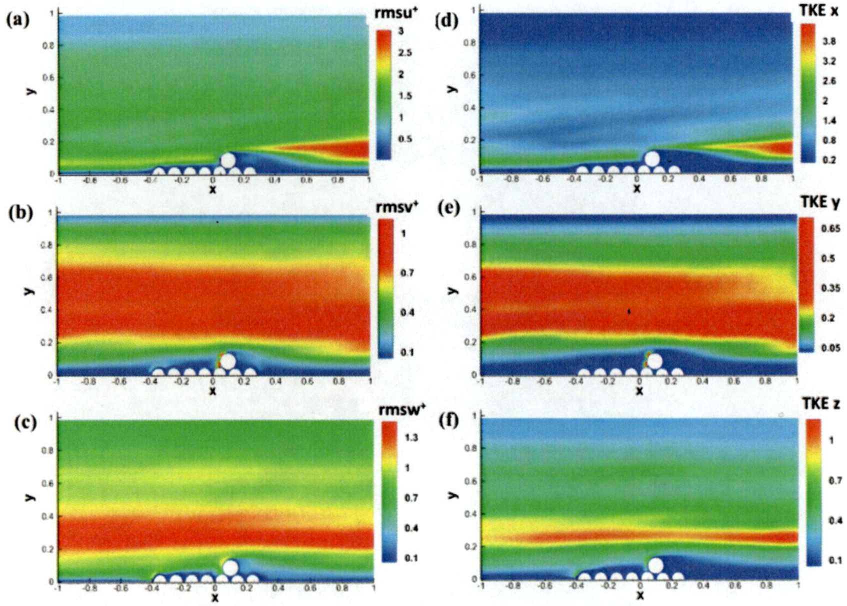

Fig. 8. Root mean square velocities (a-c) and turbulent kinetic energy in three directions (d-f). $\delta^+/d^+ = 0.0$.

number, the vortex shedding becomes stronger, and the region of turbulence augmentation in the streamwise direction locates above the attenuation region and close to the sphere, and it expands to a larger area downstream. In addition, the turbulence attenuation starts decaying at about $x = 0.7 \sim 0.8$, indicating that earlier turbulence attenuation also decays earlier. Similar phenomenon exists in the wall-normal and spanwise direction. The regions of turbulence attenuation locate in the range of about $x = 0.2 \sim 0.6$, much closer to the sphere compared to the cases of $\delta^+/d^+ = 0.0$ and $\delta^+/d^+ = 1/3$. In addition, the magnitude of augmentation is also larger.

4.2 Total Kinetic Energy

The total turbulent kinetic energy is shown in Fig. 11. As the total turbulent kinetic energy is mainly contributed by the streamwise turbulent kinetic energy, for the cases of $\delta^+/d^+ = 0.0$ and $\delta^+/d^+ = 1.0$, turbulence augmentation is observed behind the sphere. But for the case of $\delta^+/d^+ = 1/3$, turbulence augmentation is not noticed in the wake. It should be noted that the regions of turbulence augmentation are different for the cases of $\delta^+/d^+ = 0.0$ and $\delta^+/d^+ = 1.0$. In the former case, this region locates far downstream behind the sphere due to the relatively lower particle Reynolds number; while the region in the latter case is much closer to the sphere, and also decays faster.

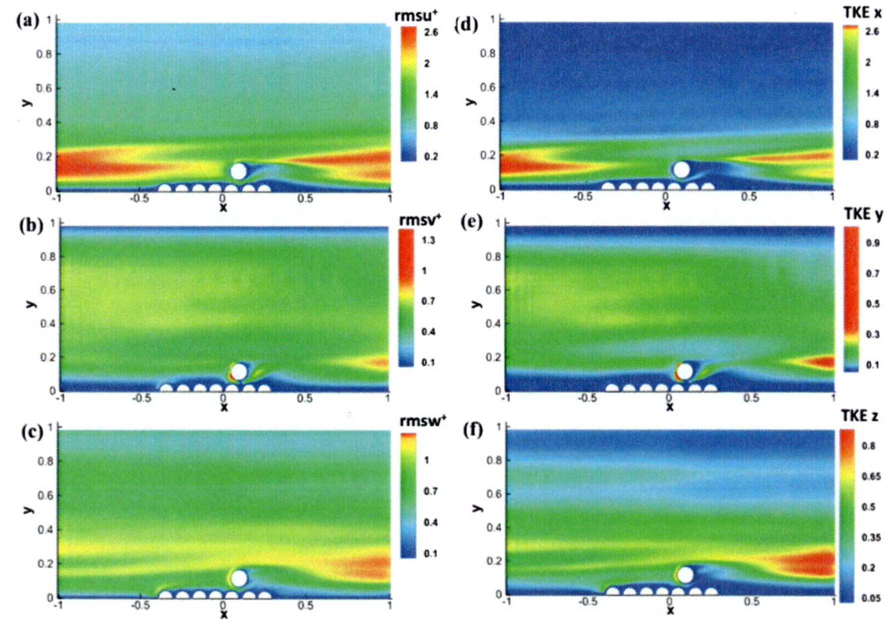

Fig. 9. Root mean square velocities (a-c) and turbulent kinetic energy in three directions (d-f). $\delta^+/d^+ = 1/3$.

5 Correlation Between the Wake and the Oncoming Flow

In this section, the interaction between the wake and the oncoming flow are examined, and their correlation is analyzed. In order to reveal the unsteady features of the wake in response to the oncoming flow, time-dependent velocity fluctuations of the wake and the oncoming flow in three directions are monitored. In addition, spectral features are also investigated to ensure the correlation between the wake and the oncoming flow.

5.1 Time-Dependent Response of the Wake to the Oncoming Fluctuation

The flow velocity fluctuations at four monitoring points behind each sphere are investigated. These points are 1.0d, 3.0d, 5.0d and 8.0d behind the sphere center, respectively. At the same time, the fluctuating velocities of the "undisturbed flow field", in which the "reference sphere" is absent while the rough bed is still included, are also recorded (this required three additional simulations). The monitoring point in the undisturbed flow field is the location of the sphere center in the disturbed flow field. Data from this point can be regarded as the oncoming flow fluctuations seen by the reference particle. Because the fluid at different monitoring points responds at different time instants to the oncoming fluctuation (e.g. the fluid at 1.0d behind the sphere reacts earlier than that at 3.0d behind the sphere), the lines of the fluctuating velocities in the disturbed field are shifted along the horizontal axis properly in order to compare them clearly. The value of the shift varies with the location of the monitoring point, but basically it is the time

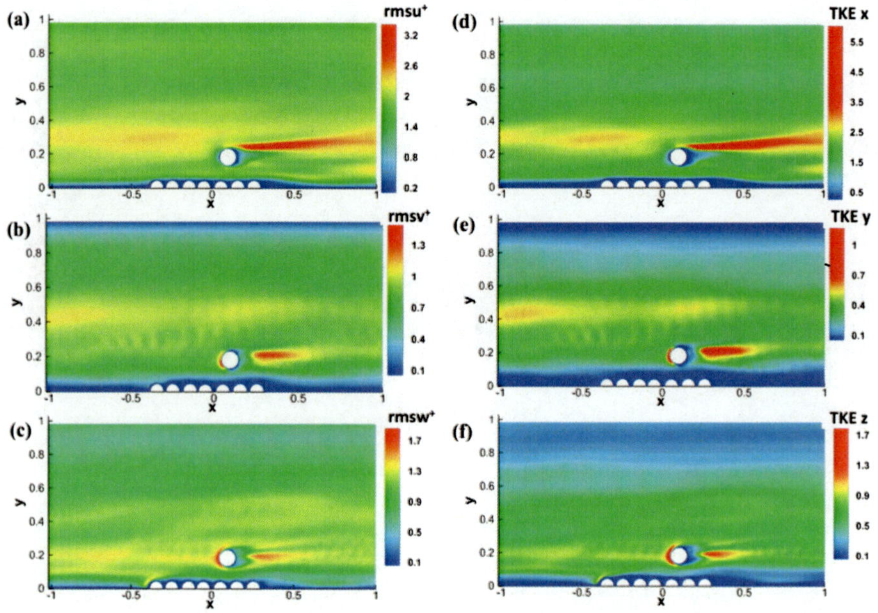

Fig. 10. Root mean square velocities (a-c) and turbulent kinetic energy in three directions (d-f). $\delta^+/d^+ = 1.0$.

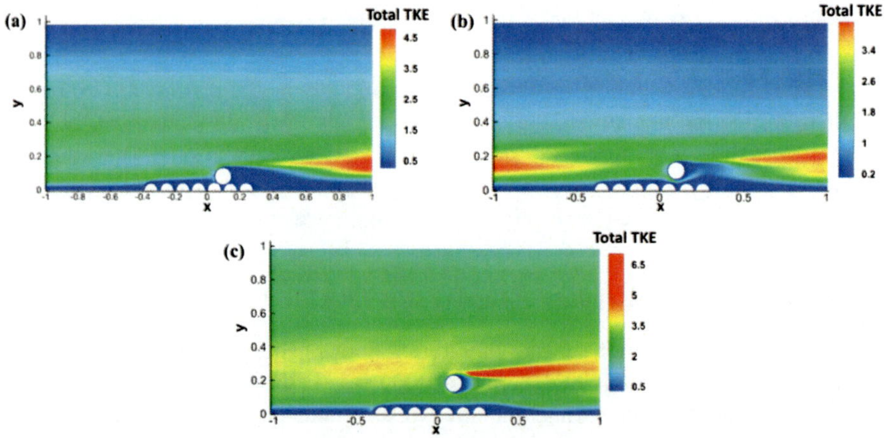

Fig. 11. Total turbulent kinetic energy on x-y plane. (a) $\delta^+/d^+ = 0.0$. (b) $\delta^+/d^+ = 1/3$. (c) $\delta^+/d^+ = 1.0$.

duration for a fluid parcel to flow from the sphere center to the streamwise location of the monitoring point. For example, for the monitoring point 8.0d behind the sphere center in Fig. 14, the shift value is about $t^+ = 13.0$. The results are shown in Fig. 12, 13 and 14.

Figure 12(a-c) shows the instantaneous fluctuating velocities of the four monitoring points behind the sphere of $\delta^+/d^+ = 0.0$, compared with that in the undisturbed flow. For this sphere, the wall-normal location of these monitoring points is at $y = 0.08165$ or $y^+ = 14.7$, the same as that of the sphere center. The first phenomenon we notice is that the wake fluctuation is inhibited near the sphere and becomes stronger downstream, which is in accordance with what is noticed in Fig. 8. Second, the fluctuation in the streamwise direction is larger than those in the other two directions. And, the wall-normal fluctuation is weaker than the spanwise one. This is also in agreement with the observations of Fig. 8. Third and most importantly, there is a correlation between the wake fluctuations and the oncoming ones. In Fig. 12(a), the wake fluctuation basically follows the oncoming one, though the magnitude is different. In comparison, the wall-normal and spanwise fluctuations show weaker correlations, as shown in Fig. 12(b) and 12(c), respectively. It should be noted in Fig. 12(a) the wake fluctuation in the vicinity of the sphere (1.0d behind the sphere) does not correlate with the oncoming flow.

For the case of $\delta^+/d^+ = 0.0$, the region of the largest turbulent kinetic energy in the wake is not located behind the sphere center, but above the four monitoring points. Therefore, the point above the fourth monitoring point (8.0d behind the sphere center) is also examined. This point is located at $y = 0.13165$ or $y^+ = 23.7$, the same vertical location as the sphere top. Thus, this point is referred to as the point behind the sphere top. In Fig. 12(d-f), the fluctuating velocities at the point 8.0d behind the sphere top is examined, compared with the oncoming fluctuations. Figure 12(d) shows the streamwise fluctuating velocities in the wake and the oncoming flow are in stronger correlation than those in Fig. 12(a). On the other hand, the wall-normal and spanwise wake fluctuations are not in such strong correlation with the oncoming velocity, as shown in Fig. 12(e) and 12(f), respectively.

For the sphere of $\delta^+/d^+ = 1/3$ shown in Fig. 13, the wake and oncoming fluctuation in the streamwise direction are in very weak correlation, which is the case for both the monitoring points behind the sphere center and that behind the sphere top. However, in the wall-normal direction, except the point very near the sphere (which locates 1.0d behind the sphere center), the correlation is stronger, especially between the wake fluctuation 8.0d behind the sphere top and the oncoming one (Fig. 13e). Though the fluctuating magnitudes are not always the same, the fluctuating directions are similar. It means the sphere amplifies or inhibits the oncoming disturbance. In the spanwise direction, the correlation is even stronger. In Fig. 13(c), it can be noticed these profiles show a very similar trend, tending to collapse onto each other. This phenomenon can also be noticed in Fig. 13(f). In addition, the turbulence augmentation exists in the wall-normal and spanwise directions, but not in the streamwise direction.

For the sphere of $\delta^+/d^+ = 1.0$ shown in Fig. 14(a), the wake fluctuation behind the sphere center and oncoming fluctuation in the streamwise direction are in weak correlation, but these monitoring points lie outside the streamwise turbulence augmentation region shown in Fig. 10(a) and 10(d). On the other hand, the wake fluctuation at 1.0d behind the sphere top shows a strong correlation with the oncoming one, as shown in Fig. 14(d). In the wall-normal direction shown in Fig. 14(b), the correlation between the oncoming fluctuation and the wake fluctuations directly behind the sphere center is much stronger, even for the point 1.0d behind the sphere. In addition, the largest fluctuation

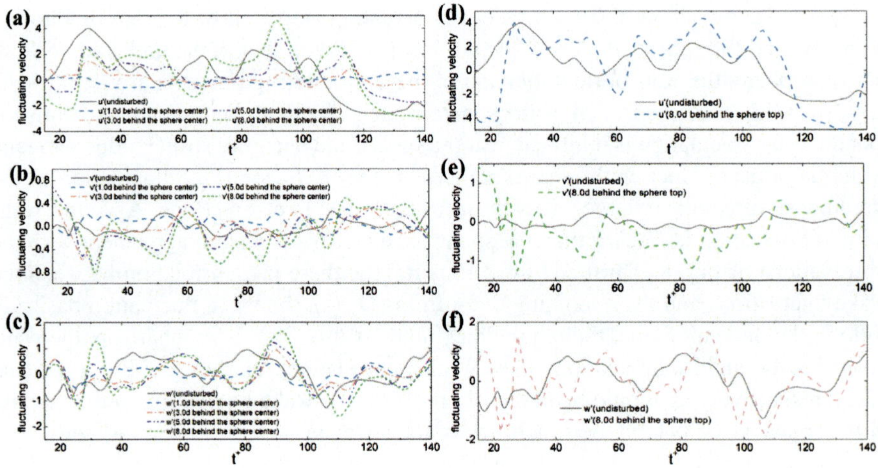

Fig. 12. Time-dependent fluctuating velocities of in the streamwise, wall-normal and spanwise directions of the sphere of $\delta^+/d^+ = 0.0$. (a-c): Time-dependent fluctuating velocities behind the sphere center ($y^+ = 14.7$). (d-f): Time-dependent fluctuating velocities behind the sphere top ($y^+ = 23.7$). (a, d): Streamwise fluctuating velocities. (b, e): Wall-normal fluctuating velocities. (c, f): Spanwise fluctuating velocities. Because the fluid at different monitoring points responds at different time instants to the oncoming fluctuation (e.g. the fluid at 1.0d behind the sphere reacts earlier than that at 3.0*d* behind the sphere), the lines of the fluctuating velocities in the disturbed field are shifted along the horizontal axis (time axis) properly in order to be shown more clearly.

magnitude lies at the second monitoring point (3.0d behind the sphere), unlike the other two cases ($\delta^+/d^+ = 0.0$ and $\delta^+/d^+ = 1/3$). This is in accordance with turbulence augmentation region shown in Fig. 10(b) and 10(e). A strong correlation also exists in the spanwise direction, except the point 1.0d behind the sphere, as shown in Fig. 14(c).

5.2 Spectral Features of the Wake and the Oncoming Flow

In order to investigate the spectral features of the wake and its relation to the oncoming flow, we analyze the amplitudes of the wake fluctuation and the oncoming fluctuation in the frequency domain. Figure 15 shows the spectra of the case of $\delta^+/d^+ = 0.0$. Fig. 15(a) shows the wake to be in strong correlation with the oncoming flow and the amplitude of the fluctuation in the wake to be generally larger than that of the oncoming flow, indicating that the oncoming fluctuation is amplified in the wake in the streamwise direction. It is interesting that the amplification ratios at different frequencies are different. However, in the other two directions shown in Fig. 15(b) and 15(c), the correlation is relatively weak, and the amplitude is not amplified.

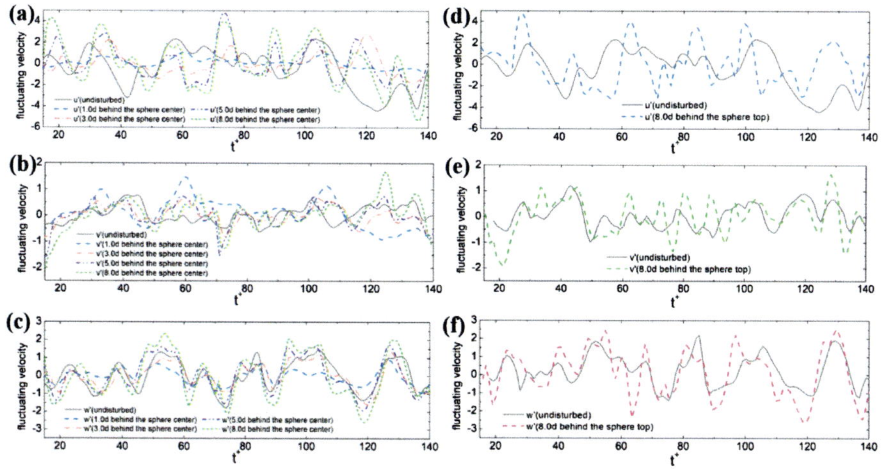

Fig. 13. Time-dependent fluctuating velocities of in the streamwise, wall-normal and spanwise directions of the sphere of $\delta^+/d^+ = 1/3$. (a-c): Time-dependent fluctuating velocities behind the sphere center ($y^+ = 20.7$). (d-f): Time-dependent fluctuating velocities behind the sphere top ($y^+ = 29.7$). (a, d): Streamwise fluctuating velocity. (b, e): Wall-normal fluctuating velocity. (c, f) Spanwise fluctuating velocity. The lines of the fluctuating velocities in the disturbed field are shifted along the horizontal axis (time axis) properly in order to be shown more clearly.

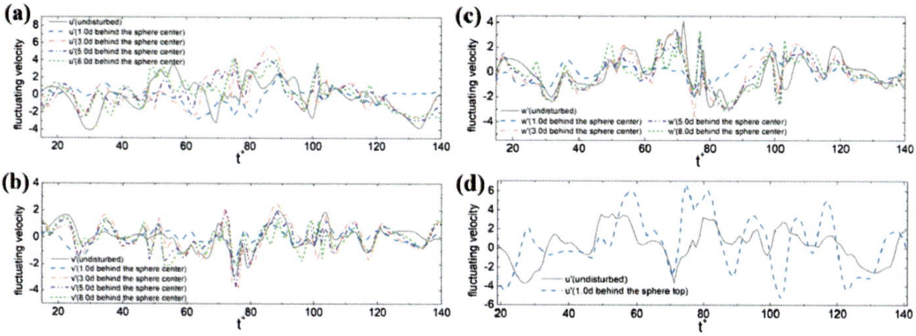

Fig. 14. Time-dependent fluctuating velocities of in the streamwise, wall-normal and spanwise directions of the sphere of $\delta^+/d^+ = 1.0$. (a-c): Time-dependent fluctuating velocities behind the sphere center ($y^+ = 23.7$). (d): Time-dependent streamwise fluctuating velocity behind the sphere top ($y^+ = 41.7$). The lines of the fluctuating velocities in the disturbed field are shifted along the horizontal axis (time axis) properly in order to be shown more clearly.

Figure 16 shows the case of $\delta^+/d^+ = 1/3$. In this case, wake fluctuations in all three directions are correlated with the oncoming ones. In addition, nearly all three components of wake fluctuations are larger than those in the oncoming flow. For the case of $\delta^+/d^+ = 0.0$, the wake of the sphere is greatly affected by the wall, and the viscous effect plays a very important role. Thus, the wake fluctuations are not so large. However, for the case of $\delta^+/d^+ = 1/3$, Re_p increases to a range of 157 ~ 280, and the wake fluctuations increase rapidly. Meanwhile, the fluctuations in the oncoming flow,

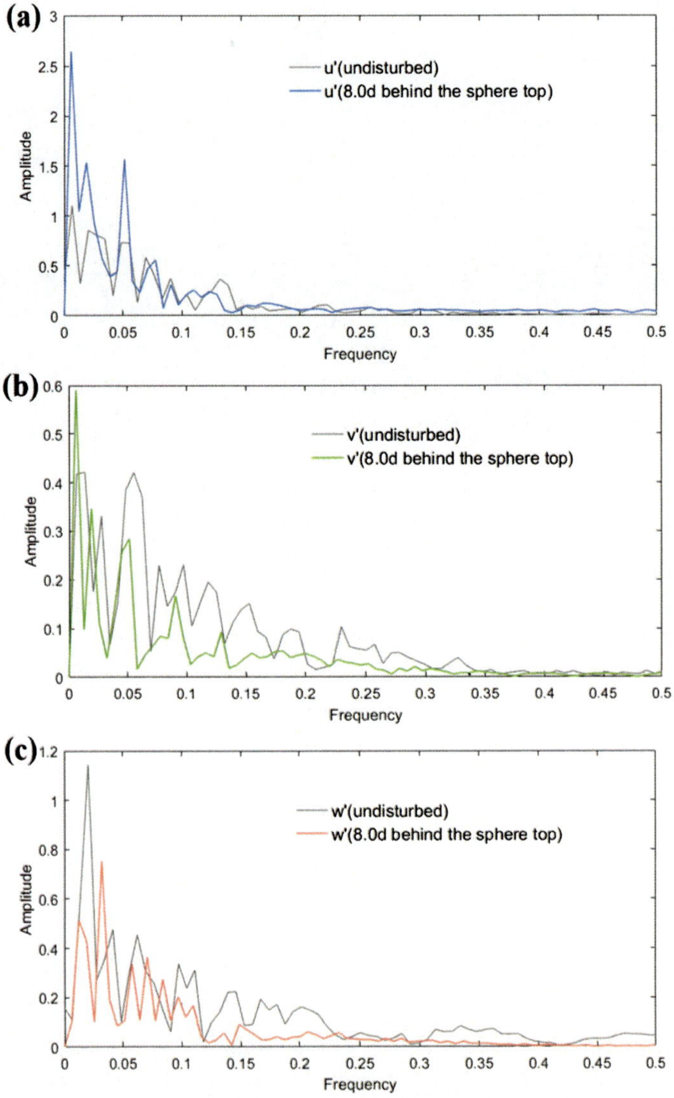

Fig. 15. Spectra of the wake fluctuations compared with those of the oncoming fluctuations for the case of $\delta^+/d^+ = 0.0$. (a) Streamwise direction; (b) Wall-normal direction; (c) Spanwise direction.

though increases slightly, still remain at low levels, especially in the wall-normal and spanwise directions. This is the reason why wake fluctuations are larger than those in the oncoming ones. As δ^+/d^+ increases to 1.0, the oncoming fluctuations experienced by the sphere increase to nearly the largest in the channel, so the wake fluctuations are not that large in comparison. This difference in the behavior of the intermediate separation case is consistent with the earlier observations. This also reveals the interesting fact that the back effect of the particle on the ambient turbulence is not monotonically dependent on either the gap between the particle and the wall or correspondingly the particle Reynolds number.

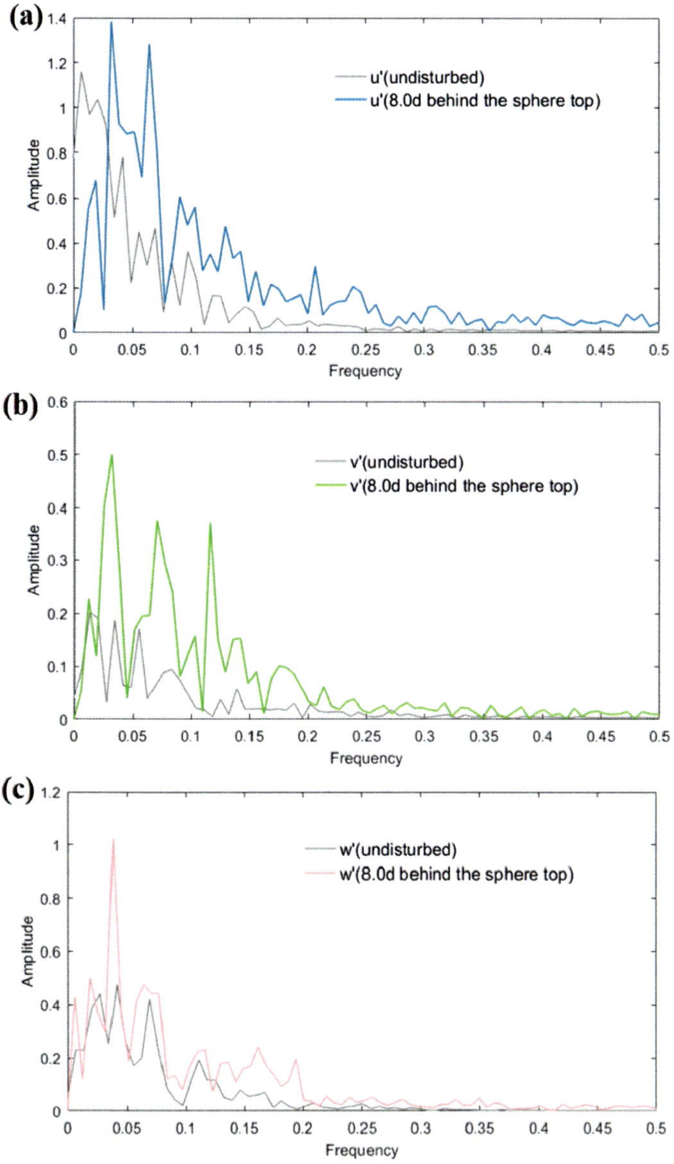

Fig. 16. Spectra of the wake fluctuations compared with those of the oncoming fluctuations for the case of $\delta^+/d^+ = 1/3$. (a) Streamwise direction; (b) Wall-normal direction; (c) Spanwise direction.

The case of $\delta^+/d^+ = 1.0$ is shown in Fig. 17. Like the previous case, strong correlations are noticed in all three components. Besides that, the most important observation is that a peak can be seen at about $f = 2.0$, both in Fig. 17(a) and 17(b). The peak is obviously taller than those near it. The height of this peak is apparently larger than that of the oncoming flow at $f = 2.0$, and the amplification ratio at this frequency is the largest. It is inferred this frequency is at which the natural vortex shedding happens. Thus, it could be inferred that the force on the particle may be mostly affected by the wake at the

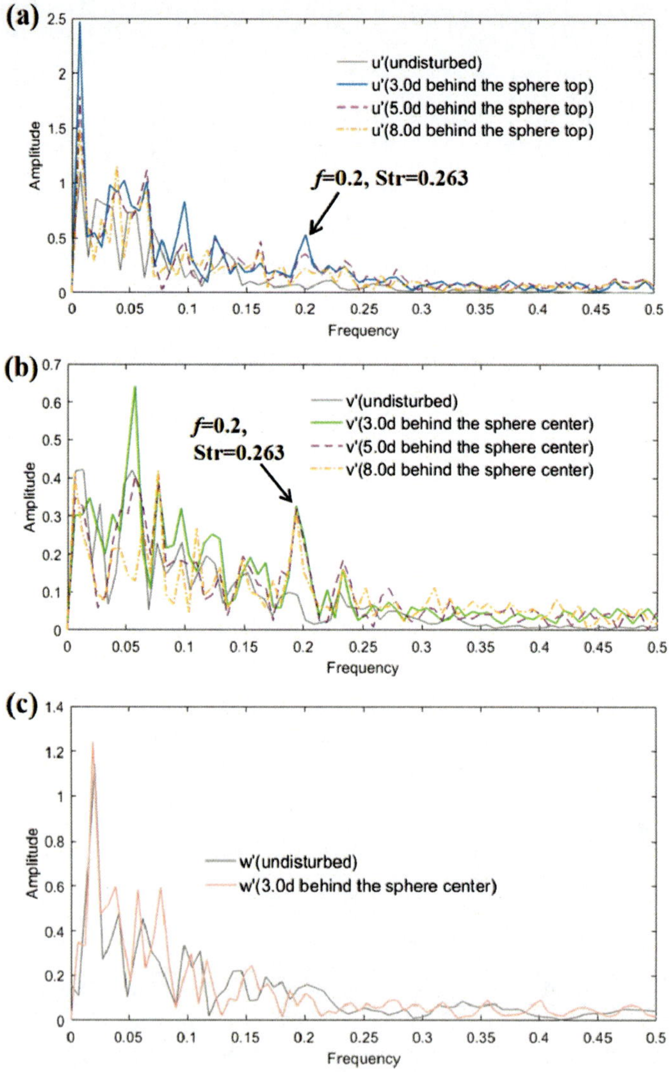

Fig. 17. Spectra of the wake fluctuations compared with those of the oncoming fluctuations for the case of $\delta^+/d^+ = 1.0$. (a) Streamwise direction; (b) Wall-normal direction; (c) Spanwise direction.

natural vortex shedding frequency. The Strouhal number (based on the particle center velocity) at $f = 2.0$ is 0.263, about twice of that of the vortex shedding behind a sphere in isotropic turbulence. Zeng et al. [25] investigated finite-sized particles in wall turbulence and found that the Strouhal number of dominant vortex shedding to be from 0.103 to 0.138 and to be in agreement with that in isotropic turbulence reported in Bagchi and Balachandar [41]. But, a peak at Strouhal number of 0.203 was also noticed, and this could be the Strouhal number of natural vortex shedding. Steward et al. [21] and Rao et al. [22] investigated the wake of a rolling sphere in a wall-bounded laminar flow and the Strouhal number was found to be around 0.12, which is close to the Strouhal number in an unbounded flow obtained by Tomboulides & Orszag [9] and Johnson & Patel [7].

6 Discussion

6.1 Correlation between the Wake and the Oncoming Flow

Our simulations show that a strong correlation between the wake fluctuation and the oncoming one exists in many cases. The oncoming fluctuation induces the shedding of vortices and the wake oscillates in the same direction as the oncoming fluctuation. This gives us the hope to establish a deterministic model for the wake-induced force on the sphere, which are urgently needed for the closure of the large eddy simulation (LES) of particle-laden flows. Nevertheless, some difficulties still exist. For example, the magnitude of wake fluctuation is quite different from the oncoming one, so the extent of force alteration by the wake is still unknown. In addition, the natural vortex shedding is found to be important in the case of $\delta^+/d^+ = 1.0$, but the relative contribution of shedding to overall force fluctuation is unclear.

A deterministic instantaneous turbulence modulation model for the wake may also be possible. The wake oscillation is dictated by the oncoming flow. Thus, the prediction of the instantaneous wake oscillation may be possible if the oncoming turbulence is resolved and the flow is moderately turbulent. The correlation is closely related to turbulence modulation. The region of turbulence augmentation is less affected by the viscous effect and is controlled mainly by the oncoming flow, and thus the correlation between turbulence augmentation region in the wake and the oncoming flow is strong. Therefore, strong correlation arises with the turbulence augmentation in the wake. On the other hand, as discussed in Sect. 6.1, the near-wall region is mainly affected by the viscous effect of the rough bed and turbulence attenuation follows. It is less affected by the oncoming flow and thus is where the correlation between the oncoming flow and the wake is weak.

6.2 Turbulence Modulation by the Wake

Turbulence modulation by the sphere wake is understandably depends on the location of the sphere due to the inhomogeneity and anisotropy of the wall-bounded flow. The region of turbulence augmentation is dictated by both the location with respect to the sphere and the sphere's location with respect to the rough wall. Basically, turbulence attenuation arises due to the strong viscous effect near the sphere surface and the wall.

Away from the sphere and the wall, turbulence augmentation is notable, especially on the upper side of the sphere. Due to the presence of the wall, vortex shedding on the lower side of the sphere is inhibited, and it is much more notable on the upper side of the sphere. At relatively low particle Reynolds number, which are the cases of $\delta^+/d^+ = 0.0$ and $\delta^+/d^+ = 1/3$, turbulence augmentation exists far downstream, even at distances more than ten times of the sphere diameter, indicating that the influence of the wake is profound and definitely not just limited to regions near the sphere, which was also suggested by Zeng et al. [28]. As the particle Reynolds number increases, the vortices are shed more rapidly, and the region of turbulence augmentation is much closer to the sphere, which is the case for $\delta^+/d^+ = 1.0$. Compared to turbulence attenuation around the particle reported in [30] at lower Reynolds numbers of $Re_p = 27 \sim 126$, the higher Re_p in the present research causes an apparent turbulence augmentation. For the case of $\delta^+/d^+ = 1/3$, turbulence augmentation in the streamwise direction near the particle does not arise like the other two cases.

The results of a sphere above a rough-wall (the case of $\delta^+/d^+ = 1/3$) and above a smooth-wall (Case 4 in Zeng et al. [28]) can be compared. In Case 4 of Zeng et al. [28], the sphere diameter is $d^+ = 24.94$ and the gap between the sphere and the wall is $\delta^+ = 5.5$. The rough-wall and the smooth-wall cases share several similarities. The one-sided vortex shedding in both cases induces more intense turbulent fluctuations on the upper side of the sphere. A region of turbulence augmentation ahead of the sphere exists in both cases. And, the patterns of the turbulent modulation away from the wall are similar. The differences lie in the near-wall region. In the smooth-wall case, the turbulent fluctuation in the gap is extremely high, even reaches the highest level in the wall-normal and spanwise components. But it is not the case for the rough-wall. The rough wall creates a larger viscous-dominated region near the wall, and the turbulent fluctuations within the gap are much lower than those in a smooth-wall case (see Fig. 9). If the gap between the sphere and the wall is large enough (e.g. $\delta^+/d^+ = 1.0$), the effect of wall roughness is minimal.

6.3 Natural Vortex Shedding of a Sphere

In the present simulations, the sphere is stationary in a rough-wall-bounded turbulent flow. The Strouhal number in the simulation is about 0.263, which is in agreement with the experimental measurement over the range 0.26 to 0.33 [42]. It is about twice of the Strouhal number of vortex shedding of a sphere in an unbounded flow [7, 9] and a rolling sphere near a wall in a uniform flow [21, 22]. The vortex shedding on the lower side of the sphere is inhibited due to the presence of the wall, so the shedding only persists on the upper side. Thus, the flow on the upper side of the sphere plays a more important role in determining the vortex shedding frequency. Since the Strouhal number is still calculated based on the particle center velocity, the Strouhal number is larger than that of vortex shedding in an unbounded flow or a rolling sphere.

The role of natural vortex shedding in the presence of ambient turbulence is still ambiguous. It depends on the turbulent intensity, the particle Reynolds number, and the size of the oncoming vortex. In our case, the length scale of the oncoming vortex is of the same order of magnitude as that of the sphere, which may be the reason why the natural vortex shedding is dominated by the oncoming vortex structure. We conjecture if the scale of the oncoming fluctuation is small enough, the natural vortex shedding may be prominent in the wake.

7 Concluding Remarks

In this paper, we performed direct numerical simulations with the immersed boundary method to investigate the interaction between the wake of a finite-sized sphere above a rough bed and the ambient wall-bounded turbulence. The flow is moderately turbulent at $Re_\tau = 180$. The sphere with the diameter of $d^+ = 18$ is placed in three vertical locations above the rough bed, which are $\delta^+/d^+ = 0.0$, $\delta^+/d^+ = 1/3$, and $\delta^+/d^+ = 1.0$, respectively. We focused on analyzing the time-dependent features of the turbulence\wake interaction and the turbulence modulation by the wake. The key findings are summarized as follows:

(1) The correlation between the velocity fluctuations in the sphere wake in regions of turbulence augmentation and the oncoming flow fluctuation is strong. The oncoming fluctuation induces the shedding of vortices and the wake oscillates in the same direction as the oncoming flow. The strongest correlation is found in the region of turbulence augmentation by the sphere wake. This phenomenon provides a possible way to build a deterministic model for the wake-induced force on the sphere, though further study is still needed for complete understanding. This kind of model can be beneficial for the closure of the large eddy simulation (LES) of particle-laden flows. In addition, a deterministic instantaneous turbulence modulation model for the wake is also possible based on this correlation.

(2) Turbulence attenuation region is found near the sphere and the wall due to the viscous effect. While the region of turbulence augmentation lies farther away from the wall and farther downstream. As the particle Reynolds number increases, the region of turbulence augmentation becomes closer to the sphere. For the case of $\delta^+/d^+ = 1/3$, the turbulence augmentation in the streamwise direction is not observed.

(3) The rough wall induces a larger viscous-dominated region than the smooth wall. Thus, the turbulence fluctuation in the gap between the sphere and the wall is weaker than that in the smooth-wall cases. Basically, the effect of roughness is prominent in the near-wall region. If the gap is large enough, the roughness plays a minimal role.

(4) Due to the presence of the wall, the vortex shedding happens on only one side of the sphere (upper side). The Strouhal number of the natural vortex shedding based on the particle center velocity is around 0.262, about twice that in an unbounded flow. This difference is partly due to the normalization by the velocity at the center of the particle while shedding mostly occurs only at the high-speed size on the top side of the particle. In the cases studied here, the natural vortex shedding is overwhelmed by the ambient turbulence. However, the natural vortex shedding can be prominent when the largest eddy of the ambient turbulence is smaller than the shedding vortices. Further study is still needed on that issue.

8 Statements and Declarations

The authors declare that they have no known competing financial interests or personal relationships that could have appeared to influence the work reported in this paper.

Acknowledgements. X.L. acknowledges the financial support from China Scholarship Council (CSC). This research is also supported by the Project funded by China Postdoctoral Science Foundation (No. 2020M683473).

References

1. Mathai, V., Prakash, V.N., Brons, J., Sun, C., Lohse, D.: Wake-driven dynamics of finite-sized buoyant spheres in turbulence. Phys. Rev. Lett. **115**(12), 124501 (2015)
2. Wu, J.S., Faeth, G.M.: Sphere wakes in still surroundings at intermediate Reynolds numbers. AIAA J. **31**(8), 1448–1455 (1993)
3. Leweke, T., Provansal, M., Ormières, D., Lebescond, R.: Vortex dynamics in the wake of a sphere. Phys. Fluids **11**(9), S12 (1999)
4. Chrust, M., Goujon-Durand, S., Wesfreid, J.E.: Loss of a fixed plane of symmetry in the wake of a sphere. J. Fluids Strut. **41**, 51–56 (2013)
5. Ern, P., Risso, F., Fabre, D., Magnaudet, J.: Wake-induced oscillatory paths of bodies freely rising or falling in fluids. Annu. Rev. Fluid Mech. **44**(1), 97–121 (2012)
6. Natarajan, R., Acrivos, A.: The instability of the steady flow past spheres and disks. J. Fluid Mech. **254**, 323–344 (1993)
7. Johnson, T.A., Patel, V.C.: Flow past a sphere up to a Reynolds number of 300. J. Fluid Mech. **378**, 19–70 (2000)
8. Ghidersa, B., Dušek, J.: Breaking of axisymmetry and onset of unsteadiness in the wake of a sphere. J. Fluid Mech. **423**, 33–69 (2000)
9. Tomboulides, A.G., Orszag, S.A.: Numerical investigation of transitional and weak turbulent flow past a sphere. J. Fluid Mech. **416**, 45–73 (2000)
10. Eshbal, L., Rinsky, V., David, T., Greenblatt, D., van Hout, R.: Measurement of vortex shedding in the wake of a sphere at Re=465. J. Fluid Mech. **870**, 290–315 (2019)
11. Eshbal, L., Kovalev, D., Rinsky, V., Greenblatt, D., van Hout, R.: Tomo-PIV measurements in the wake of a tethered sphere undergoing VIV. J. Fluids Struct. **89**, 132–141 (2019)
12. Citro, V., Siconolfi, L., Fabre, D., Giannetti, F., Luchini, P.: Stability and sensitivity analysis of the secondary instability in the sphere wake. AIAA J. **55**(11), 3661–3668 (2017)
13. Kim, H.J., Durbin, P.A.: Observations of the frequencies in a sphere wake and of drag increase by acoustic excitation. Phys. Fluids **31**(11), 3260–3265 (1988)
14. Sakamoto, H., Haniu, H.: A study on the vortex shedding from spheres in a uniform flow. Trans. ASME J. Fluids Engng. **112**, 386–392 (1990)
15. Ploumhans, P., Winckelmans, G.S., Salmon, J.K., Leonard, A., Warren, M.S.: Vortex methods for direct numerical simulation of three-dimensional bluff body flows: application to the sphere at Re=300, 500, and 1000. J. Comput. Phys. **178**(2), 427–463 (2002)
16. Yun, G., Kim, D., Choi, H.: Vortical structures behind a sphere at subcritical Reynolds numbers. Phys. Fluids **18**(1), 015102 (2006)
17. Terra, W., Sciacchitano, A., Scarano, F.: Aerodynamic drag of a transiting sphere by large-scale tomographic-PIV. Exp. Fluids **58**(7), 83 (2017)
18. David, T., Eshbal, L., Rinsky, V., van Hout, R.: Flow measurements in the near wake of a smooth sphere and one mimicking a pine cone. Phys. Rev. Fluids. **5**(7), 074301 (2020)
19. Zeng, L., Balachandar, S., Fischer, P.: Wall-induced forces on a rigid sphere at finite Reynolds number. J. Fluid Mech. **536**, 1–25 (2005)
20. Zeng, L., Najjar, F., Balachandar, S., Fischer, P.: Forces on a finite-sized particle located close to a wall in a linear shear flow. Phys. Fluids **21**(3), 033302 (2009)
21. Stewart, B.E., Thompson, C.M., Leweke, T., Hourigan, K.: Numerical and experimental studies of the rolling sphere wake. J. Fluid Mech. **643**(4), 137–162 (2010)

22. Rao, A., Passaggia, P.Y., Bolnot, H., Thompson, M.C., Leweke, T., Hourigan, K.: Transition to chaos in the wake of a rolling sphere. J. Fluid Mech. **695**, 135–148 (2012)
23. Mittal, R.: Response of the sphere wake to free-stream fluctuations. Theor. Comput. Fluid Dyn. **13**, 397–419 (2000)
24. Bagchi, P., Balachandar, S.: Response of the wake of an isolated particle to isotropic turbulent cross-flow. J. Fluid Mech. **518**, 95–123 (2004)
25. Zeng, L., Balachandar, S., Najjar, F., Fischer, P.: Interaction of a finite-sized particle with wall turbulence. J. Fluid Mech. **594**, 201–205 (2008)
26. Balachandar, S., Eaton, J.K.: Turbulent dispersed multiphase flow. Annu. Rev. Fluid Mech. **42**(1), 111–133 (2010)
27. Wu, J.S., Faeth, G.M.: Effect of ambient turbulence intensity on sphere wakes at intermediate Reynolds number. AIAA J. **33**(1), 171–173 (1995)
28. Zeng, L., Balachandar, S., Najjar, F.: Wake response of a stationary finite-sized particle in a turbulent channel flow. Intl. J. Multiphase Flow **36**(5), 406–422 (2010)
29. Naso, A., Prosperetti, A.: The interaction between a solid particle and a turbulent flow. New J. Phys. **12**(3), 033040 (2010)
30. Cisse, M., Homann, H., Bec, J.: Slipping motion of large neutrally buoyant particles in turbulence. J. Fluid Mech. **735**, R1 (2013)
31. van Hout, R., Eisma, J., Elsinga, G.E., Westerweel, J.: Experimental study of the flow in the wake of a stationary sphere immersed in a turbulent boundary layer. Phys. Rev. Fluids. **3**(2), 024601 (2018)
32. Xu, H., Bodenschatz, E.: Motion of inertial particles with size larger than Kolmogorov scale in turbulent flows. Physica D **237**, 14–17 (2008)
33. Bellani, G., Byron, M.L., Collignon, A.G., Meyer, C.R., Variano, E.A.: Shape effects on turbulent modulation by large nearly neutrally buoyant particles. J. Fluid Mech. **712**, 41–60 (2012)
34. Bellani, G., Nole, M.A., Variano, E.A.: Turbulence modulation by large ellipsoidal particles: concentration effects. Acta Mech. **224**(10), 2291–2299 (2013)
35. Yu, W., Vinkovic, I., Buffat, M.: Acceleration statistics of finite-size particles in turbulent channel flow in the absence of gravity. Flow Turb. Combust. **96**(1), 183–205 (2015)
36. Li, X., Balachandar, S., Lee, H., Bai, B.: Fully resolved simulations of a stationary finite-sized particle in wall turbulence over a rough bed. Phys. Rev. Fluids. **4**(9), 094302 (2019)
37. Lee, H., Balachandar, S.: Effects of wall roughness on drag and lift forces of a particle at finite Reynolds number. Intl. J. Multiphase Flow. **88**, 116–132 (2017)
38. Uhlmann, M.: An immersed boundary method with direct forcing for the simulation of particulate flows. J. Comput. Phys. **209**(2), 448–476 (2005)
39. Lee, H., Balachandar, S.: Critical shear stress for incipient motion of a particle on a rough bed. J. Geophys. Res. Earth Surf. **117**, F01026 (2012)
40. Lee, H., Ha, M.Y., Balachandar, S.: Work-based criterion for particle motion and implication for turbulent bed-load transport. Phys. Fluids **24**(11), 116604 (2012)
41. Bagchi, P., Balachandar, S.: Effect of turbulence on the drag and lift of a particle. Phys. Fluids **15**(11), 3496–3513 (2003)
42. Li, X., Bai, B.: Wake of a stationary sphere near a wall. CIESC J. **70**(9), 3300–3306 (2019)

Artificial Neural Network Modeling Small-Scale Turbulence of Isotropic Turbulent Flows

Jiangtao Tan[1,2] and Guodong Jin[1,2](✉)

[1] The State Key Laboratory of Nonlinear Mechanics, Institute of Mechanics, Chinese Academy of Sciences, Beijing 100190, China
tanjiangtao17@mails.ucas.ac.cn, gdjin@lnm.imech.ac.cn

[2] School of Engineering Sciences, University of Chinese Academy of Sciences, Beijing 100049, China

Abstract. Small-scale fluid motions play an important role in the relative dispersion and clustering of inertial particles in turbulent flows. In this paper, an artificial neural network (ANN) is used to recover the small-scale turbulence of isotropic turbulent flows in the context of *a prior* large-eddy simulation (LES). The nonlinear convection and pressure gradient terms in the governing equations of subgrid-scale (SGS) velocity are treated as output labels C_i, and the large-scale velocity and the velocity gradient tensor are taken as input features. The data required for training and testing of the ANN are provided by the Direct Numerical Simulation (DNS) and filtered Direct Numerical Simulation (FDNS). The optimized model highly fits the relationship between input features and output labels. Using the ANN model and large-scale flow field information, the approximate governing equation of small-scale motions is solved numerically, and the small-scale flow field can be obtained. The developed flow field can be statistically consistent with the results of DNS. The probability density function (PDF) of small-scale velocity and velocity gradient tensor are consistent with those of DNS. Our research indicates that the small-scale flow field can be calculated by combining the large-scale flow field with the ANN methods. Furthermore, we combine the ANN model with FDNS to predict the statistics of heavy particles as *a prior* LES. The results show that the *a prior* LES with ANN model can improve the prediction accuracy in clustering and relative velocity of particle pairs. This study provides a feasible method for constructing small-scale turbulence, which can reduce the amount of computation compared with DNS and used to the study of small-scale turbulent mixing.

Keywords: Large-eddy Simulation · Small-scale model · Artificial Neural Network · Inertial particles · Turbulent clustering

1 Introduction

Large eddy simulation (LES) is an important method for the numerical study of turbulence. Its basic principle is to solve only the filtered Navier-Stokes equations containing large-scale motion and establish a sub-grid scale (SGS) model to account for the influence of small-scale motion on large-scale motion and to close filtered N-S equations

X. Zheng and S. Balachandar (Eds.): IUTAM 2023, IUTAM Bookseries 41, pp. 40–57, 2024.
https://doi.org/10.1007/978-3-031-47258-9_3

[1, 2]. Small-scale turbulence is closely related to the energy dissipation of turbulence, which has the best prospect of being universal or quasi-universal [3]. The intermittency of small-scale turbulence affects the propagation of sound waves in the atmosphere and can strengthen the mixing effect on substances in combustion. In turbulent noise and turbulent two-phase flow, small-scale motion plays an important role in the relative dispersion and non-uniform preferential concentration of particles, turbulent noise high-frequency signals and the deformation of droplets or bubbles [4]. Deepening the study of small-scale turbulence can improve the accuracy of LES of flow field simulation at high Reynolds numbers.

In LES, the large scales of motion containing energy are numerically simulated, while unsolved sub-grid scales and their interactions with large scales are modeled [5]. On the one hand, SGS stresses are modeled to close the filtered N-S equations in LES. The classic SGS models include the Smagorinsky model [6], the scale similarity model [7, 8], the gradient model [9], the mixed model [10], the dynamic model [11–17], etc. The study of SGS stress has always been a hot topic in LES. Germano et al. [12] proposed a method to calculate the variation of the coefficients of the SGS eddy viscosity model with space and time. The mathematical inconsistency in the formula of this method limits the practicability of the model. Ghosal et al. [18] corrected these inconsistencies, and the new model can be used for the general non-uniform flow. On the other hand, the study of SGS velocity is also an important aspect of LES. However, the model established by the coupling of approximate deconvolution and synthetic turbulence and the stochastic differential equation method still has some shortcomings in simulating the intermittent characteristics of small-scale motion [19, 20]. It is still a difficult problem to build a high-fidelity SGS model.

The breakthrough in computer technology has led to a comprehensive improvement in computing capabilities. Therefore, algorithms such as machine learning and deep learning have played an important role in various research fields. Machine learning is a learning method that simulates humans through big data, and deep learning aims to extract intrinsic features from data. These methods have made breakthroughs in image classification, natural language processing, and face recognition. MacCulloch and Pitts [21] first proposed the concept of machine learning in their computational research on the principles of biological neural networks published in 1943. Later, with the famous "Turing Test", machine learning, an artificial intelligence research method, began to enter the field of scientific research. The modeling of machine learning mainly includes data processing, feature selection, building a framework, and optimizing parameters. Researchers in the computer field continuously propose and optimize different algorithms for different processes. For example, according to learning methods, it can be divided into supervised learning, unsupervised learning, semi-supervised learning, and reinforcement learning. The supervised learning data itself has tags, and the existing tags are used as expected results to optimize the model. The supervised learning algorithm is usually divided into classification algorithms and regression algorithms, including linear models, artificial neural networks, support vector machines, and so on. Unsupervised learning data itself does not carry labels, and machines explore the inherent relationships between the data. For example, clustering algorithms classify samples without categories based on the similarity between the samples. In semi-supervised learning, only part of the data is labeled, resulting in low learning costs and high accuracy. Reinforcement learning (RL) is an area of machine learning inspired by behaviorist psychology, concerned with how software agents ought to take actions in an environment to maximize some notion of cumulative reward. The most familiar example is the AlphaGo that blooms in the Go game. Machine learning methods are widely used, and have achieved breakthrough results in image classification, natural language processing, and face recognition. Face recognition, intelligent capture, etc. in daily life have the use of machine learning algorithms behind them. The interactive processing method of parallel computing for big data has improved the adaptability of machine learning in big data environments. Data sampling and attribute selection also provide modeling ideas for machine learning to process big data issues [22]. Deep learning allows a computing model composed of multiple processing layers to learn data with multiple levels of abstraction. These methods have greatly improved the technical level of many other fields such as speech recognition, visual object recognition, object detection, and drug discovery and genomics. Deep convolutional neural networks have made breakthroughs in processing images, video, speech, and audio, while recursive networks have brought research prospects for sequential data such as text and speech [23].

Artificial neural network (ANN) is a popular and widely used method in machine learning (ML), which was first proposed in the 1950s. ANN combines neuroscience and connects the theoretical model of biological neural networks - the McCulloch-Pitts [21] (M-P) neuron model - into a network to simulate the operation of neurons in the human brain. ANN utilizes the computing power of computers to process and analyze data and has excellent data processing and modeling capabilities. However, in the past, limited by the performance of computers, neural network methods once fell into a slump. After entering the 21st century, ANN has once again demonstrated its ability to fit data with deep learning. Deep learning, in a narrow sense, can be considered a multi-layer neural network. Data is crucial for deep learning, as too little data can easily lead to model overfitting. In the era of big data, deep learning has become more closely related to various disciplines and research fields. The ability to learn in-depth has not been disappointing and has played an extremely important role in various studies.

The application of machine learning to turbulence modeling has been increasingly discussed. The universal approximation theorem [24] states that any function may be approximated by a sufficiently large and deep neural network. Brunton et al. [25] discussed that ML is a useful technique for deriving information from data. He explained the classification of ML and analyzed several successful cases of modeling through ML. There are many opportunities and challenges in applying ML to the field of fluid dynamics. Yarlanki et al. [26] used ML algorithms to optimize the CFD model, resulting in a 35% reduction in model error compared to the k-ε model. Wang et al. [27] used random forest methods for turbulence modeling. The model was trained by Direct Numerical Simulation (DNS) data. The Reynolds stress in different flows was predicted, and the study has achieved good predictive effects. Fukami et al. [28] performed super-resolution of low-resolution turbulent flow field data with the convolutional neural network to reconstruct an accurate turbulence model, which promoted the efficient super-resolution model of fluid flow. Qu et al. [29] extracted turbulence features through convolutional neural networks, which reduced the error compared with previous work.

Gamahara and Hattori [30] used ANN to study SGS stress models, showing that ANN is a promising tool for building new SGS models. Sarghini et al. [31] used a multi-layer feedforward ANN in LES to build an SGS model, with the flow field velocity as the input and the turbulence viscosity coefficient as the output. The model was optimized by the error backpropagation algorithm. Their results showed that the computation time is shortened by 20% compared with the existing models without considering the time of neural network training, which proves the feasibility of using ANN to study turbulence dynamics. Xie et al. [32] used spatial artificial neural network (SANN) modeling to improve the prediction accuracy of SGS stress and SGS heat flow in the compressible isotropic turbulence model, which has a higher correlation and smaller error than the gradient model. The SANN model has better prediction results for speed, temperature, and instantaneous flow structure than the dynamic hybrid model in a posteriori comparison. In the study of isotropic turbulence, Zhou et al. [33] obtained training data by a spectral method, constructed a single hidden layer feed-forward ANN, took the velocity gradient tensor and filter size as input, SGS stress as output, predicted the energy transfer rate under different Reynolds numbers and filter sizes. The correlation of the training results of the ANN model was higher than 0.9, which was significantly improved compared with the traditional model. Beck et al. [34] rewrote the N-S equations for viscous compressible fluids as: $\partial U/\partial t + R(FU) = 0$. The ANN and deep learning models have been established for the $R(FU)$ terms and speed of the three dimensions. Their results showed that a model with higher stability and accuracy was obtained. Milano and Koumoutsakos [35] used an ANN to simulate and predict near-wall turbulent flow fields. Yin et al. [36] used an ANN to establish a ML model for turbulence feature quantity prediction, and the results showed that when the Reynolds stress error is small, the error of predicting the average flow is also small. Xie et al. [37] compared the modeling effect of three different ANN models on SGS stress in LES, and the correlation coefficient of SGS stress of the ANN model was improved in the prior analysis compared with the velocity gradient model, among which the relative error of the spatial ANN model and the deconvolutional artificial neural networks model was less than 15%. In a posterior verification, the prediction accuracy of the turbulence field of the three models was higher.

Based on the theory of LES, the current study modifies the governing equations of small-scale flows. The nonlinear term of small-scale velocity and pressure gradient in the equation are modeled using ANN to achieve an approximate linearization of the governing equation. The sum of the nonlinear term of small-scale velocity and pressure gradient is defined as model parameters. By numerically solving the linearized equation, velocity information of small-scale flows can be obtained. The ANN model constructed in the study uses large-scale velocities and velocity gradient tensor as inputs, and model parameters as outputs, and is optimized using an error backpropagation algorithm. The data required for training and testing of the ANN are obtained from DNS and filtered Direct Numerical Simulation (FDNS) of homogeneous isotropic turbulence. The prediction accuracy and generalization of the trained ANN model are evaluated using DNS data at various Reynolds numbers and filter widths.

The paper is organized as follows: The linearization of small-scale flow governing equations and the derivation of model parameters are introduced in Sect. 2. In Sect. 3, the theory of neural networks, error backpropagation algorithms, and network optimization schemes are introduced. In Sect. 4, the optimization results of neural networks are introduced, including the use of hyperparameters and error loss functions. In Sect. 5, we substitute the ANN model into a linearized governing equation to numerically solve small-scale flows and verify the accuracy of the model results through DNS data and FDNS data. Specifically, we tested the probability density distribution curve of velocity and velocity gradient, velocity gradient tensor joint probability density distribution, energy spectrum, etc. Section 6 gives the conclusions of this paper.

2 Governing Equations

This paper takes fully developed homogeneous isotropic turbulence as the modeling object. Isotropic turbulence is a hypothetical turbulence model, but its isotropic properties exist in the fine structures of anisotropic turbulence [38]. This study uses neural network modeling methods to study the characteristics of isotropic turbulence, which can expand the modeling ideas of isotropic turbulence and lay the foundation for the study of anisotropic turbulence. The data used in the study were derived from databases for direct numerical simulation of uniform isotropic turbulence and filtered direct numerical simulation.

The pseudo-spectral method is used to process the N-S equations and uses Fourier transform to transform the turbulent flow field from the physical field to the spectral space. After solving the N-S equations in the spectral space, the inverse Fourier transform is used to convert the velocities in spectral space back to the physical space. The periodic boundary conditions of isotropic turbulence make Fourier expansion accurate and effective [39]. FDNS uses the turbulence data obtained from DNS to filter out Fourier modes greater than the cutoff wave number in the turbulence energy spectrum through filtering processing and then inverts them back into physical space to obtain a large-scale velocity field. The small-scale velocity field is obtained by subtracting the large-scale velocity field from the full-scale velocity field [20, 40].

The modeling parameters of the ANN model are derived from the Navier-Stokes equations. The Navier-Stokes equations for incompressible flows are:

$$\frac{\partial u_i}{\partial t} + \frac{\partial u_i u_j}{\partial x_j} = -\frac{1}{\rho}\frac{\partial p}{\partial x_i} + \nu \frac{\partial^2 u_i}{\partial x_j \partial x_j}, \tag{1}$$

$$\frac{\partial u_i}{\partial x_i} = 0. \tag{2}$$

In the study of LES, the governing equation of large-scale flow is obtained by filtering the flow field.

$$\frac{\partial \overline{u}_i}{\partial t} + \frac{\partial \overline{u_i u_j}}{\partial x_j} = -\frac{1}{\rho}\frac{\partial \overline{p}}{\partial x_i} + \nu \frac{\partial^2 \overline{u}_i}{\partial x_j \partial x_j}. \tag{3}$$

From Eq. (1) and Eq. (3), the governing equation for the filtered small-scale flow can be obtained.

$$\frac{\partial u_i'}{\partial t} + \frac{\partial \left(u_i u_j - \overline{u_i u_j}\right)}{\partial x_j} = -\frac{1}{\rho}\frac{\partial p\prime}{\partial x_i} + \nu \frac{\partial^2 u_i'}{\partial x_j \partial x_j}, \tag{4}$$

where u_i denotes the full-scale velocity, p is the pressure, ρ is the fluid density, ν is the kinematic viscosity, $\overline{u}_i$ is the large-scale velocity, and u_i' is the small-scale velocity component. The nonlinear unclosed terms in Eq. (4): $\left(u_i u_j - \overline{u_i u_j}\right)$ is processed as follows:

$$u_i u_j - \overline{u_i u_j} = u_i u_j - \overline{u}_i \overline{u}_j - (\overline{u_i u_j} - \overline{u}_i \overline{u}_j) = u_i u_j - \overline{u}_i \overline{u}_j - \tau_{ij} =$$
$$(\overline{u}_i + u_i')(\overline{u}_i + u_j') - \overline{u}_i \overline{u}_j - \tau_{ij} = \overline{u}_i u_j' + u_i' \overline{u}_j + u_i' u_j' - \tau_{ij}, \tag{5}$$

where $\tau_{ij} = (\overline{u_i u_j} - \overline{u}_i \overline{u}_j)$ donates the sub-grid scale (SGS) stress. Combine Eq. (5) with Eq. (4) to obtain Eq. (6) and Eq. (7):

$$\frac{\partial u_i'}{\partial t} + \frac{\partial(\overline{u}_i u_j' + u_i' \overline{u}_j + u_i' u_j' - \tau_{ij})}{\partial x_j} = -\frac{1}{\rho}\frac{\partial p'}{\partial x_i} + \nu \frac{\partial^2 u_i'}{\partial x_j \partial x_j'}, \tag{6}$$

$$\frac{\partial u_i'}{\partial t} + \overline{u}_j \frac{\partial u_i'}{\partial x_j} + u_j' \frac{\partial \overline{u}_i}{\partial x_j} - \nu \frac{\partial^2 u_i'}{\partial x_j \partial x_j} - \frac{\partial \tau_{ij}}{\partial x_j} = C_i. \tag{7}$$

On the left side of Eq. (7) are the linear terms for small-scale velocity u_i' and large-scale velocity $\overline{u}_i$, which represent the acceleration, the convection of small-scale by the large-scale velocity $\overline{u}_i$, the effects of the shear deformation on the small scales, viscous dissipation, and sub-grid stress, respectively. The right side of the Eq. (7) represents the sum of the nonlinear convection and pressure gradient terms,

$$C_i = -\frac{1}{\rho}\frac{\partial p\prime}{\partial x_i} - \frac{\partial u_i' u_j'}{\partial x_j}, \tag{8}$$

which shall be modeled. Due to the interaction between large-scale flow and small-scale flow in the flow field, it is believed that large-scale velocity and velocity gradient influence C_i. Therefore, we used the ANN model to establish a mapping between large-scale velocity, velocity gradient tensor, and model parameter C_i.

3 ANN Modeling

In biological neural networks, neuronal cells use the specialized nature of the outer membrane to receive signals from other neurons through dendrites, integrate these signals by the cell body, and then transmit information to other neurons through axons to transmit information. Artificial Neural Network is a machine learning method that simulates biological neural networks in computer science research. Artificial neural networks mimic biological neurons, using a neuron-like M-P model (see Fig. 1) [41] as the basic unit to build the model.

In the ANN, x_j represents the input features, y_i represents the output labels, ω_{ij} denotes the weight of each input, θ_i specifies the corresponding threshold, f is the activation function (refers to the function of mapping the neuron inputs to the outputs, generally a step function or sigmoid function) to process the input values. The expression for output y_i is written as:

$$y_i = \sum\nolimits_{j=1}^{n} f(\omega_{ij} x_j - \theta_i). \tag{9}$$

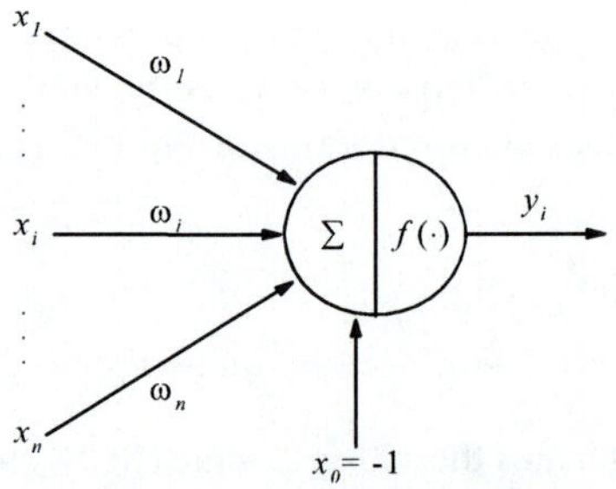

Fig. 1. M-P model of a neural cell.

Different neurons are connected by the outputs of the previous layer as the inputs of the next layer, forming an overall model of the network of neurons. In this study, the main use of feedforward neural networks - neurons are arranged in layers, and the neurons in each layer only connect with neurons in adjacent layers before and after, and neurons between the same layer and the compartment do not make connections (see Fig. 2).

In this study, the model was improved by changing the number of neuronal layers and the number of neurons, and the weight ω_{ij} and threshold θ_i in the neural network were optimized by the error Back Propagation algorithm [42]. We use the error back-propagation algorithm to optimize the model. First, we calculate the root mean square error E_k between the ANN output y and the theoretical value y_0.

$$E_k = \frac{1}{n}\sum\nolimits_{j=1}^{n}\left(y_j - y_{j0}\right)^2. \tag{10}$$

Then calculate the negative gradient of the error E_k with respect to each of the weights ω_{ij}, thresholds θ_i on the error representing their effect on the final output y.

$$\Delta\omega_{ij} = -\eta(\partial E_k/\partial\omega_{ij}), \tag{11}$$

$$\Delta\theta_i = -\eta(\partial E_k/\partial\theta_i), \tag{12}$$

where η donates the learning rate of the model, which controls the amplitude of each parameter update. A too large learning rate will cause the optimization parameters to fluctuate near the optimal value, and a too small learning rate can cause the model to be optimized too slowly. The update expression for the parameter is:

$$\omega_{ij} \leftarrow \omega_{ij} + \Delta\omega_{ij}, \tag{13}$$

$$\theta_i \leftarrow \theta_i + \Delta\theta_i. \tag{14}$$

Finally, through the continuous iteration of this method, the error between the output value of the model and the theoretical value can be gradually reduced. When the error is less than a given small number, the model training is complete.

In the study, we used a total of 12 physical quantities, including 3 components of large-scale velocity $\overline{\boldsymbol{u}}$ and 9 components of velocity gradient tensor $\nabla\overline{\boldsymbol{u}}$, as inputs to

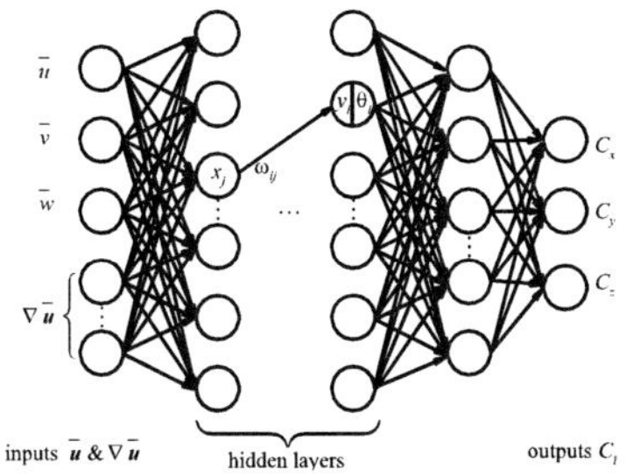

Fig. 2. Feedforward Neural Network. The inputs include large scale velocities $\overline{u}, \overline{v}, \overline{w}$ and velocity gradient tensor $\nabla\overline{\boldsymbol{u}}$. Output C_i contains components in the x, y and z directions.

the neural network. With the model parameter C_i as outputs, a fully connected neural network with 5 hidden layers and 128 network nodes per hidden layer is constructed. The flow fields with Taylor Reynolds numbers 64.43 and 128.78 were trained. Since the input has multiple features, to ensure that each feature has the same impact on the network during initial training, the maximum and minimum values of each feature are used to normalize the data.

$$x = \frac{x - x_{\min}}{x_{max} - x_{min}} \tag{15}$$

The normalized data is in the [0, 1] range, and can be restored through inverse normalization. To ensure the applicability of the model, the numerical data is randomly divided into training data sets (X_train, y_train) and test data sets (X_test, y_test). The amount of data in the training data set is larger than that in the test data set. The training dataset is brought into the neural network model to optimize the weights and threshold, and the test dataset is substituted into the optimized model to calculate the loss function size to verify the accuracy of the model.

The sigmoid function is used instead of the step function to avoid discontinuity in the substitution of the step function itself. The properties of the two are roughly similar, mainly used to indicate that they are activated when the input quantity is multiplied by the weight and the threshold value is greater than 0. The output is 1, otherwise, the output is 0.

The optimization of the network takes the loss function as a reference. This model is essentially a regression problem that uses mean square error as the loss function. The expression is written as:

$$cost = \frac{1}{n}\sum\nolimits_{j=1}^{n}\left(y_{pre} - y_0\right)^2 \tag{16}$$

The model randomly divides the data into training data sets and test data sets according to a ratio of 9:1 between the training set and the test set. The model adopts a small batch training method. Small batch training refers to taking sample values from a training dataset. Because the data exists in a unified sample, it is considered that there should be the same inherent laws in each data. Therefore, the amount of data selected should not affect the training results of the final model. On this basis, small batch training is to subdivide each round of training into n_{batch} training sessions, with each epoch randomly capturing batch_size pieces of data in the training dataset. The relationship between n_{batch}, batch_size, and the total amount of data in the training dataset training_data_size, is:

$$n_{batch} = \frac{training\_data\_size}{batch\_size} \tag{17}$$

4 Training and Validation

Model optimization is a process of continuously adjusting parameters and iterative training. Hyperparameters refer to parameters that have been set before the model starts training. The hyperparameters in this model mainly include the number of neurons in the hidden layer, initial learning rate, decay-rate, momentum size, small batch training rounds, and small batch data size. Different hyperparameters have different effects on the fitting degree and operation speed of the network. The study adopted a group of excellent super parametric combinations: the learning rate was 0.001, and the decay-rate was 0.96, which means that the learning rate decayed to 0.96 times the initial value per 1000 steps. The validation of the cost function with the training steps is shown in Fig. 3.

Initially, the weight parameter inside the model is set randomly, and the threshold parameter is 0. Therefore, the initial error is very large. With the training of the ANN model, the error backpropagation algorithm continuously updates the weights and biases, and the error decreases rapidly. The learning process is well-formulated since the training and validation losses are nearly stable after 500 global iterations. The errors in the training and validation datasets exhibit similar variations, which shows that the ANN model has been successfully trained.

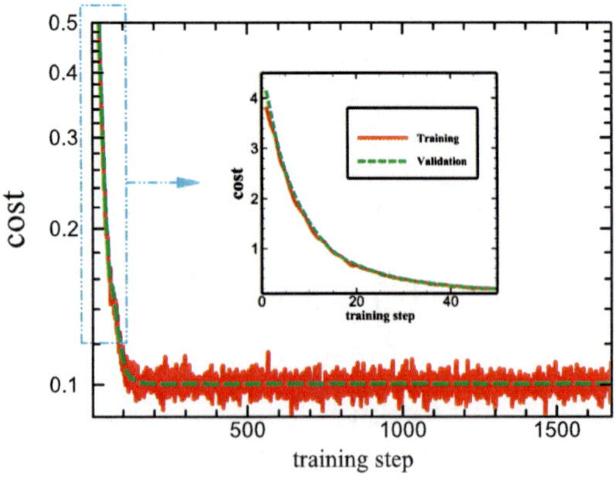

Fig. 3. Training and validation cost with the training step.

5 Results and Discussion

After training the ANN model, we can calculate the model parameter C_i by inputting the large-scale velocity and velocity gradient tensor of the initial flow field. Then, by substituting C_i into the governing equation and combining the initial value of the small-scale velocity, we can obtain the subsequent small-scale flow field numerically. Subsequent calculations only need the large-scale flow velocity from FDNS. The small-scale velocity is derived from the initial flow field, the large-scale velocity, and the governing equations.

The calculation process is shown in Fig. 4. First, input the initial flow field information - including large-scale flow and small-scale flow velocity. Then, calculate the model parameter C_i of the current flow field through the ANN model, and substitute it into the governing equation to numerically solve the small-scale flow field velocity at the next time-step. The large-scale flow at the later time-step is calculated by FDNS. In this way, the flow field information at the next time-step can be obtained. By iterating this process, the velocity of the flow field at any time can be calculated.

In this paper, we have deduced the flow field from the initial time to 500 time-steps, approximately more than 30 Kolmogorov timescales. The flow field at this time was used for data analysis to verify the effectiveness of the ANN model. We used two sets of flow field data with Reynolds numbers of 64.43 and 128.78 to evaluate the ANN model. The data are derived from DNS, FDNS, and ANN model. We compare the data obtained by the ANN model with the data calculated by DNS and FDNS.

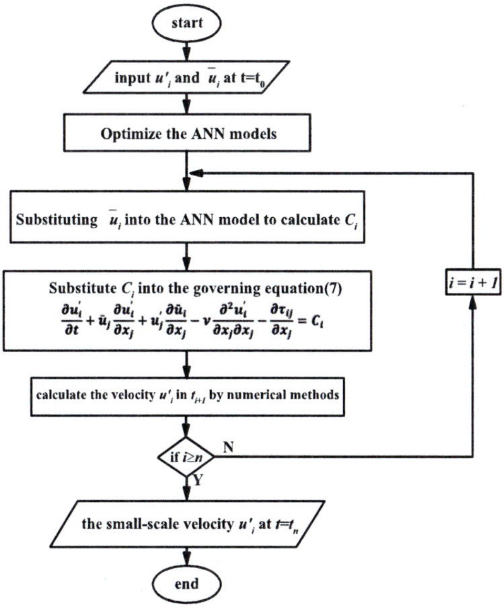

Fig. 4. Flow chart of small-scale velocity prediction.

Figure 5 shows the vortex contour for different models at the same time. In the area where the vortex structures on the right side change significantly in Fig. 5(a), the

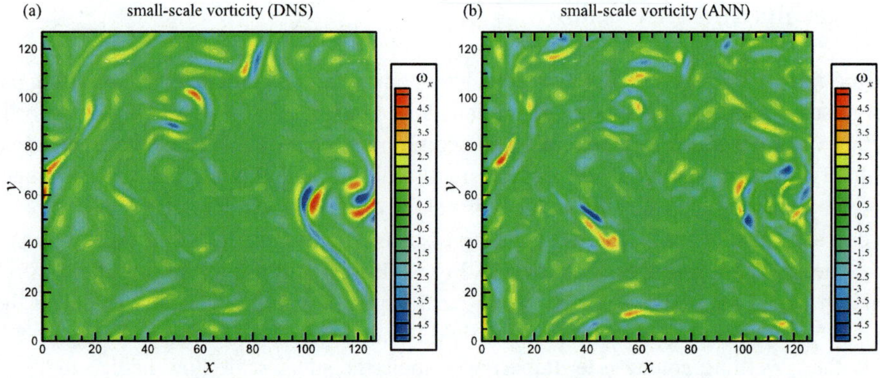

Fig. 5. Two-dimensional contours of small-scale vorticity ($Re_\tau = 64.43$). (a) the small-scale vorticity calculated by DNS and FDNS; (b) the small-scale vorticity calculated by ANN models.

structures obtained from the ANN model in Fig. 5(b) are consistent with those obtained by DNS. It can be seen from the contour that the ANN model can predict the occurrence of vortex well, and the vortex structures in the flow can be well reproduced. The flow field calculated by the ANN model can recover the structural characteristics of small-scale flow fields to a certain extent.

The probability density functions (PDFs) of small-scale velocity obtained by DNS and ANN model are shown in Fig. 6. Figure (a) shows the result at $Re_\tau = 64.43$, and Figure (b) shows the result at $Re_\tau = 128.78$. Unlike the Gaussian PDFs of the full-scale velocity, we can observe that the PDFs of small-scale velocity behaves apart from the Gaussian distribution at the tails. Furthermore, the ANN results of both sets of data can well match the DNS data. The results show that the ANN network is feasible in reconstructing the velocity distribution of small-scale flow fields.

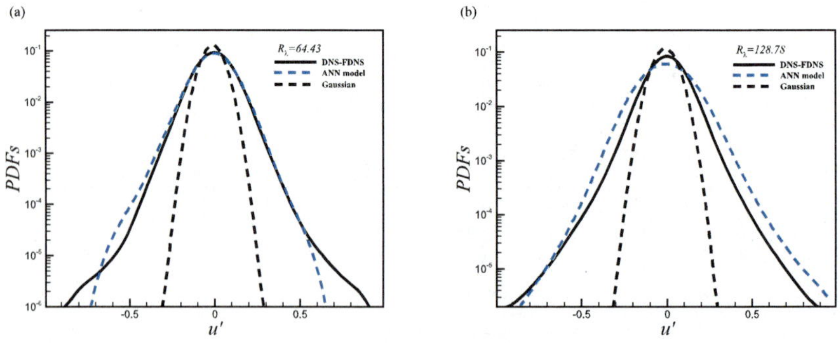

Fig. 6. The PDF of small-scale velocity. (a): Re = 64.43, (b): Re = 128.78.

Intermittency is an important property of small-scale turbulence. Statistically, intermittency is characterized by the non-Gaussian nature of the probability density functions of turbulent dissipation and velocity gradients. The non-Gaussian nature becomes increasingly evident as the Reynolds number increases. On the PDF diagram, the tails will significantly deviate from the Gaussian curve, and the probability of data at both ends will significantly increase. Figure 7 is a comparison of the PDFs of the velocity gradients from the ANN model and DNS data, where Fig. 7 (a) shows the result at $Re_\tau = 64.43$, and Fig. 7 (b) shows the result at $Re_\tau = 128.78$. Statistically, the ANN model can recover the intermittency of the small structures in turbulent flows.

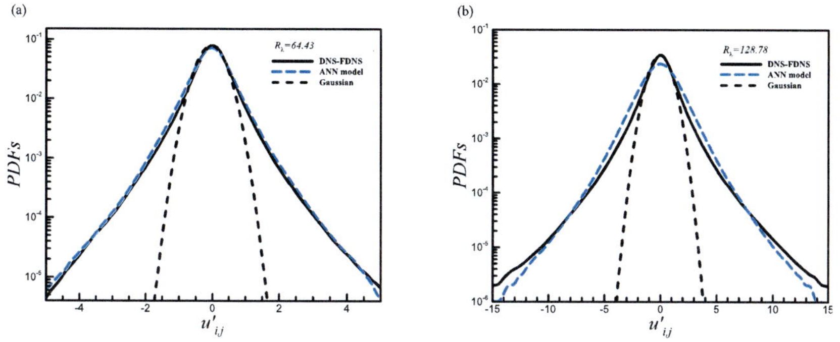

Fig. 7. The PDF of small-scale velocity gradient tensor. (a): Re = 64.43, (b): Re = 128.78.

To further examine the capability of the ANN model to recover the turbulent structures [43], the Q-R joint probability density distribution of the velocity gradient tensor was examined (see Fig. 8). Q and R are the second and third invariants of the velocity gradient tensor, respectively, $Q \equiv \frac{1}{2}\left(u_{i,i}^2 - u_{i,j}u_{j,i}\right)$ and $R \equiv Det(u_{i,j})$ [44]. According to the comparison of two sets of data, we can observe that the ANN model performs a good reconstruction of the flow structures.

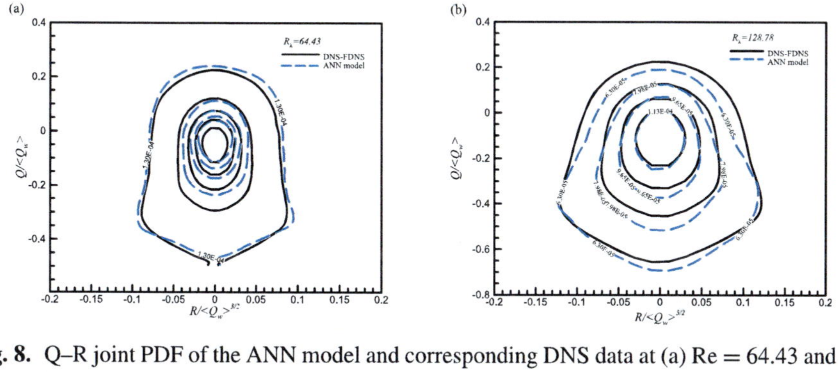

Fig. 8. Q–R joint PDF of the ANN model and corresponding DNS data at (a) Re = 64.43 and (b) Re = 128.78. Q and R are the second and third invariants of the velocity gradient tensor.

The energy spectrum of turbulent velocity is displayed in Fig. 9 for the DNS, FDNS, and ANN models. Due to the sharp-spectral filter used in FDNS, when the wave number k is greater than the cutoff wavenumber, the energy spectrum value is 0. The flow field

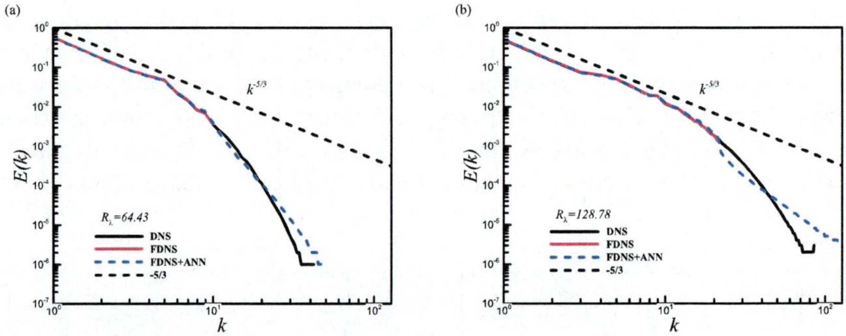

Fig. 9. Energy spectra of the isotropic turbulent flow fields obtained from DNS coupled with FDNS and the ANN model. (a) Re = 64.43, (b) Re = 128.78.

energy spectrum calculated by the ANN model can well fit the DNS curves when the wave number is smaller than the cutoff wavenumber. When the wave number is larger than the truncated wave number, the curve trend is consistent with that of DNS. According to the transformation between physical space and Fourier space, the larger the wave number, the smaller the corresponding physical spatial scale. Therefore, the ANN model can supplement small-scale information of the flow field, and the obtained energy spectrum is relatively accurate.

Next, we apply the ANN model to a prior LES of the statistics of inertial particles in isotropic turbulent flows. We calculate the motion of the inertial particles using DNS, FDNS and FDNS with the ANN model by tracking the Lagrangian trajectory of individual particles. The ANN model combined with FDNS is used to recover the missing small-scale turbulent motions in FDNS and drive the motion of particles.

The equations of motion of inertial particles are

$$\begin{cases} \frac{d\mathbf{x}_p(t)}{dt} = \mathbf{v}_p(t) \\ \frac{d\mathbf{v}_p(t)}{dt} = \frac{[\mathbf{u}^*(\mathbf{x}_p(t),t)-\mathbf{v}_p(t)]f}{\tau_p} \end{cases} \tag{18}$$

where $\mathbf{x}_p(t)$, $\mathbf{v}_p(t)$ are the particle position and velocity at time t, $\mathbf{u}^*(\mathbf{x}_p(t), t)$ is the fluid velocity seen by the particle, which can be calculated by a three-dimensional sixth-order Lagrangian interpolation scheme. $\mathbf{u}^*(\mathbf{x}_p(t), t)$ is used to represent $\mathbf{u}$ in DNS, $\tilde{\mathbf{u}}$ in FDNS and $\tilde{\mathbf{u}} + \mathbf{u}'_{ANN}$ in FDNS with the ANN model. f denotes the nonlinear drag coefficient $f(Re_p) = 1 + 0.15Re_p^{0.687}$, $Re_p = |\mathbf{u}^* - \mathbf{v}_p| d_p/\nu$ is the particle Reynolds number, $\tau_p = \rho_p d_p^2/18\rho_f \nu$ is the particle response time, and $d_p = 0.4\eta$ and η is the Kolmogorov length scale. To characterize the inertia of particles, we define the Stokes number

$$St = \frac{\tau_p}{\tau_\eta}, \tag{19}$$

where τ_η is the turbulent Kolmogorov timescale. We set St = 0.5, 1, and 2 in this study.

In order to validate the performance of the ANN model in a prior LES of particle relative motions, which is a long-lasting challenge for LES of small-scale process, we statistically analyze the Radial Distribution Function (RDF) and the Radial Relative Velocity (RRV) of particle pairs [45–47]. The Radial Distribution Function is defined as the relative probability of finding a pair of particles at a given separation distance within a unit volume, compared to the expected number of pairs in a uniform distribution of particles:

$$g(r) = \frac{Q_{p,r}/\Delta V_r}{Q_p/V} = \frac{\frac{Q_{p,r}}{\frac{4}{3}\pi\left[\left(r+\frac{1}{2}dr\right)^3-\left(r-\frac{1}{2}dr\right)^3\right]}}{\frac{\frac{1}{2}N_p(N_p-1)}{(2\pi)^3}} \tag{20}$$

where N_p is the total number of particles in the turbulent flow. The number of particle pairs is given by $Q_p = \frac{1}{2}N_p(N_p-1)$. $Q_{p,r}$ represents the average number of particles found within the spherical shell volume ΔV_r with a distance of $r = |r|$ from the detecting particle. Here, $\Delta V_r = \frac{4}{3}\pi\left[\left(r+\frac{1}{2}dr\right)^3 - \left(r-\frac{1}{2}dr\right)^3\right]$, and $V = (2\pi)^3$ represents the total volume of the flow field. The RDF g(r) is a measure of particle preferential concentration, or clustering in turbulent flows. The Radial Relative Velocity of particle pairs is defined as:

$$w_r(r) = \left[\boldsymbol{v}_{p1}\left(\boldsymbol{x}_{p1}, t\right) - v_{p2}\left(\boldsymbol{x}_{p2}, t\right)\right] \cdot \frac{\boldsymbol{l}}{|\boldsymbol{l}|} \tag{21}$$

where $\boldsymbol{v}_{p1}\left(\boldsymbol{x}_{p1}, t\right)$ and $v_{p2}\left(\boldsymbol{x}_{p2}, t\right)$ are the velocities of two particles located at $\boldsymbol{x}_{p1}$ and $\boldsymbol{x}_{p2}$, respectively, and the relative separation between the two particles is $\boldsymbol{l} = \boldsymbol{x}_{p1} - \boldsymbol{x}_{p2}$. We test the relative separation between particles, $r \in [0.4\eta, 6.0\eta]$, divided into 280 segments. For each detecting particle pairs, we search particles within a spherical shell with a thickness of 0.02η to calculate the RDF and RRV.

Figure 10 (a) (c) (e) compares the curves of RDF as a function of dimensionless distance for St = 0.5, 1.0, and 2.0, in DNS, FDNS, and FDNS + ANN. Compared to DNS, FDNS exhibits significant differences in RDF due to the loss of small-scale velocity caused by filtering. It is clearly observed that FDNS underpredicts the RDF at St = 0.5 and overpredicts the RDF at St = 2.0. The missing small-scale turbulence has little effects on the RDF at St = 1.0. With the incorporation of the ANN model, the contributions of the missing small-scale velocity to particle clustering, RDF are compensated. The RDF of FDNS + ANN is well consistent with that of DNS. Figure 10 (b) (d) (f) compares the curves of the mean of the absolute RRV as a function of dimensionless distance for DNS, FDNS, and FDNS + ANN at different St numbers. Compared to DNS, FDNS underpredicts the RRV of particles due to the loss of small-scale velocity caused by filtering at all of the 3 Stokes numbers. The inclusion of the ANN model compensates for the small-scale velocity and mitigates the RRV errors between FDNS and DNS at all of the 3 St numbers. Therefore, with the inclusion of the ANN model in FDNS, significant improvements in the statistics of particle relative motions can be achieved.

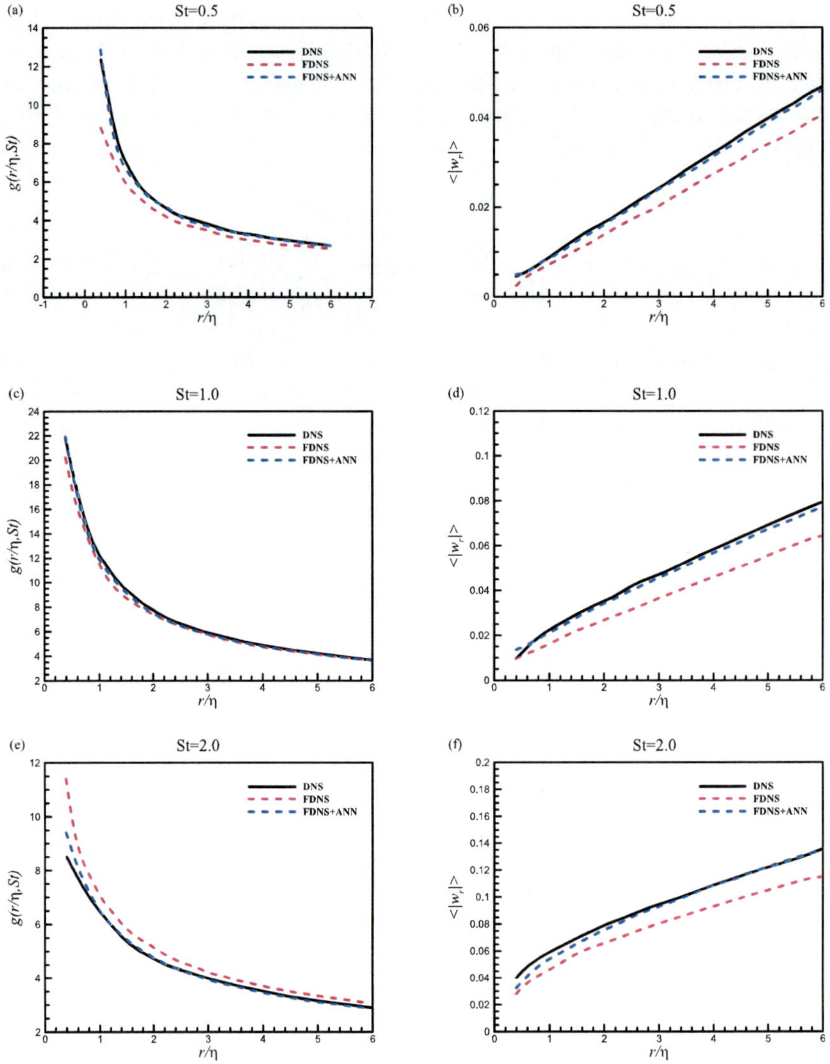

Fig. 10. Variation of the RDF [(a) (c) and (e)] and RRV [(b) (d) and (f)] with r/η in the DNS, FDNS, and FDNS with ANN model for St $= 0.5$, 1.0, and 2.0.

6 Conclusion

In this study, we have modeled the nonlinear terms and pressure gradient terms in the small-scale turbulence equations using an ANN model. The ANN model is used to calculate these two terms from large-scale information of the flow field. The equation is approximately linearized. In ANN modeling, large-scale flow velocity and velocity gradient tensor are used as inputs for optimization using an error backpropagation algorithm. Combining equations with ANN models can numerically solve turbulent small-scale flows. The study used flow fields with Reynolds numbers of 64.43 and 128.78 to test the model. By comparing the ANN and DNS data, the ANN model can well maintain the vortex structure of small-scale flow in the prediction of the flow field. Statistically, the probability density distribution of small-scale velocity and velocity gradient of the ANN model and DNS can also be well consistent. Through velocity gradient probability density distribution, ANN data can also reflect the intermittency of small-scale turbulence. Since the ANN model is constructed using FDNS data for small-scale turbulence, it can be seen from the energy spectrum that the ANN model can supplement the small-scale flow field energy filtered out by FDNS during the filtering process. The addition of the ANN model serves as a supplement to the small-scale flow, allowing FDNS to make more accurate predictions of the relative motion of inertial particles. Once the neural network model is modeled, the Reynolds number and resolution for the same flow field remain unchanged, so only one network training is required, and then it can be directly used. This makes this research method a good application prospect in engineering applications. Combining the method of large eddy simulation can not only obtain relatively accurate small-scale flow field data but also save a lot of computational time and cost compared with DNS. The results of this paper show that the method is feasible for constructing small-scale flow fields, and the calculated data are reliable. This study expands the research methods of turbulent small-scale flows and further proves the feasibility of applying ANN modeling to turbulent research.

In future work, we will continue to study how many time-steps this model can recover the missing SGS flow field while ensuring data accuracy. Our research currently focuses only on flows at Reynolds numbers of 64.43 and 128.78, flows at higher Reynolds numbers need to be studied in future work.

Acknowledgments. This work was supported by the NSFC Basic Science Center Program for "Multiscale Problems in Nonlinear Mechanics" (Grant No. 11988102), the NSFC Program (Grant No. 12272380), the National Key Project (Grant No. GJXM92579).

References

1. Lesieur, M., Metais, O.: New trends in large-eddy simulations of turbulence. Annu. Rev. Fluid Mech.. Rev. Fluid Mech. **28**, 45–82 (1996)
2. Meneveau, C., Katz, J.: Scale-invariance and turbulence models for large-eddy simulation. Annu. Rev. Fluid Mech.. Rev. Fluid Mech. **32**(1), 1–32 (2000)
3. Sreenivasan, K.R., Antonia, R.A.: The phenomenology of small-scale turbulence. Annu. Rev. Fluid Mech.. Rev. Fluid Mech. **29**, 435–472 (1997)
4. Ghate, A.S., Lele, S.K.: Gabor mode enrichment in large eddy simulations of turbulent flow. J. Fluid Mech. **903**, A-13 (2020)
5. Pierce, C.D., Moin, P.: Progress-variable approach for large-eddy simulation of non-premixed turbulent combustion. J. Fluid Mech. **504**, 73–97 (2004)
6. Smagorinsky, J.: General circulation experiments with the primitive equations: I the basic experiment. Mon. Weather Rev. **91**(3), 99–164 (1962)
7. Bardina, J., Ferziger, J., et al.: Improved subgrid-scale models for large-eddy simulation. In: Fluid and Plasma Dynamics Conference, 13th, Snowmass, CO, U.S.A. American Institute of Aeronautics and Astronautics (1980)
8. Bardina, J., Ferziger, J., et al.: Improved turbulence models based on large eddy simulation of homogeneous, incompressible turbulent flows. Ph.D. thesis (1983)

9. Clark, R.A., Ferziger, J.H., et al.: Evaluation of subgrid-scale models using an accurately simulated turbulent flow. J. Fluid Mech. **91**(01), 1–16 (1979)
10. Zang, Y., Street, R., et al.: A dynamic mixed subgrid-scale model and its application to turbulent recirculating flows. Phys. Fluids A **5**, 3186–3196 (1993)
11. Germano, M.: Turbulence: the filtering approach. J. Fluid Mech. **238**, 325–336 (1992)
12. Germano, M., Piomelli, U., et al.: A dynamic subgrid-scale eddy viscosity model. Phys. Fluids A **3**, 1760–1765 (1990)
13. Vreman, B., Geurts, B., et al.: Large-eddy simulation of the temporal mixing layer using the Clark model. Theoret. Comput. Fluid Dyn.. Comput. Fluid Dyn. **8**(4), 309–324 (1996)
14. Morinishi, Y., Vasilyev, O.V.: A recommended modification to the dynamic two-parameter mixed subgrid scale model for large eddy simulation of wall bounded turbulent flow. Phys. Fluids **13**(11), 3400–3410 (2001)
15. Yang, Z., Cui, G., et al.: A modified nonlinear sub-grid scale model for large eddy simulation with application to rotating turbulent channel flows. Phys. Fluids **24**(7), 075113 (2012)
16. Yang, Z., Cui, G., et al.: Large eddy simulation of rotating turbulent channel flow with a new dynamic global-coefficient nonlinear subgrid stress model. J. Turbul.Turbul. **13**, 1–20 (2012)
17. Ghaisas, N., Frankel, S.: Dynamic gradient models for the sub-grid scale stress tensor and scalar flux vector in large eddy simulation. J. Turbul.Turbul. **17**, 30–50 (2016)
18. Ghosal, S., Lund, T.S., et al.: A dynamic localization model for large-eddy simulation of turbulent flows. J. Fluid Mech. **286**, 229–255 (1995)
19. Zhou, Z., Wang, S., et al.: A structural subgrid-scale model for relative dispersion in large-eddy simulation of isotropic turbulent flows by coupling kinematic simulation with approximate deconvolution method. Phys. Fluids **30**(10), 105110 (2018)
20. Jin, G., He, G.: A nonlinear model for the subgrid timescale experienced by heavy particles in large eddy simulation of isotropic turbulence with a stochastic differential equation. New J. Phys. **15**(3), 1–27 (2013)
21. Mcculloch, W.S., Pitts, W.: A logical calculus of the ideas immanent in nervous activity. J. Symb. Log.Symb. Log. **5**, 115–133 (1943)
22. He, Q., Li, N., et al.: A survey of machine learning algorithms for big data. Moshi Shibie yu Rengong Zhineng/Pattern Recogn. Artif. Intell. **27**(4), 327–336 (2014)
23. LeCun, Y., Bengio, Y., et al.: Deep learning. Nature **521**(7553), 436–444 (2015)
24. Hornik, K., Stinchcombe, M., et al.: Multilayer feedforward networks are universal approximators. Neural Netw.Netw. **2**, 359–366 (1989)
25. Brunton, S., Noack, B., et al.: Machine learning for fluid mechanics. Annu. Rev. Fluid Mech.. Rev. Fluid Mech. **52**, 477–508 (2020)
26. Yarlanki, S., Rajendran, B., et al.: Estimation of turbulence closure coefficients for data centers using machine learning algorithms. In: IEEE 2012 13th IEEE Intersociety Conference on Thermal and Thermomechanical Phenomena in Electronic Systems (ITherm), San Diego, CA, USA, pp. 38–42 (2012)
27. Wang, J., Wu, J., et al.: Physics-informed machine learning approach for reconstructing Reynolds stress modeling discrepancies based on DNS data. Phys. Rev. Fluids **2**, 034603 (2017)
28. Fukami, K., Fukagata, K., et al.: Super-resolution reconstruction of turbulent flows with machine learning. J. Fluid Mech. **870**, 106–120 (2019)
29. Qu, J., Tang, Z., et al.: Turbulence feature extraction and analysis based on convolutional neural networks. In: The 10th National Conference on Fluid Mechanics (2018). (in Chinese)
30. Gamahara, M., Hattori, Y.: Searching for turbulence models by artificial neural network. Phys. Rev. Fluids **2**(5), 054604 (2016)
31. Sarghini, F., Felice, G., et al.: Neural networks based subgrid scale modeling in large eddy simulations. Comput. Fluids. Fluids **32**(1), 97–108 (2003)

32. Xie, C., Wang, J., et al.: Spatial artificial neural network model for subgrid-scale stress and heat flux of compressible turbulence. Theor. Appl. Mech. Lett.. Appl. Mech. Lett. **10**(1), 27–32 (2020)
33. Zhou, Z., He, G., et al.: Subgrid-scale model for large-eddy simulation of isotropic turbulent flows using an artificial neural network. Comput. Fluids. Fluids **195**, 104319 (2019)
34. Beck, A., Flad, D., et al.: Deep neural networks for data-driven LES closure models. J. Comput. Phys.Comput. Phys. **398**, 108910 (2019)
35. Milano, M., Koumoutsakos, P.: Neural net modeling for near wall turbulent flow. J. Comput. Phys.Comput. Phys. **182**(1), 1–26 (2002)
36. Yin, Y., Li, H., et al.: Application of machine learning assisted turbulence modeling in flow separation prediction. Acta Aerodynamica Sinica **39**(2), 23–32 (2021)
37. Xie, C., Yuan, Z., et al.: Artificial neural network-based subgrid-scale models for large-eddy simulation of turbulence (in Chinese). Lixue Xuebao/Chin. J. Theor. Appl. Mech. **53**(1), 1–16 (2021)
38. Hinze, J.O.: Turbulence: An Introduction to Its Mechanism and Theory. McGraw-Hill Book Company, inc., New York (1960)
39. Zhang, Z., Cui, G., et al.: Theory and Modeling of Turbulence. 2nd edn. Tsinghua University Press, Beijing (2017)
40. Yan, X., Li, J., et al.: The influence of sub-grid scale motions on particle collision in homogeneous isotropic turbulence. Acta Mech. Sin. **34**(1), 22–36 (2018)
41. Jiao, L., Yang, S., et al.: Seventy years beyond neural networks: retrospect and prospect. JisuanjiXuebao (Chin. J. Comput.) **39** (2016)
42. Rumelhart, D., Williams, R., et al.: Learning representations by back-propagating errors. Nature **323**, 533–536 (1986)
43. Liu, B., Tang, J., et al.: Deep learning methods for super-resolution reconstruction of turbulent flows. Phys. Fluids **32**(2), 025105 (2020)
44. Jeong, J., Hussain, F.: On the identification of a vortex. J. Fluid Mech. **285**, 69–94 (1995)
45. Zhou, Z.: Large-eddy simulation of particle-laden isotropic turbulent flows and sub-grid scale models (2019)
46. Ray, B., Collins, L.R.: A subgrid model for clustering of high-inertia particles in large-eddy simulations of turbulence. J. Turbul.Turbul. **15**(6), 366–385 (2014)
47. Sundaram, S., Collins, L.R.: Collision statistics in an isotropic particle-laden turbulent suspension .1. Direct numerical simulations. J. Fluid Mech. **335**, 75–109 (1997)

Unveiling the Dynamics of Snow Settling in Atmospheric Turbulence: A Review of Nearly a Decade of Field Research at EOLOS, MN

Jiaqi Li[1,2] and Jiarong Hong[1,2](✉)

[1] Saint Anthony Falls Laboratory, University of Minnesota, Minneapolis, MN 55414, USA
[2] Department of Mechanical Engineering, University of Minnesota, Minneapolis, MN 55455, USA
jhong@umn.edu

Abstract. This paper provides a comprehensive synthesis of a decade's advancements in field investigations into snow settling dynamics, carried out at the EOLOS field research station in Rosemount, MN, USA. The research deployed a succession of progressively sophisticated particle characterization and tracking systems, culminating in the development of a 3D particle tracking velocimetry (PTV) system. Our findings show the compelling role of preferential sweeping mechanisms, noting a consistent enhancement in snow particle settling velocities in association with turbulent eddies across multiple deployments. However, the magnitude of this turbulence-induced enhancement displayed variability, potentially modulated by the differences in turbulence and snow conditions inherent to each deployment. Additionally, intriguing patterns in snow clustering were revealed, characterized by an exponent power-law decay of -2 in cluster size distribution. This, coupled with a vertical orientation of clusters indicative of gravitational influences, adds a layer of complexity to our understanding of snow dynamics. Our research further deciphers the interplay between preferential sweeping and the preferential concentration, or clustering, of snow particles. Higher concentrations of snow particles were consistently found on the downward side of vortices, irrespective of vortex type, with an accelerated settling in these regions. Intriguingly, this disparity in average settling velocity was more pronounced in the case of prograde vortices. The advent of 3D particle tracking velocimetry (PTV) has allowed a nuanced examination of 3D settling trajectories and clustering dynamics. By providing insights into the intricate movements of snow particles as they traverse atmospheric turbulence, it uncovers the subtle variations in motion that could be overlooked in a 2D analysis. These findings offer valuable insights into the effects of snow particle morphology and turbulence on settling dynamics, promising to improve the accuracy of snowfall forecasts and ground snow accumulation predictions.

X. Zheng and S. Balachandar (Eds.): IUTAM 2023, IUTAM Bookseries 41, pp. 58–82, 2024.
https://doi.org/10.1007/978-3-031-47258-9_4

Keywords: snow settling · snow clustering · PIV · PTV · field measurement

1 Introduction

Predicting ground snow accumulation is a pivotal task, touching on diverse applications from anticipating flood risk due to snowmelt [31], providing avalanche hazard warnings [45], to ensuring traffic safety during snowfalls [37]. This prediction hinges heavily on understanding the intricacies of snow settling dynamics in the atmosphere [24], a process that is far from simple. Influenced by the morphological characteristics of snow particles and the impacts of atmospheric turbulence, it presents challenges that are not easily replicated in lab conditions or numerical simulations. Cloud conditions, including temperature and humidity, can dramatically shape the morphology of snow particles [29], thus altering their settling velocities. These velocities can fluctuate from virtually zero to up to 3 m/s [16,28], adding a dynamic element to the snow's path as it descends from cloud to ground, possibly drifting over vast distances. Atmospheric turbulence, known for its complex structures and high Reynolds numbers, introduces an additional layer of complexity to the settling process. This complexity underscores the need for more comprehensive research into the physical mechanisms underlying snow settling, to not only improve the precision of snow accumulation models but also enhance our ability to predict and prepare for the diverse impacts of snowfall.

In the previous scientific investigations, the influence of snow morphology on settling velocity has received much attention, although earlier studies did not explicitly account for the impact of atmospheric turbulence. The pioneering empirical work of Locatelli & Hobbs [28] laid the groundwork by establishing correlations between snow particle dimensions, mass, and settling velocity, each tailored to distinct snow morphologies through field measurements. Passarelli [39] examined the effect of snow collection efficiency on the formation of aggregates and their subsequent fall speed. Moreover, Mitchell, Zhang & Pitter [34] explored the snowfall rate under the influence of riming, a process in which supercooled water droplets deposit on snow crystals. The advent of the Multi-Angle Snowflake Camera (MASC) system [15] marked a significant breakthrough, providing a more nuanced approach to assessing snow morphology and settling velocity across diverse snow types. On the theoretical and modeling front, Böhm [8] proposed a universal model to estimate snow settling velocity, encapsulating the correlation among drag coefficient, particle Reynolds number, and Best number. His model ingeniously linked particle shape and settling dynamics by considering the ratio of the actual snow projected area to the elliptical area enclosing it. Building on Böhm's foundational work, Mitchell & Heymsfield [33] advanced the understanding of ice particle fall speeds across a range of shapes, refining the relationship between Reynolds number and Best number. Heymsfield & Westbrook [20] then introduced new computational methods for predicting snow fall speed. Their approach, grounded in laboratory measurements on various snow morphologies, effectively corrects the projected area ratio bias. These collective efforts underscore the pivotal role of snow morphology in understanding and predicting snow settling velocity.

Taking the study beyond morphology alone, it becomes apparent that atmospheric turbulence plays a considerable role in snow settling dynamics. The complexity of this influence is evident when considering that snow particles, given their range of shapes, sizes, and densities greater than surrounding air, effectively behave as heavy inertial particles within turbulent conditions. This behavior results in characteristic patterns, such as clustering and settling velocity modulation, strongly indicating the presence of intricate particle-turbulence interactions [6]. Influenced by factors like turbulence characteristics, represented by the Kolmogorov time scale, τ_η, and particle properties such as size (D_p), density (ρ_p), and aerodynamic characteristics. These particle properties collectively define the particle response time, τ_p [32], a key mechanism within this dynamic unfolds. This mechanism, known as preferential sweeping, leads particles to accumulate in strain-dominated regions, descending as clusters [3,12,17,55,56], and can significantly amplify the settling velocity when the Stokes number $St = \tau_p/\tau_\eta$ is around 1 [3,14,42,56]. However, it is important to highlight that much of the existing literature has been based on the simplified assumption of spherical or point particles, which overlooks the intricate diversity of snow morphologies, suggesting a crucial gap that future research needs to address in order to enhance our understanding of snow settling dynamics in turbulent conditions.

Moving beyond the common assumption of spherical particles in particle-turbulence interaction studies, several investigations have expanded their scope to consider non-spherical geometries, including disks, ellipsoids, and fibers. This shift in focus has yielded valuable insights into the settling behavior of non-spherical particles in turbulence, revealing unique mechanisms influenced by turbulence. Among these mechanisms, the preferential orientation of non-spherical particles has emerged as a significant factor, as has the presence of preferential sweeping [44]. Notably, the work by Siewert et al. [44] employed direct numerical simulation (DNS) to delve into these phenomena, providing a robust computational perspective. Voth & Soldati [54] further contributed to this field with a comprehensive review of anisotropic particles in turbulence, reinforcing the importance of considering non-spherical particles in such studies. A standout study in this area is the work by Esteban, Shrimpton, & Ganapathisubramani [11], who meticulously examined the settling dynamics of disk-shaped particles in turbulence. Using a stereoscopic vision system consisting of two digital cameras and a diffuse light source, they traced the complex three-dimensional trajectories of these particles. Their work revealed an intriguing correlation between the natural frequency of particles and the turbulent flow frequency at the particle length scale. While these investigations have advanced our understanding of non-spherical particles within turbulence, they do not fully encapsulate the realistic scenario of snow settling dynamics due to the inherent simplification in the chosen particle geometries.

Despite substantial efforts in laboratory experiments and simulations, a conspicuous gap persists in field data regarding the settling of inertial particles, such as snow, droplets, and dust, in atmospheric turbulence. This shortage in data is largely attributed to the inherent challenges in performing field measurements [43], including the need to parameterize local turbulent field conditions and the complex task of discerning the influences of particle interaction and flow Reynolds numbers on non-Stokesian particle kinematics. Notable strides in this area have been made recently [12,40,52], yet limitations persist. Incorporating particle-turbulence interaction mechanisms into predictive models at geophysical scales remains particularly challenging due to the disparity in conditions between the field and laboratory or simulation environments. The conventional models that typically utilize simplified ellipsoidal shapes for particles fall short in accurately representing the drag coefficient and wake characteristics of genuine snow particles [46,47].

Addressing the compelling challenges of field measurements, our team has implemented and advanced the snow particle imaging approach, initially applied in the study of wind turbine wake flows [1,10,21] and atmospheric boundary layer research [18,51]. This systematic refinement of these methodologies, spanning nearly a decade, has allowed us to delve into the complex dynamics of snow settling within natural environments, providing invaluable firsthand observations that incorporate both the snow morphology and the influence of turbulence. The present paper is not just a summary of previous publications but represents a distillation of nearly a decade's worth of dedicated effort in field measurements and research into snow settling dynamics. It brings together our key findings encapsulated in four landmark publications [9,25,26,36], each contributing unique insights and advancements to the study of this intricate phenomenon. Moreover, the review process has allowed us to synthesize these contributions and extract new insights. The layout of this paper is thoughtfully structured: Sect. 2 offers a comprehensive overview of our approach, detailing the progression of field experiments and data analysis techniques, specifically centered on the intersection of snow particle dynamics and atmospheric turbulence. Section 3 collates the significant findings from our sustained field campaign, drawing upon the insights from the referenced publications and previewing our ongoing research initiatives. Concluding in Sect. 4, we reflect on the valuable insights and the journey of our dedicated research into the complex process of snow settling. This final section also proposes future research avenues, acknowledges the limitations encountered in this journey, and explores potential strategies to overcome these challenges.

2 Methods

Over the span of the last decade, our research team has undertaken more than 15 field deployments at the EOLOS field research station in Rosemount, MN. These deployments have been instrumental in our ongoing exploration of snow settling dynamics. To delve into this complex phenomenon, we have harnessed a range of innovative imaging techniques, including planar particle tracking velocimetry (PTV), simultaneous planar super-large-scale particle image velocimetry

(SLPIV) and PTV, field 3D PTV, and a pioneering snow particle characterization tool that employs digital inline holography (DIH). The utilization of these advanced methodologies has propelled us into uncharted territory in the study of snow settling dynamics.

The EOLOS field station, initially established as a wind energy research facility and equipped with a 2.5 MW turbine and a 130 m meteorological tower, has proven to be a unique site for our snow settling studies. The station's relatively open terrain, with minimal canopy interference, provides an ideal setting for capturing video footage of snow settling under a wide range of conditions. The meteorological tower is fitted with a comprehensive suite of sensors that measure wind velocity, temperature, and humidity, which are essential for characterizing the atmospheric and turbulence conditions during our field experiments. Specifically, the tower hosts four sonic anemometers and temperature sensors (CSAT3, Campbell Scientific), operating at a 20 Hz sampling rate, strategically placed at elevations of 10, 30, 80, and 129 m. Additionally, six cup-and-vane anemometers, along with temperature and humidity sensors (1 Hz sampling rate), are installed at 7, 27, 52, 77, 102, and 126 m elevations. This extensive sensor setup enhances our ability to collect detailed and accurate data.

In this paper, we turn our focus to five major deployments that have been the subject of our previous publications. Each of these deployments was conducted under unique flow and snow conditions, which were analyzed to understand their impact on snow settling dynamics. To quantify the atmospheric and turbulent conditions during each deployment, we utilized the sonic anemometer at 10-m elevation and other sensors installed on the meteorological tower. Key parameters for each deployment, including atmospheric stability, integral time scale, length scale, and turbulence dissipation rate, etc., are summarized in Table 1. The atmospheric stability was estimated using the bulk Richardson number (R_b) and the Monin-Obukhov length (L_{OB}). The integral time scale (τ_L) and the length scale (L) were estimated based on the temporal autocorrelation function of the streamwise velocity. The turbulence dissipation rate (ϵ) was estimated using the second-order structure function of the streamwise velocity component,

Table 1. Estimated meteorological and turbulence conditions from the sonic anemometer at z = 10 m. See the text for the definition of the symbols. Note that Bristow et al. [9] only provides mean flow conditions for Apr. 2022 deployment not the turbulence statistics.

Dataset	U (m/s)	u_{rms} (m/s)	R_b (−)	L_{OB} (m)	L (m)	τ_L (s)	ϵ (cm^2/s^3)	η (mm)	τ_η (ms)	Re_λ (−)
Jan. 2016	1.98	0.16	0.03	811	4.9	30.6	8	1.29	126	938
Nov. 2018	1.55	0.38	0.12	1007	3.4	8.9	21	1.04	83	3548
Jan. 2019	5.95	1.18	0.03	-2643	14.6	12.9	314	0.49	20	9180
Jan. 2020	5.47	1.07	0.16	1651	6.2	5.8	355	0.51	19.4	6478
Apr. 2022	1.69	0.27	0.08	191	1.47	5.35	9.7	1.2	117	2558

with the Taylor hypothesis employed to convert the measured time series into spatial velocity variations. For a more detailed discussion of the methods and equations used in these calculations, we refer readers to our previous publications [25,26,36]. This comprehensive analysis of the atmospheric and turbulent conditions during each deployment has provided us with a deeper understanding of their implications on snow settling dynamics.

Shifting focus from atmospheric and turbulent conditions to the specifics of our snow particle measurements, it is pertinent to note the significant advancements we have achieved using the digital inline holography (DIH) technique for capturing the morphology of snow particles. The evolution of our snow analyzer technology, which has seen two distinct generations of development, is depicted in Fig. 1. Figure 1a illustrates the first-generation snow particle analyzer, a DIH sensor mounted atop a field vehicle, an effective but comparatively rudimentary system. Recognizing the need for more comprehensive data, we developed a second-generation snow particle analyzer that greatly enhanced our measurement capabilities. Shown in Fig. 1b, this new analyzer combines the DIH sensor with a high-precision scale to measure snow particle weight, thereby enabling simultaneous measurement of particle size, shape, type, and density. The improved data accuracy and expanded scope of our second-generation analyzer necessitated a more advanced data processing solution. To meet this need,

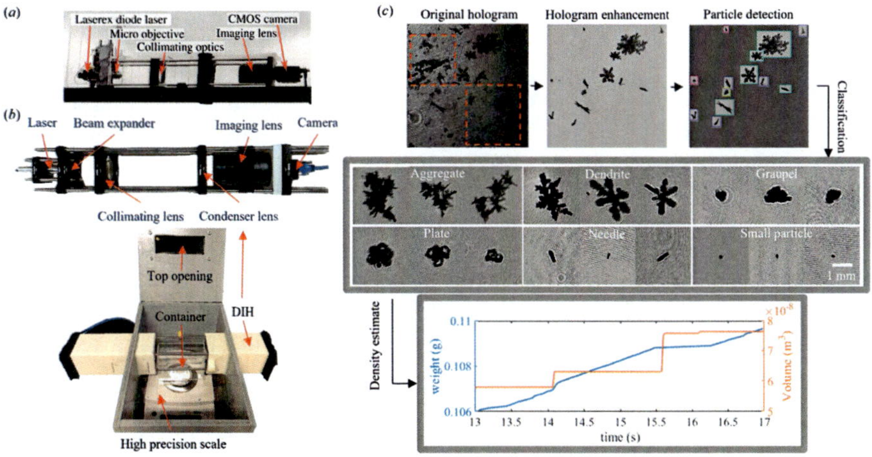

Fig. 1. (a) First-generation snow particle analyzer based on digital inline holography (DIH) sensor mounted atop a field vehicle. (b) Second-generation snow particle analyzer designed for simultaneous measurement of particle size, shape, type, and density, combining the DIH sensor and a high precision scale. (c) Real-time data processing software developed for the second-generation snow particle analyzer. The software includes image enhancement to eliminate static background (highlighted in red dashed boxes), snow particle detection and categorization using machine learning models, and particle volume and density estimation derived from weight measurements on a high-precision scale.

we developed real-time data processing software, as demonstrated in Fig. 1c. This custom software, built around the YOLOv5 machine learning model, is capable of image enhancement, snow particle detection, categorization, and estimation of particle volume and density derived from weight measurements on the high-precision scale. The software categorizes snow particles into six types according to their shape and aerodynamic response, achieving an impressive 99% particle detection rate and an average classification accuracy of 95%. For a more comprehensive understanding of the snow particle analyzer's development and operation, readers are referred to [27].

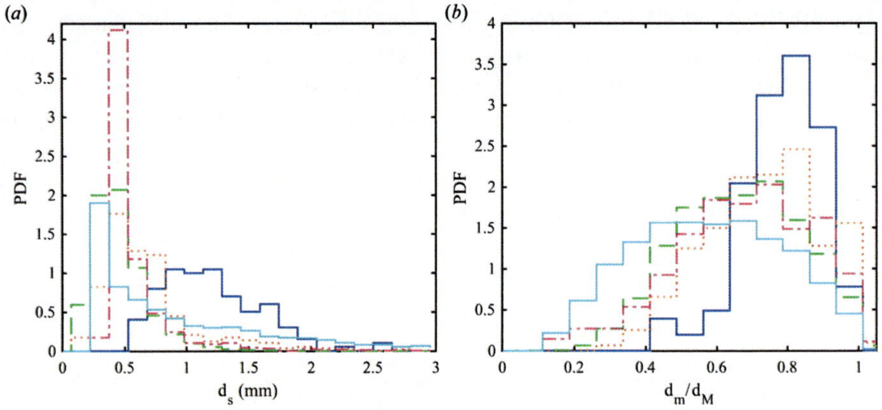

Fig. 2. The probability density functions (PDFs) of (a) size (d_s) and (b) aspect ratio (d_m/d_M) of the snow particles for January 2016 (solid blue line), November 2018 (dotted red line), January 2019 (dashed green line), January 2020 (dash-dotted magenta line), and April 2022 (solid cyan line).

With the particulars of our snow particle measurement technique established, we then focus on the resulting data. The probability distribution functions (PDFs) of snow particle size and aspect ratio are presented in Fig. 2. This figure provides a comparative analysis of data from the five deployments, offering insights into the variations in snow particle characteristics under different conditions. Complementing the graphical data in Fig. 2, Table 2 consolidates the mean values, standard deviations of particle diameter and aspect ratio, and estimates of snow number concentration and volume fraction from the different deployments.

With the advancement of our measurement techniques in quantifying turbulence and snow particle conditions, we concurrently developed a method based on light sheet-based particle tracking velocimetry (PTV). This technique, which proved vital in tracing the snow settling trajectory and concentration distribution of snow particles, was rooted in the super-large-scale particle image velocimetry (SLPIV) methods established by [21,51]. The experimental setup for this method is illustrated in Fig. 3a. In Fig. 3a, the setup integrates a 5-kW

Table 2. The snow particle properties and concentration measured using the snow analyzers from different deployments.

Dataset	Mean diameter d (mm)	Aspect ratio $d_{\text{maj}}/d_{\text{min}}$ (−)	Number concentration N (m^{-3})	Volume fraction $\phi_V \times 10^{-7}$ (−)
Jan. 2016	1.09 ± 0.45	0.73 ± 0.11	816	7.4
Nov. 2018	0.65 ± 0.41	0.65 ± 0.16	1644	6.3
Jan. 2019	0.39 ± 0.23	0.57 ± 0.17	56620	44
Jan. 2020	0.39 ± 0.29	0.62 ± 0.15	28460	38
Apr. 2022	0.97 ± 0.90	0.54 ± 0.21	933	17

searchlight paired with a curved mirror to vertically expand the beam into a light sheet. This light sheet illuminates the snow particles, and it is oriented parallel to the average wind direction to minimize out-of-plane motion. Typically, this setup is positioned 50 m upwind or downwind of the meteorological tower, thus enhancing the precision of the flow condition quantification. Along with the illumination setup, our measurement approach employs a Sony A7RII (Sony Corp.) camera to capture detailed particle motion. This camera records at a high frame rate of 120 FPS and a resolution of 720 × 1280 pixels within a targeted field of view. Table 3 details the field of view and other specifications pertinent to this camera.

To enable the elucidation of the underlying mechanism of the flow-particle interaction, we implemented a dual-camera setup, as shown in Fig. 3b. This setup complements the Sony camera with a Nikon D600 (Nikon Inc.) camera, thereby enabling simultaneous measurements of the turbulent flow field and snow particle trajectories. The Nikon camera is configured to record a broader field of view at a frame rate of 30 FPS and a resolution of 1080 × 1920 pixels, providing a comprehensive view of the overall turbulent flow field. Both cameras, referred to as PTV (Sony A7RII) and SLPIV (Nikon D600) in the subsequent sections, are mounted on tripods, with their specifications documented in Table 3.

To process the data captured through our PTV and simultaneous SLPIV/PTV systems, we utilized particle image velocimetry (PIV) analysis with the aid of LaVision Davis 8.2.0 software. We employed a multi-pass setting, with the final pass configured to 32 × 32 pixels and a 50% overlap, ensuring a comprehensive analysis of the captured data. For the extraction of particle Lagrangian trajectories, we turned to the four-frame best estimate algorithm, a method proposed by Ouellette, Xu & Bodenschatz [38]. This algorithm proved to be a solid foundation for our data extraction process. Building on this foundation, we implemented the learning-based method developed by Mallery, Shao & Hong [30] in our most recent study. This method further enhanced our ability to accurately track and analyze the trajectories of snow particles.

In our latest research endeavor, we have designed and implemented a mobile field 3D PTV system, as detailed in [9]. This system, illustrated in Fig. 3c, is engineered to capture the intricate 3D settling dynamics of snow particles. The system consists of four Wi-Fi synchronized cameras, arranged in a fan-like array spanning 90°C. These cameras surround a light cone, which is reflected and expanded by a curved mirror from the same searchlight used in our planar measurements. Each camera is positioned 5.5 m away from the searchlight and angled at 58°C, looking upward at a height of 10 m above the ground. Moreover, each camera is outfitted with an enclosed data acquisition unit, housing a board-level computer for image capturing commands, a solid-state drive for system and image storage, and a power supply. To ensure accurate measurements, we employ a calibration process using an unmanned drone equipped with two colored LEDs, set 34.5 cm apart on a wand. This calibration process is based on the wand calibration method [49]. We have developed a custom-designed camera control software that synchronizes the image capturing process across all four

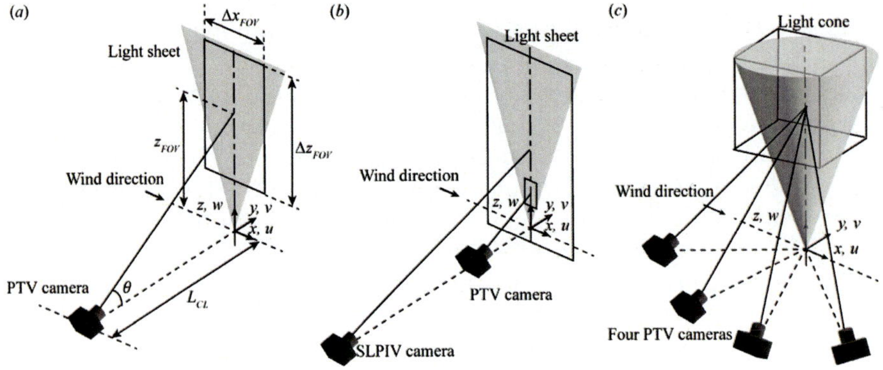

Fig. 3. Illustrations of the experimental setups used for field snow settling measurements, including (a) the planar particle tracking velocimetry (PTV) measurement setup, (b) the setup enabling simultaneous measurements using super-large-scale particle image velocimetry (SLPIV) and PTV, and (c) the ground-based multiview field 3D PTV measurement setup.

Table 3. Summary of key parameters of the super-large-scale particle image velocimetry (SLPIV), particle tracking velocimetry (PTV), and 3D field PTV measurement set-ups for the deployment dataset used in the present paper. Here FOV denotes field of view.

Dataset	SLPIV/PTV setup						DIH setup	
	Duration (min)	z_{FOV} (m)	$\Delta x_{FOV} \times \Delta z_{FOV}$ (m^2)	Resolution (mm/ pixel)	θ (deg.)	L_{CL} (m)	Resolution (μm/ pixel)	Volume (cm^3)
Jan. 2016	5	10.8	7.1 × 4.0	5.6	21.1	25	24	18.8
Nov. 2018	17	9.1	8.4 × 4.7	6.5	14.5	31	14	42
Jan. 2019	15	20.2	14.7 × 8.3	12.0	19.9	53	14	42
Jan. 2020	11	18.4	39.2 × 22.1	20.5	19.9	53	14	42
	11	7.9	5.3 × 3.0	4.2	18.4	19.1		
Apr. 2022	1	10	$4 \times 4\,(\Delta y_{FOV}) \times 6$	6.3	58	5.5	14	88

cameras. Once calibrated, the snow particle trajectories are obtained using an open-source implementation of the shake-the-box (STB) method [48]. The measurement sample volume of this setup is 4 m × 4 m × 6 m, offering a resolution of 6.3 mm per voxel, as summarized in Table 3.

3 Results

In this section, we aim to provide a succinct review of our recent advancements in the 2D field measurement of snow settling dynamics, while also offering a preview of our ongoing work on 3D PTV field measurements. The primary focus will be on the outcomes of our measurements, the comparisons drawn between different deployments, and the novel physical insights and implications that have emerged from these observations. A key aspect of this review will be the new insights gained from cross-comparison of different deployments, which go beyond the findings reported in the original papers. For a more detailed understanding of the techniques and data analysis tools that form the foundation of these findings, we recommend referring to our recent publications [9,25,26,36].

3.1 Preferential Enhancement of Snow Particle Settling

In this section, we explore the impact of turbulence on the settling velocity of snow particles, a pivotal aspect of our study. This exploration is grounded in data from several deployments, including those documented in [25,36], and [26]. Our primary objective is to compare the terminal velocity of snow particles in still air with their measured settling velocity under the influence of atmospheric turbulence.

To estimate the terminal velocity of snow particles, we first need to determine their aerodynamic response time, which is extrapolated from the probability density functions (PDFs) of normalized snow particle acceleration $(a_p/\langle a_p^2\rangle^{1/2})$. As depicted in Fig. 4a, the blue triangles correspond to data from January 2016 [36], the red diamonds to November 2018 [25], and the magenta squares to January 2020 [26]. Our findings are compared with previous experimental and simulation measurements [5,7,35]. The acceleration PDFs of January 2016 and November 2018 are well-defined, facilitating the estimation of the range of Stokes numbers (St). However, the PDF for January 2020 shows larger variability, with the probability density approaching that of November 2018 in the lower range of normalized accelerations but approaching that of January 2016 for higher values. It is important to note that while the kurtosis of the acceleration PDFs is sensitive to the Stokes number of the particles, the influence of turbulence properties, particularly the Taylor microscale Reynolds number (Re_λ), on the acceleration distribution of snow particles remains an active area of research. Existing simulations and experimental findings [22,23,53] suggest an increase in acceleration kurtosis for both fluid tracers and inertial particles with Re_λ. However, the effect of Re_λ on acceleration kurtosis might be relatively minor in the high Reynolds number regime [22]. Given these considerations, we have estimated the Stokes numbers for the snow particles based on three separate particle tracking velocimetry (PTV) deployment datasets. The estimated ranges of Stokes numbers are as follows: 0.09–0.37 for January 2016, 0.15–1.01 for November 2018, and 0.09–1.01 for January 2020. These ranges reflect the variability in the Stokes numbers of snow particles, influenced by factors such as size, shape, and density.

Building on the estimated range of Stokes numbers, we can leverage the Kolmogorov time scale of turbulence (τ_η) to calculate the particle aerodynamic response time ($\tau_p = St \cdot \tau_\eta = W_t/g$). This calculation allows us to also predict the terminal velocity of snow particles in still air (W_t). The estimated terminal velocity ranges are as follows: 0.12–0.47 m/s for January 2016, 0.12–0.82 m/s for November 2018, and 0.02–0.19 m/s for January 2020. The notably smaller terminal velocity for January 2020 can be attributed to the smallest particle size and the significantly smaller turbulence Kolmogorov time scale used for its estimation. In Fig. 4b, we compare the predicted terminal velocity (W_t) of snow particles with the settling velocity (w_p) as measured through PTV in the field. Across all three deployments, we observed an increase in settling velocity under the influence of turbulence. However, the most substantial enhancement was recorded in the January 2020 deployment. Specifically, the average settling velocity observed during the January 2020 deployment was seven times greater than the predicted terminal velocity, which was significantly higher than those derived from the January 2016 and November 2018 datasets. This pronounced enhancement in settling velocity during the January 2020 deployment may be largely attributed to the more intense turbulence conditions during this period,

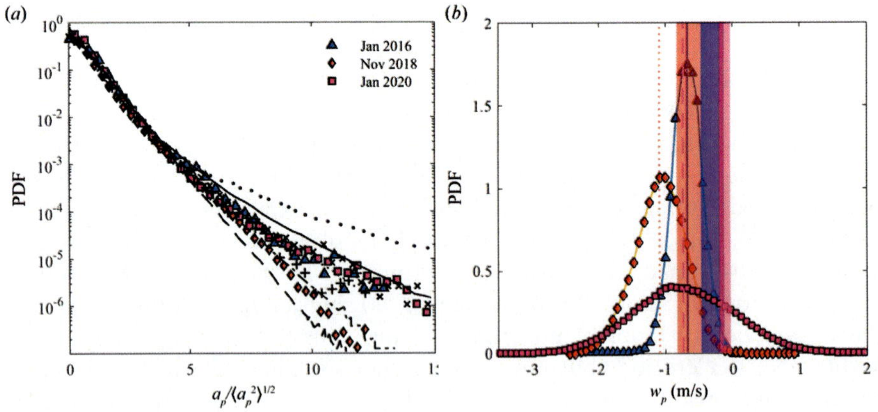

Fig. 4. (a) Probability density functions (PDFs) of normalized in-plane snow particle acceleration from different PTV deployments, compared with St = 0 from [35] (dots), [7] (St = 0.16, solid line; St = 0.37, dotted line; St = 1.01, dash-dotted line; St = 2.03, dashed line), and [5] (St = 0.09, ×; St = 0.15, +); (b) comparison of the measured distribution (same symbols as the acceleration PDF) and average (vertical lines, January 2016, blue solid line; November 2018, red dotted line; January 2020, magenta dotted line) of settling velocity with the estimated range of still-air terminal velocities (January 2016, blue region; November 2018, red region; January 2020, magenta region).

as evidenced by the highest Reynolds number (Re_λ) and dissipation rate among all the datasets. This turbulence condition brought the Stokes number closer to the critical value of '1'. This near-critical Stokes number suggests an optimal dynamic interaction between the snow particles and the turbulent air motion, leading to a heightened settling velocity. Secondarily, the January 2020 deployment had a notably higher snow particle concentration. This increased concentration intensified particle-particle interactions, which could potentially further boost the settling velocity. Though our study offers notable contributions, the conclusions are somewhat influenced by uncertainties due to simplifications in our methodology. Notably, the representation of snow particles as perfect spheres and their assumed size being smaller than the Kolmogorov scale can contribute significantly to the variance in our findings.

3.2 Snow Particle Clustering

In this section, we turn our attention to the intriguing phenomenon of snow particle clustering under intense turbulence. This occurrence, often associated with preferential concentration due to turbulence, has substantial implications for regional ground accumulation rates and interactions between particles. The data utilized for this analysis stems from multiple field deployments, as detailed in [25,26].

To delve into this investigation, we need to estimate the concentration of snow particles. This is achieved by analyzing the intensity of images captured through super-large-scale particle image velocimetry (SLPIV), as demonstrated in Fig. 5. As established by [41], there is a linear correlation between the scattered light intensity in PIV images and the number concentration of particles. Consequently, we can compute the estimated relative concentration, denoted as C^*, using the formula $C^* = I/\overline{I_{1\ \mathrm{min}}}$. Here, I represents the instantaneous image intensity, while $\overline{I_{1\ \mathrm{min}}}$ denotes the 1-minute moving average of the image intensity. Figure 5 reveals the presence of clustering during both the January 2019 and January 2020 deployments. However, a closer look reveals distinct qualitative differences between the two. The January 2020 sample exhibits larger variability in relative concentration and larger clusters. In contrast, the January 2019 sample displays clusters that resemble thin filaments, extending in the vertical direction.

Expanding upon the initial observations noted in Fig. 5, we pursued a more intricate understanding of snow particle clustering through a thorough quantitative analysis. Figure 6 offers a meticulous comparison of the statistical characteristics tied to cluster morphology across different deployments, thereby highlighting salient variations. Figure 6a focuses on the cluster area (A_C) distributions normalized by the Kolmogorov scale (η), from each deployment. Interestingly, these distributions reveal a power-law decay, echoing patterns seen in prior laboratory assessments and simulations. This decay indicates a self-similarity across clusters of diverse sizes, hinting at the prominent role turbulent eddies play in their formation. An observation of larger clusters in the January 2020 dataset seems to underscore the enhanced influence of turbulence on snow particle concentration and the broader imaging field of view (FOV). Conversely, Fig. 6b

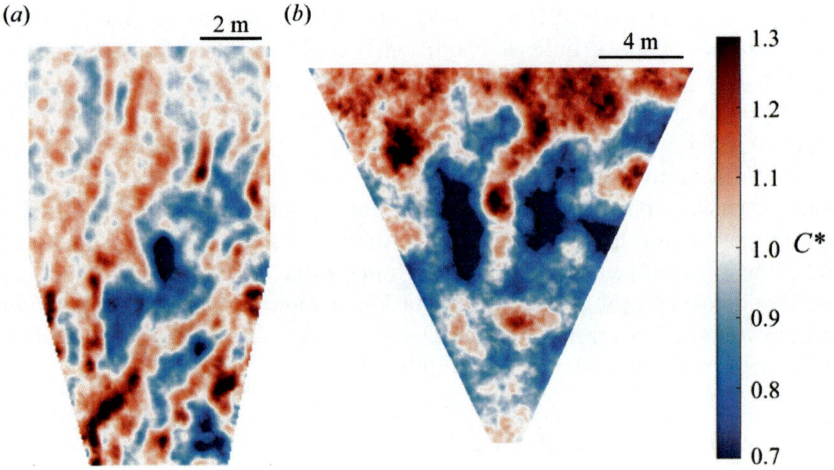

Fig. 5. Samples of instantaneous normalized snow concentration map derived from (a) January 2019 and (b) January 2020 deployments.

shifts focus towards the statistical distribution of the aspect ratio (s_2/s_1) of the clusters. The elongated shape of clusters in both datasets, most exhibiting an aspect ratio around 0.5, reflects the gravitational forces acting on them. The elongation is not a random trait but a specific characteristic of heavy inertial particles navigating turbulent flows under gravitational pull. Figure 6c offers insight into the orientation (θ_C) of the snow particle clusters, which predominantly assumes a vertical disposition-another testament to gravity's influence. Nevertheless, a notable shift in peak aspect ratio towards one and a more evenly distributed orientation are observed in the clusters from January 2020. This subtle shift indicates a reduced gravitational influence relative to turbulence when compared to the January 2019 data.

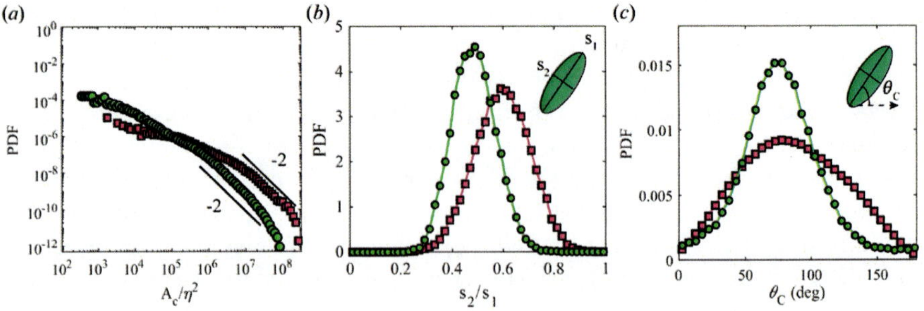

Fig. 6. (a) The probability density functions (PDFs) of snow particle cluster area (A_C) normalized by Kolmogorov length scale (η) show a power-law decay with an exponent close to -2. The PDFs of (b) the cluster aspect ratio and (c) angle orientation of the clusters. The green circles represent January 2019, and the magenta squares represent January 2020.

Overall, our analysis indicates that the consistent patterns of cluster size and morphology across the datasets signal the presence of universal physical mechanisms **(i.e., the interplay of turbulent eddies and gravitational forces in snow cluster formation)** that dictate clustering behavior. This discovery raises the possibility of constructing predictive models based on these principles. Despite these consistencies, variances across the datasets could be ascribed to the intricate interplay between atmospheric turbulence, gravity, and the unique properties of snow particles. However, considering the similar characteristics of snow particles and turbulence across the datasets, it is plausible that the prominent factor driving the difference in cluster behavior is the elevation variation between sampling zones. Specifically, the January 2020 dataset, for instance, focuses on a lower elevation and hence may capture images within the internal shear layer that is densely populated by potent prograde vortices. These vigorous coherent structures within turbulence could locally overshadow gravity's effects, diminishing cluster elongation and preferential orientation.

3.3 Connection Between Preferential Sweeping and Snow Clustering

In this section, we delve deeper into the mechanisms of the enhanced snow particle settling velocity and clustering, focusing on the preferential concentration and preferential sweeping hypothesis. To substantiate these hypotheses, we present direct measurements of the turbulent flow field and snow particle settling trajectories. These measurements, derived from the super-large-scale particle image velocimetry (SLPIV) and particle tracking velocimetry (PTV) deployment presented in [26] and the method section of the current paper, shed light on the intricate interplay between the turbulent coherent structures, snow particle concentration, and settling velocity. The results of these measurements are illustrated in Fig. 7, providing an overview of the turbulent flow field and snow particle settling behavior. Figure 7a presents the swirling strength contour of a sample instantaneous flow field with the mean velocity removed. **The swirling strength, λ_{ci}, refers to the imaginary part of the complex eigenvalues of the velocity gradient tensor, $\boldsymbol{D} = \nabla \boldsymbol{u}$, and vortices can be identified by a finite λ_{ci} value, as explained in** [2]. This visualization allows for the identification of turbulence vortices, highlighting both prograde and retrograde vortices based on their direction of rotation. Building on this, Fig. 7b superimposes a snow particle concentration color map, derived from image intensity, onto the identified vortices. This overlay reveals an interesting pattern: enhanced particle concentration on the downward moving side of vortices, suggesting a preferential concentration effect. To further explore the preferential sweeping mechanism, we combine SLPIV and PTV results to track the motion of snow particles around identified vortices. Figure 7c and d illustrate this process for a retrograde vortex and a prograde vortex, respectively. The trajectories of snow particles colored based on the dimensionless time t^*, show a clear acceleration in alignment with the downward fluid motion around both types of vortices. Conversely, on the upward moving side of these vortices, falling snow particles decelerate, even exhibiting a degree of upward motion. This direct observation provides compelling evidence for the preferential sweeping mechanism and its role in enhancing snow particle settling velocity and promoting snow clustering in turbulent flows.

We further reinforce our qualitative observations with statistical evidence, as depicted in Fig. 8. Figure 8a–c display comparative data between ensemble-averaged concentrations sourced from the SLPIV dataset within the core region of vortices and their surroundings, segmented by azimuthal locations for both prograde and retrograde vortices. A conspicuous pattern arises: higher concentrations are found on the downward side of both vortex types, suggesting a mechanism of preferential concentration. Notably, the highest localized concentration is found at the base of the downward side of retrograde vortices, contrasting with the more uniformly high concentration seen in prograde vortices. This disparity might be understood in light of the dominant presence of prograde vortices in internal shear layers, as supported by [18,57]. Their interaction with neighboring

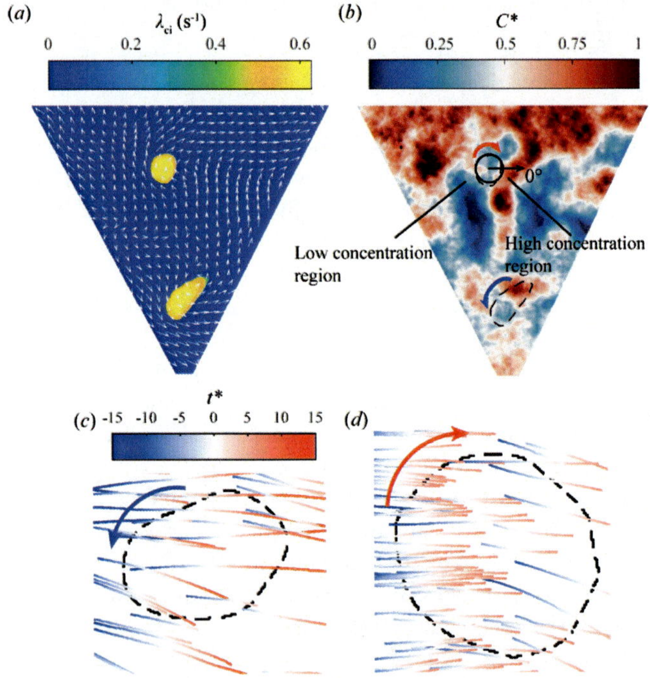

Fig. 7. (a) Swirling strength (λ_{ci}) contour of a sample of instantaneous flow field (mean velocity removed). (b) Corresponding snow particle concentration colourmap (C^* is the relative concentration) with vortices detected from (a). Samples of snow particle trajectories around (c) a retrograde vortex and (d) a prograde vortex. Black dashed lines represent vortex boundaries, and trajectories are colored based on the dimensionless time t^*, defined as the difference between the timestamps of snow particles and that of a selected vortex at one time instant normalized by the Kolmogorov time scale. Note that the figure is adapted from [26] with different samples.

vortices likely distributes particle concentration evenly along the downward side. In contrast, the relative isolation of retrograde vortices possibly fosters accumulation of snow particles at their base, likely driven by gravity and descending fluid motion. These findings constitute direct evidence of preferential concentration of snow particles in turbulent eddies, thus marking a significant stride in understanding snow particle dynamics in atmospheric turbulence. Figure 8d and e further deepen our insights by presenting the disparities in the average settling velocities of snow particles on downward and upward sides of both prograde and retrograde vortices, normalized by the average snow settling velocity. In both scenarios, it is apparent that the snow particles settle more swiftly on the downward side of the vortices. A larger disparity in average settling velocity between the downward and upward sides is seen in the case of prograde vortices compared to retrograde ones. This difference may be ascribed to the distinctive self-organization of these vortex types. Specifically, prograde vortices, typically observed in internal shear layers, may yield a compounded effect on enhancing the settling velocity of snow particles.

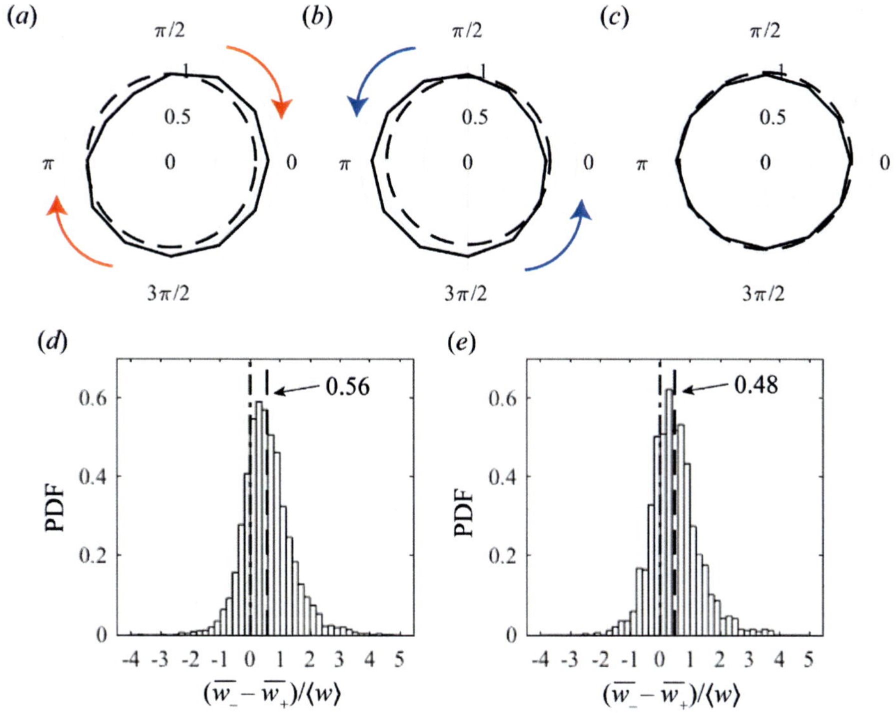

Fig. 8. Comparison between the ensemble-averaged concentration from the SLPIV dataset in the central region (dashed lines) of the vortices, and in the proximity (solid lines) for (a) prograde vortices, (b) retrograde vortices and (c) reference background. Histograms of the differences (normalized by the average snow settling velocity $\langle w \rangle$) between the settling velocities on the downward ($\overline{w_-}$) and upward ($\overline{w_+}$) sides for (d) prograde and (e) retrograde vortices.

In summary, this study presents a pioneering direct visualization and quantification of the preferential sweeping of snow particles in atmospheric turbulence. The findings underscore the significant role turbulent vortices play in shaping the interactions between snow particles and atmospheric turbulence while shedding light on the influence these vortices exert on the enhanced settling velocity of snow particles. Additionally, our analysis suggests potential intra-cluster particle interactions. Preferential sweeping might lead to increased particle collision rates within clusters, possibly causing the fragmentation or aggregation of particles. This dynamic interplay has the potential to markedly alter the physical properties of snow particles, thereby influencing their behavior within the turbulent atmospheric environment.

3.4 Snow Settling Dynamics in Three Dimensions

While planar measurements have provided significant insights into snow particle dynamics, their two-dimensional nature inherently limits our ability to fully capture the spanwise motion and three-dimensional spatial distribution of snow particles. To overcome these limitations, we have advanced our research methodologies. We have developed a novel three-dimensional (3D) field particle tracking velocimetry (PTV) system, as detailed in a recent publication by [9]. This system enhances our ability to track and analyze the motion of snow particles in a three-dimensional space. In parallel, we have incorporated a snow particle analyzer into our research toolkit, as described in [27]. This tool provides a deeper understanding of the properties of snow particles, including their size, shape, type, and density. By combining these innovative approaches, we can simultaneously monitor various properties of freshly falling snow and track particle settling trajectories in a field setting. This integrated approach not only addresses the limitations of previous methodologies but also paves the way for a more comprehensive understanding of snow particle dynamics.

Capitalizing on advancements in our research methodologies, we executed nine 3D PTV field experiments across the winter seasons of 2021/2022 and 2022/2023. These experiments utilized our novel systems and encompassed a broad spectrum of turbulence and snow conditions. The mean flow ranged from 0.5 to 8.5 m/s, and our observations captured all six prevalent snow particle types: aggregates, dendrites, graupel, plates, needles, and smaller particles. These field experiments inform our ongoing and future endeavors, primarily concentrated on the three-dimensional settling of snow particles. We are particularly keen on dissecting the role of snow morphology and the various turbulence conditions in this process. The preliminary insights gained from these deployments are gradually illuminating the complex behavior of inertial particles within atmospheric flows. A selection of these findings, specifically derived from the dataset of our April 2022 deployment, will be discussed in the section below.

By harnessing our 3D measurement capabilities, we successfully determined the 3D trajectories and corresponding kinematics of snow particles, illustrated in Fig. 9. Figure 9a showcases a selection of snow trajectories, each color-coded based on its settling velocity. This visual display offers a glimpse into the spatial variability of settling velocities among different particles. Figure 9b further highlights the spanwise acceleration of these particles along their extended trajectories. The periodic variation in spanwise acceleration suggests a meandering motion that may be linked to the dominant presence of dendrite particles in the sample. The particles also display a characteristic trait of inertial particles in turbulence - exponential tails in the acceleration PDFs, depicted in Fig. 9c. A clear asymmetry is noted in the streamwise and vertical acceleration components, which may be attributed to atmospheric flow in the positive x direction and the gravitational force acting on snow particles in the negative z direction. Additionally, our 3D measurements facilitate the extraction of compelling set-

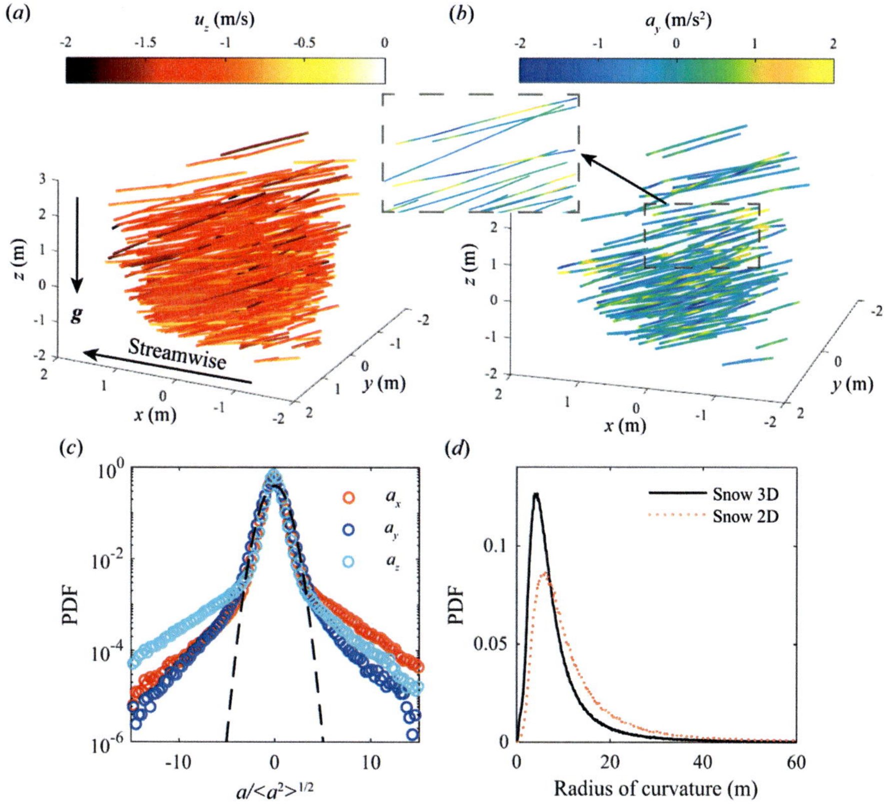

Fig. 9. (a, b) Sample snow particle trajectories, colored by (a) vertical velocity, u_z, and (b) spanwise acceleration, a_y, with the arrows indicating the directions of mean flow (streamwise) and gravity. Probability density functions (PDFs) of (c) acceleration components compared to the corresponding Gaussian distribution (black dashed line), and (d) radii of curvature, where Snow 3D and 2D represent curvature measured using x, y, z components and x, z components, respectively. The results are derived from different dataset from the same deployment as in [9].

tling kinematics, including the curvature of the trajectories. The curvature can be determined using the formula $\kappa = \frac{1}{R} = \frac{\|\boldsymbol{v} \times \boldsymbol{a}\|}{\|\boldsymbol{v}\|^3}$, with R representing the radius of curvature, and $\boldsymbol{v}$ and $\boldsymbol{a}$ denoting the Lagrangian velocity and acceleration, respectively. The L^2 norm is symbolized by $\|\cdot\|$. Figure 9d compares the PDFs of the particle trajectory curvature calculated from both full 3D and 2D $x-z$ components. The discrepancies between these distributions underscore the variable spanwise motion, also reflected in Fig. 9b.

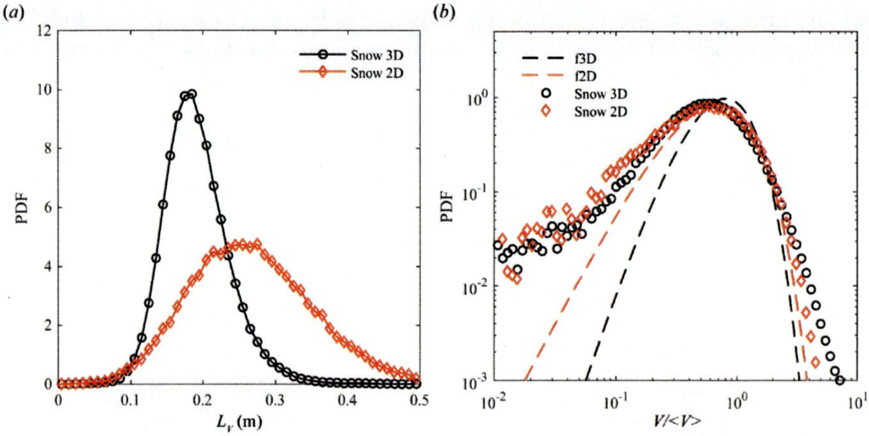

Fig. 10. Probability density functions (PDFs) of (a) Voronoi cell length scale L_V and (b) normalized cell volumes $V/\langle V\rangle$, derived from a full 3D volume (snow 3D) and the volume corresponding to a thin slice parallel to the x-z plane extracted from the entire volume (snow 2D). In (b), the results are compared against models f3D and f2D from [13] for randomly distributed particles.

In addition to our Lagrangian analysis, we have incorporated Voronoi analysis, a powerful spatial tessellation technique, to the reconstructed 3D particle field. This method offers a unique perspective on local particle concentration, as it inversely correlates with the volume of each individual tessellation cell. As illustrated in Fig. 10a, the PDFs of the length scales of both 3D and 2D Voronoi cells reveal the typical particle spacing, as indicated by the peak location. The contrast between the 3D and 2D measurements underscores the potential biases introduced by the thin light sheet used in planar measurements, thereby highlighting the value of comprehensive 3D measurements. Delving deeper into our findings, the PDFs of normalized cell volume, as depicted in Fig. 10b, show a higher probability for both small and large cell sizes compared to a random particle distribution. This pattern suggests a clustering tendency among the snow particles. The increased probabilities at both ends of the size spectrum correspond to the snow clusters and voids, respectively. Setting a size threshold for the Voronoi cells allows us to quantify these 3D snow particle clusters, which opens up new avenues for in-depth analysis. For instance, we can delve into a comparative study of particle settling dynamics within the clusters versus in the voids. This comparison could reveal significant insights into the enhancement or hindrance of settling velocity within these distinct environments. Moreover, we can explore the differences in acceleration dynamics attributable to particle-particle and particle-turbulence interactions. Comprehending these interactions offers potential illumination on the dynamics governing the observed snow particle trajectory behaviors within atmospheric turbulence, while also providing critical insights into the effects of local particle concentration on particle settling and interactions. These prospective analyses underline the remarkable potential of our 3D field imaging approach in propelling forward our knowledge of snow particle behaviors within atmospheric turbulence.

4 Conclusions and Discussion

This paper embarked on a comprehensive review, spanning nearly a decade, of field research conducted at the EOLOS Field Research Station, Rosemount, MN, USA, on the subject of snow settling. Utilizing a remarkable array of field imaging tools, we revisited and integrated the outcomes from four previous studies [9,25,26,36]. This undertaking not only served as a capstone for these works, but also spurred the generation of new insights. Our exploration of this research field was made possible through the employment and continuous refinement of a suite of measurement tools, specifically designed for characterizing snow settling dynamics. The technology deployed has evolved through three generations: planar particle tracking velocimetry (PTV), simultaneous super-large-scale particle image velocimetry (SLPIV) and PTV, and field 3D PTV. This endeavor was complemented by the development of a digital inline holography (DIH) method, custom-designed for discerning snow particle types, concentration, and size distribution.

By harnessing the power of these pioneering tools, our work uncovered several intriguing findings. Firstly, data gleaned from multiple deployments consistently point towards the preferential sweeping mechanism. Both statistical evidence and direct imaging unambiguously indicated an enhanced snow settling velocity in conjunction with turbulent eddies. The magnitude of this turbulence-induced enhancement, however, varies across different deployments, likely modulated by diverse turbulence and snow conditions. Secondly, snow clustering-a phenomena observed across different deployments-revealed intriguing trends in cluster size characteristics. An exponent power-law decay of -2 was observed in cluster size distribution across both deployments, while aspect ratio and preferential vertical orientation of clusters suggested gravitational influences. However, detailed distributions, such as mean and standard deviation, seem to be sculpted by the interplay between the significance of turbulent structures, gravity, and snow properties. Thirdly, our direct imaging painted a compelling picture of the interconnection between preferential sweeping and preferential concentration, or clustering, of snow particles. We found that higher concentrations tend to congregate on the downward side of vortices, regardless of rotation direction, and that snow particles descend more swiftly in these regions. Notably, this disparity in average settling velocity is more pronounced in the case of prograde vortices as compared to their retrograde counterparts. Finally, the deployment of 3D PTV unveiled the intricacies of 3D settling trajectories and the dynamics of snow particle clusters. It offered unprecedented insight into the complex movements of snow particles as they travel through atmospheric turbulence, unveiling subtle variations in motion that might otherwise have been obscured in a 2D analysis.

In general, the research undertaken in this study has led to a variety of valuable insights and findings regarding the interplay of turbulence and snow morphology in the process of snow settling under field conditions. Our work provides extensive field evidence and detailed characterization of the preferential sweep and concentration mechanisms that are essential for understanding snow settling dynamics. By employing novel imaging tools with field research, we have uncovered complex behavioral patterns and identified key contributing factors. However, it is crucial to acknowledge the limitations and challenges encountered in our investigation. One of the most prominent challenges we faced lies in the complexity of disentangling the influence of snow morphology from turbulence on the snow settling dynamics. More specifically, quantifying the exact influence of snow morphology on the aerodynamic response time and settling behavior of snow particles remains a significant hurdle. Our current response time estimations are based on assumptions about snow properties, which, in reality, could vary within a deployment or between the sampling regions of PTV and DIH.

Looking to the future, we are determined to tackle these challenges head-on. A promising development in our arsenal is the introduction of the snow particle analyzer [27]. This tool offers direct measurements of snow particle properties, such as size, shape, type, and notably, density. By capturing these specifics in real-time field conditions, we pave the way for an intricate and precise quantification of snow particle aerodynamics. By employing models like the Best number [8], which is dependent on particle density, ambient air density and viscosity, particle diameter, and the projected area, we can now estimate snow terminal velocity. This estimation would allow for a more detailed calculation of the aerodynamic response time, reducing the need for broad assumptions and providing us with data from actual field conditions.

Moreover, the integration of 3D PTV technology is set to revolutionize our understanding of snow particle kinematics and the complex mechanics of snow particle clustering. This technology helps us visualize the settling trajectories and clustering patterns in three dimensions. It allows us to delve into the influence of particle morphology on their behavior, elucidating complex patterns like meandering and tumbling motions influenced by turbulence-induced instabilities [4,19,54]. Notably, the combination of 3D PTV and Voronoi analysis lends us the capability to visualize and quantify 3D snow clusters, leading to further insights into the behavior of snow particles within these clusters.

With the integration of our newly introduced snow particle analyzer and 3D PTV, we are well-positioned to extract more precise information about snow properties and dynamics in the field. This allows us to conduct systematic studies under various turbulence and snow particle conditions. As a result, we can draw direct comparisons with well-controlled laboratory experiments, such as those conducted by [11,44,50], to elucidate the individual and combined effects of snow morphology and turbulence on snow settling.

The implications of these forthcoming investigations are considerable. Our efforts are likely to pave the way for the development of more robust and accurate models of snow settling dynamics. By establishing a comprehensive understanding of snow settling behaviors under realistic field conditions, we can create models that more accurately reflect the complexities of these processes. These improved models will not only serve as valuable tools for academic research but could also have practical applications in areas such as meteorology and climate science. By making predictions more precise, we can better anticipate, plan for, and respond to weather events and climate change impacts.

References

1. Abraham, A., Hong, J.: Dynamic wake modulation induced by utility-scale wind turbine operation. Appl. Energy **257**, 114003 (2020)
2. Adrian, R., Christensen, K., Liu, Z.-C.: Analysis and interpretation of instantaneous turbulent velocity fields. Exp. Fluids **29**(3), 275–290 (2000)
3. Aliseda, A., Cartellier, A., Hainaux, F., Lasheras, J.C.: Effect of preferential concentration on the settling velocity of heavy particles in homogeneous isotropic turbulence. J. Fluid Mech. **468**, 77–105 (2002)
4. Auguste, F., Magnaudet, J., Fabre, D.: Falling styles of disks. J. Fluid Mech. **719**, 388–405 (2013)
5. Ayyalasomayajula, S., Gylfason, A., Collins, L.R., Bodenschatz, E., Warhaft, Z.: Lagrangian measurements of inertial particle accelerations in grid generated wind tunnel turbulence. Phys. Rev. Lett. **97**(14), 144507 (2006)
6. Balachandar, S., Eaton, J.K.: Turbulent dispersed multiphase flow. Annu. Rev. Fluid Mech. **42**, 111–133 (2010)
7. Bec, J., et al.: Acceleration statistics of heavy particles in turbulence. J. Fluid Mech. **550**, 349–358 (2006)
8. Böhm, H.P.: A general equation for the terminal fall speed of solid hydrometeors. J. Atmos. Sci. **46**(15), 2419–2427 (1989)
9. Bristow, N., Li, J., Hartford, P., Guala, M., Hong, J.: Imaging-based 3d particle tracking system for field characterization of particle dynamics in atmospheric flows. Exp. Fluids **64**(4), 1–14 (2023)
10. Dasari, T., Wu, Y., Liu, Y., Hong, J.: Near-wake behaviour of a utility-scale wind turbine. J. Fluid Mech. **859**, 204–246 (2019)
11. Esteban, L.B., Shrimpton, J., Ganapathisubramani, B.: Edge effects on the fluttering characteristics of freely falling planar particles. Phys. Rev. Fluids **3**(6), 064302 (2018)
12. Falkinhoff, F., Obligado, M., Bourgoin, M., Mininni, P.D.: Preferential concentration of free-falling heavy particles in turbulence. Phys. Rev. Lett. **125**(6), 064504 (2020)
13. Ferenc, J.-S., Néda, Z.: On the size distribution of poisson voronoi cells. Phys. A **385**(2), 518–526 (2007)

14. Ferrante, A., Elghobashi, S.: On the physical mechanisms of two-way coupling in particle-laden isotropic turbulence. Phys. Fluids **15**(2), 315–329 (2003)
15. Garrett, T.J., Bair, E.H., Fallgatter, C.J., Shkurko, K., Davis, R.E., Howlett, D.: The multi-angle snowflake camera. In: International Snow Science Workshop, pp. 16–21 (2012)
16. Garrett, T.J., Yuter, S.E.: Observed influence of riming, temperature, and turbulence on the fallspeed of solid precipitation. Geophys. Res. Lett. **41**(18), 6515–6522 (2014)
17. Good, G., Ireland, P., Bewley, G., Bodenschatz, E., Collins, L., Warhaft, Z.: Settling regimes of inertial particles in isotropic turbulence. J. Fluid Mech. **759**, R3 (2014)
18. Heisel, M., Dasari, T., Liu, Y., Hong, J., Coletti, F., Guala, M.: The spatial structure of the logarithmic region in very-high-reynolds-number rough wall turbulent boundary layers. J. Fluid Mech. **857**, 704–747 (2018)
19. Herzhaft, B., Guazzelli, É.: Experimental study of the sedimentation of dilute and semi-dilute suspensions of fibres. J. Fluid Mech. **384**, 133–158 (1999)
20. Heymsfield, A.J., Westbrook, C.: Advances in the estimation of ice particle fall speeds using laboratory and field measurements. J. Atmos. Sci. **67**(8), 2469–2482 (2010)
21. Hong, J., et al.: Natural snowfall reveals large-scale flow structures in the wake of a 2.5-mw wind turbine. Nat. Commun. **5**(1), 4216 (2014)
22. Ireland, P.J., Bragg, A.D., Collins, L.R.: The effect of Reynolds number on inertial particle dynamics in isotropic turbulence. Part 2. Simulations with gravitational effects. J. Fluid Mech. **796**, 659–711 (2016)
23. Ishihara, T., Kaneda, Y., Yokokawa, M., Itakura, K., Uno, A.: Small-scale statistics in high-resolution direct numerical simulation of turbulence: Reynolds number dependence of one-point velocity gradient statistics. J. Fluid Mech. **592**, 335–366 (2007)
24. Khvorostyanov, V.I., Curry, J.A.: Terminal velocities of droplets and crystals: power laws with continuous parameters over the size spectrum. J. Atmos. Sci. **59**(11), 1872–1884 (2002)
25. Li, C., et al.: Settling and clustering of snow particles in atmospheric turbulence. J. Fluid Mech. **912**, A49 (2021)
26. Li, J., Abraham, A., Guala, M., Hong, J.: Evidence of preferential sweeping during snow settling in atmospheric turbulence. J. Fluid Mech. **928**, A8 (2021)
27. Li, J., Guala, M., Hong, J.: Snow particle analyzer for simultaneous measurements of snow density and morphology. J. Geophys. Res.: Atmos. **128**(16), e2023JD038987 (2023). https://doi.org/10.1029/2023JD038987
28. Locatelli, J.D., Hobbs, P.V.: Fall speeds and masses of solid precipitation particles. J. Geophys. Res. **79**(15), 2185–2197 (1974)
29. Magono, C., Lee, C.W.: Meteorological classification of natural snow crystals. J. Faculty Sci. Hokkaido Univ. Ser. 7, Geophys. 2 **4** , 321–335 (1966)
30. Mallery, K., Shao, S., Hong, J.: Dense particle tracking using a learned predictive model. Exp. Fluids **61**(10), 1–14 (2020). https://doi.org/10.1007/s00348-020-03061-y
31. Marks, D., Kimball, J., Tingey, D., Link, T.: The sensitivity of snowmelt processes to climate conditions and forest cover during rain-on-snow: a case study of the 1996 pacific northwest flood. Hydrol. Process. **12**(10–11), 1569–1587 (1998)
32. Maxey, M.R., Riley, J.J.: Equation of motion for a small rigid sphere in a nonuniform flow. Phys. Fluids **26**(4), 883–889 (1983)

33. Mitchell, D.L., Heymsfield, A.J.: Refinements in the treatment of ice particle terminal velocities, highlighting aggregates. J. Atmos. Sci. **62**(5), 1637–1644 (2005)
34. Mitchell, D.L., Zhang, R., Pitter, R.L.: Mass-dimensional relationships for ice particles and the influence of riming on snowfall rates. J. Appl. Meteorol. Climatol. **29**(2), 153–163 (1990)
35. Mordant, N., Crawford, A.M., Bodenschatz, E.: Experimental lagrangian acceleration probability density function measurement. Physica D **193**(1–4), 245–251 (2004)
36. Nemes, A., Dasari, T., Hong, J., Guala, M., Coletti, F.: Snowflakes in the atmospheric surface layer: observation of particle-turbulence dynamics. J. Fluid Mech. **814**, 592–613 (2017)
37. Ogura, T., Kageyama, I., Nasukawa, K., Miyashita, Y., Kitagawa, H., Imada, Y.: Study on a road surface sensing system for snow and ice road. JSAE Rev. **23**(3), 333–339 (2002)
38. Ouellette, N.T., Xu, H., Bodenschatz, E.: A quantitative study of three-dimensional lagrangian particle tracking algorithms. Exp. Fluids **40**, 301–313 (2006)
39. Passarelli, Jr., R.E.: Theoretical and observational study of snow-size spectra and snowflake aggregation efficiencies. J. Atmos. Sci. **35**(5), 882–889 (1978)
40. Petersen, A.J., Baker, L., Coletti, F.: Experimental study of inertial particles clustering and settling in homogeneous turbulence. J. Fluid Mech. **864**, 925–970 (2019)
41. Raffel, M., Willert, C.E., Scarano, F., Kähler, C.J., Wereley, S.T., Kompenhans, J.: Particle Image Velocimetry. Springer, Cham (2018). https://doi.org/10.1007/978-3-319-68852-7
42. Rosa, B., Parishani, H., Ayala, O., Wang, L.-P.: Settling velocity of small inertial particles in homogeneous isotropic turbulence from high-resolution dns. Int. J. Multiph. Flow **83**, 217–231 (2016)
43. Shaw, R.A.: Particle-turbulence interactions in atmospheric clouds. Annu. Rev. Fluid Mech. **35**(1), 183–227 (2003)
44. Siewert, C., Kunnen, R., Meinke, M., Schröder, W.: Orientation statistics and settling velocity of ellipsoids in decaying turbulence. Atmos. Res. **142**, 45–56 (2014)
45. Steinkogler, W., Sovilla, B., Lehning, M.: Influence of snow cover properties on avalanche dynamics. Cold Reg. Sci. Technol. **97**, 121–131 (2014)
46. Tagliavini, G., Khan, M.H., McCorquodale, M., Westbrook, C., Holzner, M.: Wake characteristics of complex-shaped snow particles: Comparison of numerical simulations with fixed snowflakes to time-resolved particle tracking velocimetry experiments with free-falling analogs. Phys. Fluids **34**(5), 055112 (2022)
47. Tagliavini, G., McCorquodale, M., Westbrook, C., Corso, P., Krol, Q., Holzner, M.: Drag coefficient prediction of complex-shaped snow particles falling in air beyond the stokes regime. Int. J. Multiph. Flow **140**, 103652 (2021)
48. Tan, S., Salibindla, A., Masuk, A.U.M., Ni, R.: Introducing openlpt: new method of removing ghost particles and high-concentration particle shadow tracking. Exp. Fluids **61**, 1–16 (2020)
49. Theriault, D.H., et al.: A protocol and calibration method for accurate multi-camera field videography. J. Exp. Biol. **217**(11), 1843–1848 (2014)
50. Tinklenberg, A., Guala, M., Coletti, F.: Thin disks falling in air. J. Fluid Mech. **962**, A3 (2023)
51. Toloui, M., et al.: Measurement of atmospheric boundary layer based on super-large-scale particle image velocimetry using natural snowfall. Exp. Fluids **55**(5), 1–14 (2014). https://doi.org/10.1007/s00348-014-1737-1
52. Tom, J., Bragg, A.D.: Multiscale preferential sweeping of particles settling in turbulence. J. Fluid Mech. **871**, 244–270 (2019)

53. Voth, G.A., La Porta, A., Crawford, A.M., Alexander, J., Bodenschatz, E.: Measurement of particle accelerations in fully developed turbulence. J. Fluid Mech. **469**, 121–160 (2002)
54. Voth, G.A., Soldati, A.: Anisotropic particles in turbulence. Annu. Rev. Fluid Mech. **49**, 249–276 (2017)
55. Wang, L.-P., Maxey, M.R.: Settling velocity and concentration distribution of heavy particles in homogeneous isotropic turbulence. J. Fluid Mech. **256**, 27–68 (1993)
56. Yang, C., Lei, U.: The role of the turbulent scales in the settling velocity of heavy particles in homogeneous isotropic turbulence. J. Fluid Mech. **371**, 179–205 (1998)
57. Zhou, J., Adrian, R.J., Balachandar, S., Kendall, T.: Mechanisms for generating coherent packets of hairpin vortices in channel flow. J. Fluid Mech. **387**, 353–396 (1999)

Numerical Study on Turbulent Dispersion of Weakly Inertial Particles under Large Temperature Differences

Chendong Shen[1,2] and Guodong Jin[1,2](✉)

[1] The State Key Laboratory of Nonlinear Mechanics, Institute of Mechanics, Chinese Academy of Sciences, Beijing 100190, China
shenchendong@imech.ac.cn, gdjin@lnm.imech.ac.cn

[2] School of Engineering Sciences, University of Chinese Academy of Sciences, Beijing 100049, China

Abstract. We study the turbulent dispersion of weakly inertial particles under large temperature differences between the particle cloud and the ambient air. The fast Eulerian method is used to describe the turbulent dispersion of inertial particles at very small Stokes numbers, the low Mach number variable density fluid model is used to simulate the large temperature differences without using the Boussinesq approximation. We simulate the dispersion process of particles under different Reynolds numbers *Re*, Grashof numbers *Gr* and Stokes numbers *St* respectively. The results show that the temperature gradient plays a major role in both horizontal and normal dispersions. The dispersion rate and the rising velocity of the particle cloud are faster for larger Grashof numbers. The incoming flow velocity affects the rates of particle dispersion by changing the temperature distribution. The normal and spanwise dispersion rates are slower, and the streamwise dispersion rate and the streamwise velocity of the particle cloud are faster for larger Reynolds numbers. In the final settling period, the settling velocity of the particle cloud is faster for larger Stokes numbers. Our work is expected to be applied to the dispersion of environmental pollutants and smoke screen under large temperature differences.

Keywords: Turbulent Dispersion · Weakly Inertial Particles · Fast Eulerian Method · Large Temperature Differences

1 Introduction

The turbulent dispersion of inertial particles is a ubiquitous phenomenon in both natural environments and industrial processes. For instance, the dispersion of air pollutants in the atmosphere has caused serious environmental pollution problems [1]. In military applications, the smoke screen, which is used to conceal and protect soldiers, heavily relies on the rapid and effective dispersion of smoke particles [2]. However, simulating and controlling the dispersion of inertial particles in the atmosphere is a highly challenging problem. On one hand, during the dispersion process, the velocity field, temperature

X. Zheng and S. Balachandar (Eds.): IUTAM 2023, IUTAM Bookseries 41, pp. 83–103, 2024.
https://doi.org/10.1007/978-3-031-47258-9_5

field, and concentration field of inertial particles are coupled with each other. Three dimensionless numbers, namely the Reynolds number *Re*, the Grashof number *Gr*, and the Stokes number *St* are used to characterize the coupling process. The Reynolds number represents the ratio of inertial force to viscous force of the fluid, the Grashof number represents the ratio of buoyancy force due to the temperature differences between the particle cloud and the ambient air to viscous force, and the Stokes number represents the ratio of particle inertia characteristic time to the turbulent characteristic time. The three dimensionless numbers can characterize the incoming velocity of the flow field, the temperature difference between the particle release temperature and the ambient temperature, and the particle inertia, respectively, all of which have a significant impact on the particle dispersion process. On the other hand, single-phase turbulence involves a large number of time scales and space scales [3]. In multiphase turbulence, the particle phase entails smaller scales, resulting in more abundant scales [4–6]. As a result, analyzing all scales is extremely challenging. Direct Numerical Simulation (DNS) and Large Eddy Simulation (LES) require a considerable amount of computational cost. Therefore, in this study, we used a fast Eulerian method based on Reynolds-Averaged Navier-Stokes (RANS) to simulate the dispersion of weakly inertial particles.

Currently, there exist several well-established dispersion models for the prediction of inertial particle dispersion, with different models being appropriate for the simulation of particles with varying Stokes numbers [7]. When the Stokes numbers approach zero, the modified density method can be used. The vanishing inertia of particles allows us to assume that the particles closely move with the fluid and are fully mixed in space, allowing the flow to be regarded as a single-phase flow. The volume fraction of particles can be used to modify the fluid density. However, the modified density method does not consider the inertia of particles and is unable to simulate the sedimentation of particles in the dispersion process. For particles at moderate Stokes numbers, the particle phase can still be viewed as a continuous phase, and the Eulerian method [8–10] can be utilized. The Eulerian method requires the solution of two sets of partial differential equations for both the fluid phase and particle phase. Using this model, Xu and Subramaniam [11] proposed a multiscale model for dilute turbulent gas-particle flows based on the equilibration of energy concept to account for the interaction of particles with a range of fluid turbulence scales. For particles at even larger Stokes numbers, the large inertia of the particles results in significant differences in velocity between particles, making it challenging to describe particle velocity using a velocity field. In this case, the particle phase can no longer be treated as a continuous phase, and the Eulerian method is no longer suitable.

Lagrangian particle tracking method can be used to describe particles at any Stokes numbers, which involves analyzing the force on a single particle and tracking the trajectory of each particle. However, this method requires a significant amount of computational cost. The Lagrangian method has been widely applied in the study of two-phase flow [12–17], and has been used by Marchioli et al. [18] to study turbulent transfer and deposition of inertial particles in vertical upward circular pipe flow. Their results support that particles exhibit a specific tendency to segregate in the near-wall region, and the instantaneous realizations of Reynolds stresses play a crucial role in determining the flux of particles toward and away from the wall. Besides, their findings confirm the relative importance of free-flight and diffusion deposition mechanisms.

The fast Eulerian method can be used to simulate the dispersion of weakly inertial particles. This method is capable of simulating particle dispersion with the particle Stokes number between that simulated by the modified density method and that simulated by the Eulerian method. The fast Eulerian method considers the weak inertia of particles, which is an advantage over the modified density method. Maxey [19] obtained the first order asymptotic expansion solution of the velocity of a weakly inertial particle affected by drag and gravity. Particle inertia can be considered by using particle velocity to replace the fluid velocity in the convection term of the original convection - diffusion equation for concentration [7, 20–23]. Therefore, this method only requires solving the concentration equation for the particle phase. Compared to the traditional Eulerian method, the fast Eulerian method provides faster solution speeds [7]. In addition, compared to the Lagrangian method, the fast Eulerian method has the advantage of requiring less computation cost. Yang et al. [24] used this method to study the large-eddy simulation and parameterization of buoyant plume dynamics in stratified flow. Shotorban and Balachandar [25] applied this method to simulate the concentration of particles with small Stokes numbers through large-eddy simulation, which showed a good agreement with the results obtained from direct numerical simulation.

When studying natural convection problems, it is common to use the Boussinesq approximation [26] to simplify the equations if the temperature difference ΔT is very small and the ratio of the temperature difference to the ambient temperature is a small quantity ($\Delta T/T_0 << 1$). This allows for the use of governing equations of incompressible fluid, where the temperature and momentum equations are decoupled and the buoyancy effect is maintained during the dynamic process. However, in some cases, such as the smoke produced by fire or the initial smokescreen formed by the explosion of smoke bombs, there is a large temperature difference between the initial particle cloud and the surrounding environment. For example, the temperature in fire can reach 700 K–800 K or even higher [27]. The temperature difference between flame temperature and ambient temperature is often much greater than the ambient temperature ($\Delta T/T_0 > 1$), and the density will change significantly. The Boussinesq approximation can lead to large errors in this case. Schatzmann et al. [28] compared the simulation results of industrial stack plumes with and without the Boussinesq approximation and found that as the initial density difference increases, the difference in the height of the plumes between the two examples also increases. When the ratio of the initial density difference to the density at infinity $(\rho\text{-}\rho_\infty)/\rho_\infty$ reaches 50%, the final height of the plumes simulated with the Boussinesq approximation is 20% larger than the simulation result without the Boussinesq approximation [28, 29]. Therefore, it is necessary to use the low Mach number variable density fluid model to calculate particle dispersion with a large temperature difference.

The aim of this paper is to address the issues of turbulent dispersion of weakly inertial particles under large temperature differences. At the initial stage of particle dispersion, the particles are subject to buoyancy due to the large temperature difference between the particle temperature and the ambient temperature, leading to an upward trend. However, after dispersion for a certain period of time, when the particle temperature approaches the ambient temperature, the weakly inertia effect of particles becomes apparent, resulting in a settling tendency under the action of gravity. We shall study the turbulent dispersion under coupling of fluid flow, large temperature difference and particle inertia. To more accurately simulate particle dispersion in this scenario, we abandon the incompressible fluid model with the Boussinesq approximation and adopt the low Mach number variable density fluid model for the fluid phase. Additionally, the fast Eulerian method is utilized to simulate weakly inertial particles.

The remainder of the paper is organized as follows. In the second part, we introduce the research methods, including the selection of governing equations, the setting of the flow field and physical parameters, and the selection of numerical methods. In the third part, we carry out the verification of the numerical calculation program, thus validating the correctness of the numerical program. In the fourth part, we use the numerical program to simulate particle dispersion under different incoming flow velocities, particle release temperatures, and particle inertia, and further analyze the influence of Grashof number *Gr*, Reynolds number *Re*, and Stokes number *St* on particle dispersion. Finally, we provide concluding remarks in the fifth part.

2 Mathematical Model

2.1 Governing Equations

In this study, we use the fast Eulerian model to simulate the dispersion process of inertial particles, while adopting the governing equation of the low Mach number variable density fluid model for the fluid phase based on RANS method. The continuity equation is given as:

$$\frac{\partial \rho}{\partial t} + \nabla \cdot (\rho \boldsymbol{u}) = 0, \tag{1}$$

where ρ represents the density of the fluid, while $\boldsymbol{u}$ denotes the velocity of the fluid. The momentum conservation equation is given as:

$$\frac{\partial (\rho \boldsymbol{u})}{\partial t} + \nabla \cdot (\rho \boldsymbol{u}\boldsymbol{u}) = \rho \boldsymbol{g} - \nabla p + \nabla \cdot \boldsymbol{\tau}. \tag{2}$$

In Eq. (2), p denotes the pressure of the fluid, while $\boldsymbol{\tau}$ is defined as:

$$\boldsymbol{\tau} = \mu_{\text{eff}}\left(\left(\nabla \boldsymbol{u} + (\nabla \boldsymbol{u})^{\mathrm{T}}\right) - \frac{2}{3}(\nabla \cdot \boldsymbol{u})\boldsymbol{I}\right), \tag{3}$$

where μ_{eff} represents the sum of molecular and turbulent dynamic viscosity. The energy conservation equation is given as:

$$\frac{\partial (\rho h)}{\partial t} + \nabla \cdot (\rho \boldsymbol{u} h) + \frac{\partial (\rho K)}{\partial t} + \nabla \cdot (\rho \boldsymbol{u} K) = \frac{\partial p}{\partial t} + \nabla \cdot (\alpha_{\text{eff}} \nabla e) + \rho \boldsymbol{u} \cdot \boldsymbol{g} + S, \tag{4}$$

where $K = \frac{1}{2}|\boldsymbol{u}|^2$ denotes the specific mechanical energy, e represents the specific internal energy, and h is the specific enthalpy. The relationship between e and h is given as:

$$e = h - \frac{p}{\rho}. \tag{5}$$

In formula (4), α_{eff} denotes the sum of molecular and turbulent thermal diffusion coefficients, while S encompasses the source term, radiation term, and other related terms, which is neglected in this study. The momentum equation and energy equation are coupled through the perfect gas equation:

$$pM = \rho RT, \tag{6}$$

where M represents the molar mass of the fluid, R is the molar gas constant with a value of 8.3145 J/(mol·K), and T denotes the temperature of the fluid. The relationship between T and h is given as:

$$\left(\frac{\partial h}{\partial T}\right)_p = c_p, \tag{7}$$

where c_p denotes the specific heat capacity at constant pressure per unit mass. In our simulation, we utilize the buoyantPimpleFoam solver in OpenFOAM to solve the aforementioned governing equations. The concentration equation that describes the dispersion of particles is given as [20]:

$$\frac{\partial \rho C}{\partial t} + \nabla \cdot (\rho \boldsymbol{v} C) = \nabla \cdot (\gamma_{\text{eff}} \nabla C). \tag{8}$$

In Eq. (8), C denotes the concentration of the particle cloud, γ_{eff} represents the sum of the molecular and turbulent concentration diffusion coefficients, and $\boldsymbol{v}$ denotes the velocity of particles. This equation takes into account the inertia of particles. It is assumed that particles are subject to both Stokes resistance and gravity, the equation of motion of particles is given as:

$$\frac{\mathrm{d}\boldsymbol{v}}{\mathrm{d}t} = \frac{1}{\tau_{\mathrm{p}}}(\boldsymbol{u} - \boldsymbol{v}) + \boldsymbol{g}. \tag{9}$$

In Eq. (9), $\boldsymbol{g}$ represents the gravitational acceleration. τ_{p} is the characteristic time of particles and is defined as:

$$\tau_{\mathrm{p}} = \frac{\rho_{\mathrm{p}} d_{\mathrm{p}}^2}{18\mu}, \tag{10}$$

where ρ_{p} represents the density of particles, d_{p} denotes the particle diameter, and μ represents the dynamic viscosity of the fluid. The relationship between $\boldsymbol{v}$ in the convection term of Eq. (8) and fluid velocity $\boldsymbol{u}$ can be obtained through the first-order asymptotic expansion solution of Eq. (9) [19, 21–23, 25]:

$$\boldsymbol{v} = \boldsymbol{u} - \tau_{\mathrm{p}} \frac{D\boldsymbol{u}}{Dt} + \tau_{\mathrm{p}} \boldsymbol{g} + o(\tau_{\mathrm{p}}) = \boldsymbol{u} - \tau_{\mathrm{p}} \left(\frac{\partial \boldsymbol{u}}{\partial t} + \boldsymbol{u} \cdot \nabla \boldsymbol{u} \right) + \tau_{\mathrm{p}} \boldsymbol{g} + o(\tau_{\mathrm{p}}). \tag{11}$$

To model turbulence, we use the k-ε model of RANS in OpenFOAM [30]:

$$\frac{\partial (\rho k)}{\partial t} + \nabla \cdot (\rho k \boldsymbol{u}) = \rho G + \nabla \cdot (\rho D_k \nabla k) - \frac{2}{3} \rho (\nabla \cdot \boldsymbol{u}) k - \rho \varepsilon, \tag{12}$$

$$\frac{\partial(\rho\varepsilon)}{\partial t} + \nabla\cdot(\rho\varepsilon\boldsymbol{u}) = C_1\rho G\frac{\varepsilon}{k} + \nabla\cdot(\rho D_\varepsilon\nabla\varepsilon) - \left(\frac{2}{3}C_1 - C_3\right)\rho(\nabla\cdot\boldsymbol{u})\varepsilon - C_2\rho\frac{\varepsilon^2}{k}. \tag{13}$$

In Eqs. (12) and (13), k denotes the turbulent kinetic energy, while ε represents the turbulent dissipation rate per unit mass. The turbulent momentum diffusion coefficient is defined as:

$$\nu_t = C_\mu\frac{k^2}{\varepsilon}. \tag{14}$$

The parameters in Eqs. (12) and (13) are defined as:

$$D_k = \nu + \frac{\nu_t}{\sigma_k}, \tag{15}$$

$$D_\varepsilon = \nu + \frac{\nu_t}{\sigma_\varepsilon}, \tag{16}$$

$$G = \nu_t\left(\nabla\boldsymbol{u} + (\nabla\boldsymbol{u})^{\mathrm{T}} - \frac{2}{3}(\nabla\cdot\boldsymbol{u})\boldsymbol{I}\right). \tag{17}$$

The constant values of the k-ε model are shown in Table 1.

Table 1. The values of the constants of the k-ε model

The constants of the k-ε model	Values
C_μ	0.09
C_1	1.44
C_2	1.92
C_3	0
σ_k	1.0
σ_ε	1.3

2.2 Settings of Flow Field

The flow field, as shown in Fig. 1, is represented in a Cartesian coordinate system, where the x-direction corresponds to the spanwise direction, the y-direction corresponds to the streamwise direction, and the z-direction corresponds to the normal direction. The total size of the flow field is set to 20 m × 120 m × 30 m. 48 uniform grids are set in the spanwise, while 720 uniform grids are set in the streamwise. The normal direction grids are divided into two parts. From 0 to 1.5 m, 40 non-uniform grids are set, and the grid size increases from the bottom to the top in an equal proportion sequence. The ratio of

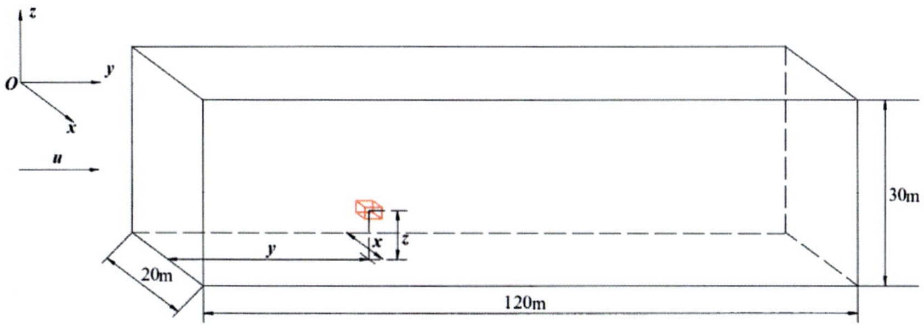

Fig. 1. The flow field diagram, in which the red box is the particle release position.

Table 2. Initial condition settings of physical quantities in the internal flow field

Physical quantities	Initial conditions
Average velocity $\boldsymbol{u}$	a fixed value
Ambient temperature T_0	a fixed value
Pressure p	100000 Pa
Concentration C	0
Turbulent kinetic energy k	0.01 m^2/s^2
Turbulent dissipation rate ε	0.0001 m^2/s^3
Turbulent momentum diffusion coefficients ν_t	0
Turbulent thermal diffusion coefficients α_t	0
Turbulent concentration diffusion coefficients γ_t	0

the largest grid size to the smallest grid size is set to 10, allowing for better analysis of the boundary layer flow. From 1.5 m to 30 m, 180 uniform grids are set.

The initial condition settings of physical quantities in the internal flow field are shown in Table 2.

The boundary conditions of physical quantities at the inlet, top, outlet, front and back are shown in Table 3. "zero gradient" and "periodic" represent the zero gradient boundary condition and the periodic boundary condition, respectively. "calculated" indicates that the value of a physical quantity needs to be calculated from other fields.

The boundary conditions of physical quantities at the ground are set as follows. The velocity is set to the no-slip condition, and the turbulence variables are set to correspond to the wall function in OpenFOAM, so that the dimensionless streamwise velocity $u_y{}^+$ and the dimensionless normal distance z^+ at the wall meet the wall function law. At the viscous sublayer ($z^+ < 5$), the relationship between $u_y{}^+$ and z^+ is given as:

$$u_y^+ = z^+. \tag{18}$$

Table 3. The boundary conditions of physical quantities

Physical quantities	Inlet	Top	Outlet	Front and Back
Average velocity $\boldsymbol{u}$	a fixed value	zero gradient	zero gradient	periodic
Ambient temperature T_0	a fixed value	zero gradient	zero gradient	periodic
Concentration C	0	zero gradient	zero gradient	periodic
Turbulent kinetic energy k	0.01 m^2/s^2	zero gradient	zero gradient	periodic
Turbulent dissipation rate ε	0.0001 m^2/s^3	zero gradient	zero gradient	periodic
Turbulent momentum diffusion coefficients ν_t	calculated	zero gradient	zero gradient	periodic
Turbulent thermal diffusion coefficients α_t	calculated	zero gradient	zero gradient	periodic
Turbulent concentration diffusion coefficients γ_t	calculated	zero gradient	zero gradient	periodic
Pressure p	zero gradient	100000 Pa	zero gradient	periodic

At the logarithmic region ($30 < z^+ < 200$), the relationship between $u_y{}^+$ and z^+ is given as:

$$u_y^+ = \frac{1}{\kappa}\ln(Ez^+). \tag{19}$$

In formula (19), κ represents the von Karman constant whose value is between 0.35 and 0.43. E is a constant related to factors such as the smoothness of the wall. The temperature is set to a fixed value, which is equal to ambient temperature T_0. The pressure and the concentration are set to the zero gradient boundary conditions.

The physical parameters in the flow field are shown in Table 4.

2.3 Numerical Methods

The numerical discretization method for each term in the governing equations is shown in Table 5.

In Table 5, the "Gauss van Leer" scheme is used for the convection term of concentration to reduce non-physical oscillations [31, 32].

The solver utilized in OpenFOAM, buoyantPimpleFoam, uses the PIMPLE algorithm, which usually adopts the prediction-correction strategy to decouple the calculation of velocity and pressure through iteration.

Table 4. Physical parameters in the flow field

Physical parameters	Values
Dynamic viscosity of air μ	1.8×10^{-5} N·s/m^2
Prandtl number Pr	0.7
Turbulent Prandtl number Pr_t	1.0
Schmidt number Sc	0.7
Turbulent Schmidt number Sc_t	1.0
Average molar mass of air M	28.9 g/mol
Particle characteristic time τ_p	0.0003 s
Specific heat capacity at constant pressure per unit mass c_p	1000 J/(kg·K)

Table 5. The numerical discretization methods used for different terms in the governing equations

Terms in the governing equations	Numerical discretization method
Time derivative term	"Euler" scheme
Gradient term	"Gauss linear" scheme
Diffusion term	"Gauss linear corrected" scheme
Convection term of velocity	"Gauss upwind" scheme
Convection term of specific enthalpy	"Gauss upwind" scheme
Convection term of turbulent kinetic energy	"Gauss upwind" scheme
Convection term of turbulent dissipation rate	"Gauss upwind" scheme
Convection term of specific mechanical energy	"Gauss linear" scheme
Convection term of concentration	"Gauss van Leer" scheme

3 Verification of Numerical Method

In order to validate the numerical methods, we reset the numerical example according to the experimental conditions [16]. The incoming flow velocity is set to 2 m/s along the y-axis direction and the ambient temperature is set to 283.15 K. We compare the subsequent dispersion process in our simulation with the experimental results. The initial release position of particles in Fig. 1 is set as follows: The coordinates of the center point of the red box are $(x, y, z) = (10$ m, 41.8 m, 5.15 m$)$, the dimensions of spanwise direction, streamwise direction and normal direction are $\Delta x = 3.6$ m, $\Delta y = 3.6$ m, $\Delta z = 0.3$ m, respectively. The particle release temperature is set to be 340 K. After the flow field becomes stable, we release the particles and compare the experimental and numerical results. Taking the smoke shielding effect into consideration, the boundary of the particle cloud is determined as 10% of the maximum particle concentration on the same plane [2]. The changes in the streamwise and normal direction sizes of the particle cloud in the first ten seconds of dispersion are recorded and compared with the experimental data [16], as shown in Fig. 2. It can be seen from Fig. 2 that the relative errors of the normal direction size and the streamwise size are mostly controlled within 15%. This indicates that the simulation results basically reflect the general law of dispersion and can reasonably predict the dispersion of particles.

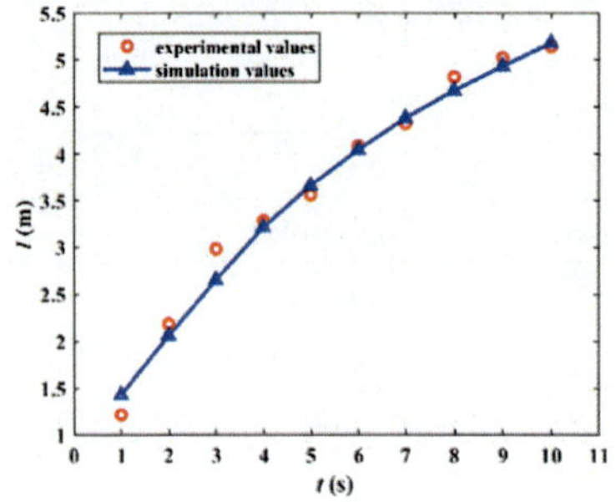

(a) The variations of the normal direction size l of the particle cloud with time t

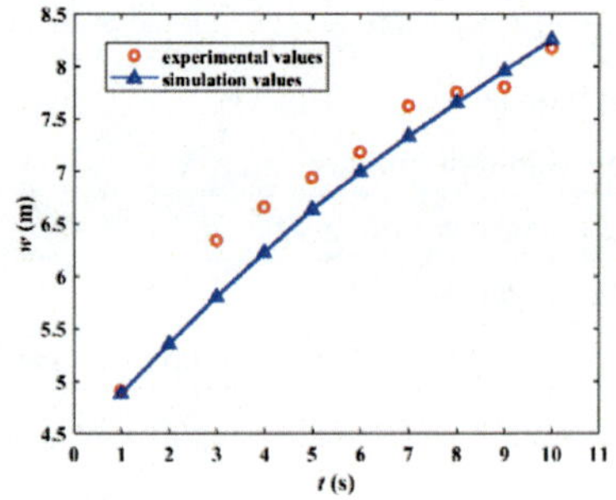

(b) The variations of the streamwise size w of the particle cloud with time t

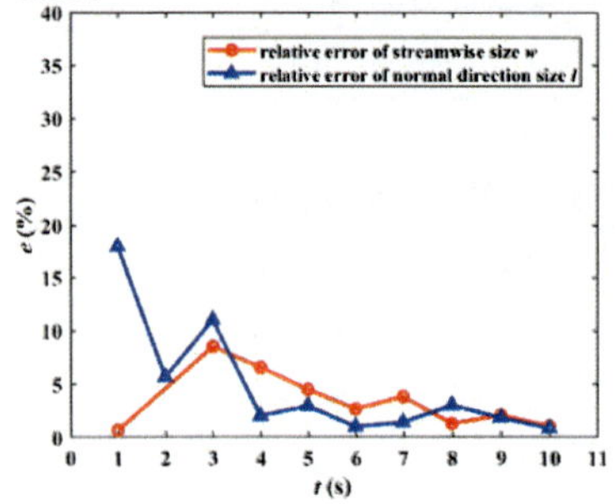

(c) The relative errors e of the normal direction size and the streamwise size with time t

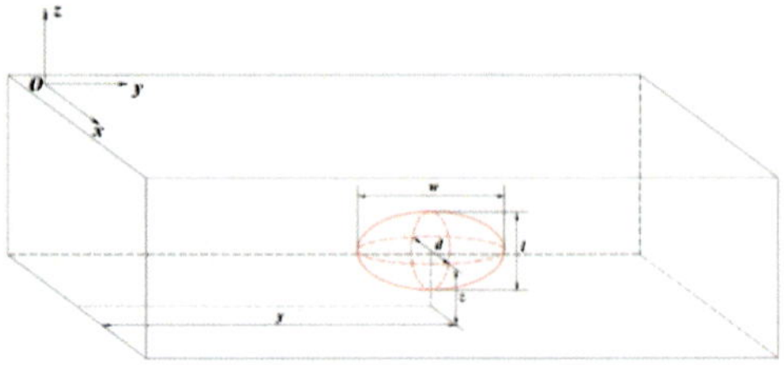

(d) The schematic diagram of each size in the process of particle dispersion

Fig. 2. In the process of particle dispersion, the simulation results are compared with the experimental results. The variations of the normal direction size l of the particle cloud and the streamwise size w of the particle cloud with time t are shown in **a** and **b** respectively. The relative errors e of the normal direction size and the streamwise size with time t is shown in **c**. **d** shows the schematic diagram of each size in the process of particle dispersion. The red ellipsoid in the figure represents the measurement boundary of the particle cloud.

4 Numerical Calculation Results and Discussion

In this study, we investigate the influences of different physical quantities on the dispersion process, including the particle release temperature T, incoming velocity is along the y-axis direction, and particle inertia represented by the characteristic time τ_p. These three physical quantities are represented by three dimensionless numbers respectively, namely the Grashof number Gr, Reynolds number Re, and Stokes number St. The dimensionless numbers are calculated using the following definitions.

The Grashof number is defined as [26]:

$$Gr = \frac{g\beta \Delta T \rho^2 V}{\mu^2}. \tag{20}$$

This dimensionless number represents the ratio of buoyant force due to the density difference caused by the temperature difference to viscous force. In formula (20), g denotes the gravitational acceleration, $\beta = 3.33 \times 10^{-3}\ \mathrm{K}^{-1}$ is the coefficient of thermal expansion, $V = 3\ \mathrm{m} \times 3\ \mathrm{m} \times 1.95\ \mathrm{m}$ denotes the volume of the initial particle cloud, $\mu = 1.8 \times 10^{-5}\ \mathrm{N{\cdot}s/m^2}$ represents the dynamic viscosity of air, and $\rho = 1.16\ \mathrm{kg/m^3}$ represents the density of air at ambient temperature. The temperature difference ΔT is defined as the difference between the particle release temperature T and the ambient temperature T_0.

The Reynolds number is defined as [26]:

$$Re = \frac{\rho U L}{\mu}. \tag{21}$$

This dimensionless number represents the ratio of inertial forces to viscous forces, In formula (21), L denotes the distance along the streamwise direction from the inlet to the location where the particles are released and the value of L is 40 m. U represents a characteristic velocity, which is the incoming flow velocity of the flow field.

The Stokes number is defined as [21, 25]:

$$St = \frac{\tau_{\mathrm{p}}}{\tau_{\eta}}. \tag{22}$$

The dimensionless number represents the ratio of the particle inertia characteristic time to the turbulent characteristic time. A smaller Stokes number indicates lower particle inertia and easier to follow the fluid, whereas a larger Stokes number indicates higher particle inertia and harder to follow the fluid. In formula (22), τ_{η} denotes the turbulent Kolmogorov time scale, which only depends on the kinematic viscosity of air ν and turbulence dissipation rate ε. Its definition is as follows:

$$\tau_\eta = \left(\frac{\nu}{\varepsilon}\right)^{0.5} \sim \left(\frac{10^{-5}}{10^{-4}}\right)^{0.5} s \sim 0.3s. \tag{23}$$

4.1 Effect of the Release Temperature of Particles on Dispersion Process

We set the ambient temperature T_0 to 300 K, the flow field inlet velocity u to 1.6 m/s, and the particle characteristic time τ_{p} to 0.0003 s. The particles are released after the flow field reaches a stable state. We release the particles at temperatures T of 400 K, 500 K and 600 K, namely the Reynolds number Re of the flow field is 4.12×10^6, Stokes number St is 10^{-3}, and Grashof numbers Gr are 2.38×10^{11}, 4.76×10^{11}, 7.14×10^{11} respectively. The initial release position of particles in Fig. 1 is set as follows: The coordinates of the center point of the red box are $(x, y, z) = (10\ \mathrm{m}, 41.5\ \mathrm{m}, 5.975\ \mathrm{m})$, the dimensions of spanwise direction, streamwise direction and normal direction are $\Delta x = 3$ m, $\Delta y = 3$ m, $\Delta z = 1.95$ m, respectively. During the first ten seconds of simulating the dispersion process, we record the streamwise moving distance y and the normal direction moving distance z of the center of the particle cloud in Fig. 3. The changes of the normal direction size l, the streamwise size w and the spanwise size d of the particle cloud are shown in Fig. 4. The sizes of the particle cloud are shown in Fig. 2d.

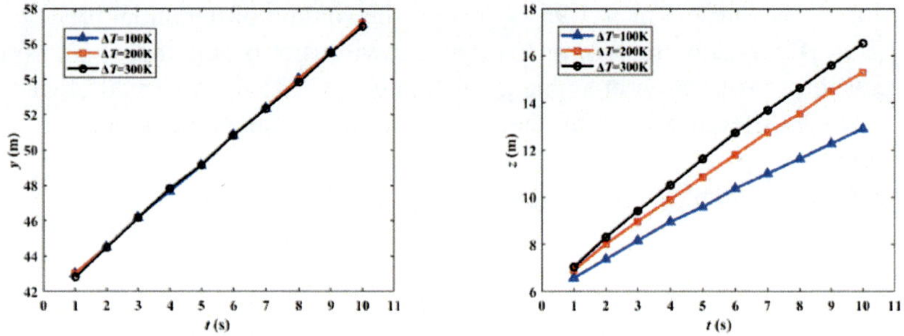

(a) The variations of the center of the particle cloud in the streamwise direction with time t

(b) The variations of the center of the particle cloud in the normal direction with time t

Fig. 3. The variations of the center of the particle cloud with time t

The particle cloud trajectories (Fig. 3) reveal that t the streamwise moving distance y and the normal direction moving distance z of the particle cloud increase with time, due to the influence of incoming flow velocity in the streamwise direction and buoyancy force caused by the temperature difference. As the initial release temperature of particles increases, the temperature difference between the particles and the ambient temperature also increases, resulting in a larger Grashof number and a stronger buoyancy force acting on the particle cloud. Consequently, the normal direction velocity increases, and the particle cloud moves farther in the normal direction at the same dispersion time. However, the streamwise moving distance of the particle cloud center remains relatively unchanged with the same incoming velocity. Furthermore, the increase in Grashof number Gr accelerates particle dispersion and leads to larger normal direction size l, streamwise size w, and spanwise size d of the particle cloud (Fig. 4a, b, c).

To investigate the reasons for the changes of particle cloud during the dispersion process, at the ninth second of particle dispersion, we analyze the distributions of temperature along three directions and the distributions of three components of velocity along their respective directions under different initial temperature differences. These distributions are shown in Fig. 5, Fig. 6, and Fig. 7, respectively. The sampling lines pass through the center of the particle cloud.

From the distributions of temperature and velocity components in the spanwise and streamwise directions, we observe that larger initial temperature differences result in higher peak values in the temperature distributions, while the widths of the normal distributions remain basically unchanged. The distributions of velocity components are antisymmetric about the center of the particle cloud, and the peak values on both sides increase with increasing initial temperature differences.

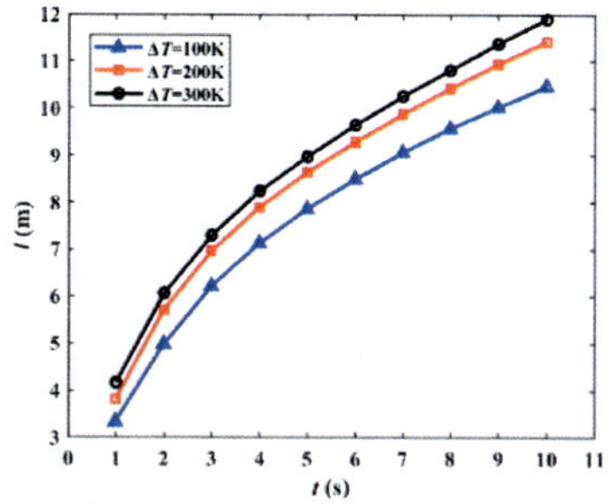

(a) The variations of the normal direction size l of the particle cloud with time t

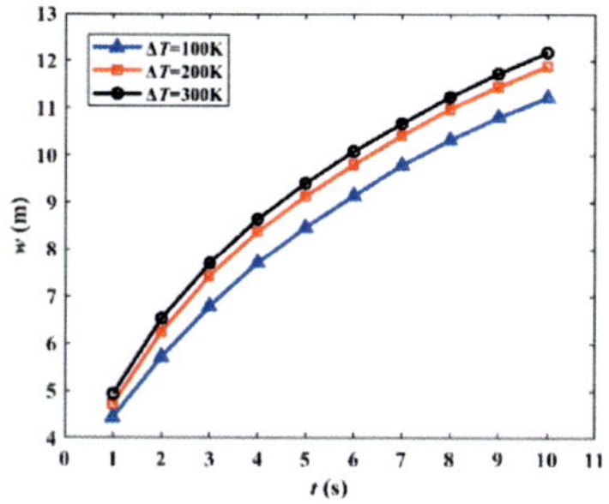

(b) The variations of the streamwise size w of the particle cloud with time t

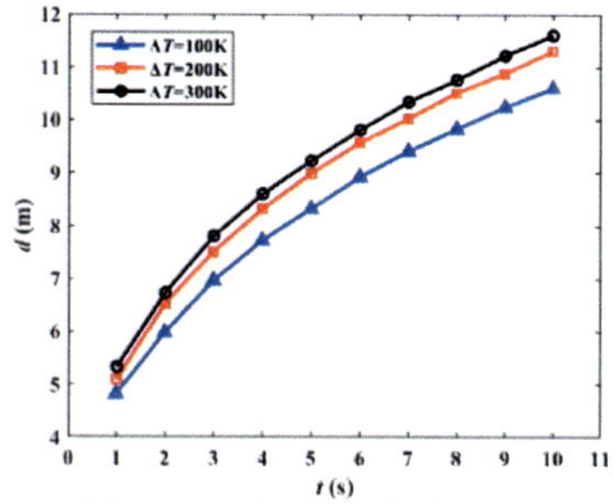

(c) The variations of the spanwise size d of the particle cloud with time t

Fig. 4. The variations of the particle cloud sizes with time t

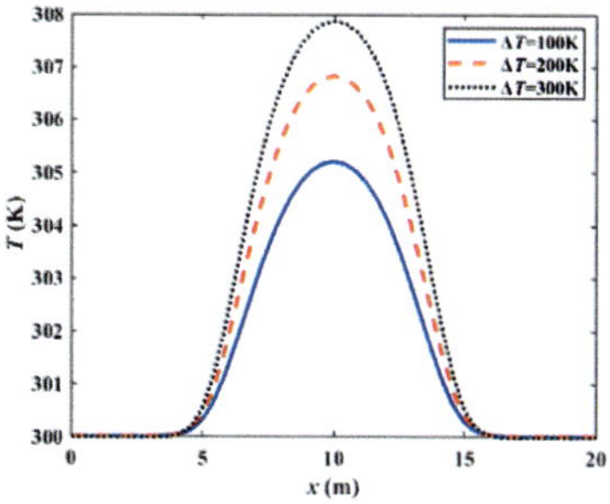

(a) The distributions of temperature of the flow field

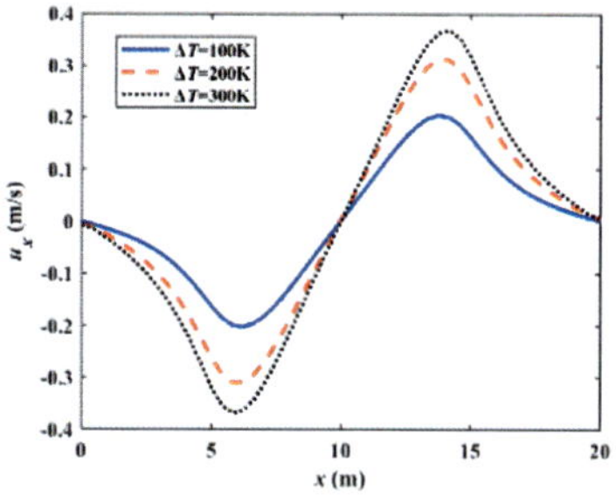

(b) The distributions of spanwise velocity of the flow field

Fig. 5. The distributions of temperature and velocity along the sampling line which passes through the center of the particle cloud along the spanwise direction

The normal direction distributions of temperature exhibit skewed distributions, with larger initial temperature differences resulting in a higher temperature peak and a more noticeable deviation from the center. The normal direction of the velocity component follows a similar trend as the temperature, where increasing initial temperature differences result in higher maximum velocities and larger displacements of peak value position from the center.

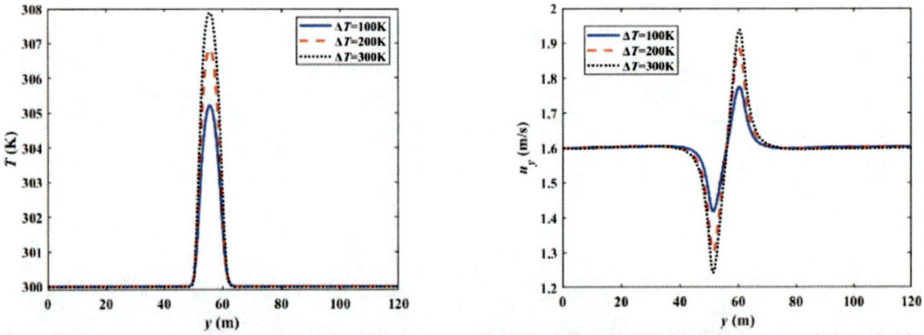

(a) The distributions of temperature of the flow field

(b) The distributions of streamwise velocity of the flow field

Fig. 6. The distributions of temperature and velocity along the sampling line which passes through the center of the particle cloud along the streamwise direction

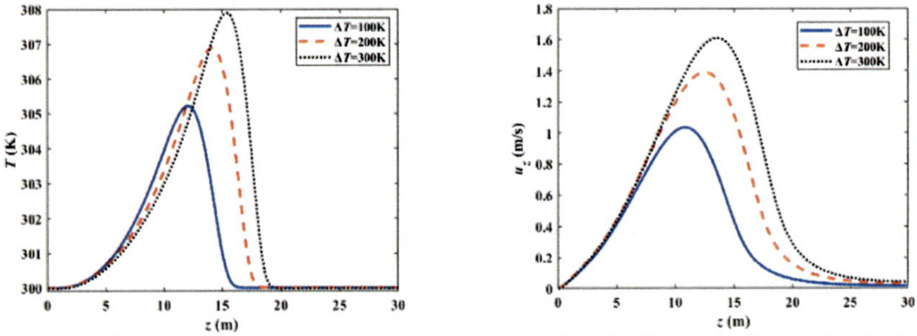

(a) The distributions of temperature of the flow field

(b) The distributions of normal velocity of the flow field

Fig. 7. The distributions of temperature and velocity along the sampling line which passes through the center of the particle cloud along the normal direction

When the initial temperature difference becomes larger, the buoyancy caused by the temperature gradient become larger, and the rising velocity of the particle cloud center becomes faster, so the position of maximum temperature and velocity becomes higher, and this is also the reason why the distributions of temperature and velocity are skewed in Fig. 7. At the same time, with the larger initial temperature difference, the normal direction moving distance z of the particle cloud becomes higher in Fig. 3b, and the three velocity components and the velocities at the boundary of the particle cloud increase, resulting in the sizes of the particle cloud becomes larger in all directions in Fig. 4.

4.2 Effect of Incoming Flow Velocity on Dispersion Process

We set the ambient temperature T_0 to 300 K, the release temperature of particles T to 500 K, particle characteristic time τ_p to 0.0003 s, the values of the incoming velocity u are taken as 1.6 m/s, 3 m/s and 4 m/s respectively, namely the Grashof number Gr is 4.76 $\times$ 10^{11}, Stokes number St is 10^{-3}, Reynolds numbers Re of flow field are respectively 4.12 $\times$ 10^6, 7.73 $\times$ 10^6, 1.03 $\times$ 10^7. The initial release position of particles in Fig. 1 is set as follows: The coordinates of the center point of the red box are $(x, y, z) = (10$ m, 41.5 m, 5.975 m), the dimensions of spanwise direction, streamwise direction and normal direction are $\Delta x = 3$ m, $\Delta y = 3$ m, $\Delta z = 1.95$ m, respectively. After the flow field is stable, we release particles. During the first ten seconds of simulating the dispersion process, we record the streamwise moving distance y and the normal direction moving distance z of the center of the particle cloud in Fig. 8. The variations of the normal direction size l, the streamwise size w and the spanwise size d of the particle cloud are shown in Fig. 9. The sizes of the particle cloud are shown in Fig. 2d.

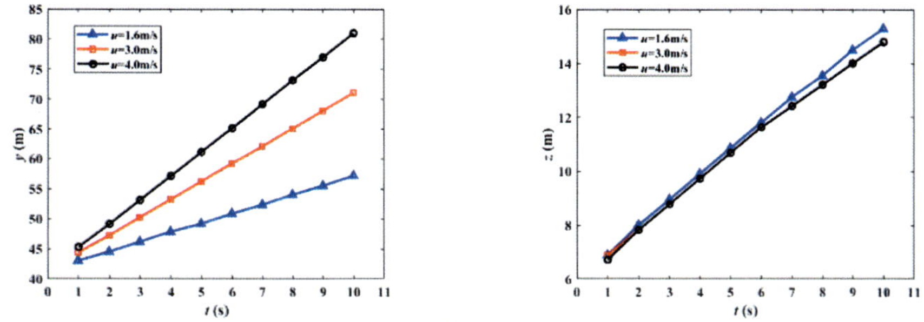

(a) The variations of the center of the particle cloud in the streamwise direction with time t

(b) The variations of the center of the particle cloud in the normal direction with time t

Fig. 8. The variations of the center of the particle cloud with time t

It can be seen from the particle cloud trajectories (Fig. 8) that the streamwise moving distance y and the normal direction moving distance z both increase with time. For the higher the incoming flow velocity, the Reynolds number is larger, and the streamwise moving velocity of particle cloud is higher (at the same dispersion time, the streamwise moving distance is larger), while the normal distance of particle cloud center movement changes slightly.

The changes in the normal direction size l, the streamwise size w and the spanwise size d of the particle cloud (Fig. 9) indicate that the normal direction size and spanwise size decrease while the streamwise size increases with the increasing Reynolds number.

We conduct further analysis on the reasons for the changes of particle cloud during the dispersion process. At the ninth second of particle dispersion, under different incoming flow velocities, we research the distributions of temperature along three directions and the distributions of three components of velocity along their respective directions. These distributions are shown in Fig. 10, Fig. 11 and Fig. 12, respectively. The sampling lines pass through the center of the particle cloud.

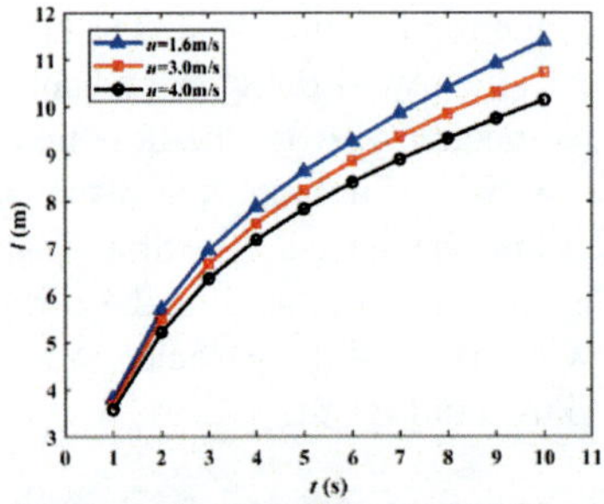

(a) The variations of the normal direction size *l* of the particle cloud with time *t*

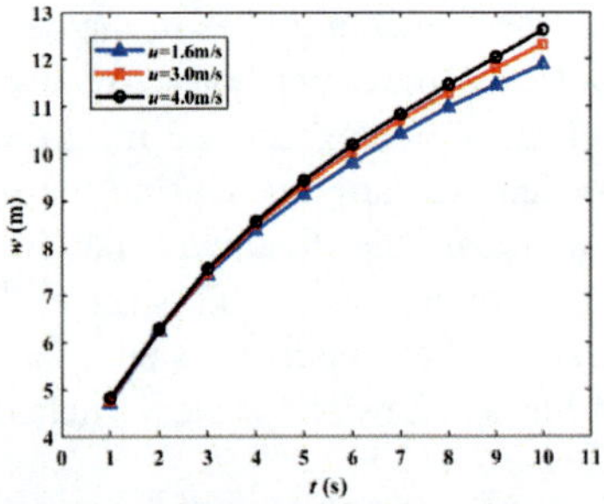

(b) The variations of the streamwise size *w* of the particle cloud with time *t*

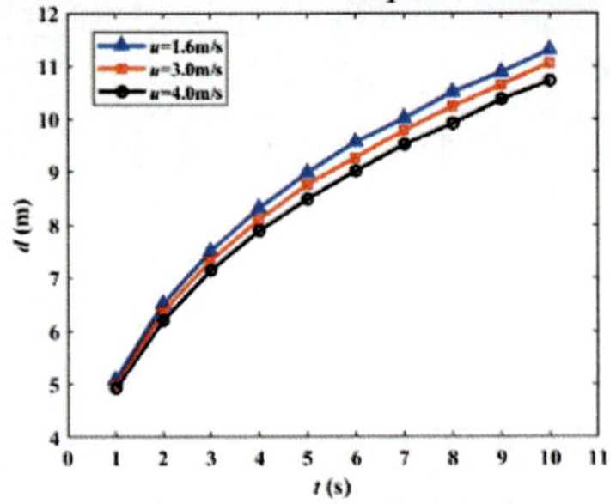

(c) The variations of the spanwise size *d* of the particle cloud with time *t*

Fig. 9. The variations of the center of the particle cloud with time *t*

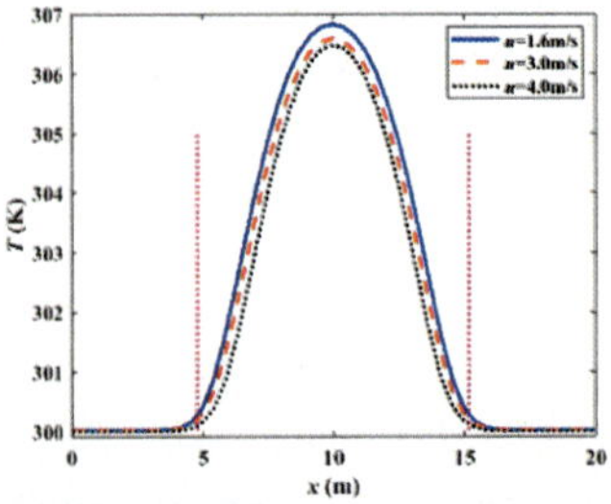

(a) The distributions of temperature of the flow field

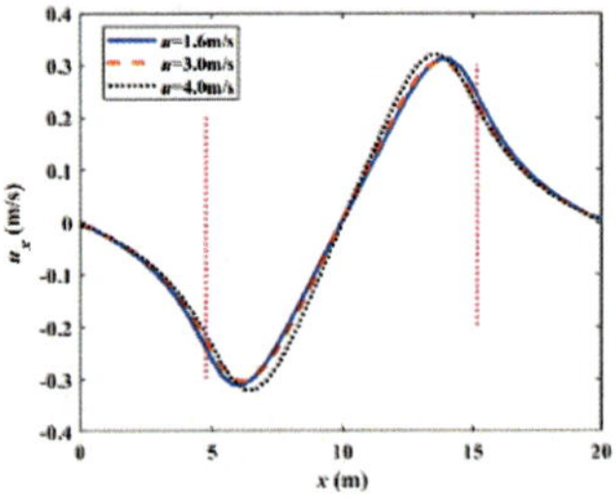

(b) The distributions of spanwise velocity of the flow field

Fig. 10. The distributions of temperature and velocity along the sampling line which passes through the center of the particle cloud along the spanwise direction are shown in **a** and **b**. The pink vertical dotted lines in the figures indicate the vicinity of the boundary of the particle cloud.

For spanwise distributions, when the incoming flow velocity becomes higher, the temperature peak becomes smaller, the value of temperature and spanwise velocity at the vicinity of the boundary of the particle cloud become smaller, so the spanwise size of the particle cloud decreases with the increase of the incoming flow velocity, as shown in Fig. 9c.

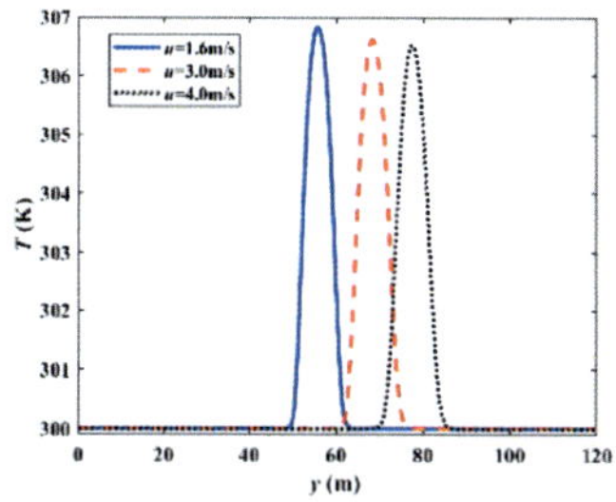

(a) The distributions of temperature of the flow field

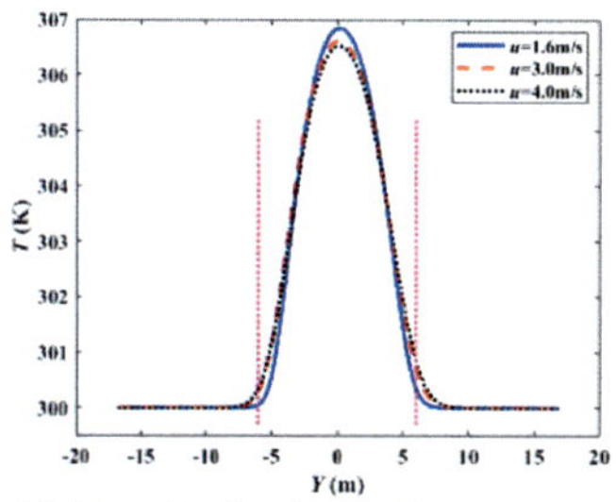

(b) The distributions of temperature in the new coordinate system

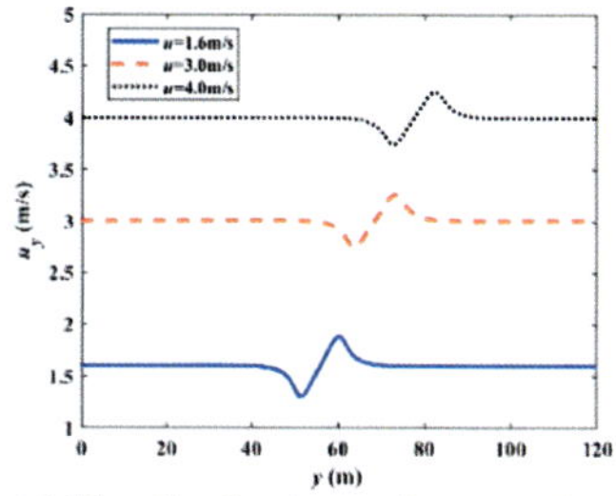

(c) The distributions of streamwise velocity of the flow field

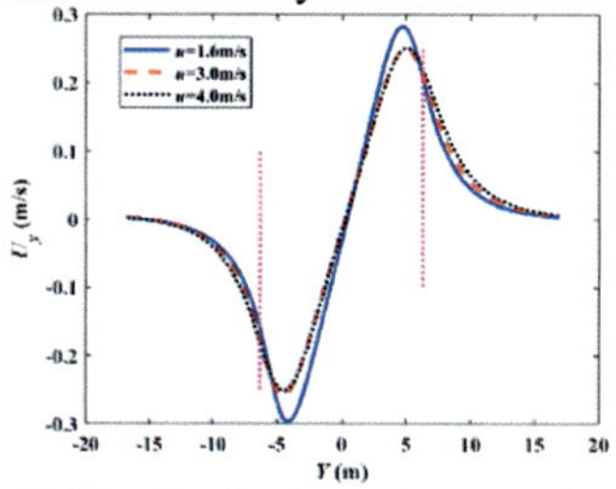

(d) The distributions of streamwise velocity in the new coordinate system

Fig. 11. The distributions of temperature and velocity along the sampling line which passes through the center of the particle cloud along the streamwise direction are shown in **a** and **c**. We move the centers of the distributions to the same position as shown in **b** and **d**. The pink vertical dotted lines in the figures indicate the vicinity of the boundary of the particle cloud.

For the streamwise distributions, under different incoming flow velocities, the centers of the temperature and streamwise velocity distributions are not in the same position, we move the centers of the distributions to the same position. Furthermore, the analysis only considers the velocity difference between the streamwise velocity and incoming velocity $U_y = u_y - u$, and takes the center of the particle cloud as the origin of the streamwise coordinate Y, as shown in Fig. 11b, d. The results indicate that with increasing incoming flow velocity, the peak value of the temperature distribution decreases slightly while the variance of the temperature distribution increases. This leads to higher temperatures near the boundary of the particle cloud, as well as higher velocities in the vicinity of the boundary, resulting in a larger streamwise size of the particle cloud, as shown in Fig. 9b.

The distributions of temperature and normal velocity in the normal direction are skewed, when the incoming flow velocity is higher, the value of temperature and normal direction velocity at the vicinity of the boundary of the particle cloud are smaller, which results in a smaller size of the particle cloud in the normal direction, as illustrated in Fig. 9a.

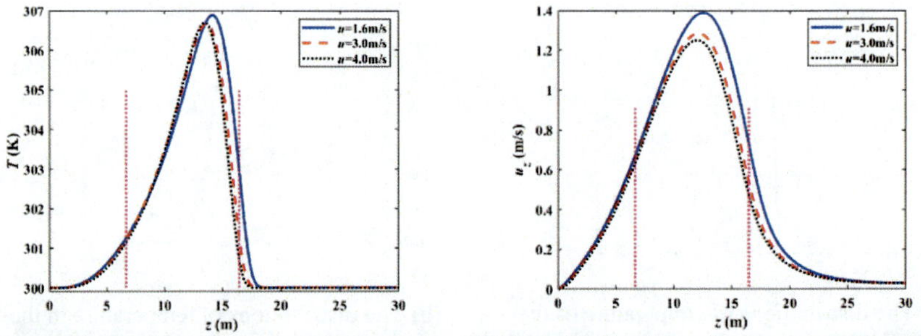

(a) The distributions of temperature of the flow field **(b)** The distributions of normal velocity of the flow field

Fig. 12. The distributions of temperature and velocity along the sampling line which passes through the center of the particle cloud along the normal direction are shown in **a** and **b**. The pink vertical dotted lines in the figures indicate the vicinity of the boundary of the particle cloud.

4.3 Effect of Particle Inertia on Dispersion Process

We set the ambient temperature T_0 to 300 K, the particle release temperature T to 300 K and the incoming flow velocity u to 1.6 m/s, particle characteristic times τ_p are taken as 0.0003 s, 0.003 s, and 0.009 s respectively, namely Reynolds number Re of flow field is 4.12×10^6, Stokes numbers St are respectively 10^{-3}, 10^{-2}, 3×10^{-2}. The settling distance Δz of the particle cloud center in the normal direction after ten seconds of dispersion is recorded for each corresponding Stokes number St, the results are presented in Fig. 13.

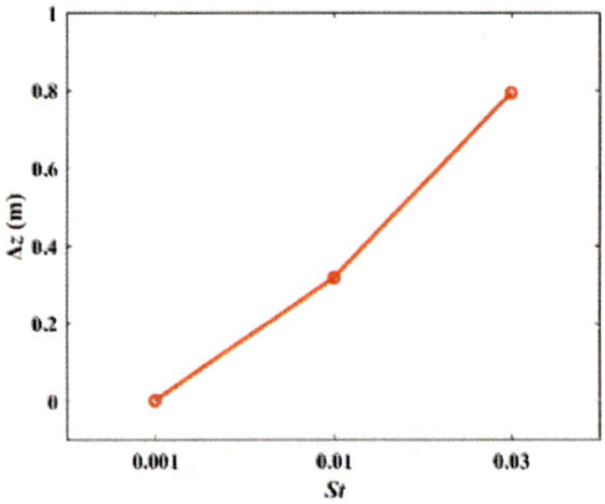

Fig. 13. The effect of particle inertia on the sedimentation process after ten seconds of dispersion

Since the release temperature of particles is equivalent to the ambient temperature, the particle cloud is not affected by buoyancy. This case corresponds to the final stage of turbulent dispersion where the buoyant effect vanishes since the temperature difference disappears. Gravity causes the particle cloud to have a downward motion. As the Stokes numbers St representing the particle inertia increases, the settling velocity of particles also increases. Figure 13 clearly illustrates that for $St = 0.001$, the particle cloud exhibits a remarkably slow settling behavior, with its height remaining almost constant during the initial ten seconds. However, as St increases, the settling velocity increases, and when $St = 0.03$, the particle cloud descends by approximately 0.8 m within the first ten seconds.

5 Conclusion

In this study, we investigate the turbulent dispersion of weakly inertial particles under large temperature differences between the particle and ambient temperature during the initial release. We use the low Mach number variable density fluid model instead of the conventional incompressible fluid model with the Boussinesq approximation. By using this model, we are able to simulate particle dispersion under large temperature differences. We use the fast Eulerian method to simulate the dispersion of the inertial of particles, which integrates the benefits of both the Eulerian and Lagrangian methods. This method replaces the fluid velocity in the convection term of the convection-diffusion equation for the concentration with the particle velocity, enabling us to capture the particle inertia and reducing computational expense through the concentration equation.

To validate the accuracy of the algorithm, we compare the numerical simulation results with the experimental data. The relative error is basically controlled within 15%, indicating that the simulation results can accurately reflect the general law of particle dispersion.

We conduct simulations of particle dispersion at different Grashof numbers and Reynolds numbers. Our results show that the temperature gradient plays a major role in the dispersion. The dispersion rate and the rising velocity of the particle cloud are faster for a large Grashof number. The incoming flow velocity affects the rates of particle dispersion by changing the temperature distribution. Differences in incoming velocity lead to changes in temperature distribution of the particle cloud in the flow field. Temperature differences near the particle cloud boundary create velocity differences, which affect dispersion rate. The normal and spanwise dispersion rates are slower, and the streamwise dispersion rate and the streamwise velocity of the particle cloud are faster for a larger Reynolds number.

We examine the impact of particle inertia (represented by Stokes number St) on the dispersion process. We set the temperature difference between the particle cloud and the ambient air to zero. The results indicate that an increase in Stokes number leads to a higher settling velocity. Furthermore, the result also shows that the effectiveness of the fast Eulerian method in accurately modeling the effect of particle inertia on dispersion.

Acknowledgements. This work was supported by the NSFC Basic Science Center Program for "Multiscale Prolems in Nonlinear Mechanics" (Grant No. 11988102), the NSFC Program (Grant No. 12272380), the National Key Project (Grant No. GJXM92579).

References

1. He, P., Zheng, B.H., Zheng, J.: Urban PM2.5 diffusion analysis based on the improved gaussian smoke plume model and support vector machine. Aerosol Air Qual. Res. **18**(12), 3177–3186 (2018)
2. Yao, L.J., Gao, J.L., Xiao, K.T., Gong, Y.G.: Theory and Testing Technique of Smoke, 1st edn. National Defense Industry Press, Beijing (2004)
3. Pope, S.B.: Turbulent Flows, 1st edn. Cambridge University Press, Cambridge (2000)
4. Balachandar, S., Eaton, J.K.: Turbulent dispersed multiphase flow. Annu. Rev. Fluid Mech.. Rev. Fluid Mech. **42**, 111–133 (2010)

5. Fox, R.O.: Large-eddy-simulation tools for multiphase flows. Annu. Rev. Fluid Mech.. Rev. Fluid Mech. **44**, 47–76 (2012)
6. Sreenivasan, K.R.: Turbulent mixing: a perspective. Proc. Natl. Acad. Sci. **116**(37), 18175–18183 (2018)
7. Ferry, J., Balachandar, S.: A fast Eulerian method for disperse two-phase flow. Int. J. Multiph. FlowMultiph. Flow **27**(7), 1199–1226 (2001)
8. Derakhti, M., Kirby, J.T.: Bubble entrainment and liquid-bubble interaction under unsteady breaking waves. J. Fluid Mech. **761**, 464–506 (2014)
9. Higuera, F.J.: Multifluid Eulerian model of an electrospray in a host gas. J. Fluid Mech. **734**, 363–386 (2013)
10. Wu, W.X., Wang, B., Xiang, G.M.: Impingement of high-speed cylindrical droplets embedded with an air/vapour cavity on a rigid wall: numerical analysis. J. Fluid Mech. **864**, 1058–1087 (2019)
11. Xu, Y., Subramaniam, S.: A multiscale model for dilute turbulent gas-particle flows based on the equilibration of energy concept. Phys. Fluids **18**(3), 033301 (2006)
12. Beetham, S., Fox, R.O., Capecelatro, J.: Sparse identification of multiphase turbulence closures for coupled fluid-particle flows. J. Fluid Mech. **914**, A11 (2021)
13. Capecelatro, J., Desjardins, O., Fox, R.O.: On the transition between turbulence regimes in particle-laden channel flows. J. Fluid Mech. **845**, 499–519 (2018)
14. Innocenti, A., Marchioli, C., Chibbaro, S.: Lagrangian filtered density function for LES-based stochastic modelling of turbulent particle-laden flows. Phys. Fluids **28**(11), 115106 (2016)
15. Marchioli, C., Fantoni, M., Soldati, A.: Orientation, distribution, and deposition of elongated, inertial fibers in turbulent channel flow. Phys. Fluids **22**(3), 033301 (2010)
16. Xu, L.C., Hao, X.Y., Xiao, K.T., Song, W.W., Chen, C.S.: Simulation study of screening efficiency of explosive smoke bomb. Acta Armamentarii **41**(7), 1299–1306 (2020)
17. Minier, J.P., Chibbaro, S., Pope, S.B.: Guidelines for the formulation of Lagrangian stochastic models for particle simulations of single-phase and dispersed two-phase turbulent flows. Phys. Fluids **26**(11), 113303 (2014)
18. Marchioli, C., Giusti, A., Salvetti, M.V., Soldati, A.: Direct numerical simulation of particle wall transfer and deposition in upward turbulent pipe flow. Int. J. Multiph. FlowMultiph. Flow **29**(6), 1017–1038 (2003)
19. Maxey, M.R.: The gravitational settling of aerosol-particles in homogeneous turbulence and random flow-fields. J. Fluid Mech. **174**, 441–465 (1987)
20. Friedlander, S.K.: Smoke, Dust, and Haze, 2nd edn. Oxford University Press, New York (2000)
21. Dodin, Z., Elperin, T.: On the motion of small heavy particles in an unsteady flow. Phys. Fluids **16**(8), 3231–3234 (2004)
22. Balkovsky, E., Falkovich, G., Fouxon, A.: Intermittent distribution of inertial particles in turbulent flows. Phys. Rev. Lett. **86**(13), 2790–2793 (2001)
23. Falkovich, G., Pumir, A.: Intermittent distribution of heavy particles in a turbulent flow. Phys. Fluids **16**(7), L47–L50 (2004)
24. Yang, D., Chen, B.C., Socolofsky, S.A., Chamecki, M., Meneveau, C.: Large-eddy simulation and parameterization of buoyant plume dynamics in stratified flow. J. Fluid Mech. **794**, 798–833 (2016)
25. Shotorban, B., Balachandar, S.: A Eulerian model for large-eddy simulation of concentration of particles with small Stokes numbers. Phys. Fluids **19**(11), 118107 (2007)
26. Kays, W.M., Crawford, M.E., Weigand, B.: Convective Heat and Mass Transfer, 4th edn. McGraw-Hill, New York (2004)
27. Chatterjee, P., Wang, Y., Meredith, K.V., Dorofeev, S.B.: Application of a subgrid soot-radiation model in the numerical simulation of a heptane pool fire. Proc. Combust. Inst. **35**, 2573–2580 (2015)

28. Schatzmann, M., Policastro, A.J.: Effects of the boussinesq approximation on the results of strongly-buoyant plume calculations. J. Climate Appl. Meteorol.Meteorol. **23**(1), 117–123 (1984)
29. Kumar, A.: Effects of the boussinesq approximation on the results of strongly-buoyant plume calculations - comments. J. Climate Appl. Meteorol.Meteorol. **24**(7), 735–739 (1985)
30. Wilcox, D.C.: Turbulence Modeling for CFD, 3rd edn. DCW Industries, California (2006)
31. Hannappel, R., Hauser, T., Friedrich, R.: A comparison of ENO and TVD schemes for the computation of shock-turbulence interaction. J. Comput. Phys.Comput. Phys. **121**(1), 176–184 (1995)
32. Shu, C.W., Osher, S.: Efficient implementation of essentially non-oscillatory shock-capturing schemes. J. Comput. Phys. **83**(1), 32–78 (1989)

Analysing and Predicting Streamwise Velocity Fluctuations in Nonstationary Atmospheric Surface Layers Using the ARMA-GARCH Model

Fei-Chi Zhang[1], Jin-Han Xie[1,2(✉)], Song Xi Chen[3], and Xiaojing Zheng[1,4]

[1] Department of Mechanics and Engineering Science at College of Engineering, and State Key Laboratory for Turbulence and Complex Systems, Peking University, Beijing 100871, People's Republic of China
zfc@stu.pku.edu.cn, jinhanxie@pku.edu.cn

[2] Joint Laboratory of Marine Hydrodynamics and Ocean Engineering, Laoshan Laboratory, Shandong 266237, People's Republic of China

[3] School of Mathematical Sciences, Guanghua School of Management and Center for Statistical Science, Peking University, Beijing 100871, People's Republic of China
csx@gsm.pku.edu.cn

[4] Center for Particle-Laden Turbulence, Lanzhou University, Lanzhou 730000, People's Republic of China
xjzheng@lzu.edu.cn

Abstract. Obtaining reliable statistical results from nonstationary data remains a big challenge in understanding the atmospheric surface layer (ASL). This work employs autoregressive moving average (ARMA) and generalized autoregressive conditional heteroskedasticity (GARCH) models to analyze and predict nonstationary streamwise velocities of the ASL data obtained from the Qingtu Lake Observation Array site in Minqin District, Gansu Province, China. The ARMA process captures temporal autocorrelation, and the GARCH process describes the varying turbulent kinetic energy induced by the mean flow. The ARMA-GARCH model allows for estimating instantaneous variance and correlation for short intervals, which is advantageous over a direct moving average by avoiding an artificially defined averaging window. In addition, we use the phase space spanned by the GARCH coefficients to characterize ASL flows for clear-air and sand-laden situations, and find that sand enhances temporal correlation. We predict the velocity fluctuation by combining an empirical linear relation between the GARCH process and the mean velocity. Even though the current simple model only captures the second-order statistics, it helps extract information from nonstationary data and benefits ASL modelling.

Keywords: atmospheric surface layer · nonstationary turbulence · statistical time series model

X. Zheng and S. Balachandar (Eds.): IUTAM 2023, IUTAM Bookseries 41, pp. 104–116, 2024.
https://doi.org/10.1007/978-3-031-47258-9_6

1 Introduction

Field measurements of the atmospheric surface layer (ASL) facilitate the understanding and investigation of high Reynolds number boundary-layer turbulence [1,2]. However, due to the complexity and uncontrollability of the actual atmospheric flows, it is often difficult to maintain uniform flow conditions over long time intervals. As a result, the mean streamwise velocity undergoes three stages of evolution roughly, namely the rising stage, the stationary stage, and the declining stage. Most studies on ASL have focused on the stationary stage, mainly due to the lack of a theory of nonstationary turbulence and the difficulty of obtaining ensemble averages for nonstationary flows. Consequently, nonstationary observed data has received less attention. To analyze nonstationary turbulence, several data-based methods have been employed, such as wavelet analysis [3], short-time Fourier transform [4], the Hilbert-Huang transform [5], and empirical mode decomposition [6,7].

Besides, there are also approaches that apply statistical time series models to nonstationary turbulence. For instance, by combining the autoregressive moving average (ARMA) model with the generalized autoregressive conditional heteroskedasticity (GARCH) model, the second-order statistics of nonstationary turbulence can be well modeled [8]. The ARMA model divides the dynamic of series into autoregressive and stochastic components, and can be used to discriminate different flows [9]. Meanwhile, the GARCH model captures the data volatility and acts as a nonlinear process in the series. In contrast to data-based methods, statistical models assume a specific form for the coherent structures (correlations) and the stochastic process. Furthermore, the extension to multivariate cases makes the statistical model more advantageous in dealing with complex turbulent flows.

In ASL flow, boundary-layer turbulence is generated and maintained by a combination of shear and thermal convection, due to daily sunrise and sunset, as well as other weather impacts. Under extreme meteorological conditions, sand and dust particles can be carried by severe winds, leading to the formation of sandstorms. Among all meteorological condition factors, the mean wind affects the occurrence of sandstorms significantly [10], and injects energy into turbulent fluctuations. Previous studies have explored and summarized the turbulent characteristics, such as large-scale coherent structures and turbulent kinetic energy fraction, of stationary sand-laden ASL turbulence [11]. And in nonstationary sand-laden ASL turbulence the fraction of streamwise turbulent kinetic energy occupied by large-scale and very large-scale coherent structures has been revealed in different flow stages [7]. The importance of streamwise flow motivates us to apply a statistical model, i.e., the ARMA-GARCH model, to streamwise ASL flow to extract more information on nonstationary turbulence.

In this study, we apply the ARMA-GARCH model on the nonstationary streamwise fluctuating velocity of ASL flow for both clean-air and sand-laden cases. For short intervals, we expect that the ARMA-GARCH model obtains an instantaneous statistical average, which is believed to be better than the moving average, where artificial parameters exist. Over daily intervals, we apply the ARMA-GARCH model to capture the nonstationary statistics for streamwise fluctuating velocities. For a fixed temporal separation, the model helps to extract the mean autocorrelation for one day. It is interesting that some key parameters in the statistical model are almost constant, so they potentially capture the key information of boundary layer dynamics, such as the characteristic time scale. Thus, given mean velocity we can predict the streamwise fluctuating velocities.

The rest of this paper is organized as follows. In Sect. 2 some basic properties of the ARMA-GARCH model are introduced, and the details of the ASL data used in this work are displayed in Sect. 3. In Sect. 4, we use the ARMA-GARCH model to analyze ASL data for short and long intervals, and predict fluctuating velocities. Finally, some remarkable results are summarized in Sect. 5.

2 ARMA-GARCH Model

The long-time correlation in turbulence necessitates the consideration of memory effects in the model. Thus for stationary case the ARMA(p, q) model is a suitable model that incorporates the memory and random effect in a linear process:

$$u(t) = \sum_{i=1}^{p} \phi_i u(t-i) + \epsilon(t) + \sum_{j=1}^{q} \theta_j \epsilon(t-j), \tag{1}$$

where ϕ_i with i ranging from 1 to p and θ_j with j ranging from 1 to q are model coefficients. The integer values of p and q represent the order of the model. The term $\epsilon(t) = \sigma_\epsilon w(t)$ denotes the random component of the model and is an independent identically distributed process, where $w(t)$ is a standard white noise. The model orders p and q are important parameters in characterizing the memory depth of the process $u(t)$ with higher values of p or q corresponding to longer-term dependencies in the data. These orders can also serve as criteria to distinguish different flows, as higher model orders may deviate more from the behaviour predicted by Kolmogorov turbulence [9]. The constraints for model coefficients under the stationary case and more details can be found in Ref. [12]. The model Eq. (1) can be rewritten as

$$u(t) = \sum_{i=0}^{\infty} \varphi_i \epsilon(t-i), \tag{2}$$

where the Green function φ is defined as

$$\varphi_i = \begin{cases} 1, & i = 0, \\ H(q-i)\theta_i + \sum\limits_{j=1}^{\min\{i,p\}} \phi_j \varphi_{i-j}, & i \geq 1, \end{cases} \tag{3}$$

with H the Heaviside function. The autocorrelation for time step k can be calculated as

$$\langle u(t)u(t+k) \rangle = \left\langle \sum_{i=0}^{\infty} \varphi_i \epsilon(t-i) \sum_{i=0}^{\infty} \varphi_i \epsilon(t+k-i) \right\rangle = \sigma_\epsilon^2 \sum_{i=0}^{\infty} \varphi_i \varphi_{i+k}, \tag{4}$$

where the angle brackets denote the average.

To determine the order of the ARMA model, one may use the principle of minimum Bayesian information criterion (BIC). The BIC is a statistical criterion that measures the balance between the goodness of fit and the complexity of the model and is defined as

$$\mathrm{BIC} = N_p \ln N_s - 2 \ln L, \tag{5}$$

where N_p is the number of parameters estimated by the model, N_s is the number of data points in series, and L represents the maximum likelihood of the model tested [12]. The first term on the right side of Eq. (5) penalizes the number of model parameters, while the second term measures the goodness of fit of the model. A lower BIC value indicates a better model fit and more details can be found in Ref. [12].

The previously mentioned methods for determining the model order apply to stationary time series data. However, for nonstationary data, such as the ASL streamwise velocity, the nonstationarity is reflected both in the variation of the mean velocity and the change of the fluctuating velocity variance, namely volatility, in addition to the autocorrelation. To address this issue, we remove the mean velocity, which transforms the fluctuating velocity into a zero-mean time series with changing variance. Several methods, such as moving average, adaptive wavelet transform, and empirical mode decomposition, are used to extract the time-varying mean value in previous studies [7]. Here we adopt the moving average method to remove the mean values for convenience. To capture the volatility, we use the GARCH model, which is defined as

$$\sigma_\epsilon^2(t) = \omega + \sum_{i=1}^{r} \alpha_i \epsilon^2(t-i) + \sum_{j=1}^{s} \beta_j \sigma_\epsilon^2(t-j). \tag{6}$$

Here, ω is a constant, and α_i and β_j are model coefficients. By defining $\epsilon(t) = \sigma_\epsilon(t) w(t)$, $\epsilon(t)$ is temporal uncorrelated and the GARCH model captures the temporal variation of the random noise $\epsilon(t)$. The GARCH model consists of an autoregressive term that is similar to the ARMA model and a quadratic term that captures nonlinearity. The GARCH coefficients are positive, and a stationarity parameter Υ can be defined as

$$\Upsilon = \sum_{i=1}^{r} \alpha_i. \tag{7}$$

A smaller Υ corresponds to less volatility, and the flow is more stationary.

Combing Eqs. (1) and (6), the ARMA(p, q)-GARCH(r, s) model can explain autocorrelation and volatility of univariate series. The variance changes described by the GARCH model are involved in the ARMA process by altering the random noise strength. The orders p, q, r, and s can be determined by minimizing the BIC. Thus, we strike a balance between model complexity and accuracy, and obtain a parsimonious model that captures the essential features of the data.

Similar to Eq. (4), the autocorrelation of the ARMA-GARCH model is calculated as

$$\langle u(t)u(t+k)\rangle = \sum_{i=0}^{\infty} \varphi_i \varphi_{i+k} \sigma_\epsilon^2(t-i). \tag{8}$$

In the turbulence studies, the nondimensional autocorrelation of streamwise fluctuating velocity R_{uu} is defined as

$$R_{uu}(\tau) = \frac{\langle u(t)u(t+\tau)\rangle}{\sqrt{\langle u^2(t)\rangle \langle u^2(t+\tau)\rangle}}, \tag{9}$$

and we can express R_{uu} for nonstationary case by the ARMA-GARCH model coefficients as:

$$R_{uu}(t,\tau) = \frac{\sum_{i=0}^{\infty} \varphi_i \varphi_{i+\tau/t_s} \sigma_\epsilon^2(t-i)}{\sqrt{\sum_{i=0}^{\infty} \varphi_i^2 \sigma_\epsilon^2(t-i) \sum_{i=0}^{\infty} \varphi_i^2 \sigma_\epsilon^2(t+\tau-i)}}, \tag{10}$$

where t_s is the sample time of series and τ is an integer multiple of t_s.

3 ASL Data

The ASL data used in this work were obtained from the Qingtu Lake Observation Array (QLOA) built on a dry flatbed of Qingtu Lake in Minqin (China). The data at different heights comes from the main tower, on which eleven three-component sonic anemometers with a sampling frequency of 50 Hz are distributed in logarithmic separation. And aerosol monitors are installed to measure PM10 (particulate matter with size below 10 μm) concentrations with a measurement range of 0–40 mg/m^3 and a sampling frequency of 1 Hz. More details on the QLOA experimental setup can be found in Ref. [2].

During field measurements, the wind direction always changes, so the raw data needs to be adjusted according to the angle ξ between the actual wind direction and the streamwise direction of the observation array:

$$u = u_m \cos\xi + v_m \sin\xi, \quad v = v_m \cos\xi - u_m \sin\xi, \tag{11}$$

where u and v are the adjusted actual streamwise and spanwise velocities, and u_m and v_m are the measured streamwise and spanwise velocities. Since there is no need to guarantee stationarity, we use all-day data for clean-air and sand-containing situations. Because the boundary layer thickness δ is not measured directly, we use 150m for estimation when δ is needed. The friction velocity u_τ is approximated by the peak of $\sqrt{\langle -uw\rangle}$ and the air kinematic viscosity ν is obtained at a standard atmospheric pressure corresponding to mean temperature. At stationary stages the friction Reynolds number Re_τ can reach the order of 10^6. The mean velocity U is extracted by taking a 1-minute moving average of the streamwise velocity.

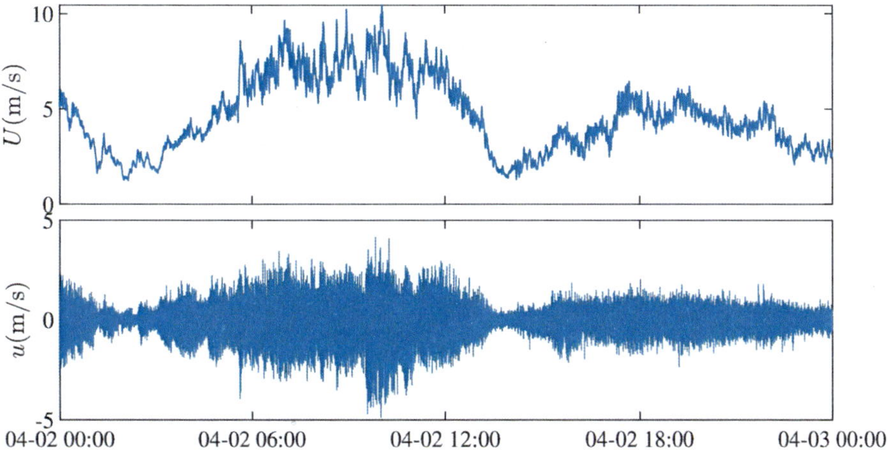

Fig. 1. Mean velocity U and fluctuating velocity u at $z = 5\,\text{m}$ on April 2, 2016.

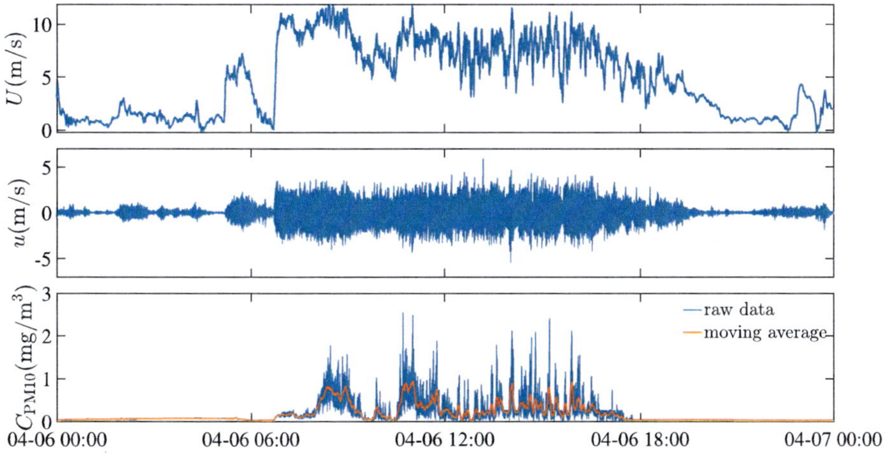

Fig. 2. Mean velocity U, fluctuating velocity u, and PM10 concentration C_{PM10} at $z = 5\,\text{m}$ on April 6, 2016.

An example of ASL data for clean-air flow is shown in Fig. 1, and for sand-laden flow in Fig. 2. In both cases, the variance of the fluctuating velocity u varies approximately with the mean velocity U. During the rising and stationary stages of sandstorms, the PM10 concentration trends almost coincide with the mean velocity. And it is clearly seen that during sandstorms, the mean velocity of the rising stage varies much faster than the clean-air flow.

4 Model Application

In this section, we apply the ARMA-GARCH model to ASL data. Due to the large amount of data throughout one day, we focus only on the rising and declining stages for the original 50 Hz series using the ARMA-GARCH model, and analyze the all-day 1 Hz series obtained by equally separated downsampling.

4.1 Short-Interval Analysis

To present the non-stationary data, we select the rising stage for a clean-air case on April 2, 2016, at 4:00-6:00 as an example. With a segment length of 2 min for the series, we can calculate instantaneous statistics according to Eqs. (8) and (10). For the clean-air case, the mean velocity U, the streamwise turbulent energy $\langle u^2 \rangle$, and the streamwise autocorrelation for 0.02 s at different heights during the rising stage are shown in Fig. 3. The results indicate that as U increases, $\langle u^2 \rangle$ increases at different heights. Although U is slowly increasing and R_{uu} is also slowly increasing, R_{uu} shows a local trend opposite to the mean wind speed, i.e., the peak of U corresponds to a valley of R_{uu}. In addition, the values of $\langle u^2 \rangle$ and R_{uu} both have a clear monotonic pattern with height z.

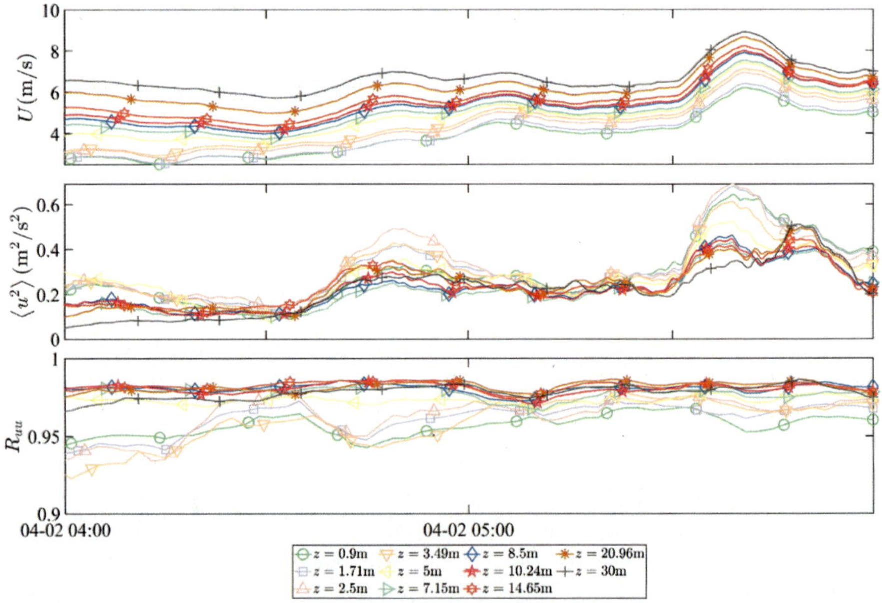

Fig. 3. Mean velocity U, streamwise turbulent energy $\langle u^2 \rangle$, and autocorrelation for 0.02 s during rising stage for clean-air case.

To verify the result of $\langle u^2 \rangle$ calculated bt the ARMA-GARCH model, we compare the model result with a direct moving-average method with a 1-minute

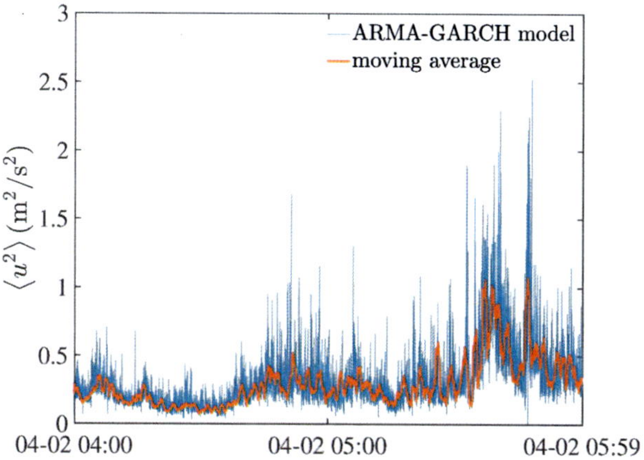

Fig. 4. Streamwise turbulent energy $\langle u^2 \rangle$ calculated using the ARMA-GARCH model and the moving average method with a 1-min window at $z = 0.9$ m.

window. As shown in Fig. 4, the $\langle u^2 \rangle$ calculated by the ARMA-GARCH model and the moving average method show similar values and trends. Besides, for stationary ASL flow the probability density function (PDF) of streamwise fluctuating velocity shows a near-Gaussian behavior. In the rising stage, we find that the PDF of fluctuating velocity normalized by the calculated standard deviation obtained from the ARMA-GARCH model is closer to Gaussian than that obtained from a simple moving average, implying the consistency of the ARMA-GARCH model in dealing with the non-stationary flow.

4.2 Long-Interval Analysis

In order to get the statistical results of the changes of some quantities, such as R_{uu}, during the whole day, we extract the 1 Hz series from the original series and apply the ARMA-GARCH model for every 20mins segment series. The mean velocity U, the streamwise turbulent energy $\langle u^2 \rangle$, the stationarity parameter Υ that characterizes the stationarity, and the streamwise autocorrelation R_{uu} for 1 s for the clean-air case is displayed in Fig. 5. It can be noticed that the value of Υ is larger when $\langle u^2 \rangle$ changes significantly throughout the day, which indicates that Υ is a good parameter to measure the flow stationarity and does not require segmentation of the considered interval like the traditional methods. Moreover, it can be seen that the value of R_{uu} does not change greatly during one day. And the same happens in the sand-containing case (cf. Fig. 6), which means one model may be applied to the whole-day data, even though it is rather crude. We thereby make predictions for the fluctuating velocities.

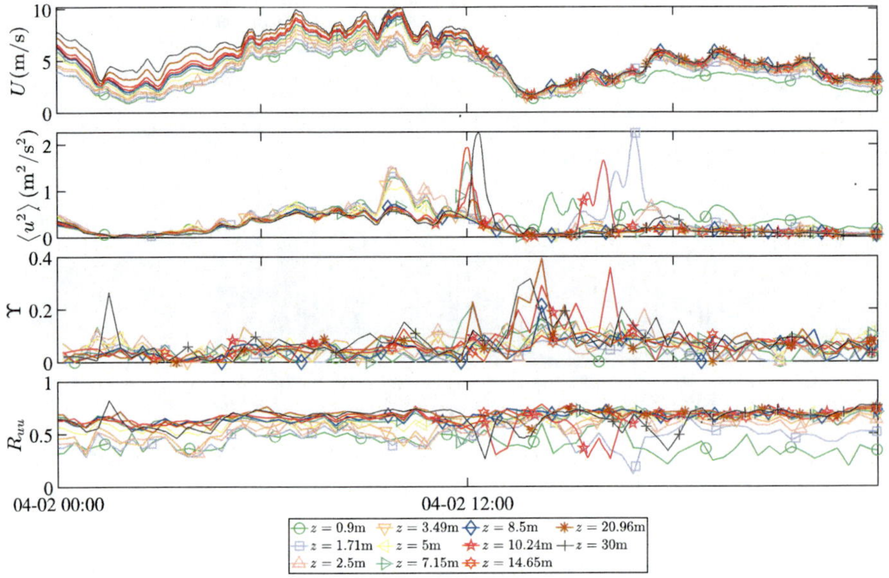

Fig. 5. Mean velocity U, streamwise turbulent energy $\langle u^2 \rangle$, stationarity parameter Υ, and streamwise autocorrelation R_{uu} for 1 s for clean-air case during April 2, 2016.

4.3 Prediction

Since the autocorrelation R_{uu} for a fixed time separation fluctuates in one day, which inspires us to fit one ARMA-GARCH model for the whole day to capture the mean value of R_{uu}. And if for each day's data R_{uu} is approximately constant, then we can use the same ARMA-GARCH coefficients to make predictions for fluctuating velocities. The autocorrelation for one time step (1 s) can be expressed as

$$\begin{aligned}
\tilde{R}_{uu} &= \frac{\langle u(t)u(t+1)\rangle}{\sqrt{\langle u^2(t)\rangle \langle u^2(t+1)\rangle}} \\
&= \frac{\sum\limits_{i=0}^{\infty} \varphi_i \varphi_{i+1} \sigma_\epsilon^2(t-i)}{\sqrt{\sum\limits_{i=0}^{\infty} \varphi_i^2 \sigma_\epsilon^2(t-i) \sum\limits_{i=0}^{\infty} \varphi_i^2 \sigma_\epsilon^2(t+1-i)}} \\
&\approx \frac{\sum\limits_{i=0}^{\infty} \varphi_i \varphi_{i+1} \sigma_\epsilon^2(t-i)}{\sum\limits_{i=0}^{\infty} \varphi_i^2 \sigma_\epsilon^2(t-i)} \approx \frac{\sum\limits_{i=0}^{N} \varphi_i \varphi_{i+1}}{\sum\limits_{i=0}^{N} \varphi_i^2},
\end{aligned} \tag{12}$$

where N is a cutoff number for calculation. Then the mean autocorrelation $\tilde{R}_{uu}$ for one day can be obtained from the ARMA-GARCH model coefficients. For

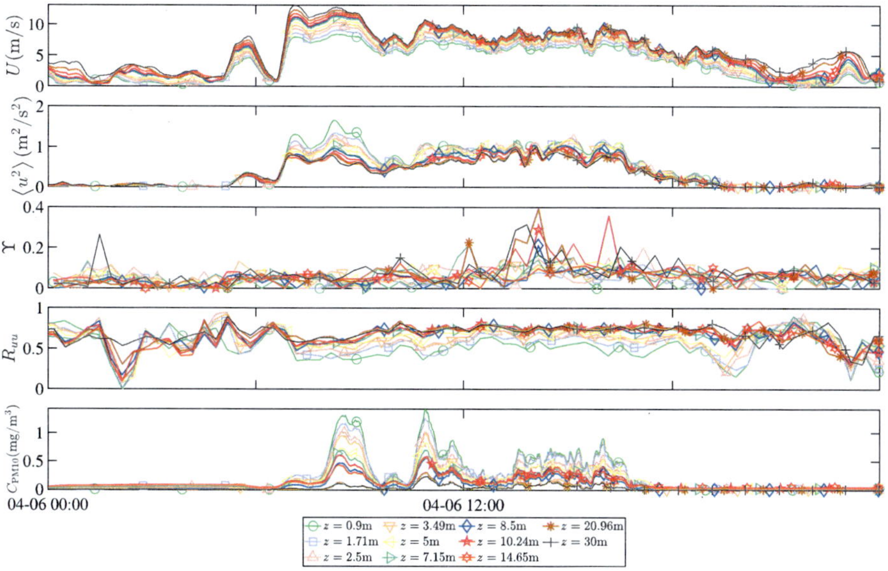

Fig. 6. Mean velocity U, streamwise turbulent energy $\langle u^2 \rangle$, stationarity parameter Υ, streamwise autocorrelation R_{uu} for 1 s, and PM10 concentration for sand-laden case during April 6, 2016.

different days 1 Hz data (five clean-air days and fifty five sand-laden days), we fit the ARMA-GARCH model and summarize the φ_1 in Fig. 7. We find that, on average, for the one-day ASL flow, sand and dust particles near the wall increase

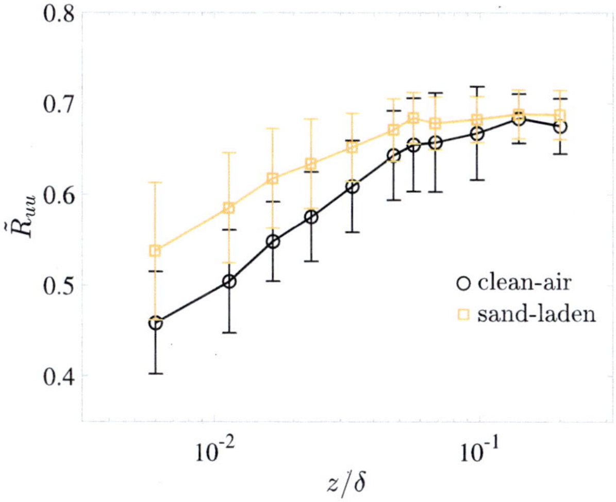

Fig. 7. Mean autocorrelation $\tilde{R}_{uu}$ for 1 s obtained using the ARMA-GARCH model during one day at different heights for clean-air and sand-laden cases.

the streamwise autocorrelation function of 1 s. And far away from the wall, the autocorrelation of 1 s is almost the same for the two cases. Considering that the sand concentrates near the wall, the sand enhances the autocorrelation, by stronger coherent structure, which remains to be studied.

In the ARMA-GARCH model an important variable is $\sigma_\epsilon(t)$, which is related to $\langle u^2 \rangle$:

$$\langle u^2(t) \rangle = \sum_{i=0}^{\infty} \varphi_i^2 \sigma_\epsilon^2(t-i) \approx \sigma_\epsilon^2(t) + \varphi_1^2 \sigma_\epsilon^2(t-1), \tag{13}$$

where other terms can be neglected because φ has a power decay form (cf. Eq. (3)). Thus $\langle u^2(t) \rangle$ calculated by the ARMA-GARCH model has an approximately linear relation with σ_ϵ^2. Besides, $\sigma_\epsilon(t)$ can be seen as the driving force of $u(t)$, and the main source of turbulent kinetic energy $\langle u^2(t) \rangle$ is from the mean flow, so roughly $\sigma_\epsilon(t)$ should relate to $U(t)$, which can be seen from Fig. 8(a). We find that there exists an approximately linear relation between

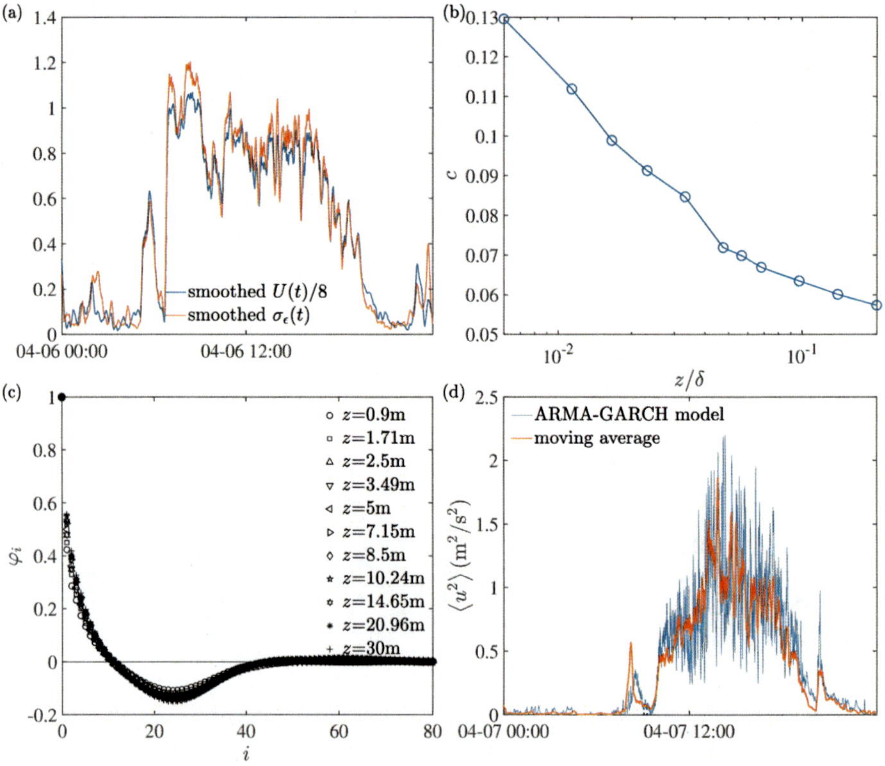

Fig. 8. (a) Linear relation between $U(t)$ and $\sigma_\epsilon(t)$. The smoothing is done using a 10-minute window for comparison purposes. (b) Constant $c = \langle \sigma_\epsilon(t)/U(t) \rangle$ for different heights. (c) Green function φ of ARMA process for different heights. (d) Comparison of the predicted streamwise turbulent kinetic energy $\langle u^2 \rangle$ and $\langle u^2 \rangle$ calculated by moving average.

$U(t)$ and $\sigma_\epsilon(t)$. Thus, by combining one ARMA-GARCH model and the linear relation between $\sigma_\epsilon(t)$ and $U(t)$, the prediction of fluctuating velocities can be achieved by the following steps: (i) For one day 1 Hz data, the ARMA-GARCH model is fitted and we obtain φ and $\sigma_\epsilon(t)$, and φ varies slightly at different heights (cf. Fig. 7 and 8(c)). (ii) Then the constant $c = \langle \sigma_\epsilon(t)/U(t) \rangle$ can be calculated (cf. Fig. 8(b)). (iii) Known the mean velocity $\tilde{U}(t)$ in next day, we take the approximation of $\tilde{\sigma}_\epsilon(t)$ as $c\tilde{U}(t)$ and obtain the prediction $\tilde{u}(t) = \sum_{i=0}^{\infty} \varphi_i \tilde{\sigma}_\epsilon(t-i) = c \sum_{i=0}^{\infty} \varphi_i \tilde{U}(t-i)$. As shown in Fig. 8(d), the trend of the predicted $\langle u^2 \rangle$ and the moving average calculated $\langle u^2 \rangle$ are in good agreement, which justifies the simple linear relation between U and σ_ϵ. This positive correlation also indicates that the mean velocity is the main source of turbulent kinetic energy, which is consistent with previous studies saying that dynamic turbulence dominates in severe sandstorms [13]. Although c may have a variation for different days, for turbulent flows (here we take $\langle u^2 \rangle > \max(\langle u^2 \rangle)/10$ in one day as a simple criterion for defining turbulence), the absolute percent error of the results predicted using the previous day's c ranges from 15% to 35%. The error increases with height, which may be due to the fact that the model ignores the effect of temperature, which is stronger at higher locations.

5 Summary and Discussion

The ARMA-GARCH model applied in the ASL flow can resolve the nonstationary flow well. For short intervals, we obtain the variance and short-time streamwise temporal autocorrelation. For long intervals, the ARMA-GARCH model captures the mean temporal autocorrelation, which is significantly different near the wall for clean-air and sand-laden flows. And we find that sand enhances the autocorrelation for a specific temporal separation of one second. Further statistical analysis of the autocorrelation for more temporal separations can provide information about characteristic length scales in sand-laden ASL turbulence [14].

In the ARMA-GARCH model, the GARCH part that captures the nonstationarity represents the forcing signal for the fluctuating velocity u. The GARCH signal σ_ϵ is found to have an approximately linear dependence on the mean velocity U. Since the standard deviation of streamwise fluctuating velocity σ_u and σ_ϵ are also linearly related (cf. Eq. (13)), we obtain that σ_u and U are linearly dependent approximately, which is intuitive but rarely reported. Through linear relation between σ_ϵ and U we achieve the prediction of u from known U, providing a means to model ASL fluctuating signals. One potential drawback is that the ARMA model lacks long-time correlation, making the large-scale statistics biased, which could be improved by introducing other statistical models. And the thermal effect in ASL needs to be considered in further analysis.

References

1. Hutchins, N., Chauhan, K., Marusic, I., Monty, J., Klewicki, J.: Towards reconciling the large-scale structure of turbulent boundary layers in the atmosphere and laboratory. Boundary-Layer Meteorol. **145**(2), 273–306 (2012)
2. Wang, G., Zheng, X.: Very large scale motions in the atmospheric surface layer: a field investigation. J. Fluid Mech. **802**, 464–489 (2016)
3. Treviño, G., Andreas, E.L.: On wavelet analysis of nonstationary turbulence. Boundary-Layer Meteorol. **81**(3–4), 271–288 (1996)
4. Treviño, G., Andreas, E.L.: Averaging intervals for spectral analysis of nonstationary turbulence. Boundary-Layer Meteorol. **95**, 231–247 (2000)
5. Vincent, C., Giebel, G., Pinson, P., Madsen, H.: Resolving nonstationary spectral information in wind speed time series using the Hilbert-Huang transform. J. Appl. Meteorol. Clim. **49**(2), 253–267 (2010)
6. Xu, Y.L., Chen, J.: Characterizing nonstationary wind speed using empirical mode decomposition. J. Struct. Eng. **130**(6), 912–920 (2004)
7. Liu, H., Shi, Y., Zheng, X.: Evolution of turbulent kinetic energy during the entire sandstorm process. Atmos. Chem. Phys. **22**(13), 8787–8803 (2022)
8. Huang, Z., Gu, M.: Characterizing nonstationary wind speed using the ARMA-GARCH model. J. Struct. Eng. **145**(1), 04018226 (2019)
9. Faranda, D., et al.: Modelling and analysis of turbulent datasets using auto regressive moving average processes. Phys. Fluids **26**(10), 105101 (2014)
10. Li, W.Y.: Statistical quantification of the local daily surface meteorological condition's impact properties on dust storm occurrence: style, intensity, significance, contribution, and decisiveness, taking North and Northwest China as an example. Theor. Appl. Climatol. **143**, 403–428 (2020)
11. Liu, H., Zheng, X.: Large-scale structures of wall-bounded turbulence in single-and two-phase flows: advancing understanding of the atmospheric surface layer during sandstorms. Flow **1**, E5 (2021)
12. Brockwell, P.J., Davis, R.A.: Time Series: Theory and Methods. Springer, New York (2009). https://doi.org/10.1007/978-1-4899-0004-3
13. Zhang, H., Zhu, H., Peng, Y., Kang, L., Chen, J., Park, S.U.: Experiment on dust flux during duststorm periods over desert area. Acta Meteorol. Sin. **22**(2), 239–247 (2008)
14. Wang, G., Gu, H., Zheng, X.: Large scale structures of turbulent flows in the atmospheric surface layer with and without sand. Phys. Fluids **32**(10), 106604 (2020)

Turbulence Models for Compressible Disperse Multiphase Flows

Rodney O. Fox[1,2](✉)

[1] Department of Chemical and Biological Engineering, Iowa State University, Ames, IA 50011-1098, USA
rofox@iastate.edu

[2] Center for Multiphase Flow Research and Education, Iowa State University, Ames, IA 50011-1096, USA

Abstract. Multiphase turbulence driven by coupling between the phases can arise in the absence of mean shear. For example, due to gravity (or other body forces) in disperse multiphase flows (e.g., particle-laden and bubbly flows), the mean-slip velocity between phases and spontaneous cluster formation generate strong turbulence that is nearly one dimensional. Such complex flows can be modeled using Eulerian–Eulerian two-fluid models. Here, starting from a compressible two-fluid model, a Reynolds-averaged turbulence model is derived for disperse multiphase flows driven by gravity. Due to the diagonal form of the Reynolds-stress tensors, the model equations are hyperbolic and thus require time-dependent solutions and dedicated numerical solvers.

Keywords: disperse multiphase flow · turbulence modeling · hyperbolic two-fluid model

1 Introduction

Multiphase turbulence takes many different forms and arises in many different contexts [1–10]. Here, we will consider only disperse multiphase flows with a discrete phase (e.g., solid particles, droplets or bubbles) and a continuous fluid phase. Furthermore, for simplicity, we will focus on cases where the disperse-phase elements are identical, i.e., have the same material density and geometry. In other words, we will consider only monodisperse multiphase flows. Even with these restrictions, disperse multiphase turbulent flows can be further categorized into two broad classes: (i) flows where the turbulence originates in the continuous phase with the discrete phase modulating the small scales of the turbulence [11–17], and (ii) flows where the turbulence arises due to the coupling between the discrete and continuous phases. The former is often associated with particle-laden turbulent flows and its study is mainly focused on how the discrete phase modifies the classical turbulence structures seen in single-phase flows. The latter, which is the topic of this work, can be observed in gas–particle flows when the

X. Zheng and S. Balachandar (Eds.): IUTAM 2023, IUTAM Bookseries 41, pp. 117–140, 2024.
https://doi.org/10.1007/978-3-031-47258-9_7

mass of the particles is of the same order or greater than that of the gas phase; or in bubbly flows when the bubble volume fraction is high enough to lead to buoyancy-driven instabilities.

The remainder of the work is organized as follows. First, we present the Eulerian–Eulerian (EE) two-fluid model for compressible disperse multiphase flows developed in [18,19] (but not considering pseudoturbulence nor particle internal energy for the sake of clarity). Next, we describe the Reynolds-averaged turbulence model that results from the EE model after averaging and applying closures. This turbulence model is our main result, and care is taken to ensure that it is globally hyperbolic (as described in [18]). As an application, we then reduce the model equations for the case of statistically stationary, homogeneous turbulence driven by gravity (e.g., cluster-induced turbulence (CIT) [20] and buoyancy-driven turbulence (BDT) [21]). The paper concludes with some general remarks concerning multiphase turbulence models and open questions.

2 Compressible Two-Fluid Model

The starting point for formulating the turbulence model is the inviscid, compressible two-fluid model in Table 1. As discussed in detail in [18,19], this model accounts for added mass by allowing a portion of the fluid with volume fraction $\alpha_a = \alpha_p^\star - \alpha_p$ to move with the particle-phase velocity $\mathbf{u}_p$. The particle material density ρ_p is constant, while the fluid density ρ_f varies due to compression/expansion. The particle-phase volume fraction is α_p, and $\alpha_f = 1 - \alpha_p$ is that of the fluid phase.

2.1 Mass

In the mass balances, the source term S_a models fluid exchange between the particle wakes and the bulk fluid. The latter has velocity $\mathbf{u}_f$ and volume fraction $\alpha_f^\star$. As demonstrated in [18], treating the added mass is this manner is exactly equivalent to the virtual-mass force used in two-fluid modeling, but has the advantage of ensuring that the model is hyperbolic for arbitrary material-density ratios $Z = \rho_f/\rho_p$. Thus, the model in Table 1 is applicable to gas–particle flows with $\rho_p \gg \rho_f$, as well as to bubbly flows with $\rho_p \ll \rho_f$ or suspensions with $\rho_p \approx \rho_f$. In order to allow for large-density fluids (e.g., water or other liquids), the fluid equation of state (EOS) is modeled as a stiffened gas:

$$p_f = \rho_f(\gamma_f - 1)\Theta_f - \gamma_f p_f^o, \tag{1}$$

but any other EOS could be used as well. The conserved variables in the mass balances are

$$(X_1, X_2, X_3) = (\alpha_p, \rho_e \alpha_p^\star, \rho_f \alpha_f^\star) \tag{2}$$

where ρ_e is the effective density of the particle with its added mass. Thus, the volume fractions and fluid density are found from the conserved variables by

$$(\alpha_p, \rho_f, \alpha_a) = \left(X_1, \frac{X_3 + X_2 - \rho_p X_1}{\alpha_f}, \frac{X_2 - \rho_p X_1}{\rho_f} \right). \tag{3}$$

Finally, to enforce the limit $0 \leq \alpha_a \leq \alpha_f$, a new variable c_m is introduced such that $\alpha_a = c_m \alpha_f \alpha_p$. In the very dilute limit, c_m corresponds to the virtual-mass constant, whose value is 0.5 for spherical particles. The latter is imposed in the model for S_a using the parameter $c_m^\star$. In summary, the two-fluid model in Table 1 contains three mass balances in order to account for added mass while remaining globally hyperbolic [18,19].

2.2 Momentum

The two momentum balances in Table 1 use the two conserved variables

$$(X_4, X_5) = (\rho_e \alpha_p^\star \mathbf{u}_p, \rho_f \alpha_f^\star \mathbf{u}_f), \tag{4}$$

which correspond, respectively, to the momentum of the particles with the added mass, and to the momentum of the bulk fluid.[1] The phase velocities are related to the conserved variables by

$$(\mathbf{u}_p, \mathbf{u}_f) = \left(\frac{X_4}{X_2}, \frac{X_5}{X_3} \right). \tag{5}$$

In addition to the fluid pressure p_f and the particle pressure p_p, a pressure contribution due to the slip-velocity tensor R appears for the particle phase. This particle-fluid-particle (pfp) pressure tensor P_{pfp} arises in the kinetic theory derivation of a binary system [22], and is nonzero whenever the slip velocity $\mathbf{u}_{fp}$ is nonzero. As shown in [22], it plays an important role in gas–particle flows to ensure that the system is hyperbolic. The particle pressure p_p is the sum of two contributions. The first is the usual collisional pressure that depends on particle-phase granular temperature Θ_p. The second is the frictional or elastic pressure for close-packed particles ($\alpha_p > 0.63$) that depends on α_p only. The parameter p_c is set to a large value to resist increases in α_p once the particles are in contact. The latter is controlled by the tanh function, which provides a sharp transition near $\alpha_p = 0.6$. Note that this pressure contribution is present in dense flows such as fluidized beds, but will be absent for riser flows where $\alpha_p \ll 0.63$. For frictional particles, by including a interparticle-friction dependence and a saltation model in the particle pressure and viscous terms (which are neglected here), good agreement with discrete particle simulations of dense liquid–particle flows has been reported in the literature [23].

The source terms on the right-hand sides of the momentum balances represent, respectively, fluid drag; buoyancy; forces due to fluid density gradients, expansion/contraction, and mean shear (e.g., lift); momentum transfer due to added mass; and gravity. The drag coefficient C_D depends on Re_p and α_p and a correlation [24] must be provided to complete the model, e.g., $C_D Re_p = 24$

[1] As discussed in [18], the total fluid velocity $\mathbf{v}_f$ is related to the bulk fluid velocity by $\alpha_f \mathbf{v}_f = \alpha_a \mathbf{u}_p + \alpha_f^\star \mathbf{u}_f$. Thus, depending on the value of α_a, the turbulence statistics will be different for the bulk and total fluid. In most two-fluid simulations [21], the statistics are computed using $\mathbf{v}_f$.

Table 1. Inviscid, compressible two-fluid model for monodisperse elastic particles in a fluid modeled as a stiffened gas with parameters γ_f and p_f^o [18]. The drag coefficient, C_D, depends on Re_p and α_p. The slip-velocity tensor R is the exact expression from kinetic theory and defines the pfp-pressure tensor P_{pfp}.

Mass

$$\begin{aligned}
\partial_t \alpha_p + \partial_{\mathbf{x}} \cdot (\alpha_p \mathbf{u}_p) &= 0 \\
\partial_t (\rho_e \alpha_p^\star) + \partial_{\mathbf{x}} \cdot (\rho_e \alpha_p^\star \mathbf{u}_p) &= S_a \\
\partial_t (\rho_f \alpha_f^\star) + \partial_{\mathbf{x}} \cdot (\rho_f \alpha_f^\star \mathbf{u}_f) &= -S_a
\end{aligned}$$

Momentum

$$\begin{aligned}
\partial_t (\rho_e \alpha_p^\star \mathbf{u}_p) + \partial_{\mathbf{x}} \cdot (\rho_e \alpha_p^\star \mathbf{u}_p \mathbf{u}_p + p_p \mathsf{I} + \mathsf{P}_{pfp}) &= K \mathbf{u}_{fp} - \alpha_p^\star (\partial_{\mathbf{x}} p_f + \mathbf{F}_{pf}) + \mathbf{S}_{fp} + \rho_e \alpha_p^\star \mathbf{g} \\
\partial_t (\rho_f \alpha_f^\star \mathbf{u}_f) + \partial_{\mathbf{x}} \cdot (\rho_f \alpha_f^\star \mathbf{u}_f \mathbf{u}_f + p_f \mathsf{I}) &= K \mathbf{u}_{pf} + \alpha_p^\star (\partial_{\mathbf{x}} p_f + \mathbf{F}_{pf}) - \mathbf{S}_{fp} + \rho_f \alpha_f^\star \mathbf{g}
\end{aligned}$$

Total Energy

$$\begin{aligned}
&\partial_t (\rho_e \alpha_p^\star E_p) + \partial_{\mathbf{x}} \cdot (\rho_e \alpha_p^\star E_p \mathbf{u}_p + p_p \mathbf{u}_p + \mathsf{P}_{pfp} \cdot \mathbf{u}_p) = \mathsf{P}_{pfp} : \partial_{\mathbf{x}} \mathbf{u}_p \\
&\quad - K(3\Theta_p + \mathbf{u}_p \cdot \mathbf{u}_{pf}) - \alpha_p^\star \mathbf{u}_p \cdot (\partial_{\mathbf{x}} p_f + \mathbf{F}_{pf}) + S_E + \rho_e \alpha_p^\star \mathbf{u}_p \cdot \mathbf{g}
\end{aligned}$$

$$\begin{aligned}
&\partial_t (\rho_f \alpha_f^\star E_f) + \partial_{\mathbf{x}} \cdot [\rho_f \alpha_f^\star E_f \mathbf{u}_f + (\alpha_f^\star \mathbf{u}_f + \alpha_p^\star \mathbf{u}_p) p_f] = -\mathsf{P}_{pfp} : \partial_{\mathbf{x}} \mathbf{u}_p \\
&\quad + K(3\Theta_p + \mathbf{u}_p \cdot \mathbf{u}_{pf}) + \alpha_p^\star \mathbf{u}_p \cdot (\partial_{\mathbf{x}} p_f + \mathbf{F}_{pf}) - S_E + \rho_f \alpha_f^\star \mathbf{u}_f \cdot \mathbf{g}
\end{aligned}$$

where

$$\alpha_f + \alpha_p = 1 \quad \alpha_f^\star + \alpha_p^\star = 1 \quad \alpha_p^\star = \alpha_p + \alpha_a \quad \rho_e = \frac{\rho_p \alpha_p + \rho_f \alpha_a}{\alpha_p + \alpha_a}$$

$$\mathbf{u}_{fp} = -\mathbf{u}_{pf} = \mathbf{u}_f - \mathbf{u}_p \quad K = \frac{3\rho_e \alpha_p^\star C_D Re_p}{4\tau_p} \quad \tau_p = \frac{\rho_e d_p^2}{\mu_f} \quad Re_p = \frac{\alpha_f^\star}{\alpha_f} \frac{d_p u_{fp}}{\nu_f}$$

$$S_a = \frac{\rho_f}{\tau_a}(c_m^\star \alpha_f \alpha_p - \alpha_a) \quad \mathbf{S}_{fp} = \max(S_a, 0)\mathbf{u}_f + \min(S_a, 0)\mathbf{u}_p$$

$$S_E = \max(S_a, 0)\frac{1}{2}u_f^2 + \min(S_a, 0)E_p \quad \tau_a = \frac{4d_p^2}{3\nu_f C_D Re_p} \quad c_m^\star = \frac{1}{2}\min(1 + 2\alpha_p, 2)$$

$$\mathbf{F}_{pf} = \mathsf{R} \cdot \partial_{\mathbf{x}} \rho_f - \rho_f(\gamma_f - 1) tr(\mathsf{\Gamma}) \mathbf{u}_{fp} + \frac{4}{5}\rho_f \mathsf{S} \cdot \mathbf{u}_{fp} \quad \mathsf{R} = \frac{1}{5}u_{fp}^2 \mathsf{I} + \frac{2}{5}\mathbf{u}_{fp}\mathbf{u}_{fp}$$

$$\mathsf{\Gamma} = \frac{1}{2}\left[\partial_{\mathbf{x}} \mathbf{u}_f + (\partial_{\mathbf{x}} \mathbf{u}_f)^t\right] \quad \mathsf{S} = \mathsf{\Gamma} - \frac{1}{3}tr(\mathsf{\Gamma})\mathsf{I}$$

$$p_f = \rho_f(\gamma_f - 1)\Theta_f - \gamma_f p_f^o \qquad \mathsf{P}_{pfp} = C_{pfp}\rho_f \alpha_p^\star \mathsf{R} \quad C_{pfp} = c_m^\star$$

$$p_p = \rho_e \alpha_p^\star \Theta_p (1 + 4\alpha_p g_0) + p_c \alpha_p g_0 \frac{1}{2}\left[1 + \tanh\left(\frac{\alpha_p - 0.6}{0.01}\right)\right] \quad g_0 = \frac{1 + \alpha_f}{2\alpha_f^3}$$

$$\Theta_f = E_f - \frac{1}{2}u_f^2 \quad \Theta_p = \frac{2}{3}\left(E_p - \frac{1}{2}u_p^2\right)$$

for Stokes drag. As shown in [18], P_{pfp} is crucial for ensuring that the system is globally hyperbolic for bubbly flows and for gas–particles flows when, e.g., the granular temperature Θ_p is negligible. Note that P_{pfp} scales like $\alpha_p^\star$ (i.e., it is first order in particle concentration[2]) and corresponds to particle–particle interactions mediated by the fluid. For global hyperbolicity for any Z, the parameter C_{pfp} is set equal to $c_m^\star$, but smaller values can be used in gas–particle flows.

2.3 Total Energy

The total energy balances in Table 1 are written in terms of the two conserved variables

$$(X_6, X_7) = (\rho_e \alpha_p^\star E_p, \rho_f \alpha_f^\star E_f), \tag{6}$$

and thus the specific total energies are related to the conserved variables by

$$(E_p, E_f) = \left(\frac{X_6}{X_2}, \frac{X_7}{X_3} \right). \tag{7}$$

For the particle phase (including the added mass), the internal energy is not included in the total energy, implying that the phase is adiabatic. This is possible when the particle–particle collisions are assumed elastic. When inelastic collisions are required (e.g., high-speed gas–particle flows [24]), then the energy dissipation term in the particle phase would appear as a source term in the particle-phase internal energy balance [19]. The granular temperature Θ_p corresponds to the particle velocity fluctuations relative to the mean velocity $\mathbf{u}_p$. In contrast, the fluid-phase thermodynamic temperature is proportional to its internal energy Θ_f. The relationship between these temperatures and the total energies are given in Table 1 where γ_f is the heat-capacity ratio for the fluid.

As shown in [22], kinetic theory provides closed forms for the energy fluxes. The total energy flux for the particle phase has the classical form of an ideal gas, but with an additional contribution due to the pfp-pressure tensor. For the fluid phase an additional energy-flux term appears due to the presence of the particle phase. Note that $(\alpha_f^\star \mathbf{u}_f + \alpha_p^\star \mathbf{u}_p)$ is the volume-average velocity of the fluid–particle mixture, which sees the continuous phase pressure p_f. The source terms on the right-hand sides of the total energy balances represent, respectively, the work done by the fluid on the particles to lower α_p due to the pressure P_{pfp},[3] fluid drag, buoyancy, work done by forces $\mathbf{F}_{pf}$, potential energy, and mass transfer. Many of these terms represent energy exchange from the particle phase to the fluid phase (i.e., fluid heating and pseudoturbulence [18,19]). Here, we neglect pseudoturbulence and focus on large-scale multiphase turbulence accompanied by fluctuations in α_p (i.e., cluster-induced and buoyancy-driven turbulence). A two-fluid model that includes heat transfer and pseudoturbulence can be found in [19].

---

2 A second-order model is also analyzed in [18].

3 Using the mass balance for α_p, $\partial_{\mathbf{x}} \cdot \mathbf{u}_p$ can be replaced by the rate of change of α_p as it moves with velocity $\mathbf{u}_p$.

For systems with inelastic collisions and/or strong fluid drag, the granular temperature can be very small and care must be taken to ensure that numerical errors do not result in negative Θ_p when it is found by subtracting the mean kinetic energy $\frac{1}{2}u_p^2$ from E_p. Alternatively, a non-conservative equation for Θ_p:

$$\frac{3}{2}\partial_t(\rho_e\alpha_p^\star\Theta_p) + \frac{3}{2}\partial_\mathbf{x}\cdot(\rho_e\alpha_p^\star\Theta_p\mathbf{u}_p) = -p_p\partial_\mathbf{x}\cdot\mathbf{u}_p - 3K\Theta_p \tag{8}$$

could be employed. Note that this equation does not depend on P_{pfp}, and is valid for elastic collisions and $\alpha_p \ll 0.63$. For inelastic collisions or larger α_p, dissipation terms appear that change the particle-phase internal energy [19]. In practice [19], when the conservative form is used to solve for E_p, it suffices to clip Θ_p to non-negative values when evaluating terms on which it depends. This is possible because when Θ_p is negligible, the eigenvalues of system are real-valued due to P_{pfp} and, hence, the fluxes are non-degenerate. In most applications, the eigenvalues associated with p_f will be much larger than those associated with the particle phase, especially if the fluid phase is a liquid. For reference, when $Z \approx 0$ (e.g., gas–particle flows), the particle-phase speed of sound is $\sqrt{5\Theta_p/3}$ [18]. Thus, the particle-phase Mach number can correspond to hypersonic flow when Θ_p is small. This is also approximately true for larger values of Z where other physical mechanisms in addition to drag are important (e.g., buoyancy).

2.4 Summary

In summary, the two-fluid model in Table 1 is the starting point for developing a turbulence model for compressible disperse multiphase flows. Aside from the body force $\mathbf{g}$, these equations conserve mass, momentum and total energy for the mixture, i.e., by summing together the two equations representing each phase. As shown in [18, 19], this model is globally hyperbolic and can be solved using numerical methods for hyperbolic conservation laws [25]. Nonetheless, the wide range of timescales appearing in the fluxes (especially for liquids) make the numerical solution challenging from the point of view of their cost. In high-speed multiphase flows, the fluid-phase Mach number will be very large, implying that shocks will be present in the solutions. As in single-phase flows, shock-resolving solutions require fine grids (at least locally near the shock). In contrast, in gravity-driven flows the fluid phase usually has a low Mach number and the fluid density ρ_f can be treated as constant, eliminating the need for the fluid-phase total energy balance [18].

The principal objective when developing a turbulence model will be to reduce the grid-resolution requirement due to 'turbulent viscosity/diffusion' and by decreasing the particle-phase Mach number. Indeed, as the eigenvalues of the fluid phase will not be significantly changed (or removed by taking ρ_f constant) in the turbulence model, a coarser grid will allow for larger time steps while still satisfying the CFL condition. Nonetheless, even after phase averaging, the system must remain hyperbolic and large temporal variations in the volume fraction will not be damped out. Thus, the model will usually not exhibit stable time-independent solutions unless the particle-phase Mach number is very small.

3 Multiphase Turbulence Model

The turbulence model in Table 2 is found by phase averaging the two-fluid model in Table 1 following the method presented in [26,27]. During this process, numerous unclosed terms are introduced that require physics-based closures. The turbulence model in Table 2 is the final closed form, and care has been taken to ensure that the closures for the fluxes do not change the hyperbolicity of the system. We remind the reader that only terms involving space and time derivatives influence the hyperbolicity. Thus, the closures for point-wise source terms do not affect the eigenvalues and, hence, more latitude is available when choosing their mathematical forms. In general, we have chosen forms that attempt to capture the behavior of cluster-induced (CIT) and buoyancy-driven (BDT) turbulence [27]; however, turbulence arising due to mean shear is also included. At this stage, we consider only the simplest forms for the Reynolds-stress transport equations corresponding to linear relaxation. Likewise, the spatial transport of scalar quantities due to turbulent fluctuations is modeled using a gradient-transport model [28]. For clarity and simplicity, we drop the subscripts introduced in [26] to denote phase averages, and give the final closed forms for the turbulence model developed in [27].

One important point that is often underestimated in multiphase turbulence models is that for CIT and BDT the phasic Reynolds stresses are nearly one-dimensional turbulence [28]. In other words, the shear stress components are very small, implying that turbulent transport in the directions normal to gravity is small (or even null). In this case, the Reynolds stresses do not add a 'turbulent' viscosity to the momentum balances because the shear stress is null. Consequently, the mean volume fraction $\langle \alpha_p \rangle$ will be time-dependent in fully developed flows because the 'molecular' viscosity is not large enough to smooth out velocity fluctuations and the compression/expansion of the disperse phase. In short, even a Reynolds-averaged model like the one in Table 2 will exhibit large-scale fluctuations characteristic of single-phase large-eddy simulations [29] if the CIT/BDT forces are large enough.[4] For systems with low mass loading or very dilute systems, there is no mechanism (besides mean shear) to sustain the turbulence in spatially homogeneous flows and thus it will decay in time as seen in single-phase turbulence. Such flows have been widely studied in the literature, and thus are not the target application for the turbulence model described in this work. As shown recently for particle-laden channel flows [30], there is a clear separation between CIT and classical channel flow that occurs for a mass loading near unity. The former lacks 'turbulent' viscosity giving a hyperbolic system, while the latter has strong 'turbulent' viscosity leading to a parabolic system with a time-independent statistics.

[4] For CIT, the mass loading of the particle phase is the key parameter [30], while in BDT the key parameter is $\langle \alpha_p \rangle$ [21].

3.1 Mass

The mass balances in Table 2 use the conserved variables

$$(X_1, X_2, X_3) = (\langle \alpha_p \rangle, \langle \rho_e \alpha_p^\star \rangle, \langle \rho_f \alpha_f^\star \rangle), \tag{9}$$

which are chosen such that the phase-averaged velocities are defined by $\langle \mathbf{u}_p \rangle$ and $\langle \mathbf{u}_f \rangle$. Closures relating (X_1, X_2, X_3) to Reynolds-averaged quantities, e.g., $\langle \rho_f \rangle$, $\langle \alpha_a \rangle$ and $\langle c_m \rangle$, are provided in the table. These formulas simplify significantly when ρ_f can be treated as constant. Because the conserved variables in the mass balances, as well as their fluxes, have the same mathematical form as in the original two-fluid model, the hyperbolicity will not be affected by these choices.

3.2 Momentum

The momentum balances in Table 2 use the conserved variables

$$(X_4, X_5) = (\langle \rho_e \alpha_p^\star \rangle \langle \mathbf{u}_p \rangle, \langle \rho_f \alpha_f^\star \rangle \langle \mathbf{u}_f \rangle), \tag{10}$$

and, hence,

$$(\langle \mathbf{u}_p \rangle_p, \langle \mathbf{u}_f \rangle_f) = \left(\frac{X_4}{X_2}, \frac{X_5}{X_3} \right). \tag{11}$$

The momentum fluxes on the left-hand sides of the momentum balances have an additional contribution due to the phasic Reynolds stresses R_p and R_f. At first glance, one might conclude that the eigenvalues associated with these terms would increase (i.e., the effective pressure increases). However, this is not the case because the total energy is fixed so that increasing R_f results in decreasing $\langle \Theta_f \rangle$ (see definitions of phase-average temperatures in Table 3). Nonetheless, the *effective* particle-phase pressure can increase, especially when the granular temperature is small. In contrast, the closure for $\langle \mathsf{R} \rangle$ can change the eigenvalues. This closure is given in Table 3 and contains a 'laminar' contribution $\langle \mathsf{R}_1 \rangle$ and a turbulent contribution $\langle \mathsf{R}_2 \rangle$. In general, the latter will modify the eigenvalues (e.g., increase their magnitude) due to the difference in turbulence intensities in each phase. Note that the Reynolds-stress tensors are symmetric, positive definite, and therefore the symmetric, non-negative square-root $(\mathsf{R}_f \mathsf{R}_p + \mathsf{R}_p \mathsf{R}_f)^{1/2}$ is uniquely defined using the positive square roots of its (real-valued) eigenvalues.[5] Rescaling this matrix by one-half its trace yields the matrix R_{fp}. In the hyperbolicity analysis, $tr(\langle \mathsf{R} \rangle)$ increases the eigenvalues and $tr(\langle \mathsf{R}_2 \rangle) = 2(k_f^{1/2} - k_p^{1/2})^2$ is non-negative. For the other flux terms, the 'laminar' closures given in Table 3 are retained to ensure that the turbulence model is hyperbolic.

For CIT and BDT, the most important turbulence closure in the momentum balances is the drag modification represented by $(1 - C_d)$ [26]. If the particles remained uniformly distributed in space, then $C_d = 0$; thus, $0 < C_d \le 1/2$

[5] For CIT and BDT, the Reynolds-stress tensors are diagonal, and hence $\mathsf{R}_f \mathsf{R}_p$ is diagonal and non-negative. However, for mean-shear flows this is not the case, so care must be taken when defining R_{fp} to be sure that the eigenvalues are real-valued.

Table 2. Turbulence model for mean statistics found from two-fluid model in Table 1. The deformation $\langle \Gamma \rangle$ and strain-rate $\langle S \rangle$ tensors use the phase-average fluid velocity $\langle \mathbf{u}_f \rangle$. The pressure tensors are defined in Table 3. The Reynolds stresses R_p and R_f and turbulent viscosities $\nu_{p,t}$ and $\nu_{f,t}$ are defined in Table 4.

Mass

$$\begin{aligned} \partial_t \langle \alpha_p \rangle + \partial_{\mathbf{x}} \cdot (\langle \alpha_p \rangle \langle \mathbf{u}_p \rangle) &= 0 \\ \partial_t \langle \rho_e \alpha_p^\star \rangle + \partial_{\mathbf{x}} \cdot (\langle \rho_e \alpha_p^\star \rangle \langle \mathbf{u}_p \rangle) &= \langle S_a \rangle \\ \partial_t \langle \rho_f \alpha_f^\star \rangle + \partial_{\mathbf{x}} \cdot (\langle \rho_f \alpha_f^\star \rangle \langle \mathbf{u}_f \rangle) &= - \langle S_a \rangle \end{aligned}$$

Momentum

$$\begin{aligned} &\partial_t(\langle \rho_e \alpha_p^\star \rangle \langle \mathbf{u}_p \rangle) + \partial_{\mathbf{x}} \cdot (\langle \rho_e \alpha_p^\star \rangle \langle \mathbf{u}_p \rangle \langle \mathbf{u}_p \rangle + \langle \rho_e \alpha_p^\star \rangle \mathsf{R}_p + \langle p_p \rangle \mathsf{I} + \langle \mathsf{P}_{pfp} \rangle) = \\ &\quad (1 - C_d)\langle K \rangle \langle \mathbf{u}_{fp} \rangle - \langle \alpha_p^\star \rangle \partial_{\mathbf{x}} \langle p_f \rangle - \langle \alpha_p^\star \rangle \langle \mathbf{F}_{pf} \rangle + \langle \mathbf{S}_{fp} \rangle + \langle \rho_e \alpha_p^\star \rangle \mathbf{g} \end{aligned}$$

$$\begin{aligned} &\partial_t(\langle \rho_f \alpha_f^\star \rangle \langle \mathbf{u}_f \rangle) + \partial_{\mathbf{x}} \cdot (\langle \rho_f \alpha_f^\star \rangle \langle \mathbf{u}_f \rangle \langle \mathbf{u}_f \rangle + \langle \rho_f \alpha_f^\star \rangle \mathsf{R}_f + \langle p_f \rangle \mathsf{I}) = \\ &\quad (1 - C_d)\langle K \rangle \langle \mathbf{u}_{pf} \rangle + \langle \alpha_p^\star \rangle \partial_{\mathbf{x}} \langle p_f \rangle + \langle \alpha_p^\star \rangle \langle \mathbf{F}_{pf} \rangle - \langle \mathbf{S}_{fp} \rangle + \langle \rho_f \alpha_f^\star \rangle \mathbf{g} \end{aligned}$$

Total Energy

$$\begin{aligned} &\partial_t(\langle \rho_e \alpha_p^\star \rangle \langle E_p \rangle) + \partial_{\mathbf{x}} \cdot [\langle \rho_e \alpha_p^\star \rangle \langle E_p \rangle \langle \mathbf{u}_p \rangle + \langle \rho_e \alpha_p^\star \rangle \mathsf{R}_p \cdot \langle \mathbf{u}_p \rangle + (\langle p_p \rangle \mathsf{I} + \langle \mathsf{P}_{pfp} \rangle) \cdot \langle \mathbf{u}_p \rangle] = \\ &\quad \partial_{\mathbf{x}} \cdot (\langle \rho_e \alpha_p^\star \rangle \nu_{p,t} \partial_{\mathbf{x}} \langle E_p \rangle) + \langle \mathsf{P}_{pfp} \rangle : \partial_{\mathbf{x}} \langle \mathbf{u}_p \rangle \\ &\quad - \langle K \rangle \left[3\langle \Theta_p \rangle + 2k_p - 2(k_f k_p)^{1/2} + (1 - C_d)\langle \mathbf{u}_p \rangle \cdot \langle \mathbf{u}_{pf} \rangle - C_b \langle u_{fp} \rangle^2 \right] \\ &\quad - \langle \alpha_p^\star \rangle \langle \mathbf{u}_p \rangle \cdot \partial_{\mathbf{x}} \langle p_f \rangle - \langle \alpha_p^\star \rangle \langle \mathbf{u}_p \rangle \cdot \langle \mathbf{F}_{pf} \rangle + \langle S_E \rangle + \langle \rho_e \alpha_p^\star \rangle \langle \mathbf{u}_p \rangle \cdot \mathbf{g} \end{aligned}$$

$$\begin{aligned} &\partial_t(\langle \rho_f \alpha_f^\star \rangle \langle E_f \rangle) + \partial_{\mathbf{x}} \cdot [\langle \rho_f \alpha_f^\star \rangle \langle E_f \rangle \langle \mathbf{u}_f \rangle + \langle \rho_f \alpha_f^\star \rangle \mathsf{R}_f \cdot \langle \mathbf{u}_f \rangle + (\langle \alpha_f^\star \rangle \langle \mathbf{u}_f \rangle + \langle \alpha_p^\star \rangle \langle \mathbf{u}_p \rangle) \langle p_f \rangle] = \\ &\quad \partial_{\mathbf{x}} \cdot (\langle \rho_f \alpha_f^\star \rangle \nu_{f,t} \partial_{\mathbf{x}} \langle E_f \rangle) - \langle \mathsf{P}_{pfp} \rangle : \partial_{\mathbf{x}} \langle \mathbf{u}_p \rangle \\ &\quad + \langle K \rangle \left[3\langle \Theta_p \rangle + 2k_p - 2(k_f k_p)^{1/2} + (1 - C_d)\langle \mathbf{u}_p \rangle \cdot \langle \mathbf{u}_{pf} \rangle - C_b \langle u_{fp} \rangle^2 \right] \\ &\quad + \langle \alpha_p^\star \rangle \langle \mathbf{u}_p \rangle \cdot \partial_{\mathbf{x}} \langle p_f \rangle + \langle \alpha_p^\star \rangle \langle \mathbf{u}_p \rangle \cdot \langle \mathbf{F}_{pf} \rangle - \langle S_E \rangle + \langle \rho_f \alpha_f^\star \rangle \langle \mathbf{u}_f \rangle \cdot \mathbf{g} \end{aligned}$$

where

$$\langle \alpha_f \rangle + \langle \alpha_p \rangle = 1 \quad \langle \alpha_f^\star \rangle + \langle \alpha_p^\star \rangle = 1$$

$$\langle \rho_f \alpha_a \rangle = \langle \rho_e \alpha_p^\star \rangle - \rho_p \langle \alpha_p \rangle \quad \langle \rho_f \alpha_f \rangle = \langle \rho_f \alpha_f^\star \rangle + \langle \rho_f \alpha_a \rangle$$

$$\langle \rho_f \rangle = \frac{\langle \rho_f \alpha_f^\star \rangle + \langle \rho_e \alpha_p^\star \rangle - \rho_p \langle \alpha_p \rangle}{\langle \alpha_f \rangle} \quad \langle \alpha_a \rangle = \frac{\langle \rho_e \alpha_p^\star \rangle - \rho_p \langle \alpha_p \rangle}{\langle \rho_f \rangle} \quad \langle c_m \rangle = \frac{\langle \rho_f \alpha_a \rangle}{\langle \rho_f \alpha_f \rangle \langle \alpha_p \rangle}$$

$$\langle K \rangle = \frac{3\langle \rho_e \alpha_p^\star \rangle C_D Re_p}{4\tau_p} \quad Re_p = \frac{\langle \alpha_f^\star \rangle}{\langle \alpha_f \rangle} \frac{d_p \langle u_{fp} \rangle}{\nu_f} \quad c_m^\star = \frac{1}{2} \min\left(1 + 2\langle \alpha_p \rangle, 2\right)$$

$$\langle S_a \rangle = \frac{1}{\tau_a} \langle \rho_f \alpha_f \rangle \langle \alpha_p \rangle (c_m^\star - \langle c_m \rangle) \quad \langle \mathbf{S}_{fp} \rangle = \max(\langle S_a \rangle, 0)\langle \mathbf{u}_f \rangle + \min(\langle S_a \rangle, 0)\langle \mathbf{u}_p \rangle$$

$$\langle S_E \rangle = \max(\langle S_a \rangle, 0) \frac{1}{2} \langle u_f \rangle^2 + \min(\langle S_a \rangle, 0)\langle E_p \rangle \quad g_0 = \frac{1 + \langle \alpha_f \rangle}{2\langle \alpha_f \rangle^3}$$

$$\langle \mathbf{F}_{pf} \rangle = \langle \rho_f \rangle \left[C_c\, tr(\langle \Gamma \rangle) \mathsf{I} + C_l \langle \mathsf{S} \rangle \right] \cdot \langle \mathbf{u}_{fp} \rangle \quad \langle \mathbf{u}_{pf} \rangle = -\langle \mathbf{u}_{fp} \rangle = \langle \mathbf{u}_p \rangle - \langle \mathbf{u}_f \rangle$$

$$\langle u_{fp} \rangle^2 = \langle \mathbf{u}_{fp} \rangle \cdot \langle \mathbf{u}_{fp} \rangle \quad \langle u_f \rangle^2 = \langle \mathbf{u}_f \rangle_f \cdot \langle \mathbf{u}_f \rangle \quad \langle u_p \rangle^2 = \langle \mathbf{u}_p \rangle \cdot \langle \mathbf{u}_p \rangle$$

Table 3. Pressure tensors for turbulence model in Table 2. Reynolds stresses R_p and R_f are defined in Table 4. All pressure tensors are non-negative and symmetric. $\langle \mathsf{R}_2 \rangle$ is a closure for the turbulent contribution to the pfp-pressure with trace $2(k_f^{1/2} - k_p^{1/2})^2$.

Slip-Velocity Tensor

$$\langle \mathsf{R} \rangle = \langle \mathsf{R}_1 \rangle + \langle \mathsf{R}_2 \rangle$$

$$\langle \mathsf{R}_1 \rangle = \frac{1}{5} \langle u_{fp} \rangle^2 \mathsf{I} + \frac{2}{5} \langle \mathbf{u}_{fp} \rangle \langle \mathbf{u}_{fp} \rangle$$

$$\langle \mathsf{R}_2 \rangle = \frac{2}{5} (k_f^{1/2} - k_p^{1/2})^2 \mathsf{I} + \frac{2}{5} \left(\mathsf{R}_f - 2(k_f k_p)^{1/2} \mathsf{R}_{fp} + \mathsf{R}_p \right)$$

where the symmetric tensor $\mathsf{R}_{fp} = r_{fp}(\mathsf{R}_f \mathsf{R}_p + \mathsf{R}_p \mathsf{R}_f)^{1/2}$ has $tr(\mathsf{R}_{fp}) = 2$, which fixes r_{fp} given the Reynolds-stress tensors.

Particle-Fluid-Particle-Pressure Tensor

$$\langle \mathsf{P}_{pfp} \rangle = C_{pfp} \langle \rho_f \rangle \langle \alpha_p^\star \rangle \langle \mathsf{R} \rangle$$

Fluid Pressure

$$\langle p_f \rangle = \langle \rho_f \rangle (\gamma_f - 1) \langle \Theta_f \rangle - \gamma_f p_f^o$$

Granular Pressure

$$\langle p_p \rangle = \langle \rho_e \alpha_p^\star \rangle \langle \Theta_p \rangle (1 + 4 \langle \alpha_p^\star \rangle g_0) + p_c \langle \alpha_p \rangle g_0 \frac{1}{2} \left[1 + \tanh \left(\frac{\langle \alpha_p \rangle - 0.6}{0.01} \right) \right]$$

where

$$\langle \Theta_f \rangle = \langle E_f \rangle - k_f - \frac{1}{2} \langle u_f \rangle^2$$

$$\langle \Theta_p \rangle = \frac{2}{3} \left(\langle E_p \rangle - k_p - \frac{1}{2} \langle u_p \rangle^2 \right)$$

represents reduced drag due to particle clustering. In general, C_d depends on many parameters such as $\langle \alpha_p \rangle$ and Re_p [31]. However, here we simply take $C_d = 0.5$ as the default value, knowing that in most flows it will be smaller. For the other source terms in the momentum balances, most of which affect the hyperbolicity, we use the 'laminar' closure and introduce parameters C_c (compression) and C_l (lift) with default values of unity. These parameters could be fit to canonical cases, for example, involving shocks passing over particle beds (C_c) and mean shear flow (C_l). For the various pressure tensors, use of the 'laminar' forms is justified because they are usually small [18] and will mostly be important for low-turbulent-Reynolds-number flows.

3.3 Total Energy

The total energy balances in Table 2 use the conserved variables

$$(X_6, X_7) = (\langle \rho_e \alpha_p^\star \rangle \langle E_p \rangle, \langle \rho_f \alpha_f^\star \rangle \langle E_f \rangle), \tag{12}$$

and thus

$$(\langle E_p \rangle, \langle E_f \rangle) = \left(\frac{X_6}{X_2}, \frac{X_7}{X_3} \right). \tag{13}$$

The flux closures are analogous to the 'laminar' model but with the additional contribution of the Reynolds stress and the gradient-diffusion model involving turbulent viscosity ν_t (see Table 4). As discussed earlier, one role of the turbulent viscosity is to 'thicken' the shock-like structures that arise due to the high Mach number of the particle phase.

The source terms on the right-hand sides of the energy balances in Table 2 are analogous to the 'laminar' model, but with additional terms that depend on the turbulent kinetic energies (TKE) k_f, k_p and the mean slip velocity $\langle u_{fp} \rangle^2$. Recall that, aside from gravity $\mathbf{g}$, the sum of the source terms for the two phases must be zero due to conservation on energy. Because in most gas–particle flows $k_p < k_f$, the transfer terms move TKE from the particle phase to the fluid phase where it ultimately dissipates to increase $\langle \Theta_f \rangle$. The source terms depending on the mean slip velocity also depend on the material density ratio $Z = \langle \rho_f \rangle / \rho_p$ such that $C_d = C_b + C_g$. As a default closure, we propose the relation given in Table 5, but acknowledge that it is a crude approximation for large Z (i.e., bubbly flows). Physically, C_b corresponds to the particle-phase TKE generated by buoyancy (e.g., BDT) [21], while C_g corresponds to the fluid-phase TKE generated by gravity (e.g., CIT) [26]. In fully developed flow, the terms involving buoyancy and drag are much larger than all of the other source terms. It is interesting to note that the dependence of C_b and C_g on Z implies that in homogeneous bubbly flows $k_f \leq k_p$ [21], while in gas–particle flows $k_p \leq k_f$ [32]. Naturally, for homogeneous suspensions with $Z \approx 1$, $k_p \approx k_f$. Finally, we must remind the reader that the proposed models for C_b and C_g are based on limited data from EE and EL simulations. As is true for C_d, further work is needed to account for the other flow parameters.

3.4 Multiphase Turbulence Statistics

The proposed multiphase turbulence model uses the Reynolds-stress model given in Table 4. By construction, the balances for the Reynolds stresses reduce to those of the TKE, e.g., $2k_f = tr(\mathsf{R}_f)$. The turbulent length and time scales are found using the turbulent dissipate rates (TDR) ε. The models in Table 4 for the TKE and the TDR were first proposed in [26] as generalization of the classical k–ε model [28]. There are two coupling terms, one involving the fluid drag $\langle K \rangle$ and the difference in the statistics in each phase (i.e., the drag-exchange term), and the other is a production term due to the mean-slip velocity. Because the TKE is part of the total energy, the same source terms appear (along with several others) in the total energy balances. Observe that the drag-exchange terms (in isolation) work to equalize the TKE in both phases. On the other hand, the mean-slip-production terms depend on C_b and C_g and inject TKE extracted from the mean kinetic energy. Of these two terms, C_g is important for gas–particle flows, while C_b is important in bubbly flows. At present, only C_g has been studied in detail [32]. The scaling in terms of Z in the drag-production term involving C_b is unknown, but likely must increase with Z in bubbly flows. Unlike for the total energy of the mixture, the sources terms for TKE of the mixture need not be conservative. In fact, the drag-exchange terms are strictly dissipative for the mixture [26]. At very large mass loading, the magnitudes of the drag-exchange and mean-slip-production terms are much larger than the others and thus determine the values of the TKE and TDR.

The TDR in the TKE balances has the usual form as in single-phase turbulence models, i.e., it dissipates k due to the turbulent energy flux from large to small scales. The constant $C_{\varepsilon,2}$ controls the turbulence decay rate in homogeneous systems [28]. As discussed in [26], the granular temperature equation has the form

$$\begin{aligned}\tfrac{3}{2}\partial_t(\langle\rho_e\alpha_p^\star\rangle\langle\Theta_p\rangle) + \tfrac{3}{2}\partial_{\mathbf{x}}\cdot(\langle\rho_e\alpha_p^\star\rangle\langle\Theta_p\rangle\langle\mathbf{u}_p\rangle) &= \tfrac{3}{2}\partial_{\mathbf{x}}\cdot(\langle\rho_e\alpha_p^\star\rangle\nu_{p,t}\partial_{\mathbf{x}}\langle\Theta_p\rangle) \\ &- \langle p_p\rangle\partial_{\mathbf{x}}\cdot\langle\mathbf{u}_p\rangle - 3\langle K\rangle\langle\Theta_p\rangle + \langle\rho_e\alpha_p^\star\rangle\varepsilon_p \end{aligned} \tag{14}$$

where the final term represents production of uncorrelated granular energy due the turbulent dissipation of k_p. Because we use the conservative form of the model with the total energy, this balance equation is not needed. However, one can clearly observed that $\langle\Theta_p\rangle$ is only dissipated by drag (i.e., viscous heating of the fluid). In other words, in the particle phase there is a cascade of energy from large to small scales: $\frac{1}{2}\langle u_p\rangle^2 \to k_p \to \langle\Theta_p\rangle \to \langle\Theta_f\rangle$.

The balance equations for TDR are described in detail in [26]. Again, the two new terms are due to drag exchange and mean-slip production. These terms have the same forms as for the TKE, but there are subtle differences from classical closures used for multiphase turbulence. First, instead of multiplying the drag-exchange term from the TKE by a turbulence time scale, here we define the drag-exchange term using ε (which automatically has the correct units). As shown in [26], by adjusting $C_{\varepsilon,3}$ this simple model reproduces the decay of homogeneous multiphase turbulence in the absence of a mean-slip velocity [11]. The mean-slip-production term uses the timescale τ_p modified by the constant $C_{\varepsilon,4}$. The latter is likely a function of Z and/or mass loading $\varphi = \langle\rho_p\alpha_p\rangle/\langle\rho_f\alpha_f\rangle$. In general, for fixed Z, the value of $C_{\varepsilon,2}/C_{\varepsilon,4}$ controls the magnitude of the steady-state TKE in systems with no production due to mean shear.

Turning now to the balance equations for the Reynolds-stress tensors in Table 4, it is easily verified that their traces reduce to the TKE balances as expected. For the second-order tensors appearing in the balances, the drag-exchange terms are written in terms of the symmetric matrix R_{fp}, defined by rescaling the square-root matrix found from the symmetric form $\mathsf{R}_f\mathsf{R}_p + \mathsf{R}_p\mathsf{R}_f$ such that the trace is equal to two. To the author's knowledge, this simple form has not been used previously and has yet to be tested for flows with mean shear (i.e., away from one-dimensional turbulence). The mean-slip-production terms involve the tensor P_{fp}, which is the dyadic product of the mean-slip velocity. It is this simple (closed) form that leads to diagonal Reynolds stresses for gravity-driven flows in the absence of mean shear. For the other terms, the isotropization tensors $\boldsymbol{\Phi}$ have the classical linear form used in single-phase turbulence models

Table 4. Closures for higher-order statistics in the multiphase turbulence model in Table 2. Default values of the model constants are provided in Table 5.

Turbulent Kinetic Energy

$$\begin{aligned}\partial_t(\langle\rho_f\alpha_f^\star\rangle k_f) + \partial_{\mathbf{x}}\cdot(\langle\rho_f\alpha_f^\star\rangle k_f\langle\mathbf{u}_f\rangle) &= \partial_{\mathbf{x}}\cdot(\langle\rho_f\alpha_f^\star\rangle\nu_{f,t}\partial_{\mathbf{x}}k_f) - \langle\rho_f\alpha_f^\star\rangle\varepsilon_f\\ &- \langle\rho_f\alpha_f^\star\rangle\mathsf{R}_f : \partial_{\mathbf{x}}\langle\mathbf{u}_f\rangle_f + 2\langle K\rangle\left[(k_pk_f)^{1/2} - k_f\right] + C_g\langle K\rangle\langle u_{fp}\rangle^2\end{aligned}$$

$$\begin{aligned}\partial_t(\langle\rho_e\alpha_p^\star\rangle k_p) + \partial_{\mathbf{x}}\cdot(\langle\rho_e\alpha_p^\star\rangle k_p\langle\mathbf{u}_p\rangle) &= \partial_{\mathbf{x}}\cdot(\langle\rho_e\alpha_p^\star\rangle\nu_{p,t}\partial_{\mathbf{x}}k_p) - \langle\rho_e\alpha_p^\star\rangle\varepsilon_p\\ &- \langle\rho_e\alpha_p^\star\rangle\mathsf{R}_p : \partial_{\mathbf{x}}\langle\mathbf{u}_p\rangle + 2\langle K\rangle\left[(k_pk_f)^{1/2} - k_p\right] + C_b\langle K\rangle\langle u_{fp}\rangle^2\end{aligned}$$

Turbulent Dissipation Rate

$$\begin{aligned}\partial_t(\langle\rho_f\alpha_f^\star\rangle\varepsilon_f) + \partial_{\mathbf{x}}\cdot(\langle\rho_f\alpha_f^\star\rangle\varepsilon_f\langle\mathbf{u}_f\rangle) &= \partial_{\mathbf{x}}\cdot(\langle\rho_f\alpha_f^\star\rangle\nu_{f,t}\partial_{\mathbf{x}}\varepsilon_f) - C_{\varepsilon,2}\langle\rho_f\alpha_f^\star\rangle\tfrac{\varepsilon_f}{k_f}\varepsilon_f\\ - C_{\varepsilon,1}\tfrac{\varepsilon_f}{k_f}\langle\rho_f\alpha_f^\star\rangle\mathsf{R}_f : \partial_{\mathbf{x}}\langle\mathbf{u}_f\rangle &+ 2C_{\varepsilon,3}\langle K\rangle\left[(\varepsilon_p\varepsilon_f)^{1/2} - \varepsilon_f\right] + C_{\varepsilon,4}\tfrac{1}{\tau_p}C_g\langle K\rangle\langle u_{fp}\rangle^2\end{aligned}$$

$$\begin{aligned}\partial_t(\langle\rho_e\alpha_p^\star\rangle\varepsilon_p) + \partial_{\mathbf{x}}\cdot(\langle\rho_e\alpha_p^\star\rangle\varepsilon_p\langle\mathbf{u}_p\rangle) &= \partial_{\mathbf{x}}\cdot(\langle\rho_e\alpha_p^\star\rangle\nu_{p,t}\partial_{\mathbf{x}}\varepsilon_p) - C_{\varepsilon,2}\langle\rho_e\alpha_p^\star\rangle\tfrac{\varepsilon_p}{k_p}\varepsilon_p\\ - C_{\varepsilon,1}\tfrac{\varepsilon_p}{k_p}\langle\rho_e\alpha_p^\star\rangle\mathsf{R}_p : \partial_{\mathbf{x}}\langle\mathbf{u}_p\rangle &+ 2C_{\varepsilon,3}\langle K\rangle\left[(\varepsilon_p\varepsilon_f)^{1/2} - \varepsilon_p\right] + C_{\varepsilon,4}\tfrac{1}{\tau_p}C_b\langle K\rangle\langle u_{fp}\rangle^2\end{aligned}$$

Reynolds-Stress Tensor

$$\begin{aligned}\partial_t(\langle\rho_f\alpha_f^\star\rangle\mathsf{R}_f) + \partial_{\mathbf{x}}\cdot(\langle\rho_f\alpha_f^\star\rangle\langle\mathbf{u}_f\rangle\mathsf{R}_f) &= \partial_{\mathbf{x}}\cdot(\langle\rho_f\alpha_f^\star\rangle\nu_{f,t}\partial_{\mathbf{x}}\mathsf{R}_f) - \langle\rho_f\alpha_f^\star\rangle\mathbf{\Phi}_f - \langle\rho_f\alpha_f^\star\rangle\epsilon_f\\ &+ \langle\rho_f\alpha_f^\star\rangle\mathsf{P}_f + 2\langle K\rangle\left((k_fk_p)^{1/2}\mathsf{R}_{fp} - \mathsf{R}_f\right) + 2C_g\langle K\rangle\mathsf{P}_{fp}\end{aligned}$$

$$\begin{aligned}\partial_t(\langle\rho_e\alpha_p^\star\rangle\mathsf{R}_p) + \partial_{\mathbf{x}}\cdot(\langle\rho_e\alpha_p^\star\rangle\langle\mathbf{u}_p\rangle\mathsf{R}_p) &= \partial_{\mathbf{x}}\cdot(\langle\rho_e\alpha_p^\star\rangle\nu_{p,t}\partial_{\mathbf{x}}\mathsf{R}_p) - \langle\rho_e\alpha_p^\star\rangle\mathbf{\Phi}_p - \langle\rho_e\alpha_p^\star\rangle\epsilon_p\\ &+ \langle\rho_e\alpha_p^\star\rangle\mathsf{P}_p + 2\langle K\rangle\left((k_fk_p)^{1/2}\mathsf{R}_{fp} - \mathsf{R}_p\right) + 2C_b\langle K\rangle\mathsf{P}_{fp}\end{aligned}$$

where $(\mathsf{Y})^\dagger = \frac{1}{2}(\mathsf{Y} + \mathsf{Y}^t)$, $\mathsf{R}_{fp} \propto (\mathsf{R}_f\mathsf{R}_p + \mathsf{R}_p\mathsf{R}_f)^{1/2}$ with $tr(\mathsf{R}_{fp}) = 2$,

$$\tau_p = \frac{\langle\rho_e\alpha_p^\star\rangle}{\langle K\rangle} \qquad \nu_{f,t} = C_{\mu,f}\frac{k_f^2}{\varepsilon_f} \qquad \nu_{p,t} = C_{\mu,p}\frac{k_p^2}{\varepsilon_p} \qquad \mathsf{P}_{fp} = \langle\mathbf{u}_{fp}\rangle\langle\mathbf{u}_{fp}\rangle \quad P_{fp} = tr(\mathsf{P}_{fp})$$

$$\mathsf{P}_f = -2(\mathsf{R}_f\cdot\partial_{\mathbf{x}}\langle\mathbf{u}_f\rangle)^\dagger \quad P_f = tr(\mathsf{P}_f) \qquad \mathsf{P}_p = -2(\mathsf{R}_p\cdot\partial_{\mathbf{x}}\langle\mathbf{u}_p\rangle)^\dagger \quad P_p = tr(\mathsf{P}_p)$$

$$\mathbf{\Phi}_f = C_{1f}\frac{\varepsilon_f}{k_f}\left(\mathsf{R}_f - \frac{2}{3}k_f\mathsf{I}\right) + C_{2f}\left(\mathsf{P}_f - \frac{1}{3}P_f\mathsf{I}\right) + C_{fp}\frac{2C_g\langle K\rangle}{\langle\rho_f\alpha_f^\star\rangle}\left(\mathsf{P}_{fp} - \frac{1}{3}P_{fp}\mathsf{I}\right)$$

$$\mathbf{\Phi}_p = C_{1p}\frac{\varepsilon_p}{k_p}\left(\mathsf{R}_p - \frac{2}{3}k_p\mathsf{I}\right) + C_{2p}\left(\mathsf{P}_p - \frac{1}{3}P_p\mathsf{I}\right) + C_{fp}\frac{2C_b\langle K\rangle}{\langle\rho_e\alpha_p^\star\rangle}\left(\mathsf{P}_{fp} - \frac{1}{3}P_{fp}\mathsf{I}\right)$$

$$\epsilon_f = \left(C_f\frac{1}{k_f}\mathsf{R}_f + \frac{2}{3}(1 - C_f)\mathsf{I}\right)\varepsilon_f \qquad \epsilon_p = \left(C_p\frac{1}{k_p}\mathsf{R}_p + \frac{2}{3}(1 - C_p)\mathsf{I}\right)\varepsilon_p$$

Table 5. Default values of the turbulence model constants with $Z = \langle\rho_f\rangle/\rho_p$. For water, $\gamma_f = 6.1$ and $p_f^o = 4 \times 10^8$ Pa. For air, $\gamma_f = 1.4$ and $p_f^o = 0$ Pa. Except perhaps for $Z \leq 1$, the expressions for C_g and C_b are rough approximations and should not be used for $1 \ll Z$ (where $C_g \approx 0$) as explained in the main text. More generally, the dissipation model parameters must be fit to data for statistically homogeneous flows, especially for large Z (bubbly flows).

Constant	Value
c_f	0.2
C_c	1
C_l	1
C_g	$[\langle\alpha_f\rangle - \langle\alpha_a\rangle \min(Z, 1)]C_d$
C_b	$[\langle\alpha_p\rangle + \langle\alpha_a\rangle \min(Z, 1)]C_d$
C_d	0.5
$C_{\mu,f}$, $C_{\mu,f}$	0.09
C_{1f}	1.5
C_{1p}	1.5
C_{2f}	0.6
C_{2p}	0.6
C_{fp}	0
C_f, C_p	0
$C_{\varepsilon,1}$	1.44
$C_{\varepsilon,2}$	1.92
$C_{\varepsilon,3}$	1.92
$C_{\varepsilon,4}$	1.92

[28], and the Rotta model is used for the turbulent dissipation tensors ϵ. Obviously, other forms of these terms are possible, and likely desirable, to capture the temporal and spatial variations of the Reynolds stresses in real disperse multiphase flows.

Finally, we should note that in the hyperbolicity analysis the eigenvalues associated with the TKE and TDR balances are just the mean velocities. Thus, the hyperbolicity of the turbulence model is not compromised. Nonetheless, as mentioned earlier in the discussion of Table 2, the magnitude of some of the eigenvalues will increase with increasing TKE. In other words, the speed of some processes will become faster when turbulence is present. However, the largest eigenvalues (which control the numerical time step [18]) are usually fixed by the speed of sound in the fluid ($\sqrt{\gamma_f p_f^o}$). If the fluid phase were treated as incompressible [18], then the speed of sound for the particle phase (which depends on $\sqrt{6k_p}$) would determine the time step.

3.5 Summary

In addition to the seven conserved variables X_i corresponding to the mean statistics, the turbulence model introduces four additional conserved quantities:

$$(X_8, X_9, X_{10}, X_{11}) = (\langle \rho_e \alpha_p^\star \rangle \mathsf{R}_p, \langle \rho_f \alpha_f^\star \rangle \mathsf{R}_f, \langle \rho_e \alpha_p^\star \rangle \varepsilon_p, \langle \rho_f \alpha_f^\star \rangle \varepsilon_f). \quad (15)$$

The Reynolds stresses contain six variables each; however, in homogeneous CIT and BDT, only two of the six are independent. The default values for the model constants are provided in Table 5. Nonetheless, further research is needed to determine whether (1) the simple linear-relaxation model used for the Reynolds stresses is sufficient to handle a wide range of flows, (2) whether the default values of the constants need to be adjusted or made to depend on flow parameters such as Re_p and $\langle \alpha_p \rangle$, and (3) what additional terms beyond the 'laminar' terms are needed in the flux and source-term closures.

Finally, we should remind the reader that the starting two-fluid model in Table 1 neglects viscous stresses in the fluid phase, assumes elastic collisions for the particles, and treats all particles as adiabatic spheres with identical properties. While all of these additional effects can easily be included in the turbulence model, at present the model in Table 2 is applicable only to high-turbulent-Reynolds-number flows. For many applications, the main effect of the turbulence model will be to modify the phasic speeds of sound due to the TKE. For example, gas–particle flows with $Z = 0$ and negligible granular temperature have a particle-phase speed of sound equal to $\sqrt{6k_p}$. In principle, if k_p were large enough to make the particle phase subsonic, then a much coarser grid could be employed to resolve the particle phase.

Another important consideration when applying the turbulence model to real flows are the boundary conditions. For example, in particle-laden channel flows with significant mass loading the fluid-phase turbulence structures are altered by the presence of the clusters, which changes the behavior of the turbulent boundary layer [17,30]. Although some initial work on developing boundary conditions for channel flows has been reported [33], much work remains to accommodate more general flow geometries. Thus, in the example applications below, we will focus on homogeneous systems for which boundary conditions are not required.

4 Application to Statistically Homogeneous Flows

As an example application, we consider statistically homogeneous flows that are controlled by the drag-exchange and mean-slip-production terms. Two specific examples of such flows are (i) cluster-induced turbulence with $Z \approx 0$ [20] and (ii) buoyancy-driven turbulence with $1 \ll Z$ [21].

4.1 Mass Balances

For statistically stationary and spatially homogeneous flows, the turbulence model reduces to algebraic relations. For example, $\langle \alpha_p \rangle$ is a fixed constant, so that $\langle c_m \rangle = c_m^\star = \frac{1}{2} + \langle \alpha_p \rangle$ and $\langle \alpha_a \rangle = \langle c_m \rangle \langle \alpha_p \rangle \langle \alpha_f \rangle$ are constants. Furthermore, compressibility plays no role in these flows, so that ρ_f is constant and

$$\frac{\langle \rho_f \alpha_f^\star \rangle}{\langle \rho_e \alpha_p^\star \rangle} = \frac{Z \langle \alpha_f^\star \rangle}{\langle \alpha_p \rangle + Z \langle \alpha_a \rangle} \tag{16}$$

where $Z = \rho_f / \rho_p$ and the second term in the denominator is the contribution due to added mass. In general, any averaged variable involving the volume fractions and/or densities is time independent for spatially homogeneous flows.

4.2 Momentum Balances

The mean-slip velocity $\langle \mathbf{u}_{fp} \rangle$ is with respect to the *bulk* fluid velocity $\langle \mathbf{u}_f \rangle$ and not the fluid velocity $\langle \mathbf{v}_f \rangle$. As shown in [18], they differ by a factor of $\langle \alpha_f \rangle / \langle \alpha_f^\star \rangle$, i.e., $\langle \alpha_f^\star \rangle \langle \mathbf{u}_{pf} \rangle = \langle \alpha_f \rangle \langle \mathbf{v}_{pf} \rangle$. In most simulations of statistically homogeneous flows, $\langle \mathbf{v}_f \rangle$ is set to zero by controlling the fluid pressure gradient. For the fluid velocity to be null, we must have $\langle \rho_f \alpha_f^\star \rangle \langle \mathbf{u}_f \rangle = -\langle \rho_f \alpha_a \rangle \langle \mathbf{u}_p \rangle$, i.e., the bulk fluid velocity is opposite the particle velocity and proportional to the added mass. The mean-slip velocity is related to the mean particle velocity by

$$\langle \mathbf{u}_{pf} \rangle = \frac{\langle \alpha_f \rangle}{\langle \alpha_f^\star \rangle} \langle \mathbf{u}_p \rangle. \tag{17}$$

Thus, $\langle \alpha_f^\star \rangle d \langle \mathbf{u}_{pf} \rangle / dt = \langle \alpha_f \rangle d \langle \mathbf{u}_p \rangle / dt$ for statistically homogeneous flows.

We can take gravity to be $\mathbf{g} = g(t) \mathbf{e}_x$ where $g(t)$ is a possibly time-dependent magnitude. From the mixture momentum balance, we then find

$$\partial_{\mathbf{x}} \langle p_f \rangle + \rho_p \langle \alpha_p \rangle \partial_t \langle \mathbf{u}_p \rangle = (\rho_p \langle \alpha_p \rangle + \rho_f \langle \alpha_f \rangle) g(t) \mathbf{e}_x. \tag{18}$$

The mean-slip velocity is then set by $\langle \mathbf{u}_{pf} \rangle = \langle u_{pf} \rangle \mathbf{e}_x$, which is determined from the balance between gravity, buoyancy and fluid drag. Using (18) to eliminate the fluid-phase pressure gradient, the particle-phase momentum balance reduces to

$$\frac{\rho_p \langle \alpha_p \rangle \langle \alpha_f^\star \rangle + \rho_f \langle \alpha_a \rangle}{\langle K \rangle} \frac{d \langle u_p \rangle}{dt} = -(1 - C_d) \langle u_{pf} \rangle + \frac{\langle \alpha_f^\star \rangle \rho_p \langle \alpha_p \rangle (1 - Z)}{\langle K \rangle} g(t). \tag{19}$$

For a steady-state system, $g(t) = g$ is constant. With Stokes drag, $\langle K \rangle$ is

$$\langle K \rangle = \langle \rho_e \alpha_p^\star \rangle \frac{18 \mu_f}{\rho_p d_p^2} = \langle \rho_e \alpha_p^\star \rangle \frac{1}{\tau_p} \tag{20}$$

and the mean-slip velocity is

$$\langle u_{pf} \rangle = \frac{\langle \alpha_f^\star \rangle (1 - Z) \tau_p g}{(1 - C_d)(1 + Z \langle c_m \rangle \langle \alpha_f \rangle)}. \tag{21}$$

For CIT, $Z \approx 0$ so that $\langle u_{pf} \rangle$ has the same direction as gravity. The opposite is observed with BDT.

Recall that the factor $(1 - C_d)$ accounts for clustering, i.e., the magnitude of the mean-slip velocity increases compared to uniform flows. In the very dilute limit ($\langle \alpha_p \rangle \to 0$) there are no clusters, so that the mean-slip velocity for Stokes flow is

$$\langle u_{pf} \rangle_0 = \frac{1 - Z}{1 + Z \langle c_m \rangle} \tau_p g. \tag{22}$$

For finite Re_p, the drag coefficient will depend on Re_p and the volume fraction. Nevertheless, the mean-slip velocity is uniquely determined from (19). By defining $\tau_p^\star$ as

$$\tau_p^\star = \frac{\langle \rho_e \alpha_p^\star \rangle}{\langle K \rangle}, \tag{23}$$

the terminal velocity $V_t = \tau_p^\star g |1 - Z| / (1 + Z \langle c_m \rangle)$ is defined for arbitrary Re_p.

4.3 Granular Energy Balance

The total energy equation for the fluid provides no new information. For the particle phase, the steady-state granular temperature is found from (14):

$$\langle \Theta_p \rangle = \frac{\langle \rho_e \alpha_p^\star \rangle}{3 \langle K \rangle} \varepsilon_p = \frac{1}{3} \tau_p^\star \varepsilon_p, \tag{24}$$

and thus depends on the value of $\tau_p^\star \varepsilon_p$. In the model for CIT and BDT, the granular temperature does not affect any of the other variables. However, direct-numerical simulations of uniform systems suggest that it may modify the drag coefficient C_D [34]. At least for CIT, the amount of total energy represented by $\langle \Theta_p \rangle$ is relatively small as compared to k_p (i.e., approximately 10% of the turbulent kinetic energy [32]).

4.4 Turbulent Kinetic Energy Balances

The TKE and TDR are found by solving four differential equations:[6]

$$\frac{\rho_f \langle \alpha_f^\star \rangle}{\langle \rho_e \alpha_p^\star \rangle} \frac{d\hat{k}_f}{d\hat{t}} = 2(\hat{k}_p \hat{k}_f)^{1/2} - 2\hat{k}_f + \hat{C}_g \gamma(\hat{t}) - \frac{\rho_f \langle \alpha_f^\star \rangle}{\langle \rho_e \alpha_p^\star \rangle} \hat{\varepsilon}_f, \tag{25}$$

$$\frac{d\hat{k}_p}{d\hat{t}} = 2(\hat{k}_p \hat{k}_f)^{1/2} - 2\hat{k}_p + \hat{C}_b \gamma(\hat{t}) - \hat{\varepsilon}_p, \tag{26}$$

$$\frac{\rho_f \langle \alpha_f^\star \rangle}{\langle \rho_e \alpha_p^\star \rangle} \frac{d\hat{\varepsilon}_f}{d\hat{t}} = 2C_{\varepsilon,3} [(\hat{\varepsilon}_p \hat{\varepsilon}_f)^{1/2} - \hat{\varepsilon}_f] + C_{\varepsilon,4} \hat{C}_g \gamma(\hat{t}) - \frac{\rho_f \langle \alpha_f^\star \rangle}{\langle \rho_e \alpha_p^\star \rangle} C_{\varepsilon,2} \hat{\omega}_f \hat{\varepsilon}_f, \tag{27}$$

[6] Only algebraic equations are required at steady state, but the differential equations provide information concerning the stability and relaxation behavior of the solution.

$$\frac{d\hat{\varepsilon}_p}{d\hat{t}} = 2C_{\varepsilon,3}[(\hat{\varepsilon}_p\hat{\varepsilon}_f)^{1/2} - \hat{\varepsilon}_p] + C_{\varepsilon,4}\hat{C}_b\gamma(\hat{t}) - C_{\varepsilon,2}\hat{\omega}_p\hat{\varepsilon}_p \tag{28}$$

where $\hat{C}_g = C_g/C_d$ and $\hat{C}_b = C_b/C_d$ (which are non-negative, depend on Z, and sum to unity) modify the split of the TKE between the phases. The dimensionless variables are defined by

$$\hat{t} = \frac{t}{\tau_p^\star}, \quad \hat{k} = \frac{k}{C_d V_t^2}, \quad \hat{\varepsilon} = \frac{\tau_p^\star \varepsilon}{C_d V_t^2}, \quad \hat{\omega} = \frac{\hat{\varepsilon}}{\hat{k}}, \quad \gamma(\hat{t}) = \frac{\langle u_{pf}\rangle^2}{V_t^2} \tag{29}$$

where $\tau_p^\star$ is defined by (23). The forcing term $\gamma(\hat{t})$ is constant at steady state, but can be made to vary with time by making $g(t)$ time dependent in (19), e.g., with periodic forcing $g(t) = g\cos(\omega_g \hat{t})$.

The steady-state solution depends on the parameter ratios $R_3 = C_{\varepsilon,3}/C_{\varepsilon,2}$ and $R_4 = C_{\varepsilon,4}/C_{\varepsilon,2}$. In general, if $R_3 = 1$ then the steady-state solution gives $\hat{\omega}_p = \hat{\omega}_f$, and hence $\hat{\varepsilon}_f = \hat{\omega}\hat{k}_f$ and $\hat{\varepsilon}_p = \hat{\omega}\hat{k}_p$ where $\hat{\omega} = R_4$ is the shared frequency. Then, for $Z = 1$, if $\hat{C}_g(1) = \langle\alpha_f^\star\rangle$ and $\hat{C}_b(1) = \langle\alpha_p^\star\rangle$, the steady-state solution will be $\hat{k}_f = \hat{k}_p = 1/R_4$. In general, C_d is used to control the mean-slip velocity, thus the magnitude of the TKE can controlled using R_4.[7] Then the TKE ratio $R_{TKE} = \hat{k}_f/\hat{k}_p$ for other values of Z is controlled by the model for $\hat{C}_g(Z)$ (given that $\hat{C}_b(Z) = 1 - \hat{C}_g(Z)$). For example, for $1 \ll Z$ it is known that $R_{TKE} < 1$ but nearly constant for fixed $\langle\alpha_p\rangle$, e.g., $R_{TKE} \approx 0.696$ for $\langle\alpha_p\rangle = 0.5$ and $Z = 1000$ [21]. On the other hand, for $Z \approx 0$, $R_{TKE} > 1$, e.g., $R_{TKE} \approx 1.66$ for $\langle\alpha_p\rangle = 0.01$ and $Z = 0.001$ [32].

In summary, the default dissipation parameters are $C_{\varepsilon,3} = C_{\varepsilon,2} = 1.92$ and $C_{\varepsilon,4} = R_4 C_{\varepsilon,2}$ where R_4 can be set to control the level of the TKE. The parameters $C_b(Z)$ and $C_g(Z)$ have the property $C_b + C_g = C_d$ where C_d fixes the mean-slip velocity and depends on Z, Re_p and $\langle\alpha_p\rangle$. The ratio $\hat{C}_g(Z)$ must have the properties $\hat{C}_g(1) = \langle\alpha_f^\star\rangle$ and $\hat{C}_g(0) = 1$. For other values of Z, $\hat{C}_g(Z)$ is chosen to fix the TKE ratio R_{TKE}, but must always be non-negative and less than unity. Also, for simplicity, the drag-exchange term for TKE and TDR is written here with the correlation constants βk and β_ε (see [26]) equal to one. At least for CIT with $\hat{C}_g = 1$, other values may be needed for these parameters to get the correct R_{TKE}. As in other dynamical systems, it would be informative to compare the model to Euler–Lagrange or two-fluid simulations with periodic forcing, i.e., $g(t) = g\cos(\omega_g \hat{t})$, for different values of the forcing frequency ω_g. For a given Z, the dynamical response of R_{TKE} would provide valuable information concerning R_3 and R_4 that is indiscernible from the steady-state solution.

[7] Adjusting R_4 is physically equivalent to adjusting the timescale for the cluster-induced turbulence relative to τ_p as defined in (23). For bubbly flow with $1 \ll Z$ the predicted value of k_f is very small when $R_4 = 1$. However, since part of the fluid travels with the particles, k_f is less than the TKE of fluid reported in [21].

4.5 Reynolds Stresses

Turning now to the Reynolds stresses, in the coordinate system with the x-axis aligned opposite of $\mathbf{g}$ (i.e., $\mathbf{e}_x \cdot \mathbf{g} = -g$), the stress tensor will be diagonal with $R^\star_{xx} > R^\star_{yy} = R^\star_{zz}$. Thus, for each phase there are only two independent components and $2k^\star = R^\star_{xx} + 2R^\star_{yy}$. The mean-slip-production tensor P_{fp} has only the xx-component nonzero and is equal to $\langle u_{fp}\rangle^2$. The dimensionless isotropization tensors are traceless and have the form

$$\hat{\Phi}_f = 2C_{1f}\hat{\varepsilon}_f\left(\tilde{\mathsf{R}}_f - \frac{1}{3}\mathsf{I}\right) + 2C_{fp}\hat{C}_g\gamma(\hat{t})\frac{\langle\rho_e\alpha^\star_p\rangle}{\rho_f\langle\alpha^\star_f\rangle}\left(\mathbf{e}_x\mathbf{e}_x - \frac{1}{3}\mathsf{I}\right) \tag{30}$$

$$\hat{\Phi}_p = 2C_{1p}\hat{\varepsilon}_p\left(\tilde{\mathsf{R}}_p - \frac{1}{3}\mathsf{I}\right) + 2C_{fp}\hat{C}_b\gamma(\hat{t})\left(\mathbf{e}_x\mathbf{e}_x - \frac{1}{3}\mathsf{I}\right) \tag{31}$$

where $\tilde{\mathsf{R}} = \mathsf{R}^\star/(2k^\star)$ is diagonal with unit trace. The first term reorients TKE from the xx-component towards the other diagonal components, while the second term makes the mean-slip-production term more isotropic. The dimensionless dissipation tensors are

$$\hat{\epsilon}_f = 2\left(C_f\tilde{\mathsf{R}}_f + \frac{1}{3}(1-C_f)\mathsf{I}\right)\hat{\varepsilon}_f \quad \hat{\epsilon}_p = 2\left(C_p\tilde{\mathsf{R}}_p + \frac{1}{3}(1-C_p)\mathsf{I}\right)\hat{\varepsilon}_p \tag{32}$$

where $tr(\hat{\epsilon}) = 2\hat{\varepsilon}$. As in single-phase turbulence, the term involving C_f (C_p) in (32) can be combined with the C_1 term in Φ, hence we can let $C_f = C_p = 0$.[8]

With these definitions, the Reynolds stresses are found from

$$\frac{\rho_f\langle\alpha^\star_f\rangle}{\langle\rho_e\alpha^\star_p\rangle}\frac{d\hat{\mathsf{R}}_f}{d\hat{t}} = 2(\hat{k}_f\hat{k}_p)^{1/2}\mathsf{R}_{fp} - 2\hat{\mathsf{R}}_f + 2\hat{C}_g\gamma(\hat{t})\mathbf{e}_x\mathbf{e}_x - \frac{\rho_f\langle\alpha^\star_f\rangle}{\langle\rho_e\alpha^\star_p\rangle}(\hat{\Phi}_f + \hat{\epsilon}_f) \tag{33}$$

$$\frac{d\hat{\mathsf{R}}_p}{d\hat{t}} = 2(\hat{k}_f\hat{k}_p)^{1/2}\mathsf{R}_{fp} - 2\hat{\mathsf{R}}_p + 2\hat{C}_b\gamma(\hat{t})\mathbf{e}_x\mathbf{e}_x - (\hat{\Phi}_p + \hat{\epsilon}_p) \tag{34}$$

where $tr(\hat{\mathsf{R}}) = 2\hat{k}$ and R_{fp} is defined such that $tr(\mathsf{R}_{fp}) = 2$:

$$\mathsf{R}_{fp} = 2(\tilde{R}^{1/2}_{xx} + 2\tilde{R}^{1/2}_{yy})^{-1}diag(\tilde{R}^{1/2}_{xx}, \tilde{R}^{1/2}_{yy}, \tilde{R}^{1/2}_{yy}), \tag{35}$$

and $\tilde{R}_{xx}$, $\tilde{R}_{yy}$ are the two independent components of the diagonal matrix $\tilde{\mathsf{R}}_f\tilde{\mathsf{R}}_p$. Given the solution for the TKE and TDR, the xx-component of the relations in (33) and (34) can be solved to find $\hat{R}_{xx}$ for each phase. Note that the drag-exchange terms couple the two phases and work to equalize the anisotropy between phases. In contrast, the mean-slip-production term is highly anisotropic and would drive the system to $\tilde{R}_{xx} = 1$ if not for the isotropization term $\hat{\Phi}$. For example, with $C_{fp} = 1$ the production terms will be isotropic so that $\tilde{R}_{xx} = 1/3$.

[8] For particle-laden flows, the particle-pressure tensor is highly anisotropic [35]. In general, if a full second-order model is used for the particle phase [36], nonzero C_p is required to predict this anisotropy [32].

For the isotropization term, one can first set C_1 to the standard value for single-phase turbulence [28], and then use $0 \leq C_{fp} < 1$ to achieve the correct value for $\tilde{R}_{zz}$. This statement assumes that with $C_{fp} = 0$, $\tilde{R}_{xx}$ is too large when the standard value of C_1 is used. If this is not the case, then C_1 must be decreased to get the desired value for $\hat{R}_{xx}$.

Table 6. Steady-state predictions from the turbulence model with default values of parameters in Table 5. The particle volume fractions are chosen to correspond to CIT [32] and BDT [21]. The statistics $\tilde{R}$ are the xx component of the Reynolds-stress tensor scaled by $2\hat{k}$. For the default parameters $\hat{\varepsilon} = \hat{k}$ and dimensionless mean-slip velocity is unity.

Z	$\langle \alpha_p \rangle$	$\hat{k}_f$	$\hat{k}_p$	R_{TKE}	$\tilde{R}_f$	$\tilde{R}_p$
0.001	0.01	1.307	0.588	2.223	0.871	0.750
1	0.1	0.154	0.154	1	0.778	0.778
1000	0.5	0.403	0.569	0.708	0.752	0.797
1000	0.1	0.057	0.102	0.559	0.774	0.818

4.6 Example Results

To show predictions from the multiphase turbulence model, we will use the default parameters and three values for Z, namely, 0.001, 1, and 1000. The first corresponds to gas–particle flow, the second to buoyancy-neutral flow, and the third to bubbly flow. For the default parameters, the steady-state frequencies are $\hat{\omega}_p = \hat{\omega}_f = 1$. Steady-state results for the other statistics are given in Table 6. As expected, for $Z = 1$ both phases have the same values for all statistics. We should mention that for $Z = 1$, we find $V_t = 0$ so that the dimensional statistics are null. As expected, for gas–particle flow $\hat{k}_f \gg \hat{k}_p$ and $\tilde{R}_f$ is near 0.9 [32]. However, R_{TKE} is larger than observed in [32]. Likewise, for bubbly flow $\hat{k}_f < \hat{k}_p$ [21]. Here, R_{TKE} for $\langle \alpha_p \rangle = 0.5$ is close to the value observed in [21]. These trends validate the structure of the mathematical model and its dependence on Z. Quantitative agreement with direct-simulation data can be obtained by optimizing the parameters around their default values in Table 5. For bubbly flow, no turbulence is observed for $\langle \alpha_p \rangle \leq 0.3$ in direct simulations [21]. The turbulence model predicts a much lower, but nonzero, value for the dimensionless TKE. However, recall that $C_d = 0$ when the flow is non-turbulent (i.e., when only pseudoturbulence present [37]) and, hence, the dimensional TKE $k^\star$ will be null. In other words, the transition to turbulence must be incorporated into the drag closure for C_d.

An example of the time-dependent turbulence model predictions for gas–particle flow with periodic forcing $g(\hat{t}) = \cos(\omega_g \hat{t})$ and $\omega_g = \pi/3$ is shown in Fig. 1. In the top panel, u_p is scaled such that the steady-state value is unity. At the chosen frequency, the mean-slip velocity lags behind the forcing and has maximum magnitude of approximately 0.5. In the second panel, as expected, the TKE oscillates at twice the frequency of $g(\hat{t})$ and its magnitude is significantly lower than the steady-state value in Table 6. In the third panel, the TKE and TDR ratios are seen to vary significantly with time, as is the case for $\hat{\omega}$ in the bottom panel. It is noteworthy that the phasic $\hat{\omega}$ are not equal, and most of the time are below the steady-state value of unity. These differences are due to the fact that the characteristic timescale for the fluid from (16) depends on the value of Z. The results for $Z \gg 1$ follow similar trends to those in Fig. 1. From the point of view of fitting the model parameters, the periodic-forcing data are much richer than the steady-state data. It would therefore be useful to obtain such data from direct simulations for model validation. Finally, we can note that the dissipation parameter $C_{\varepsilon,2}$ can be fit to direct-simulation data by starting from steady state and observing the decay rate of $\hat{\varepsilon}$ with $g = 0$.

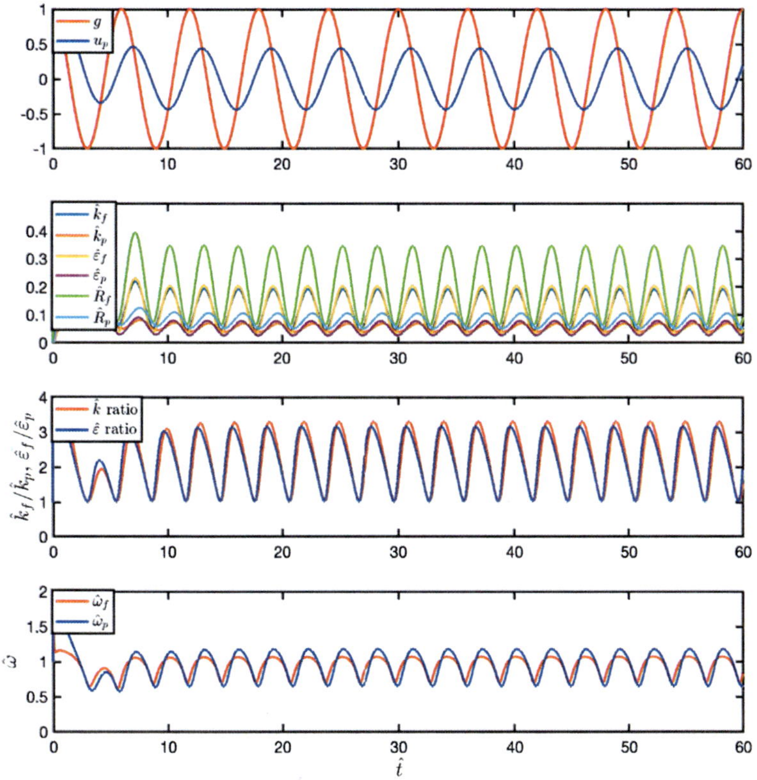

Fig. 1. Example predictions from multiphase turbulence model for CIT with $Z = 0.001$, $\langle \alpha_p \rangle = 0.01$, $\omega_g = \pi/3$, and default parameters in Table 5. All model variables are dimensionless and R_f and R_p are the xx-component of the Reynolds stresses. **top** $g(\hat{t})$ and $u_p(\hat{t})$. **middle-top** $\hat{k}_f(\hat{t})$, $\hat{k}_p(\hat{t})$, $\hat{\varepsilon}_f(\hat{t})$, $\hat{\varepsilon}_p(\hat{t})$, $\hat{R}_f(\hat{t})$, $\hat{R}_p(\hat{t})$. **middle-bottom** $R_{TKE}(\hat{t})$, $R_{TDR}(\hat{t})$. **bottom** $\hat{\omega}_f(\hat{t})$, $\hat{\omega}_p(\hat{t})$

5 Conclusions

Starting from a hyperbolic two-fluid model for monodisperse multiphase flow consisting of a compressible fluid and incompressible spherical particles, a multiphase turbulence model is proposed based on Reynolds averaging. In order to allow for arbitrary material-density ratios, the two-fluid model includes the added mass due to the fluid in particle wakes, which travels with the velocity of the particles. The fluid phase is thus divided into a bulk fluid and the added mass, which travel at different velocities. As shown in previous work [18], this two-fluid formulation involving seven equations is globally hyperbolic. By carefully formulating the closures in the turbulence model, the latter retains the same hyperbolicity characteristics as the original two-fluid model. However, the phasic speeds of sound are altered due to the 'turbulent pressure' associated with the phasic turbulent kinetic energies.

In addition to the seven conserved variables, the turbulence model includes four additional balance equations for the fluid- and particle-phase Reynolds stresses and turbulence dissipation rates. In the target application—gravity-driven turbulence—the Reynolds-stress tensors are diagonal but anisotropic. Thus, they do not produce 'turbulent-viscosity' terms in the mean momentum balances as in classical mean-shear flows. For this reason, the characteristic timescales in the hyperbolic multiphase turbulence model are similar to the original two-fluid model (albeit with a higher speed of sound due to the turbulent kinetic energy), and solutions to the multiphase turbulence model for real applications—such as particle-laden riser flow—will not be time independent. Nonetheless, 'turbulent diffusion' of turbulent kinetic energy and its dissipation will be present. Together with the higher speed of sound in the particle phase, this should make it possible to achieve accurate time-dependent solutions on coarser grids than would be the case with the original two-fluid model.

The closures employed in the balances for turbulent kinetic energy were, for the most part, originally proposed in [26]. Both drag and mean-shear production are included in these balances, and the form of the drag-exchange term has been discussed in detail in [26]. The balances for the turbulent dissipation rates follow the same formulation, but we have introduced a slight (but important) change in the drag-production term. Contrary to what was done in [26], the characteristic timescale for drag production in the dissipation equation is τ_p. This change was made to allow for stationary solutions for homogeneous gravity-driven turbulence, and is consistent with the treatment of pseudoturbulence in [18]. We have also shown that the dissipation parameters must be made to depend on Z (i.e., the material-density ratio) in order to reproduce the behavior seen in CIT and BDT (and the anticipated equality of the turbulence statistics for $Z = 1$). Nevertheless, further work is needed to determine how these parameters depend on the other properties of the flow (e.g., mass loading, particle Reynolds number, etc.).

For the Reynolds-stress balances, the closures for the isotropization tensors are the simplest linear-relaxation models from the literature. This choice was made because data are lacking to validate more complex closures involving nonlinear terms. The proposed closures for the drag-exchange terms for the Reynolds stresses are novel. First, their trace yields the correct expression for the drag-exchange model for the turbulent kinetic energies. Second, their mathematical forms are analogous to the latter (e.g., linear terms keep the same form but replace the scalar with the tensor). Thus, the most speculative closure is the tensor R_{fp} towards which the phasic Reynolds stresses relax due to drag coupling. Obviously, this tensor is not unique, but its form (e.g., symmetric, positive-definite) can be tested for gravity-driven and mean-shear flows in future work.

In summary, we have presented a 'basic' multiphase turbulence model for compressible disperse multiphase flows. This model has favorable mathematical properties (e.g., hyperbolic, realizable, etc.) that will make its numerical implementation in compressible flow codes very robust. Future work will be devoted to validating the closures and extending them as needed to account for multiphase turbulence physics in systems with both gravity-driven and mean-shear turbulence production.

References

1. Theofanous, T.G., Sullivan, J.: J. Fluid Mech. **116**, 343 (1982)
2. Besnard, D.C., Harlow, F.H.: Int. J. Multiphase Flow **14**(6), 679 (1988)
3. Champney, J.M., Dobrovolskis, A.R., Cuzzi, J.N.: Phys. Fluids **7**(7), 1703 (1995)
4. Bi, H.T., Ellis, N., Abba, I.A., Grace, J.R.: Chem. Eng. Sci. **55**, 4789 (2000)
5. Sundaresan, S.: AIChE J. **46**(6), 1102 (2000)
6. Krause, M.: New Astron. Rev. **51**, 174 (2007)
7. Doronzo, D.M., de Tullio, M.D., Pascazio, G.: Comput. Fluids **44**, 56 (2011)
8. Colombo, M., Fairweather, M.: Int. J. Multiphase Flow **77**, 222 (2015)
9. Besagni, G., Guédon, G.R., Inzoli, F.: Nucl. Eng. Des. **331**, 222 (2018)
10. Sun, Z., Zhu, J.: AIChE J. **65**(9), 1 (2019)
11. Février, P., Simonin, O., Squires, K.D.: J. Fluid Mech. **533**, 1 (2005)
12. Bosse, T., Kleiser, L., Meiburg, E.: Phys. Fluids **18**, 027102 (2006)
13. Loisel, V., Abbas, M., Masbernat, O., Climent, E.: Phys. Fluids **25**, 123304 (2013)
14. Picano, F., Breugem, W.P., Brandt, L.: J. Fluid Mech. **764**, 463 (2015)
15. Huck, P.D., Bateson, C., Volk, R., Cartellier, A., Bourgoin, M., Aliseda, A.: J. Fluid Mech. **846**, 1059 (2018)
16. Schneiders, L., Fröhlich, K., Meinke, M., Schröder, W.: J. Fluid Mech. **875**, 520 (2019)
17. Muramulla, P., Tyagi, A., Goswami, P.S., Kumaran, V.: J. Fluid Mech. **889**, A28 (2020)
18. Fox, R.O., Vié, A., Laurent, F.: J. Fluid Mech. **903**, A5 (2020)
19. Boniou, V., Fox, R.O., Laurent, F.: J. Computat. Phys. (2023). HAL preprint https://hal.science/hal-04037062
20. Capecelatro, J., Desjardins, O., Fox, R.O.: J. Fluid Mech. **747**, R2 (2014)
21. Panicker, N., Passalacqua, A., Fox, R.O.: Chem. Eng. Sci. **216**, 115546 (2020)
22. Fox, R.O.: J. Fluid Mech. **877**, 282 (2019)
23. Chassagne, R., Bonamy, C., Chauchat, J.: J. Fluid Mech. **964**, A27 (2023)
24. Boniou, V., Fox, R.O.: Int. J. Multiphase Flow **169**, 104591 (2023)
25. Toro, E.F.: Riemann Solvers and Numerical Methods for Fluid Dynamics: A Practical Introduction, 3rd edn. Springer, Heidelberg (2009)

26. Fox, R.O.: J. Fluid Mech. **742**, 368 (2014)
27. Fox, R.O.: Multiphase turbulence. In: Advanced Approaches in Turbulence. Theory, Modeling, Simulation, and Data Analysis for Turbulent Flows, pp. 307–371. Elsevier (2021)
28. Pope, S.B.: Turbulent Flows. Cambridge University Press, Cambridge (2000)
29. Fox, R.O.: Annu. Rev. Fluid Mech. **44**, 47 (2012)
30. Capecelatro, J., Desjardins, O., Fox, R.O.: J. Fluid Mech. **845**, 499 (2018)
31. Schneiderbauer, S.: AIChE J. **63**(8), 3544 (2017)
32. Capecelatro, J., Desjardins, O., Fox, R.O.: J. Fluid Mech. **780**, 578 (2015)
33. Baker, M.C., Fox, R.O., Kong, B., Capecelatro, J., Desjardins, O.: Phys. Rev. Fluids **5**, 074304 (2020)
34. Mehrabadi, M., Tenneti, S., Garg, R., Subramaniam, S.: J. Fluid Mech. **770**, 210 (2015)
35. Capecelatro, J., Desjardins, O., Fox, R.O.: Phys. Fluids **28**(3), 033306 (2016)
36. Capecelatro, J., Desjardins, O., Fox, R.O.: Phys. Fluids **28**(3), 033307 (2016)
37. Risso, F.: Annu. Rev. Fluid Mech. **50**, 25 (2018)

Persistence of Logarithmic Temperature Profile in Unstably Stratified Atmospheric Boundary Layers with and Without Sand

Junning Wang[1], Jin-Han Xie[1,2(✉)], and Xiaojing Zheng[1,3]

[1] Department of Mechanics and Engineering Science at College of Engineering and State Key Laboratory for Turbulence and Complex Systems, Peking University, Beijing, China
jnwang@pku.edu.cn

[2] Joint Laboratory of Marine Hydrodynamics and Ocean Engineering, Laoshan Laboratory, Shandong 266237, People's Republic of China
jinhanxie@pku.edu.cn

[3] Center for Particle-Laden Turbulence, Lanzhou University, Lanzhou 730000, People's Republic of China
xjzheng@xidian.edu.cn

Abstract. Monin-Obukhov similarity theory (MOST) is a widely used framework in atmospheric boundary layer (ASL) research. MOST predicts that the mean potential temperature profile deviates from the logarithmic law in buoyancy-dominated turbulence. However, recent studies show that the logarithmic profile remains intact in unstably stratified atmospheric boundary layers. Using data from the Qingtu Lake Observation Array(QLOA) with Reynolds number up to O(106), we investigate the similarity functions and mean temperature profile with different stability parameters and sand-bearing conditions. By assessing the accuracy of the logarithmic profile and the MOST-based empirical expressions obtained by Hogstrom and Wilson, we discover that the logarithmic law is a better fit than the MOST expressions. The Von Kármán constant of the potential temperature profile has a power function dependences on the stability parameter. In sand-laden ASLs, it is remarkable to find that the logarithmic law still holds, and the error of the MOST expressions amplifies. The von Kármán constant of potential temperature increases in sandiness conditions. Still, a quantitative theory that describes the sand effect on the mean temperature profile remains to be studied.

Keywords: Monin-Obukhov similarity theory · unstable boundary layer · potential temperature profile

1 Introduction

The logarithmic mean velocity profile in the near-wall region of boundary-layer turbulence has been observed in laboratory and field observations and numerical

X. Zheng and S. Balachandar (Eds.): IUTAM 2023, IUTAM Bookseries 41, pp. 141–151, 2024.
https://doi.org/10.1007/978-3-031-47258-9_8

simulations [16,19,20]. Similarly, the logarithmic profile for mean temperature was also observed [15,25]. Kader [13] pointed out that the average temperature profile in wall turbulence satisfies the law of the wall. However, when buoyancy affects the flow field, the universal log law is insufficient for accurately describing the mean velocity [11,26] and temperature profiles [7,17,24,27].

Monin-Obukhov similarity theory (MOST) [21,23] is frequently used to account for buoyancy effects. MOST suggests that four parameters govern the quasi-steady turbulence structure in the Atmospheric Surface Layer (ASL): the distance z from the surface, the friction velocity u_τ (which is the square root of the magnitude of the mean kinematic surface stress), the mean temperature flux at the surface Q_0, and the buoyancy parameter g/θ_0, where θ_0 is a characteristic potential temperature. Combining the dependent variables yields dimensionless parameters. The Monin-Obukhov length is defined as $L = -(u_\tau)^3 \theta_0/\kappa g Q_0$ [22], where κ is the von Kármán constant. Thus, z/L is defined as the stability parameter. MOST implies that the mean velocity and potential temperature gradients are expressed as:

$$\frac{\kappa z}{u_\tau}\frac{\partial U}{\partial z} = \phi_m\left(\frac{z}{L}\right), \tag{1a}$$

$$-\frac{\kappa z u_\tau}{Q_0}\frac{\partial \Theta}{\partial z} = \frac{\kappa z}{\theta_\tau}\frac{\partial \Theta}{\partial z} = \phi_h\left(\frac{z}{L}\right), \tag{1b}$$

where U is the mean velocity and Θ is the mean potential temperature at the same height. ϕ_m and ϕ_h are Monin-Obukhov similarity functions. Many different forms of similarity functions have been proposed [3,4,9,14,31], and they depart from the logarithmic profiles. Particularly, Hogstrom [9] suggested:

$$\phi_h = \begin{cases} 1 + 7.8\dfrac{z}{L}, & \dfrac{z}{L} > 0, \\ \left(1 - 12\dfrac{z}{L}\right)^{-1/2}, & \dfrac{z}{L} < 0. \end{cases} \tag{2}$$

Wilson [31] proposed

$$\phi_h = \left(1 + 7.9\left|\frac{z}{L}\right|^{2/3}\right)^{-1/2} \tag{3}$$

for unstable conditions. The expression of law of the wall [28] is

$$\frac{u}{u_\tau} = \frac{1}{\kappa}\ln\frac{z}{\delta_\nu} + C, \tag{4}$$

where $\delta_\nu = \nu/u_\tau$ is the viscous length , which defines the Kármán constant.

However, recent studies find the logarithmic profile remains intact in unstably stratified atmospheric boundary layers [1,2,8,10]. It has been shown that in turbulent flows driven by both shear and buoyancy, the mean temperature profile may deviate from the traditional power law observed in Rayleigh-Bénard Convection [1,2] and natural convection [10]. Instead, a logarithmic temperature profile with varying slopes has been reported to be more prevalent than the velocity log law, and they are observed even in extremely buoyant flows [1]. These findings motivate further examination of the existence of a temperature log law in such turbulent flows. Johansson et al. [12] found that the similarity functions also depend on z_i/L, where z_i is the convective boundary layer, by performing large eddy simulations and studying field measurements. Cheng et al. [5] investigates the existence and slope of the logarithmic temperature profile in turbulent flows driven by both shear and buoyancy, such as the unstably stratified atmospheric boundary layer. Their results suggest that the buoyancy force does not modify the logarithmic nature of the mean potential temperature profile but modulates its slope instead.

Inspired by this work, we verified this conclusion under different Reynolds numbers and sediment conditions compared to theirs, examining the variables that contribute to the expression of temperature profiles. We compared the accuracy of the logarithmic profile and the MOST-based empirical expressions obtained by Hogstrom and Wilson based on the data of QLOA [18,29,30].

This paper is arranged as follows. Section 2 introduces the observation facility, data selection and pretreatment. Section 3 contains the similarity functions, mean velocity and potential temperature profiles, comparison of logarithmic profile and MOST expressions, and the slopes of the profiles vary with different abscissas. Section 4 gives the conclusions.

2 Experimental Facility and Data Details

This study utilizes data collected from the QLOA site situated on the flat and arid Qingtu Lake in western China. The observation array comprises of 21 towers, with the main tower standing at a height of 32 m and 20 smaller towers at 5 m height, all capable of measuring the three components of wind speed and dust concentration simultaneously. To measure wind speed and temperature, eleven acoustic anemometers (Campbell scientific, CSAT-3B) were mounted on the main tower between 0.9 and 30 m. The temperature measurement range is from −40°C to 60°C, with a minimum resolution of 1°C and less than 1°C absolute error.

Due to the complexity and unpredictability of environmental conditions during field observations, specific selection and preprocessing of original data were performed to ensure accurate representation of the turbulent boundary layer at high Reynolds numbers. The pretreatment involved stratification stability judgment and classification, wind direction adjustment, and long-term trend filtering. Additionally, quality verification was conducted to ensure that the wind data were statistically stable. The non-stationary index was calculated based on the study by Foken et al. [6] to identify high-quality data with $\mathrm{IST}_u < 30\%$ and $\mathrm{IST}_\theta < 30\%$. The definition of IST_u and IST_θ is

$$\mathrm{IST}_u = \left|\left(\mathrm{CV}_{um} - \mathrm{CV}_{u1h}\right)/\mathrm{CV}_{u1h}\right| \times 100\%, \tag{5}$$

where $\mathrm{CV}_{um} = \sum_{i=1}^{12} \mathrm{CV}_{ui}/12, \mathrm{CV}_{u1}, \mathrm{CV}_{u2}, \ldots, \mathrm{CV}_{u12}$ are the streamwise velocity variances every 5 min in 1 h, and CV_{u1h} is the total streamwise velocity variance for the hour.

$$\mathrm{IST}_\theta = |(\mathrm{CV}_{\theta m} - \mathrm{CV}_{\theta 1\,h})/\mathrm{CV}_{\theta 1\,h}| \times 100\%, \tag{6}$$

with $\mathrm{CV}_{\theta m} = \sum_{i=1}^{12} \mathrm{CV}_{\theta i}/12, \mathrm{CV}_{\theta 1}, \mathrm{CV}_{\theta 2}, \ldots, \mathrm{CV}_{\theta 12}$ are the potential temperature variances every 5 min in 1 h, and CV_{u1h} is the total potential temperature variance for the hour.

Table 1. Main parameters of the selected representative wall flow data intervals. The Re_τ of the flow is $O(10^6)$. ν is the kinematic viscosity coefficient and α is the thermal diffusion coefficient. The z/L here is calculated with the minimum experimental height 0.9m. The PM10 measured at 5m is used to reflect the amount of sediment concentration.

No	Time and date	u_τ (ms^{-1})	θ_τ (K)	Re_τ $(\times 10^6)$	ν $(\times 10^{-5} m^2 s^{-1})$	α $(\times 10^{-5} m^2 s^{-1})$	z/L $(\times 10^{-3})$	PM10 $(mg \cdot m^{-}3)$
1	2016.3.19 12:00–13:00	0.43	−0.31	4.37	1.47	2.05	−20.6	–
2	2017.3.27 17:00–18:00	0.46	−0.15	4.67	1.47	2.04	−9.4	0.023
3	2016.3.28 10:00–11:00	0.27	−0.23	2.81	1.46	2.03	−21.3	0.075
4	2016.3.28 17:00–18:00	0.16	−0.34	1.64	1.48	2.06	−90.6	0.060
5	2016.5.14 8:00–9:00	0.48	−0.12	4.96	1.45	2.02	−3.3	0.432
6	2017.3.27 10:00–11:00	0.34	−0.21	3.54	1.44	2.01	−12.9	0.050

High-quality data with $\mathrm{IST}_u < 30\%, \mathrm{IST}_\theta < 30\%$ are selected in the work. Unstable stratified data is mainly studied. The used data were measured from 2015 to 2017 and can be classified into two types: sand-laden flow data observed during sandstorm weather and clean air data measured in clean air with little sand.

As shown in Table 1, several intervals are selected to analyze the data. No.1 is a clean air data and N0.2–N0.6 are data measured in sandstorm. $\theta_\tau = Q_0/u_\tau$ is the friction temperature. Negative z/L and θ_τ corresponds to an unstable situation. The kinetic viscosity coefficient and thermal diffusion coefficient of them have little difference. The key distinction is the stability parameter and sediment concentration. The measured data of 0.9m and 10.24m is applied in the calculation of z/L and PM10 separately.

From Fig. 1, the kinetic flux and potential temperature flux are relatively steady with elevation change, which indicates the defined characteristic parameters u_τ and θ_τ can be applicable to each height for both sets of data. No significant change is observed due to the presence of sediment.

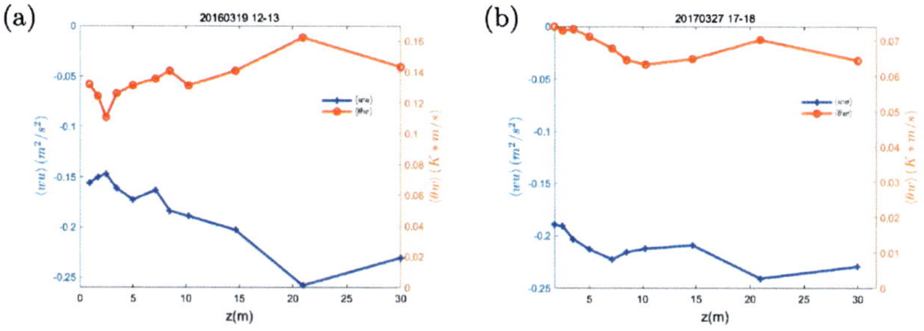

Fig. 1. $\langle wu \rangle$ and $\langle w\theta \rangle$ of clean air and sand-laden flow. Q_0 equals $\langle w\theta \rangle$ on the surface. (a) The kinetic surface flux and potential temperature flux of different heights in clean air with u_τ and θ_τ on the top right corner The blue line is $\langle wu \rangle$ while the red line is $\langle w\theta \rangle$. (b) The flux in a sand-laden flow.

3 Results

3.1 Similarity Functions

Figure 2 shows the similarity functions of the measured data and verifies the expressions of Businger [3] and Hogstrom [9]. Businger suggests the approximate expression is:

$$\phi_h = \begin{cases} 0.74 + 4.7\dfrac{z}{L}, & \dfrac{z}{L} > 0, \\ 0.74\left(1 - 9\dfrac{z}{L}\right)^{-1/2}. & \dfrac{z}{L} < 0. \end{cases} \tag{7}$$

This work focus on the unstable condition, so only the negative z/L data is plotted. In Fig. 2, measured value of ϕ_h at several heights and different time intervals are plotted. ϕ_h of different heights at the same time is close to each other. Hogstrom's formula exhibits better performance than Businger's, but discrepancies with our experimental data still exists. The predicted values are widely underestimated but the general trend is consistent with the empirical formula. The similarity functions for the sand-laden and clean air cases are essentially equivalent.

3.2 The Potential Temperature Profile

Similar to Eq. (4), the potential temperature profile expression is:

$$\frac{\Theta - \Theta_0}{\theta_\tau} = \frac{1}{\kappa_\theta} \ln \frac{z}{z_r}, \tag{8}$$

where Θ is the mean potential temperature, Θ_0 is the mean potential temperature at the lowest height, κ_θ is the potential temperature Kármán constant, and z_r is the reference height.

Figure 3 is the potential temperature profile. The horizontal axis is $z^+ = z/\delta_\nu = zu_\tau/\nu$ and the vertical axis is $\Theta^+ = (\Theta - \Theta_0)/\theta_\tau$. The predicted profile

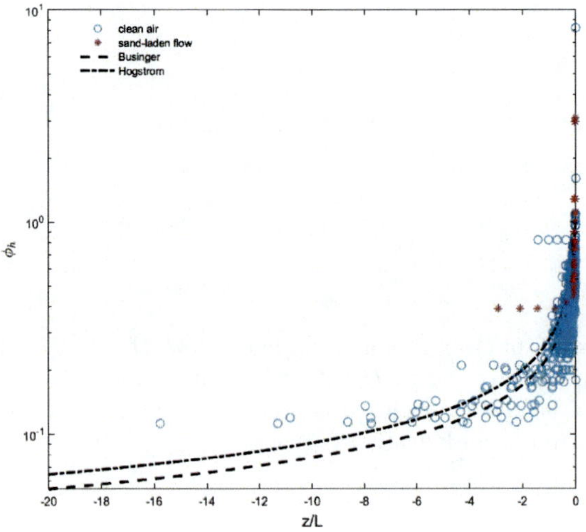

Fig. 2. Similarity function ϕ_h of clean air and sand-laden flow. The circles are the clean air data and the asterisks are the sand-laden data. The dotted line is Businger's (1971) expression and the dot dash line is Hogstrom's(1988) expression. The figure has a logarithmic ordinate.

of Hogstrom and Wilson deviates more from the measured data than the clean air situation.

In unstable conditions, the mean temperature grows with height. It can be seen from the left figure that both the three estimates of the profile under the clear wind flow are in good agreement with the actual measured data. Under the sand-laden flow, the measured data has a larger fluctuation and the two empirical expressions deviate more from the logarithmic profile. The potential temperature profiles in sand-laden flow have larger fluctuation than clean air, but they still satisfy the logarithmic profile in general.

3.3 The Root-Mean-Square Error (RMSE) Comparison of Log Law and MOST-Based Empirical Equations

To assess the accuracy of the logarithmic law equation and the MOST-based empirical equations, we computed the root mean square error (RMSE) between the measured temperature (Θ_i) and the estimated temperature ($\hat{\Theta}_i$). The RMSE is defined as:

$$\sigma_{l,H,W} = \sqrt{\frac{1}{n}\sum_{i=1}^{n}(\Theta_i - \hat{\Theta}_i)^2}, \tag{9}$$

where n represents the total number of data points. The subscripts l, H, and W correspond to the logarithmic law expression, Hogstrom's expression, and Wilson's expression, respectively. A smaller value of σ indicates a lower error, indicating a better fit between the estimated and measured temperatures.

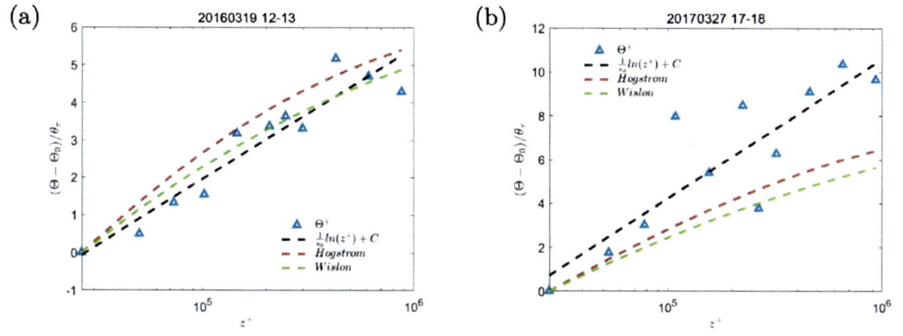

Fig. 3. The potential temperature profiles of clean air and sand-laden flow. The titles represent the time of data measurement. The triangular scatters are the measured potential temperature. The black dashed line is the log law line. The red and green lines are the MOST-based empirical results. Subfigure (a) is the clean air while (b) is the sand-laden flow.

From Fig. 4, the logarithmic coordinate plot, it can be seen that error in unstable conditions is unexpectedly smaller than in near-neutral conditions. σ has a straight line tendency in the logarithmic graph and the log law equation always has the minimum error.

The presence of sand particles does not significantly affect the error, as there is no significant difference of RSME between the sand-laden flow and clean air scenarios. The error of two empirical formulas is consistently higher than the log law equation.

The conclusion remains in the sand-laden flow. The log-law expression has the least error in all cases. Strong unstable conditions have smaller errors than weak unstable conditions. The PM10 concentration's influence is not remarkable (Fig. 5).

3.4 Slope of the Potential Temperature Profile

This section shows the dependents of κ_θ on different parameters.

We discover that the logarithmic profile remains in even strong unstable conditions when the buoyancy term plays an important role in turbulence. Different stability conditions mainly impact the slope of the mean temperature profile. The slope of the profile equals $\theta_\tau/\kappa_\theta$. The Kármán constants are power functions of $-\delta_\tau/L$, $-u_\tau^2/gL$, $-(\nu^2/g)^{1/3}/L$ and $-(\alpha^2/g)^{1/3}/L$ from Fig. 6. The abscissa employed distinct dimensionless characteristic lengths from the boundary layer flow.

Based on the fitting results of Fig. 6, we have:

$$\kappa_\theta = C_1 \left(-\frac{\nu}{u_\tau L}\right)^{0.251} \approx C_1 \left(-\frac{\nu}{u_\tau L}\right)^{4/15}, \tag{10a}$$

$$\kappa_\theta = C_2 \left(-\frac{u_\tau^2}{gL}\right)^{0.642} \approx C_2 \left(-\frac{u_\tau^2}{gL}\right)^{2/3}, \tag{10b}$$

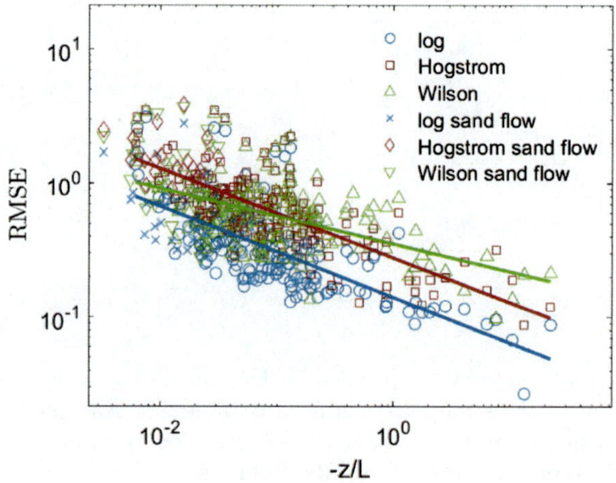

Fig. 4. RMSE of different expressions. Different color and different marker styles represent different equations' RMSE. The straight lines denotes the trends.

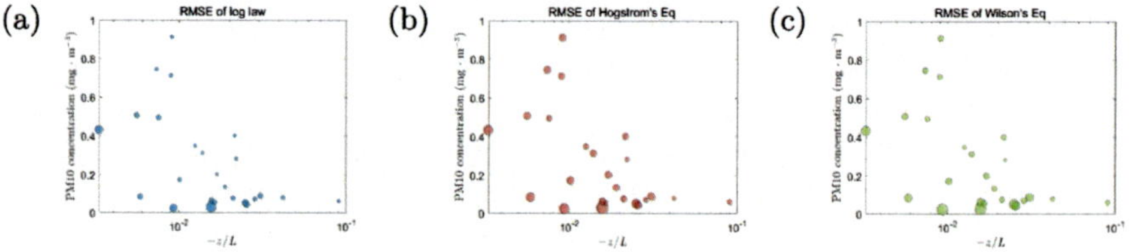

Fig. 5. RMSE of sand-laden flow. The color and size represent the value of RMSE. A larger size indicates that RMSE is larger.

$$\kappa_\theta = C_3 \left(-\frac{\nu^{2/3}}{g^{1/3} L} \right)^{0.347} \approx C_3 \left(-\frac{\nu^{2/3}}{g^{1/3} L} \right)^{1/3}, \tag{10c}$$

$$\kappa_\theta = C_4 \left(-\frac{\alpha^{2/3}}{g^{1/3} L} \right)^{0.338} \approx C_4 \left(-\frac{\alpha^{2/3}}{g^{1/3} L} \right)^{1/3}, \tag{10d}$$

where $C_{1,2,3,4}$ are constants. As Table 1 shows, ν and α vary slightly at different time intervals and can be approximated as constants.

Equation (10b) and Eq. (10c) can be rewritten as:

$$u_\tau = g^{1/2}(-L)^{1/2} \left(\frac{\kappa_\theta}{C_2} \right)^{3/4}, \quad \nu = g^{1/2}(-L)^{3/2} \left(\frac{\kappa_\theta}{C_3} \right)^{9/2}. \tag{11}$$

Substituting them in Eq. (10a), we get:

$$C_1 = \frac{C_3^{6/5}}{C_2^{1/5}}, \tag{12}$$

which shows that the scalings are consistent. The Kármán constant in the sand-laden situation has little difference with clean air conditions. A larger PM10 give rise to a larger bias from the trend line. The sand-laden data has same slope but larger κ_θ in the logarithmic plots than clean air data. A rough estimate gives that κ_θ in sand-laden data is 20–40% higher compared to clean air. However, the number of sandstorm data point is quite small. To verify the conclusion, more data from different experiment condition should be used for verification in the future.

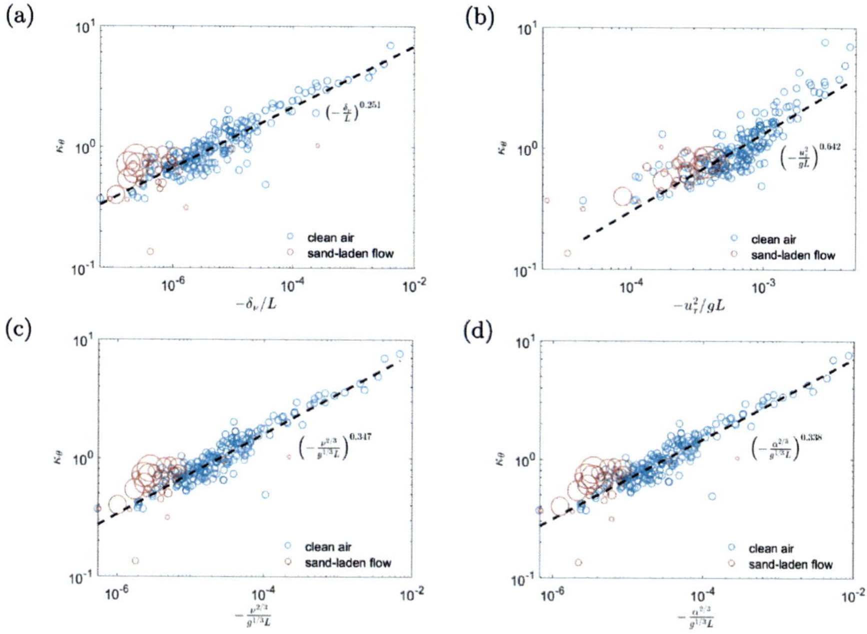

Fig. 6. The velocity and potential temperature Kármán constant of clean air and sand-laden flow. The clean air data markers are blue and the sand-laden data markers are red. Larger red circles represent larger PM10 concentrations. The dotted lines are the trend lines calculated by logarithmic polynomial fitting. Different horizontal coordinates are used in the subfigures.

4 Conclusion and Discussion

We verified the logarithmic law remains valid even in the presence of unstably stratified atmospheric conditions by investigating QLQA data with and without sand. By studying the similarity functions, potential temperature profile, root-mean-square error and slope of the profiles, we find that when buoyancy effects become significant, the empirical equations derived from the Monin-Obukhov similarity theory do not exhibit superior performance compared to the logarithmic law. We focus on the impact of stability parameters on the Kármán constant of the potential temperature profile, κ_θ. A power expression of κ_θ about

u_τ and Obukhov length L is found. $\kappa_\theta = C_1\left(-\frac{\nu}{u_\tau L}\right)^{4/15}$, $\kappa_\theta = C_2\left(-\frac{u_\tau^2}{gL}\right)^{2/3}$ and $\kappa_\theta = C_3\left(-\frac{\nu^{2/3}}{g^{1/3}L}\right)^{1/3}$. Flows containing sand have approximately 20 – 40% higher κ_θ values compared to sand-free flows.

Due to the limited data, we do only provide qualitative impact of sand on the mean temperature profile, and the quantitative effect should be studied in the future. Also, we need theoretical explanations to the scalings for the Kármán constant of the potential temperature profile.

References

1. Ahlers, G., et al.: Logarithmic temperature profiles in turbulent Rayleigh-Bénard convection. Phys. Rev. Lett. **109**(11), 114501 (2012)
2. Ahlers, G., Bodenschatz, E., He, X.: Logarithmic temperature profiles of turbulent Rayleigh–Bénard convection in the classical and ultimate state for a prandtl number of 0.8. J. Fluid Mech. **758**, 436–467 (2014)
3. Businger, J.A., Wyngaard, J.C., Izumi, Y., Bradley, E.F.: Flux-profile relationships in the atmospheric surface layer. J. Atmos. Sci. **28**(2), 181–189 (1971)
4. Carl, D.M., Tarbell, T.C., Panofsky, H.A.: Profiles of wind and temperature from towers over homogeneous terrain. J. Atmos. Sci. **30**(5), 788–794 (1973)
5. Cheng, Yu., Li, Q., Li, D., Gentine, P.: Logarithmic profile of temperature in sheared and unstably stratified atmospheric boundary layers. Phys. Rev. Fluids **6**(3), 034606 (2021)
6. Foken, T., Göockede, M., Mauder, M., Mahrt, L., Amiro, B., Munger, W.: Post-field data quality control. In: Lee, X., Massman, W., Law, B. (eds.) Handbook of Micrometeorology. Atmospheric and Oceanographic Sciences Library, vol. 29, pp. 181–208. Springer, Dordrecht (2004). https://doi.org/10.1007/1-4020-2265-4_9
7. Garai, A., Kleissl, J., Sarkar, S.: Flow and heat transfer in convectively unstable turbulent channel flow with solid-wall heat conduction. J. Fluid Mech. **757**, 57–81 (2014)
8. Grossmann, S., Lohse, D.: Logarithmic temperature profiles in the ultimate regime of thermal convection. Phys. Fluids **24**(12), 125103 (2012)
9. Högström, U.L.F.: Non-dimensional wind and temperature profiles in the atmospheric surface layer: a re-evaluation. Top. Micrometeorol. A Festschrift Arch Dyer, pp. 55–78 (1988)
10. Hölling, M., Herwig, H.: Asymptotic analysis of the near-wall region of turbulent natural convection flows. J. Fluid Mech. **541**, 383–397 (2005)
11. Iida, O., Kasagi, N.: Direct numerical simulation of unstably stratified turbulent channel flow (1997)
12. Johansson, C., Smedman, A.-S., Högström, U., Brasseur, J.G., Khanna, S.: Critical test of the validity of Monin–Obukhov similarity during convective conditions. J. Atmos. Sci. **58**(12), 1549–1566 (2001)
13. Kader, B.A.: Temperature and concentration profiles in fully turbulent boundary layers. Int. J. Heat Mass Transf. **24**(9), 1541–1544 (1981)
14. Kader, B.A., Yaglom, A.M.: Mean fields and fluctuation moments in unstably stratified turbulent boundary layers. J. Fluid Mech. **212**, 637–662 (1990)
15. Kawamura, H., Abe, H., Matsuo, Y.: DNS of turbulent heat transfer in channel flow with respect to Reynolds and Prandtl number effects. Int. J. Heat Fluid Flow **20**(3), 196–207 (1999)

16. Lee, M., Moser, R.D.: Direct numerical simulation of turbulent channel flow up to. J. Fluid Mech. **774**, 395–415 (2015)
17. Li, D., Luo, K., Fan, J.: Buoyancy effects in an unstably stratified turbulent boundary layer flow. Phys. Fluids **29**(1), 015104 (2017)
18. Liu, H., Wang, G., Zheng, X.: Amplitude modulation between multi-scale turbulent motions in high-Reynolds-number atmospheric surface layers. J. Fluid Mech. **861**, 585–607 (2019)
19. Marusic, I., Monty, J.P., Hultmark, M., Smits, A.J.: On the logarithmic region in wall turbulence. J. Fluid Mech. **716**, R3 (2013)
20. McNaughton, K.G.: Turbulence structure of the unstable atmospheric surface layer and transition to the outer layer. Bound.-Layer Meteorol. **112**, 199–221 (2004)
21. Monin, A.S., Obukhov, A.K.: Basic laws of turbulent mixing in the surface layer of the atmosphere. Contrib. Geophys. Inst. Acad. Sci. USSR **151**(163), e187 (1954)
22. Obukhov, A.: Trudy inta teoret. Geofiz. Akad. Nauk. SSSR **1**, 95 (1946)
23. Obukhov, A.: Turbulence in thermally inhomogeneous atmosphere. Trudy Inst. Teor. Geofiz. Akad. Nauk SSSR **1**, 95–115 (1946)
24. Pirozzoli, S., Bernardini, M., Verzicco, R., Orlandi, P.: Mixed convection in turbulent channels with unstable stratification. J. Fluid Mech. **821**, 482–516 (2017)
25. Redjem-Saad, L., Ould-Rouiss, M., Lauriat, G.: Direct numerical simulation of turbulent heat transfer in pipe flows: effect of Prandtl number. Int. J. Heat Fluid Flow **28**(5), 847–861 (2007)
26. Scagliarini, A., Einarsson, H., Gylfason, Á., Toschi, F.: Law of the wall in an unstably stratified turbulent channel flow. J. Fluid Mech. **781**, R5 (2015)
27. Sid, S., Dubief, Y., Terrapon, V.: Direct numerical simulation of mixed convection in turbulent channel flow: on the Reynolds number dependency of momentum and heat transfer under unstable stratification. In: 8th International Conference on Computational Heat and Mass Transfer (2015)
28. Von Kármán, T.: Mechanical similitude and turbulence. Number 611. National Advisory Committee for Aeronautics (1931)
29. Wang, G., Zheng, X.: Very large scale motions in the atmospheric surface layer: a field investigation. J. Fluid Mech. **802**, 464–489 (2016)
30. Wang, G., Zheng, X., Tao, J.: Very large scale motions and pm10 concentration in a high-re boundary layer. Phys. Fluids **29**(6), 061701 (2017)
31. Wilson, D.K.: An alternative function for the wind and temperature gradients in unstable surface layers. Boundary-Layer Meteorol. **99**, 151–158 (2001)

Large Eddy Simulation of the Mixing Process of Gaseous Pollutants Using a Lagrangian Stochastic Model

Shuming Jia[1,2], Xiaolan Yu[3], and Zhengshi Wang[3](✉)

[1] College of Soil and Water Conservation Science and Engineering, Northwest A&F University, Yangling 712100, Shaanxi, China
jiashm2020@nwafu.edu.cn

[2] State Key Laboratory of Soil Erosion and Dryland Farming on the Loess Plateau, Institute of Soil and Water Conservation, Northwest A&F University, Yangling 712100, Shaanxi, China

[3] College of Natural Resources and Environment, Northwest A&F University, Yangling 712100, Shaanxi, China
{yxl123,wangzs2020}@nwafu.edu.cn

Abstract. The mixing process of reacting gaseous pollutants in a turbulent flow is essential to understand the formation of resultant aerosols. Unfortunately, this issue is only rarely investigated. In this study, a large eddy simulation (LES) model is combined with a Lagrangian stochastic model to explore the mixing process of two gaseous pollutants in an atmospheric turbulent boundary layer. Simulation results show that the atmospheric turbulences drive the pollutant parcels to form obvious spatial structures, and the mass concentration exhibits an inhomogeneous distribution. The growth and dissipation rate of the mixing zone sizes both show approximately linear behaviors, and the variation rate of the latter is almost twice of the former. Surprisingly, the distance between two emission sources generally does not affect the formation and dissipation time of the mixing zone with the same mass concentration. The mean volume of mixing zone with a certain mass concentration decreases linearly with distance of two emission sources under the same wind condition, while it decreases exponentially with increasing free stream velocity if the distance of two sources is fixed. Thus, the increasing wind velocity can reduce the mixing zone more effectively compared with that of inter-sources distance. From this study, a critical distance between two emission sources and their emission strategies can be suggested if the wind condition is known. This research can also provide an effective way to determine an optimized siting of various emission sources.

Keywords: Gaseous pollutants · Large eddy simulation (LES) · Lagrangian stochastic (LS) model · Atmospheric turbulent boundary layer · Mixing process

1 Introduction

Aerosol particles are direct source of material of haze, which not only endanger people's health, but also may affect the climate system through absorbing the solar radiation [1, 2]. Only small part of the haze pollution caused by the primary particulate matter (PM) that

X. Zheng and S. Balachandar (Eds.): IUTAM 2023, IUTAM Bookseries 41, pp. 152–163, 2024.
https://doi.org/10.1007/978-3-031-47258-9_9

comes from the direct emission of fossil fuel burning and automobile exhaust [3, 4], while about 70% comes from the secondary PM that is contributed by the oxidizing reaction of various gaseous pollutants [5–7]. The secondary PM essentially accounts for a large proportion during this process, which also requires a thorough mixing state of various chemical components [8, 9]. Thus, a thoroughly investigation on the mixing features of various gaseous pollutants is essential to understand the formation and evolution of dust-haze.

The diffusion process of pollutants has been a long topic, and the numerical models can be divided into two major aspects. One is the scalar diffusion (SD) model that describes the diffusion process by solving the scalar transport equation [10, 11], the other is the Lagrangian stochastic (LS) model that tracks the trajectory of marked fluid particle (pollutant parcel) [12, 13]. The SD model is poor to address the diffusion of the point source (e.g., unrealistic initialization and diffusion velocity), especially for the larger grid. However, the transport and space distribution of airborne contaminant described by the LS model is independent of the grid resolution, and thus is more suitable for the diffusion of point sources under bigger mesh spacing.

As a matter of fact, the chemical component of effluent from one source is relative unitary, thus, the mixing process of various chemical components is the basic for the formation of secondary PM. Although the mixing processes of two or more-point source emissions have been investigated by a large number of researchers [11, 14], the mixing process of various gaseous components has been rarely investigated based on large eddy simulation (LES), especially under the actions of atmospheric turbulence. In this way, an accurate simulation of the atmospheric turbulent boundary layer is required. The LES has become one of the most effective approaches to model atmospheric turbulences with different scales [15–17], which can provide an accurate fluid field for the pollutants diffusion.

In this study, the mixing features of two kinds of gaseous pollutants in an atmospheric turbulent boundary layer produced by a LES model are investigated, and the evolution process of gaseous pollutants is described by a Lagrangian stochastic model. The structures of pollutant micelles in the turbulence flow are presented, and the distribution characteristics and the evolution of the mixing zones under different model configurations are analyzed.

2 Model and Method

2.1 Lagrangian Stochastic Model

The Lagrangian stochastic (LS) model calculates trajectories of marked fluid particles based on the generalized Langevin equation, in which each fluid particle represents a scalar parcel [18, 19]:

$$dx_{pi} = \left(\tilde{u}_i + u'_{pi}\right)dt \tag{1}$$

where x_{pi} is the position coordinate of the particle, t is time, $\tilde{u}_i$ stand for the resolved filtered velocity obtained by the later Eqs. (2) and (3), u'_{pi} is the sub-grid scale (SGS) fluctuating velocity of the fluid; The subscript $i = 1, 2, 3$ indicates the streamwise, spanwise and vertical direction, respectively.

The filtered velocity of the fluid is calculated with the mesoscale large eddy simulation model of Advanced Regional Prediction System (ARPS, version 5.3.3) [20]. The filtered mass and momentum conservation equations can be written as:

$$\frac{\partial \rho}{\partial t} + \frac{\partial}{\partial x_i}(\rho \tilde{u}_i) = 0 \tag{2}$$

$$\frac{\partial (\rho \tilde{u}_i)}{\partial t} + \frac{\partial (\rho \tilde{u}_i \tilde{u}_j)}{\partial x_j} = -\frac{\partial \tilde{p}^*}{\partial x_i} + B\delta_{i3} - \frac{\partial \tau_{ij}}{\partial x_j} \tag{3}$$

where '~' represents filter variables, $\rho = p\Lambda / R_d T$ is the density of humid air, in which p is the air pressure, R_d is the gas constant of dry air, T is the air temperature, $\Lambda = \left(1 - q_v / (\lambda + q_v)\right)(1 + q_v)$ is a parameter related to the specific humidity of air q_v, and $\lambda = R_d / R_v$ is the ratio of gas constant between dry air and water vapor. $p^* = p' - \alpha \nabla(\rho \mathbf{u})$ contains the pressure perturbation and damping terms, in which α is the damping coefficient; δ_{ij} is the Kronecker delta, and $B = -g\rho' / \rho$ is the buoyancy caused by air density perturbation ρ', in which g is gravitational acceleration; τ_{ij} is the SGS stress, and can be expressed as [14]:

$$\tau_{ij} = -2\nu_t \tilde{S}_{ij} \tag{4}$$

where $\tilde{S}_{ij} = 0.5(\partial \tilde{u}_i / \partial x_j + \partial \tilde{u}_j / \partial x_i)$ is the strain rate tensor, $\nu_t = \left(C_s \tilde{\Delta}\right)^2 \left|\tilde{S}_{ij}\right|$ is the turbulent eddy viscosity, $\tilde{\Delta}$ and $\left|\tilde{S}_{ij}\right|$ are the filter scale and magnitude of $\tilde{S}_{ij}$, respectively; The Smagorinsky coefficient C_s is determined by the Lagrangian dynamic SGS model of Meneveau et al. [16].

The SGS stochastic model for fluid particle dispersion that coupled with large eddy simulation is adopted to model the SGS fluid particle velocity u'_{pi} [21], which can be read as:

$$du'_{pi} = a_i dt + b_{ij} d\eta_i(t) \tag{5}$$

where $d\eta_i$ is the increment of a vector-valued Wiener process, and its mean and variance are 0 and dt, respectively; a_i and b_{ij} are coefficients to be determined, and can be expressed as:

$$a_i = \left(-\frac{1}{T_L} + \frac{1}{2\tilde{E}}\frac{d\tilde{E}}{dt}\right)u'_{pi} \text{ and } b_{ij} = \sqrt{\frac{4\tilde{E}}{3T_L}} \tag{6}$$

where $T_L = 4\tilde{E} / 3C_0\tilde{\varepsilon}$ is the Lagrangian correlation time scale, $\tilde{E}$ is the SGS turbulent kinetic energy (TKE), C_0 is the Lagrangian constant, and $\tilde{\varepsilon} = C_e \tilde{E}^{3/2} / \tilde{\Delta}$ is the turbulent dissipation rate, C_e is the dispersion constant. Besides, a transport equation for $\tilde{E}$ is solved to obtain the TKE [15]:

$$\frac{\partial \tilde{E}}{\partial t} = -\frac{\partial \tilde{u}_j \tilde{E}}{\partial x_j} + 2\frac{\partial}{\partial x_j}\left(\nu_t \frac{\partial \tilde{E}}{\partial x_j}\right) + \frac{1}{3}\nu_t \frac{g}{\theta_0}\frac{\partial \tilde{\theta}}{\partial x_3} + 2\nu_t \tilde{S}_{ij}^2 + \tilde{\varepsilon} \tag{7}$$

where $\widetilde{\theta}$ and θ_0 are potential temperature (PT) and reference PT, respectively.

2.2 Concentration Statistic Model

The traced pollutant parcel also increases in volume under the dispersing effect of turbulence smaller than the parcel itself [22]. A Puff-particle (PP) model can describe the evolution of the pollutant parcel, and its concentration is generally assumed to obey a Gaussian distribution with different standard deviations along different directions. According to P Hurley [23], The standard deviation σ_i can be evaluated from the function:

$$\sigma_i^2(t+\Delta t)=\sigma_i^2(t)+2\sigma_{ui}^2 T_L\left[1-\exp\left(-\frac{t+\Delta t/2}{T_L}\right)\right]\Delta t \quad (8)$$

where σ_{ui} is the standard deviation for the i-th wind velocity component. In this simulation, σ_{ui} is determined by the wind velocity information in the fully developed turbulent boundary layer. Besides, Δt is set to be the fluid timestep.

The emitted pollutant parcel is assumed to be spherical, and its initial size σ_0 can be expressed as:

$$\frac{4}{3}\pi\sigma_0^3 c_{ini}=m_e \quad (9)$$

where $m_e = q_{emit}/N_{per}$ is the mass of each emitted parcel, q_{emit} and N_{per} are the emission rate of pollutant and the number of released parcels per unit time from the emission source, respectively, c_{ini} is the initial mass concentration.

In this way, the local concentration c can be calculated from the statistic equation [24, 25]:

$$\begin{aligned} c=&\frac{1}{(2\pi)^{1.5}}\sum_{p=1}^{N}\frac{m_e}{\sigma_1\sigma_2\sigma_3}\exp\left[-0.5\frac{(x_{p1}-x_1)^2}{\sigma_1^2}\right]\times\exp\left[-0.5\frac{(x_{p2}-x_2)^2}{\sigma_2^2}\right]\\ &\times\left\{\exp\left[-0.5\frac{(x_{p3}-x_3)^2}{\sigma_3^2}\right]+\exp\left[-0.5\frac{(x_{p3}+x_3)^2}{\sigma_3^2}\right]\right\} \end{aligned} \quad (10)$$

2.3 Simulation Settings

A neutral boundary layer is simulated and the initial depth of the boundary layer is set as $H_0 = 1000$ m. The initial potential temperature (PT) θ_0 is 300 K within the boundary layer, and the PT profile has a gradient of $\partial\tilde{\theta}/\partial z = 0.005\,\text{Km}^{-1}$ above the boundary layer. The initial wind profile is logarithmic in the boundary layer with a free stream velocity of U_e and a surface roughness of z_0 [26]. Atmospheric turbulences are induced by random temperature perturbations near the surface with a maximum magnitude of 0.5 K, and sustained by a constant heat flux of q_0 [16]. The evolution time for an atmospheric turbulent boundary layer is 2 h.

As shown in Fig. 1, the dimensions of the computational domain are 20 H_0, 8 H_0 and 5 H_0 in x, y and z directions, respectively. The uniform grid size of 100 m is adopted in horizontal directions. The vertical grid is densified near the surface, corresponding to

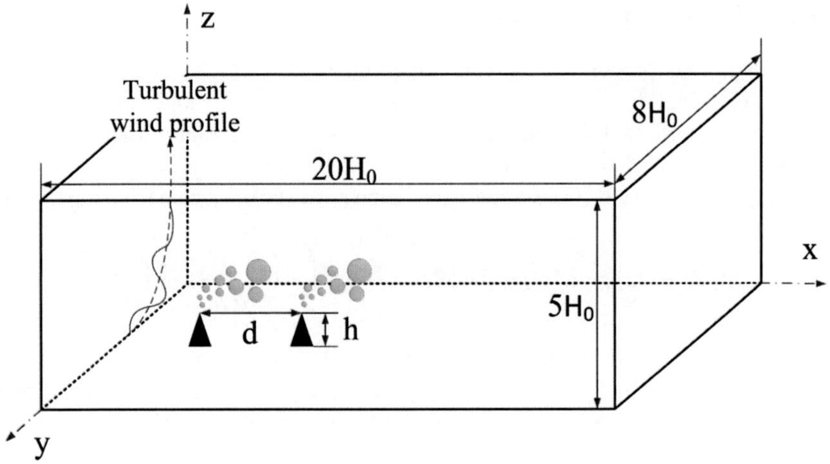

Fig. 1. Sketch of the computational domain.

a mean grid size of 100 m and a smallest grid size of 10 m. Periodic boundary conditions are applied at lateral boundaries. The bottom is set as a rigid wall, and the top uses an open radiation boundary with a Rayleigh damping layer of 1000 m in depth.

In this simulation, two emission sources with a constant height of h are placed on the ground to simulate the mixing process of the pollutants from two individual sources, and $h = 100$ m is approximately the common height of industrial chimneys. The coordinate of the first emission source (the upstream one) is (2 H_0, 4 H_0, h), and the second emission source is located at the downstream direction with a distance of d away from the first one. Besides, the maximum distance d is chosen to be 8 km to meet the common

Table 1. Parameters used in this simulation.

Parameter	Description	Unit	Value
R_d	Gas constant for dry air	$m^2s^{-2}K^{-1}$	287.0
R_v	Gas constant for water vapor	$m^2s^{-2}K^{-1}$	461.5
α	Damping coefficient	–	0.6
H_0	Thickness of the boundary layer	m	1000
q_{emit}	Emission rate of pollutant from the emission source	gs^{-1}	1000
N_{per}	Number of released parcels per unit time from the emission source	s^{-1}	2
c_{ini}	Initial mass concentration	gm^{-3}	10.0
C_0	Lagrangian constant	–	2.1
C_e	Dispersion constant	–	0.41
h	Height of the emission source	m	100
z_0	Surface roughness	m	0.01
q_0	Surface heat flux	Wm^{-2}	50.0

dimensions of a large industrial park. Besides, the two emission sources are aligned upwards considering that this condition corresponds the most severe pollutants mixing under the same wind speed and d. The detailed calculation parameters of the simulation model are given in Table 1.

3 Results and Discussion

3.1 Turbulent Characteristics in the Atmospheric Boundary Layer

Turbulence characteristics in the boundary layer are analyzed firstly, since the wind field is the foundation of pollutants transportation. Figure 2(a) shows contour surfaces of updraft ($u_3 = 0.5$ ms^{-1}) and downdraft ($u_3 = -0.5$ ms^{-1}) in a fully-developed atmospheric turbulent boundary layer, and the free stream velocity U_e is 10.0 ms^{-1}. It can be seen that numerous large scale turbulent structures are generated within the boundary layer. Figure 2(b) shows a z-slice of the vertical wind speed at z = 500 m. It can be seen that updraft and downdraft with different sizes and shapes exhibit interval distribution in space. In addition, the intensity of updraft is obvious stronger than that of the downdraft, mainly because heat fluxes are supplied from the bottom.

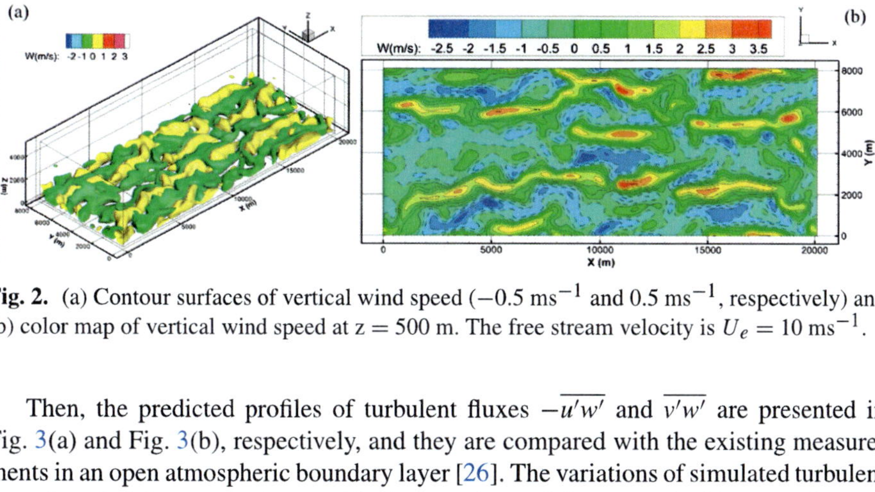

Fig. 2. (a) Contour surfaces of vertical wind speed (-0.5 ms^{-1} and 0.5 ms^{-1}, respectively) and (b) color map of vertical wind speed at z = 500 m. The free stream velocity is $U_e = 10$ ms^{-1}.

Then, the predicted profiles of turbulent fluxes $-\overline{u'w'}$ and $\overline{v'w'}$ are presented in Fig. 3(a) and Fig. 3(b), respectively, and they are compared with the existing measurements in an open atmospheric boundary layer [26]. The variations of simulated turbulent flux along height are consistent with the experimental data exactly. However, the predicted turbulent flux $\overline{v'w'}$ is slightly larger than the measurements. This is mainly because the given heat flux may be deviated from the real situation due to the lack of observations, thus, the vertical flow motions are underestimated to a certain extent.

In order to observe the structures of pollutant parcels in the turbulent flow directly, the locations of marked fluid particles at various moments are shown in Fig. 4(a), in which different colors represent the pollutants from different emission sources. Figure 4(b) shows the top view at the moment of $t = 4$ h. In addition, the horizontal cloud map of mass concentration at a height of z = 100 m is presented in Fig. 4(c), and the moment is $t = 4$ h. Seen from these figures, pollutant particles spread around during they travel to the downwind. At the same time, particles also aggregate locally under the actions

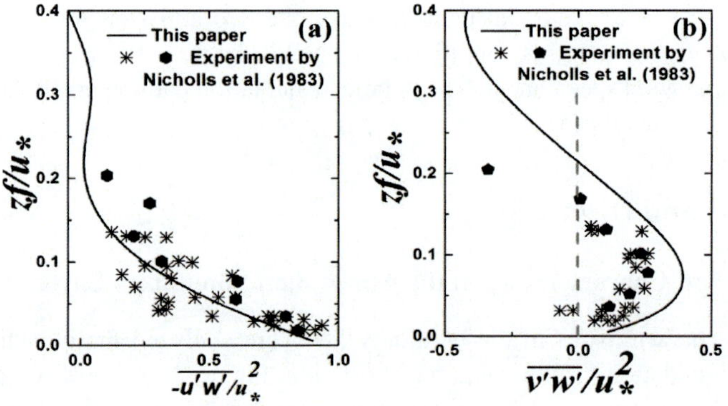

Fig. 3. Comparison of the simulated turbulent flux profiles (lines) and experimental data (scatters): (a) $-\overline{u'w'}$ and (b) $\overline{v'w'}$.

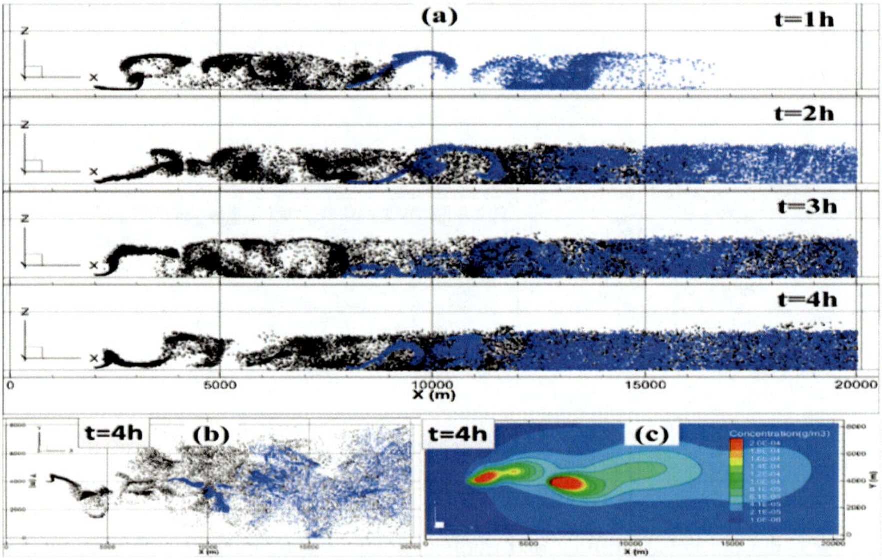

Fig. 4. (a) Lateral (x-z) and (b) top view (x-y) of structures of the pollutant particles in the turbulent flow at various moments; and (c) the mass concentration (x-y slice) at height of 100 m. The distance between two emission sources is $d = 6.0$ km, and the free stream velocity U_e is 2.0 ms^{-1}.

of large-scale turbulent vortexes and form obvious spatial structures, which also induce an asymmetrical distribution of the mass concentration of pollutant (see Fig. 4(c)). As time goes on, two kinds of the pollutant particles are jointed, and begin to mix with each other. The space distributions of particles become more and more uniformly when they transport towards the downstream direction.

3.2 Mixing Process of Pollutants from Two Emission Sources

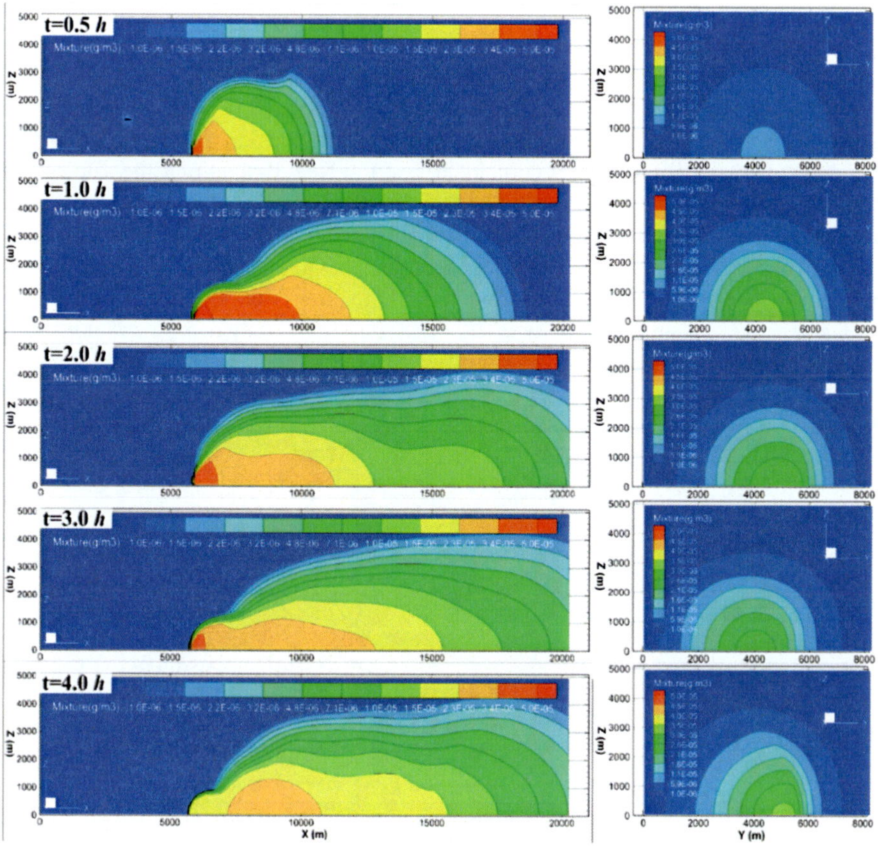

Fig. 5. Color map of the pollution level χ at various moments. The x-z slice and y-z slice are at $y = 4000$ m and $x = 10000$ m, respectively. And the free stream velocity U_e is 4.0 ms^{-1}.

During the mixing process of two kinds of gaseous pollutants that can be reacted chemically, the one with lower concentration generally controls the quantity of the resultant aerosols. Thus, the smaller value of the concentrations of pollutant at a grid point can represent the pollution levels to a certain extent. Here, we use $\chi = \min(c_1, c_2)$ to represent the pollution level, and the reaction between two pollutants is neglected. Figure 5 shows the evolution of the mixing zones with different pollution levels, in which the distance between the two emission sources is $d = 4.0$ km, and the free stream velocity is $U_e = 4.0\ \text{ms}^{-1}$. In order to explore the dissipation of the pollutant, emission sources are turned off at $t = 3$ h.

As shown in Fig. 5, an obvious transport process towards the downwind can be observed. The mixing zones generally exhibit a slender shape, and the maximum value appears just behind the second emission source. The dimension of the domain is large enough to explore the evolution of the main mixing zone (e.g., $\chi > 1.0 \times 10^{-5}$ gm^{-3}). At the same time, the contaminated zones sway from side to side during they move forward in the turbulent flow, and thus the concentration also shows unsymmetrical distribution along spanwise.

The variations of the volume of mixing zone (VMZ) versus time under various distances d are shown in Fig. 6, and the free stream velocity is $U_e = 4.0$ ms^{-1}. From this figure, although the size of the main mixing zone ($\chi > 1.0 \times 10^{-5}$ gm^{-3}) decreases with inter-sources distance d, however, the development time to form a steady state mixing zone under various distances d is coincident, and thus the dissipation time, for a given mass concentration. The reason could be that the free stream velocity is the same, and thus the pollutant parcels pass through the whole domain with almost a constant period of time. The VMZ also goes up and down with time due to the existence of atmospheric turbulences. The VMZs still maintain for over an hour after shutting the emission sources down under the free stream velocity of $U_e = 4.0$ ms^{-1}, and they decrease rapidly then. The dissipation rate of the VMZ is almost twice of the growth rate.

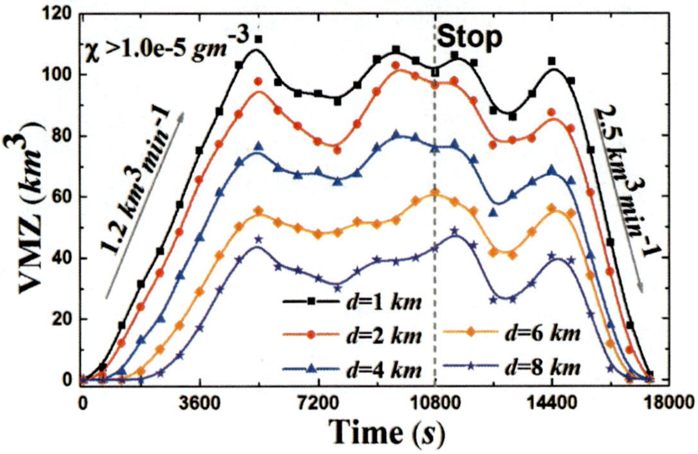

Fig. 6. Volume of mixing zone (VMZ) versus emission time at different distance d between two emission sources.

In addition, the evolution of the mean VMZ versus inter-sources distance d is shown in Fig. 7. It can be seen that the mean VMZ decreases linearly with increasing distance d, which may also reduce to zero for higher pollution levels. This mainly because the pollution zone of a single emission source is almost constant under the same wind speed, leading to the mixing zone of the two decreases linearly with the increase of the distance. For example, the critical distance for the pollution level larger than 5.0×10^{-5} gm^{-3} is approximately $d = 4$ km under this wind condition. At the same time, the mean VMZ increases with the decreasing pollution level (see Fig. 7).

Finally, the variation of the mean VMZ versus free stream velocity is shown in Fig. 8, and the distance between two emission sources is $d = 6$ km. The mean VMZ is also almost decreases exponentially with increasing free stream velocity on account of the approximately exponential relation between the mass concentration of pollutant and

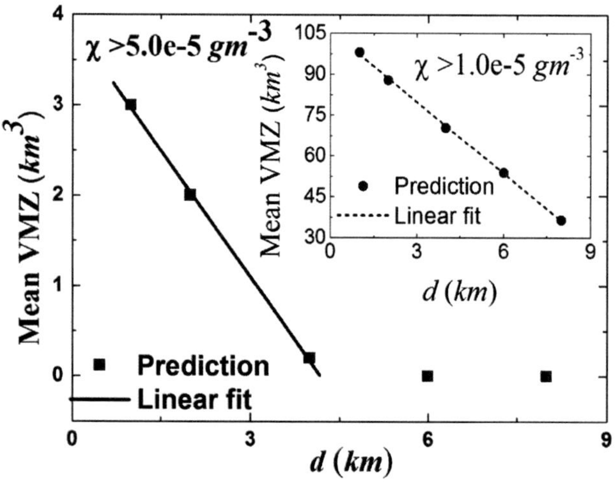

Fig. 7. Variation of mean VMZ with changing distance d at pollution level of $\chi > 5.0 \times 10^{-5}$ gm^{-3}. Inset shows the results of $\chi > 1.0 \times 10^{-5}$ gm^{-3}.

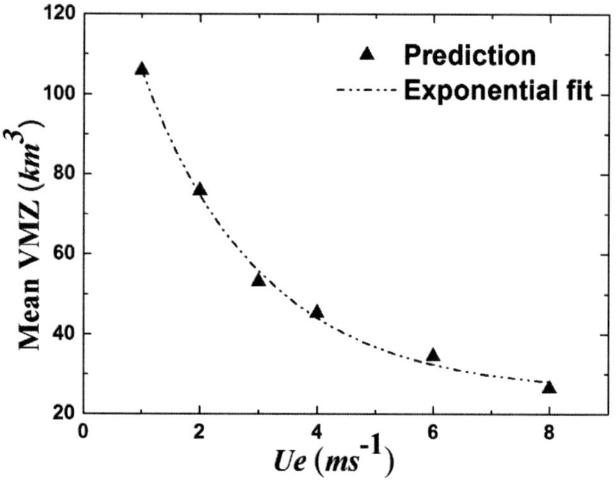

Fig. 8. Variation of the mean VMZ versus free stream velocity U_e at the d of 6 km.

turbulent wind (Eq. 8). That is, compared with the inter-sources distance d, larger wind velocity can reduce the VMZ more effectively.

4 Conclusions

In this study, the mixing features of two kinds of gaseous pollutants in an atmospheric turbulent boundary layer are investigated based on the large eddy simulation (LES) and the Lagrangian stochastic model. A fully developed neutral atmospheric boundary layer is produced, and the obtained turbulent fluxes profiles are consistent with the existing observations. Airborne contaminants are driven by the atmospheric turbulent motions and form obvious spatial structures, which results in a swing wake towards the downwind.

In addition, a mixing zone in slender shape is generally formed behind of the downstream emission source, and the pollution level decreases with their distance. The mean volume of mixing zone (VMZ) increases in a linear way once the two chemical components come into contact with each other, which also exhibits a fluctuating state after achieving a thoroughly mixing state due to the existence of turbulence. The dissipative stage of the VMZ also shows a linear behavior, and its rate is almost twice of the growth stage. For a given mass concentration, the formation and dissipation time of the steady state mixing zone seems independent of the distance between two emission sources.

Furthermore, the mean VMZ decreases linearly with their distance under the same free stream velocity. However, the mean VMZ decreases exponentially with the free stream velocity if their distance is fixed. In this way, when siting a emission source, the wind condition should be considered prior of its distance with other sources. In general, a critical distance can be determined by known wind conditions based on the simulation model in this work.

Absolutely, this is still a simplified simulation for pollutant transportation and mixture, the local terrain and surface characteristics should also be included for a real situation run.

Acknowledgements. This work was supported by the National Natural Science Foundation of China (12102364, 11902341), the China Postdoctoral Science Foundation (2021M702679), the Young Talent Fund of University Association for Science and Technology in Shaanxi, China (20210503), and the Chinese Universities Scientific Fund (2452021048, 2452020266). Open-source code of Advanced Regional Prediction System (ARPS) is available online (http://www.caps.ou.edu/ARPS/).

References

1. Li, Z.: East Asian studies of tropospheric aerosols and their impact on regional climate (EAST-AIRC): an overview. J. Geophys. Res. Atmos. **116**(4), 220–237 (2011)
2. Li, W.: A review of single aerosol particle studies in the atmosphere of East Asia: morphology, mixing state, source, and heterogeneous reactions. J. Clean. Prod. **112**, 1330–1349 (2016)
3. Chen, S.: Direct observations of organic aerosols in common wintertime hazes in North China: insights into direct emissions from Chinese residential stoves. Atmos. Chem. Phys. **17**(2), 1–32 (2017)
4. Li, W.: Integrated evaluation of aerosols from regional brown hazes over northern China in winter: concentrations, sources, transformation, and mixing states. J. Geophys. Res. Atmos. **116**(D9), D09301 (2011)
5. Du, Z.: Gasoline direct injection vehicles exceed port fuel injection ones in both primary aerosol emission and secondary aerosol formation. Atmos. Chem. Phys. **776**, 1–31 (2017)
6. Limbeck, A.: Secondary organic aerosol formation in the atmosphere via heterogeneous reaction of gaseous isoprene on acidic particles. Geophys. Res. Lett. **30**(30), 379–394 (2003)
7. Huang, R.J.: High secondary aerosol contribution to particulate pollution during haze events in China. Nature **514**(7521), 218–222 (2014)

8. Jacobson, M.Z.: Development of mixed-phase clouds from multiple aerosol size distributions and the effect of the clouds on aerosol removal. J. Geophys. Res. Atmos. **108**(D8), D84245 (2003)
9. Yang, F.: Evolution of the mixing state of fine aerosols during haze events in Shanghai. Atmos. Res. **104**(1), 193–201 (2012)
10. Manceau, D.: A diffusion model for preannounced products. Cust. Needs Solut. **1**(1), 77–89 (2014)
11. Kaplan, H.: A lagrangian particle transport and diffusion model for non-passive scalars. Bound.-Layer Meteorol. **156**(1), 1–16 (2015)
12. Gao, Z.: Estimating gas emissions from multiple sources using a backward Lagrangian stochastic model. J. Air Waste Manag. Assoc. **58**(11), 1415–1421 (2008)
13. Pope, S.B.: A stochastic Lagrangian model for acceleration in turbulent flows. Phys. Fluids **14**(7), 2360–2375 (2002)
14. Smagorinsky, J.: General circulation eeperiments with the primitive equations. Mon. Weather Rev. **91**(3), 99–164 (1963)
15. Deardorff, J.W.: Stratocumulus-capped mixed layers derived from a three-dimensional model. Bound.-Layer Meteorol. **18**(4), 495–527 (1980)
16. Meneveau, C.: A Lagrangian dynamic subgrid-scale model of turbulence. J. Fluid Mech. **319**(319), 353–385 (1996)
17. Fang, J.: Large-eddy simulation of very-large-scale motions in the neutrally stratified atmospheric boundary layer. Bound.-Layer Meteorol. **155**(3), 397–416 (2015)
18. Thomson, D.J.: Criteria for the selection of stochastic models of particle trajectories in turbulent flows. J. Fluid Mech. **180**, 529–556 (1987)
19. Rodean, H.C.: Stochastic Lagrangian Models of Turbulent Diffusion, vol. 26, 1st edn. American Meteorological Society Boston (1996)
20. Xue, M.: The advanced regional prediction system (ARPS) – a multi-scale nonhydro-static atmospheric simulation and prediction tool. Part II: Model physics and applications. Meteorol. Atmospheric Phys. **76**(3–4), 143–165 (2001)
21. Vinkovic, I.: Large-eddy simulation of the dispersion of solid particles in a turbulent boundary layer. Bound.-Layer Meteorol. **121**(2), 283–311 (2006)
22. De Haan, P.: A novel approach to atmospheric dispersion modelling: the Puff-Particle model. Q. J. R. Meteorol. Soc. **124**(552), 2771–2792 (1998)
23. Hurley, P.: Partpuffa lagrangian particle-puff approach for plume dispersion modeling applications. J. Appl. Meteorol. **33**(2), 285–294 (1994)
24. Raza, S.S.: A 3-D Lagrangian stochastic model for the meso-scale atmospheric dispersion applications. Nucl. Eng. Des. **208**(1), 15–28 (2001)
25. Yamada, T.: Development of a nested grid, second moment turbulence closure model and application to the 1982 ASCOT brush creek data simulation. J. Appl. Meteorol. **27**, 562–578 (1988)
26. Nicholls, S.: The structure of the turbulent atmospheric boundary layer and discussion. Philos. Trans. R. Soc. B Biol. Sci. **308**(1503), 307–309 (1983)

Rheology of Granular Flow of Platonic Solid-Shaped Particles

Jiahui Hao and Yu Guo(✉)

Zhejiang University, Hangzhou 310027, China
yguo@zju.edu.cn

Abstract. An investigation of frictional spherical and Platonic solid-shaped particle flow on inclined plane is provided by Discrete Element simulations and the effect of particle shape on rheological properties are studied. The results indicate that as the particle sphericity increases, the tangential stress along the particle bed and the solid volume fraction corresponding to the rapid increase of coordination number and tangential stress of granular flow are reduced. The friction and dilatancy laws of spherical and polyhedral particle flows are significantly different due to the distinction in particle rotation ability, similar flow features to the Platonic solid-shaped particle flows can be achieved by applying rolling friction to spherical particles. Rescaling by a power of dimensionless granular temperature, the variation of bulk friction coefficient μ versus inertial number I of granular flows of different Platonic solid-shaped particles can be described by a master curve.

Keywords: DEM · granular flow · Platonic solid-shaped particles · constitutive law

1 Introduction

Granular matter is ubiquitous in the natural and industrial environment, the granular flow in reality is a complex nonlinear system contains a large number of discrete particles differ in size, shape, density or some surface parameters [1–4], particle interaction makes granular assemblies have special physical and mechanical properties. Thus, assessing the flow behavior and constitutive law of granular system, exploring a unified theory to characterize the flow of particles with varied properties is of paramount importance and attracts more research attentions.

Particle shape is considered as one of most crucial parameters which modifies the microstructure and macro behavior of granular system [5–8]. In DEM simulations, 2D or 3D spherical particles are commonly implemented shape for the less computation cost, particularly in the process of contact detection [9]. In recent years, various models like multi-sphere particle, hyper-ellipsoid, cylinder or fiber, convex or concave polyhedron are considered to explore the granular flow with non-spherical particles. Both spherical particle and pinacoids obey the flow start and stop laws [10] when flowing down a rough inclined plane, while compared with spheres, the start and stop tilt angles of shear flow of pinacoids are much larger due to the influence of angularity [11]. In the hopper

X. Zheng and S. Balachandar (Eds.): IUTAM 2023, IUTAM Bookseries 41, pp. 164–177, 2024.
https://doi.org/10.1007/978-3-031-47258-9_10

flow of cylinders and Platonic solid-shaped particles, the decrease of particle sphericity will reduce the flow rate and make the particles flow unevenly [12]. Especially in the hexahedral system, stacks are easier to form and particles may arch soon after flow starts [13, 14]. Moreover, strong particle alignment of cubes can be observed in the twist shearing [15] and solid volume fraction increases monotonically in the process of twists. For the irregular polyhedral particle system [16], the shear strength grows with particle angularity on account of the increase of fabric and force anisotropies, then converges when the face number is less than 46. In the triaxial compression simulation of superballs [17], it was found that a larger particle asphericity tends to enlarge normal contact force as a result of the higher proportion of sliding contacts and strengthen the particle interlock. For the granular flow of large aspect ratio ellipsoids, force chains are easier to form than spheres and behaves like elastic flow at small particle concentrations [3]. Also, the dispersion in velocity and stresses distributions of ellipsoidal shear flow are comparatively small [18].

In the work of Pouliquen [19], the well-known rheological theory, $\mu(I)$ theory has been proposed as a rheological model which associates the bulk friction coefficient μ with inertial number I to describe the dense spherical granular flow under different boundary conditions [20, 21]:

$$\mu = \mu_s + \frac{\mu_m - \mu_s}{1 + I_0/I},$$

where the model parameters depend on the particle characteristics [22] (e.g., shape, polydispersity or surface parameters). The increasing particle friction coefficient [23] and angularity [24] lead to higher value of the minimum Bulk friction coefficient μ_s, which is related to the angle of repose [25]. In later studies, this model is proved to be extended to non-spherical or polydisperse particle systems. The constitutive laws of both frictional and frictionless flow of superballs conform with $\mu(I)$ theory, and the bulk friction coefficient rises when particle shape gradually varies from sphere to cube, i.e., becomes more aspherical [26]. In the shear flow of triangular nonconvex particles [27], the rheological model is established by introducing particle nonconvexity to the $\mu(I)$ theory. Similarly, the characteristic diameter affected by inclination particle angle in steady state can also be introduced to the inertial number to express the flow of elongated particles [28]. Besides, in the rod-like particle shear flow, the stresses of frictionless poly-sized system can be obtained as a sum of the volume-weighted stresses of the mono-sized particle components involved in the mixture [29]. In the frictional system, the jamming volume fraction will decrease with the increasing percentage of cylinders with large aspect ratio [30]. Besides, the influence of aspect ratio on bulk friction coefficient is non monotonic when the interparticle friction coefficient μ_{pp} is lower than 0.4, while for a higher μ_{pp}, the bulk friction coefficient monotonically increases with the aspect ratio [31].

The rheology and constitutive law of different particle flows has been widely discussed in the previous research, while it is still unclear how particle shape affects the flow of polyhedral particles. In this paper, using the Discrete Element Method (DEM), flows of spherical particles and Platonic solid-shaped particles on an inclined plane have been simulated, rheological curves and the $\mu(I)$, $\phi(I)$ correlations are compared to investigate the effect of particle shape on the on the mechanical properties and flow feature of granular system. Finally, it is proposed that the shear flow of polyhedral particles can be simulated using rotational restricted spherical particles.

2 Simulations and Methods

2.1 Discrete Element Method

In this study, we utilize the software RockyDEM™. In the Discrete Element Method (DEM), The particle motion is described via Newton's second law:

$$m_i \frac{d\boldsymbol{v}_i}{dt} = \boldsymbol{F}_i, \tag{1}$$

$$\boldsymbol{I}_i \frac{d\boldsymbol{\omega}_i}{dt} - (\boldsymbol{I}_i \cdot \boldsymbol{\omega}_i) \times \boldsymbol{\omega}_i = \boldsymbol{T}_i, \tag{2}$$

where $\boldsymbol{v}_i$ and $\boldsymbol{\omega}_i$ are the particle translational and angular velocities, $\boldsymbol{F}_i$ is the resultant contact force exerted on the particle i and $\boldsymbol{T}_i$ is the moment acting on the particle mass center produced by the contact forces with the other particles.

When particles contact, the Linear spring dashpot model is used for the normal contact force:

$$\boldsymbol{F}_n^c = K_n \delta_n \boldsymbol{n} - \beta_n \boldsymbol{v}_n, \tag{3}$$

wherein the normal contact stiffness is calculated as:

$$K_n = \frac{E_i d_i \cdot E_j d_j}{E_i d_i + E_j d_j}, \tag{4}$$

d_i and d_j are the equivalent volume diameters of two contacting particles i and j. E_i and E_j are Young's moduli for contact partners.

The normal damping coefficient $\beta_n = 2\eta\sqrt{m^* K_n}$, here η is the damping ratio and $m^* = \frac{m_i m_j}{m_i + m_j}$ is the reduced mass for contacting objects i and j. δ_n represents the normal contact overlap and $\dot{\delta}_n$ is the time derivative of overlap.

The tangential contact force at time t is given by Linear spring Coulomb limit model:

$$\boldsymbol{F}_\tau^{c,t} = \min\left(\left|\boldsymbol{F}_{\tau,e}^{c,t}\right|, \mu\left|\boldsymbol{F}_n^{c,t}\right|\right) \frac{\boldsymbol{F}_{\tau,e}^{c,t}}{\left|\boldsymbol{F}_{\tau,e}^{c,t}\right|}, \tag{5}$$

$$\boldsymbol{F}_{\tau,e}^{c,t} = \boldsymbol{F}_\tau^{c,t-\Delta t} - K_\tau \boldsymbol{v}_\tau \Delta t - \beta_\tau \boldsymbol{v}_\tau, \tag{6}$$

In which μ is the particle-particle or particle-boundary friction coefficient, $\boldsymbol{F}_n^{c,t}$ is the normal contact force at time t, $\boldsymbol{F}_\tau^{c,t-\Delta t}$ is the tangential contact force at previous time step, K_τ is the tangential contact stiffness. $-\beta_\tau \boldsymbol{v}_\tau$ is a dissipation term, $\boldsymbol{v}_\tau$ is the relative tangential velocity vector at the contact point, and β_τ is the tangential damping coefficient.

Besides, the contact between polyhedral particles is more complex than spherical particles and the size of overlap δ_n should be considered, the contact detection algorithms in this study is based on the shortest-link method [32].

For the spherical particles that are restricted from rolling, the rolling friction is expressed by a constant rolling torque resistance $\boldsymbol{M}_r$,

$$\boldsymbol{M}_r = -\mu_r |R_r| F_n \frac{\boldsymbol{\omega}}{|\boldsymbol{\omega}|} \tag{7}$$

where μ_r is defined as the rolling resistance coefficient, the rolling radius R_r is the vector connecting the particle mass center and the contact point, $\boldsymbol{\omega}$ is the particle angular velocity vector.

2.2 Numerical Set-Up of Shear Flow

In the present study, granular flows of frictional spherical and Platonic solid-shaped particles (tetrahedron, cube, octahedron, dodecahedron and icosahedron) are simulated by means of the above-mentioned DEM method. The granular system consists of slightly polydisperse particle mixture of mean equivalent volume diameter d_v to avoid crystallization (10% of the average radii) and a rough bottom plane. As shown in Fig. 1a, the lengths of simulation domain in x and z (flow and transverse) directions are both $20d_v$. Periodic boundary conditions are assigned to x and z directions in order to exclude side wall effects, the inclined bumpy plane formed by randomly placed spherical particles with diameter $d = 3$ mm is used as a non-sliding boundary. In Fig. 1b, spherical and five kinds of Platonic solid-shaped particles are used to study the effect of particle shape. All the particle systems in this work have the same mean equivalent volume diameter d_v. Detailed parameters used in the DEM simulation are summarized in Table 1.

Table 1. Parameters used in DEM simulations

Parameters	Value
Mean equivalent volume diameter, d_v (mm)	3
Particle density, ρ	2500
Young's modulus, E (Pa)	3×10^7
Poisson's ratio, ζ	0.3
Coefficient of restitution, e	0.6
Particle sliding friction coefficient, μ_{sld}	0.5
Particle-boundary friction coefficient, μ_{pb}	0.5
Particle number, N	10200, 20400

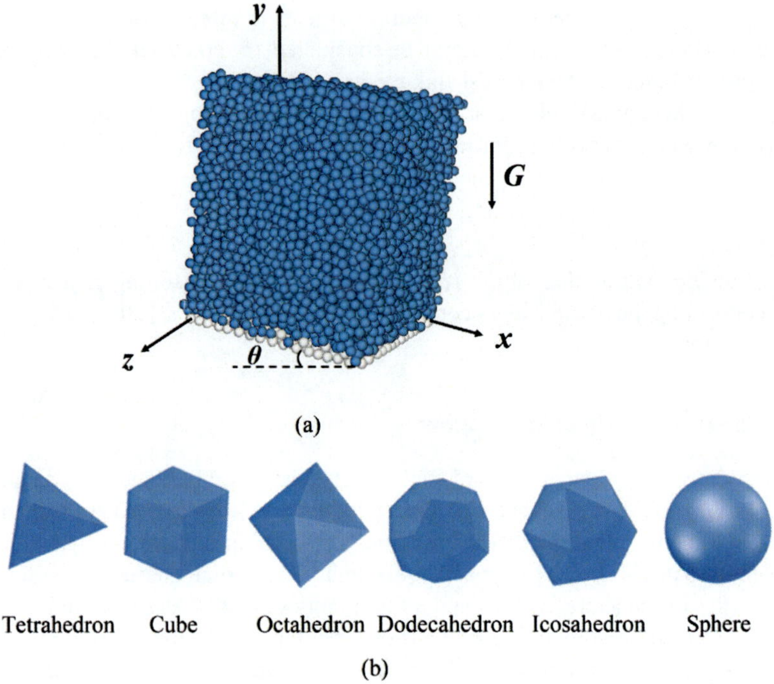

Fig. 1. (a) Snapshot of the spherical particles flowing over a bumpy inclined bottom plane and (b) the particles used in the simulations.

In the simulation, the particle bed is created by uniformly generating particles with friction coefficient $\mu_{pp} = 0.5$ in a rectangular domain with the same width and length as the simulation domain, then particles fall down under gravity in y (vertical) direction and remain stationary long enough to reach a nearly static state with inappreciable velocities. The dumpy bottom plane generated by randomly arranged particles can effectively prevent crystallization, then the granular flow is initiated by gravity G and varying flow states are reached by adjusting the inclination angle θ. In the shear process, the state of steady flow of granular system is reached when the interparticle contact number, stress components tend to converge, the data involved in the following discussion are taken from the steady flow state.

The shear rate of granular flow is $\dot{\gamma} = |dv_x/dy|$, v_x is the mean particle velocity profile. For the calculation of solid volume fraction and stress, the particle bed is divided into several layers in y direction (6 layers for $y = 10200$ and 12 layers for $y = 20400$), particles near the bottom wall and the free surface are excluded to avoid the influence of boundary.

The solid-phase stress tensor σ of a bin in the computational domain is a sum of kinetic component and σ_{kin} collisional component σ_{col}:

$$\boldsymbol{\sigma}_{col} = \sum\nolimits_{c=1}^{N_c} (\boldsymbol{F}_c \otimes l_c)/V_{bin}, \tag{8}$$

V_{bin} is the volume of representation bin, N_c is the total contact number in this bin, $\boldsymbol{F}_c$ is the contact force of a contact pair between particle i and j, and $\boldsymbol{l}_c$ is the vector connecting mass centers of two particles in contact.

$$\boldsymbol{\sigma}_{kin} = \sum_{i}^{N_p} m_i(\boldsymbol{c}_i\boldsymbol{c}_i - \boldsymbol{v}\boldsymbol{v})/V_{bin} \quad (9)$$

In which N_p represents the particle number in a bin, m_i and $\boldsymbol{c}_i$ are the mass and velocity of particle i, $\boldsymbol{v}$ is the local time-averaged particle velocity in the bin. The normal stress is the trace of stress tensor $p = (\sigma_{xx} + \sigma_{yy} + \sigma_{zz})/3$, and tangential stress is $\tau_{xy} = |\sigma_{xy}|$.

3 Results and Discussion

3.1 Flow Behavior and Stresses

We carry out simulations of spheres and Platonic solid-shaped particles, the velocities profile along y direction for granular systems with N = 10200 are illustrated in Fig. 2 for various inclination angles. The mean velocities in the flow direction (v_x) of all cases increase with the inclination angle and no slip happens near the substrate, the profile nearly match the Bagnold velocity profile and there is no steep shear band in the particle bed. Compared to spherical particles, Platonic solid-shaped particles start to flow at a larger inclination angle, because the angularity leads to stronger particle interlock and reduced fluidity. In some flow states of cubic particles, e.g., $\theta = 36°, 37°$, a "dead zone" with a height of $6d_v$ appears near the rough bottom because of the local particle stacking and ordered arrangement, this layer of particles in the flow is in a quasi-static state.

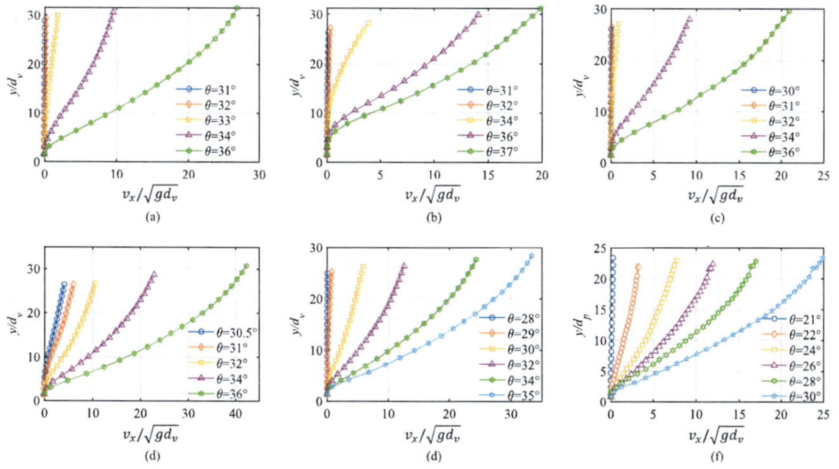

Fig. 2. Velocity profiles for granular flows of (a) tetrahedra (b) cubes, (c) octahedra, (d) dodecahedra, (e) icosahedra and (f) spheres ($N = 10200$).

Tangential stresses of spherical and Platonic solid-shaped particle flows in Fig. 3 and Fig. 4 profile linearly towards the inclined bottom along the flow layer and increase with the inclination angle. According to the momentum balance equations:

$$\rho g sin\theta = \frac{\partial \tau_{xy}}{\partial y}, \tag{10}$$

tangential stress of particle flow with different shapes on the inclined plane can be obtained by [22]:

$$\tau_{xy} = -g sin\theta \int_{y}^{H} \rho dy, \tag{11}$$

which are only related to the particle shape and height of the particle bed and declines when the particle face number increases.

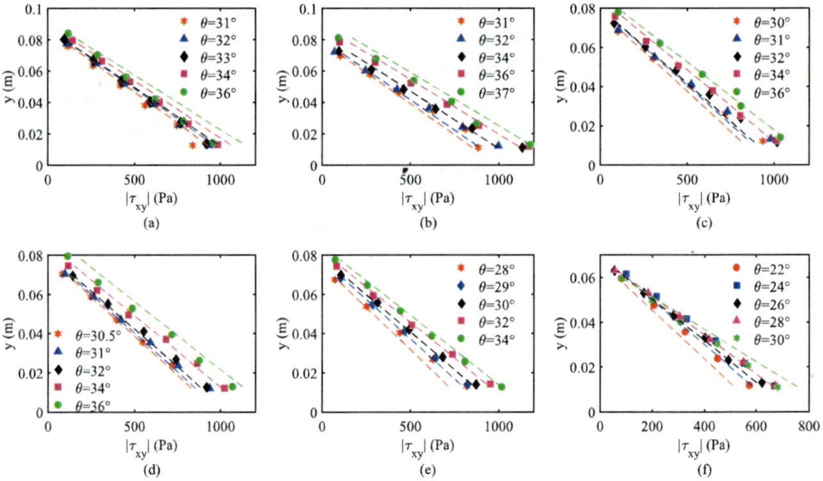

Fig. 3. Profiles of tangential stress (τ_{xy}) for granular flows of (a) tetrahedra (b) cubes, (c) octahedra, (d) dodecahedra, (e) icosahedra and (f) spheres ($N = 10200$). Solid symbols are data from simulation and dashed lines denote the prediction from Eq. (11).

3.2 The Effect of Solid Volume Fraction

For the granular shear flows of spherical and Platonic solid-shaped particles, the normalized tangential stresses, $\tau_{xy}/\rho d_v^2 \gamma^2$, are presented as a function of solid volume fraction φ in Fig. 5(a). In the frictional spherical granular flows with the $\mu_p = 0.5$, two states can be observed: inertial state and elastic state. In the inertial state when the solid volume fraction is smaller than 0.55, so that the stress slowly increases with the increase of solid volume fraction. In the elastic state with a solid volume fraction higher than 0.55, particle movement is not that active and the granular system approaches the quasi-static state.

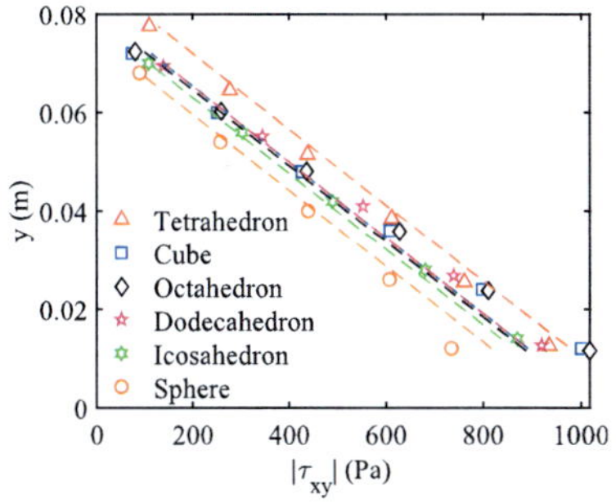

Fig. 4. Profiles of tangential stress (τ_{xy}) of particles with different shapes when $\theta = 32°$ ($N = 10200$). Solid symbols are data from simulation and dashed lines denote the prediction from Eq. (11).

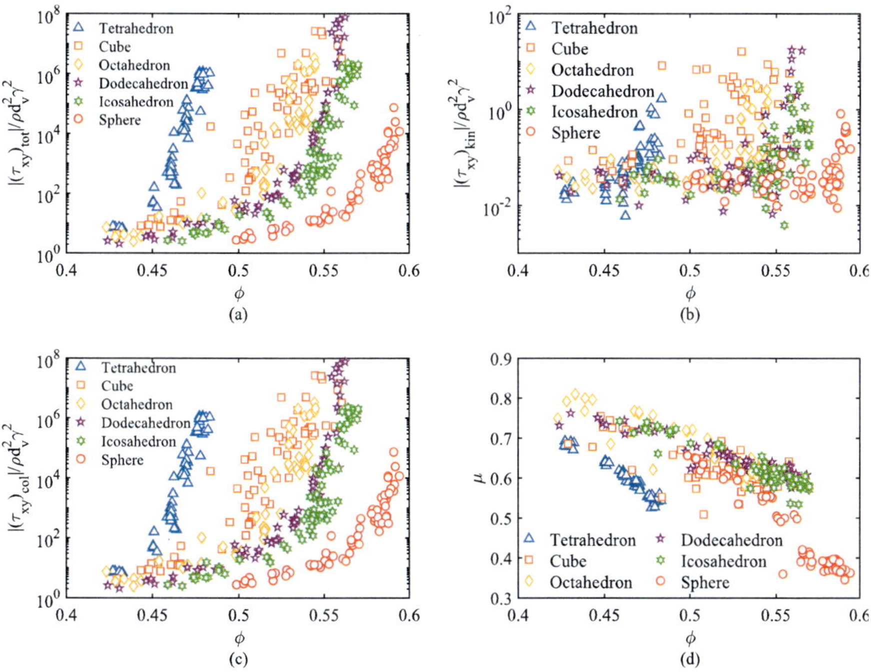

Fig. 5. Normalized (a) tangential stresses and the (b) kinetic components and (c) collisional components as a function of solid volume fraction.

The strong force chain network is formed in the granular system, and the continuous and long-lasting contact between particles results in steep stress growth with a small increase in solid volume fraction. Moreover, the kinetic components of tangential stress in Fig. 5(b) are several orders of magnitude smaller than the collision components in Fig. 5(c), which means that the stress is dominated by contact between particles.

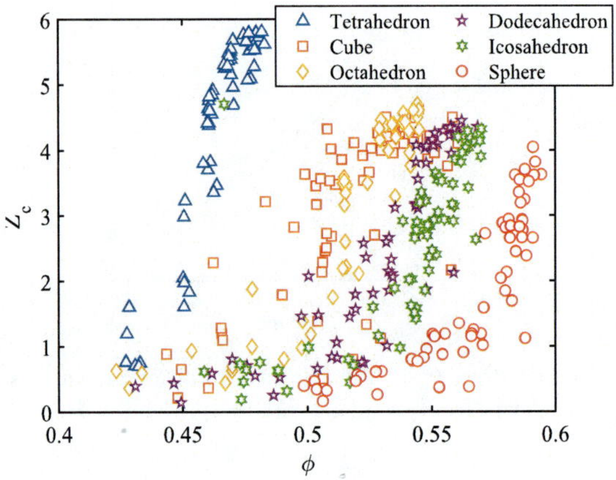

Fig. 6. Variation of coordination number Z_c with solid volume fraction.

As compared in Fig. 5(a–c), the tangential stress and stress components of Platonic solid-shaped granular flow are affected by particle shape, but follow trends similar to that of spherical particles. When the particle face number become less (particles become more aspherical), the particle interlock is stronger under the effect of angularity, the particle rotation is impaired and contact are enhanced.

As illustrated in Fig. 6, the crucial parameter to describe particle flow, coordination number Z_c of granular flow with different particle shapes has a similar tendency to that of stresses. Coordination numbers are very small in the inertial states and the solid volume fraction corresponding to the sharp increase of coordination number and stress is the same. Thus, the dense space-spanning force chain networks are more likely to form and the granular flow transfers from inertial state to elastic state at smaller solid volume fraction in granular flow of polyhedrons with less faces.

In particle system, the bulk friction coefficient μ is calculated as the ratio of tangential stress to normal stress inside the assembly. Figure 5(d) demonstrates that in the dense particle packing, μ declines linearly with the increasing φ, because the normal stress increases faster than the tangential stress in this region, this is consistent with the phenomenon observed in cylindrical particle systems [30], and the particle shape shows no noticeable influence on the $\mu - \varphi$ relationship.

3.3 The Bulk Friction Coefficient and Dilatancy

The inertial number $I = \frac{\dot{\gamma} d_v}{\sqrt{p/\rho}}$, is a dimensionless parameter to describe the flow state dense particle system. Figure 7(a) plots the bulk friction coefficient versus inertial number for spheres with different particle friction coefficient and Platonic solid-shaped particles, data near the rough bottom and free surface is removed. The bulk friction coefficients of polyhedral particle flows are much larger than that of spheres, yet the particle face number has no obvious impact on the $\mu - I$ data in granular flow of Platonic solid-shaped particles. When the sliding friction coefficient μ_{sld} increases from 0.5 to 0.8, the $\mu - I$ relationship stays almost unchanged, which is consistent with previous work. While by adding rolling friction coefficient $\mu_{rol} = 0.1$ to spheres to restrict the particle rotation, the bulk friction coefficients increase greatly and numerically close to those of Platonic solid-shaped particles. This also confirms that the difference in rheological properties between Platonic solids and spheres is led by the weakened particle rotation. So that it is reasonable to simulate flow of polyhedral particles with angularity by using spheres with rolling friction.

As in Fig. 7(a), the variation of bulk friction coefficient versus inertial number of Platonic solid-shaped particles can be captured by $\mu - I$ rheology. As an example, the parameters for granular system with $N = 10200$ of various particle shapes are detailed in Table 2. For the particles with large sphericity (dodecahedron, icosahedron, sphere), particle rolling will be weakened by corners and edges, thus the reduced sphericity gives rise to larger theoretical angle of repose θ_r for the initial state and the critical yield friction coefficient for $I \to 0$, μ_s. While for particles with large angularity (tetrahedron, cube, octahedron), μ_s and θ_r decrease with the increasing particle sphericity.

The dilatancy data obtained from the present simulations are in Fig. 7(b), the solid volume fraction of granular flow on an inclined wall increases with the particle sphericity. The solid volume fraction ϕ can be developed as a function of inertial number by $\phi = \phi_{max} - aI^b$, ϕ_{max} is the theoretical maximum random packing fraction when the granular flow transfer to the quasi-static regime of $I \to 0$. As in Table 2, ϕ_{max} grows with particle sphericity except for cubes, since particle stacks can form in the cubic granular flows due to the local ordered structure. Meanwhile, the $\varphi - I$ relationship of spherical particle flows with rolling friction is similar to those of Platonic solid-shaped particles.

As the inertial number increases, the dilatancy of the particle bed results in a decline of coordination number Z_c, as revealed by the Fig. 7(c). The angularity has little effect on coordination number when particle face number is more than 6, and coordination number increases significantly only in the granular system with tetrahedra, corresponding to a stronger force transmission network.

As proposed by Kamrin [33], in spherical granular flows, the $\mu - I$ data from different simulations can be gathered together onto one curve by the power law of dimensionless granular temperature $\Theta \equiv \rho T/P$, where the granular temperature T is calculated as $T = \frac{T_x+T_y+T_z}{3}$, $T_i = (u_i - u_i)^2$, with $i = \{x, y, z\}$, $u_i - u_i$ is the velocity fluctuation in i direction. In Fig. 7(b), similar to the spheres, the granular flow of Platonic solid-shaped particles can be predicted by combing the effect of dimensionless granular temperature Θ and the bulk friction coefficient μ. The $\mu(I, \Theta)$ relation of spherical regimes can be divided into 2 stages by the range of inertial number. In the regimes of polyhedral, the dependence of the exponent of inertial number I in the function on particle shape is insignificant (Fig. 8).

$$\mu\Theta^{1/6} = f(I) \sim \begin{cases} I^{0.5} \text{ for sphere, } I > 0.1 \\ I^{0.3} \text{ for sphere, } I < 0.1 \\ I^{0.35} \text{ for polyhedron} \end{cases} \quad (12)$$

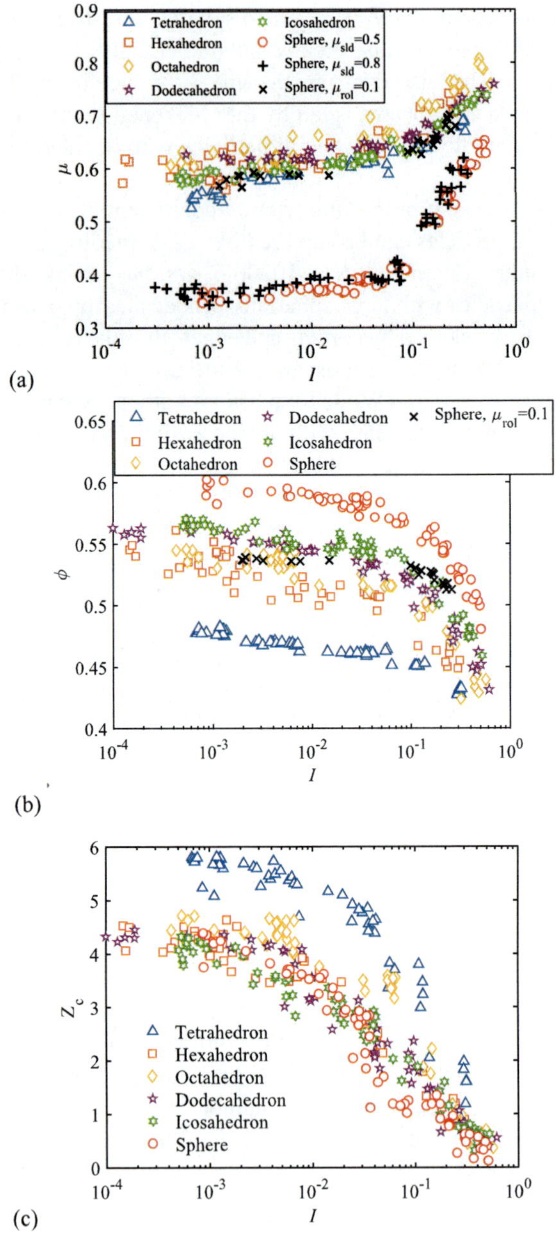

Fig. 7. Variation of (a) the bulk friction coefficient μ, (b) solid volume fraction φ and (c) coordination number Z_C versus inertial number I.

Table 2. Constitutive parameters for granular system with different particle shapes.

Shape	Sphericity, ϵ	μ_s	μ_m	I_0	Angle of repose, $\theta_r(°)$	μ_{max}
Tetrahedron	0.671	0.5583	0.6913	0.05068	29.175	0.4808
Cube	0.806	0.6004	0.8146	0.1785	30.981	0.5566
Octahedron	0.846	0.6158	0.855	0.1953	31.625	0.5414
Dodecahedron	0.910	0.6181	0.852	0.4215	31.72	0.558
Icosahedron	0.939	0.5888	0.8092	0.2394	30.49	0.5639
Sphere	1	0.3513	0.948	0.4914	19.356	0.5935

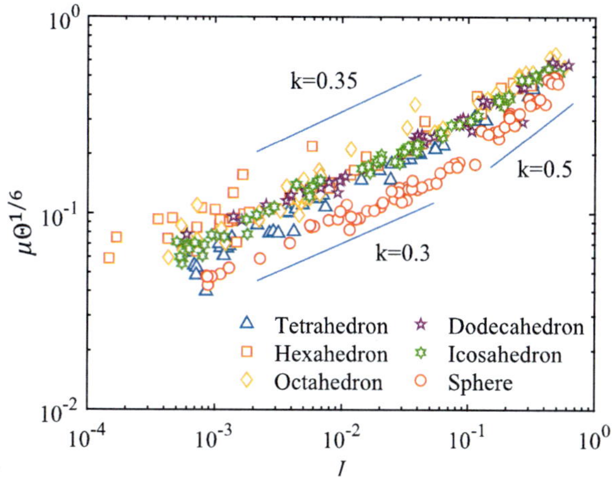

Fig. 8. Relations between bulk friction coefficient μ and inertial number I.

4 Conclusion

In summary, the flows of frictional spherical and Platonic solid-shaped particles on an inclined plane are simulated, the influence of particle shape on the rheological properties and constitutive laws are elucidated.

Compared with the spherical particles, Platonic solid-shaped particles begin to flow at a larger inclination angle θ in the granular flow on inclined plane. The particle velocity increases with the inclination angle, the tangential stress profiles linearly along the particle bed and decreases when the particle face number increases.

The coordination number Z_c and normalized tangential stress $\tau_{xy}/\rho d_v^2 \gamma^2$ of granular system both raise with the solid volume fraction φ, when particle edges become sharper, and the tangential stress will increase abruptly at a smaller solid volume fraction. Meanwhile, the bulk friction coefficients μ of Platonic solid-shaped particle flows are considerably larger than those of spheres, but the influence of particle shape on the $\mu - I$ relations of a regular polyhedron is not significant.

The bulk friction coefficient of Platonic solid-shaped particle flows of various shapes can be rescaled by a power law of granular temperature. Besides, it would be rational to simulate the shear flow of polyhedral particles by introducing rolling resistance to spheres.

References

1. Guo, Y., Wassgren, C., Ketterhagen, W., et al.: A numerical study of granular shear flows of rod-like particles using the discrete element method. J. Fluid Mech. **713**, 1–26 (2012)
2. Guo, Y., Wassgren, C., Hancock, B., et al.: Granular shear flows of flat disks and elongated rods without and with friction. Phys. Fluids (2013)
3. Campbell, C.S.: Elastic granular flows of ellipsoidal particles. Phys. Fluids (2011)
4. Binaree, T., Preechawuttipong, I., Azéma, E.: Effects of particle shape mixture on strength and structure of sheared granular materials. Phys. Rev. E (2019)
5. Guo, Y., Curtis, J.S.: Discrete element method simulations for complex granular flows. Ann. Rev. Fluid Mech. (2015)
6. Zhao, S., Zhang, N., Zhou, X., et al.: Particle shape effects on fabric of granular random packing. Powder Technol. (2017)
7. Cwalina, C.D., Harrison, K.J., Wagner, N.J.: Rheology of cubic particles suspended in a Newtonian fluid. Soft Matter (2016)
8. Malmir, H., Sahimi, M., Rahimi Tabar, M.R.: Statistical characterization of microstructure of packings of polydisperse hard cubes. Phys. Rev. E (2017)
9. Zhao, S., Zhou, X., Liu, W.: Discrete element simulations of direct shear tests with particle angularity effect. Granular Matter (2015)
10. Pouliquen, O.: Scaling laws in granular flows down rough inclined planes. Phys. Fluids (1999)
11. Azéma, E., Descantes, Y., Roquet, N., et al.: Discrete simulation of dense flows of polyhedral grains down a rough inclined plane. Phys. Rev. E (2012)
12. Höhner, D., Wirtz, S., Scherer, V.: A study on the influence of particle shape on the mechanical interactions of granular media in a hopper using the Discrete Element Method. Powder Technol. (2015)
13. Höhner, D., Wirtz, S., Scherer, V.: Experimental and numerical investigation on the influence of particle shape and shape approximation on hopper discharge using the discrete element method. Powder Technol. (2013)
14. Fraige, F.Y., Langston, P.A., Al-Khatib, L.A.: Polyhedral particles hopper flowrate predictions using discrete element method. Chem. Prod. Process Model. (2011)
15. Asencio, K., Acevedo, M., Zuriguel, I., et al.: Experimental study of ordering of hard cubes by shearing. Phys. Rev. Lett. (2017)
16. Azema, E., Radjai, F., Dubois, F.: Packings of irregular polyhedral particles: strength, structure, and effects of angularity. Phys. Rev. E Stat. Nonlinear Soft Matter Phys. (2013)
17. Zhao, S., Zhou, X.: Effects of particle asphericity on the macro- and micro-mechanical behaviors of granular assemblies. Granular Matter (2017)
18. Hossain, M., Zhu, H.P., Yu, A.B.: Numerical investigation on effect of particle aspect ratio on the dynamical behaviour of ellipsoidal particle flow. J. Phys. Condensed Matter (2021)
19. Pouliquen, O., Cassar, C., Jop, P., et al.: Flow of dense granular material: towards simple constitutive laws. J. Stat. Mech. Theory Experiment (2006)
20. Li, L., Andrade, J.E.: Identifying spatial transitions in heterogenous granular flow. Granular Matter (2020)
21. Azéma, E., Radjaï, F.: Internal structure of inertial granular flows. Phys. Rev. Lett. (2014)

22. Tripathi, A., Khakhar, D.V.: Rheology of binary granular mixtures in the dense flow regime. Phys. Fluids (2011)
23. Diego, B., Dalila, V.: Different singularities in the functions of extended kinetic theory at the origin of the yield stress in granular flows. Phys. Fluids (2015)
24. Azéma, É., Radjaï, F., Roux, J.-N.: Inertial shear flow of assemblies of frictionless polygons: rheology and microstructure. Eur. Phys. J. E (2018)
25. Trulsson, M.: Rheology and shear jamming of frictional ellipses. J. Fluid Mech. (2018)
26. Salerno, K.M., Bolintineanu, D.S., Grest, G.S., et al.: Effect of shape and friction on the packing and flow of granular materials. Phys. Rev. E (2018)
27. Hossain, M., Zhu, H., Yu, A.: Microdynamic analysis of ellipsoidal particle flow in a shear cell. In: IV International Conference on Particle-based Methods (2015)
28. Hidalgo, R.C., Szabó, B., Gillemot, K., et al.: Rheological response of nonspherical granular flows down an incline. Phys. Rev. Fluids (2018)
29. Hao, J., Li, Y., Guo, Y., et al.: The effect of polydispersity on the stresses of cylindrical particle flows. Powder Technol. (2019)
30. Hao, J., Li, Y., Liu, Y., et al.: Jamming in granular shear flows of frictional, polydisperse cylindrical particles. Adv. Powder Technol. (2021)
31. Nagy, D.B., Claudin, P., Börzsönyi, T., et al.: Flow and rheology of frictional elongated grains. New J. Phys. (2020)
32. Nezami, E.G., Hashash, Y., Zhao, D.: Shortest link method for contact detection in discrete element method. Int. J. Numer. Anal. Methods Geomech. (2006)
33. Kim, S., Kamrin, K.: Power-law scaling in granular rheology across flow geometries. Phys. Rev. Lett. (2020)

Simulation of Aerodynamic Entrainment Process of Soil Crust Based on Discrete Element Method

Wentao Hu[1,2], Zhengshi Wang[3], and Shuming Jia[1,2](✉)

[1] College of Soil and Water Conservation Science and Engineering, Northwest A&F University, Yangling 712100, Shaanxi, China
{huwt,jiashm2020}@nwafu.edu.cn

[2] State Key Laboratory of Soil Erosion and Dryland Farming On the Loess Plateau, Institute of Soil and Water Conservation, Northwest A&F University, Yangling 712100, Shaanxi, China

[3] College of Natural Resources and Environment, Northwest A&F University, Yangling 712100, Shaanxi, China
wangzs2020@nwafu.edu.cn

Abstract. Wind erosion is one of the main causes of dust storms and land desertification in arid and semi-arid areas, and the existence of soil crust significantly increases resistance to wind erosion. However, the aerodynamic entrainment mechanism of crusted soil, a typical continuous-discrete coupled system, requires further investigation. In this study, the discrete sand particles and the continuous soil crust characterized by the spring damping model combined with the parallel bonding model were established in discrete element method (DEM). And the effects of crust thickness and strength on aerodynamic entrainment were studied. Simulation results showed that in comparison to the discrete sand, soil crust increased the threshold friction velocity to a factor of two to seven. The linear increase of particle take-off velocity and take-off angle with friction velocity was unaffected, and the thickness of the crust did not affect the distribution law of take-off velocity, or angle, but it did lower the take-off angle of the particles by about fifty percent. These results provided new insights into the mechanisms underlying the effects of crust on wind erosion.

Keywords: Soil Crust · Discrete Element Method (DEM) · Aerodynamic Entrainment

1 Introduction

The wind-blown sand movement is an important process that affects sediment transport, landform evolution, and climate change on Earth, Mars, and other terrestrial planets [1, 2]. As the first stage of the aeolian transport, the aerodynamic entrainment directly determines the threshold of sand particle motion and its subsequent development, and it also plays a crucial role in the unsaturated wind-sand flows [3, 4]. In arid and semi-arid regions, the soil crust was a key factor in reducing wind erosion and was more valuable than other factors [5].

X. Zheng and S. Balachandar (Eds.): IUTAM 2023, IUTAM Bookseries 41, pp. 178–187, 2024.
https://doi.org/10.1007/978-3-031-47258-9_11

Soil physical crust is a compact structure formed on the soil surface by non-biological processes such as raindrop splash and runoff erosion [6]. As a network of interconnected particle groups, the soil crust provides physical shields for soil particles, resisting the abrasion of the salting grains [7]. The rate of wind erosion on non-crusted surfaces was six to forty times that of crusted surfaces [7, 8]. As for the mechanism of aerodynamic entrainment, Anderson and Haff proposed a linear scaling law for aerodynamic entrainment based on momentum conservation, which has been extensively utilized in simulations of wind-blown sand movement [9]. Jia and Wang proposed a complete scheme for aerodynamic entrainment based on the discrete element method (DEM), and discovered that the cohesion force of particles increased the threshold friction velocity of aerodynamic entrainment, while almost not affecting the distribution of take-off angles and velocities [10].

Previous numerical simulations of aerodynamic entrainment have investigated the sand particle, but only a few on the continuum of the soil crust, which was commonly found in arid and semi-arid regions. Furthermore, the aerodynamic entrainment of the crust particle in a continuous-discrete coupled system (i.e., crust-sand) has not been studied. Although DEM simulations on the formation of the physical crust have been provided [11], additional research is required to determine how crust properties affect aerodynamic entrainment of particles.

Therefore, this study firstly introduces a parallel bonding model characterizing the continuum of crust and combines it with a spring damping model characterizing the discrete of sand. Secondly, a continuous-discrete coupled model of crust-sand sediment bed is developed. Thirdly, this study examines the evolution of particle diameter as a function of threshold friction velocity and the response time of aerodynamic entrainment under crusted conditions. Finally, the effects of crust strength and thickness on the aerodynamic entrainment of the crust particle are investigated.

2 Model and Method

2.1 Discrete Element Method of Particle Motion

The simulation was performed with the Discrete Element Method (DEM) developed by Cundall and Strack [12]. The governing equations of particle motion in DEM simulation are Newton's second laws, and the equations of translation and rotation are as follows

$$m\frac{dv_p}{dt} = F_c + F_d + mg \tag{1}$$

$$I\frac{d\omega_p}{dt} = M_c + M_d \tag{2}$$

where m and I stand for the particle mass and moment of inertia, v_p and ω_p are the particle velocity and angular velocity, respectively, $F_c = \sum_{i=1}^{n} F_{p_i}$ is the resultant contact force of all normal and frictional contacts, F_d means the fluid drag force, g denotes the gravitational acceleration, M_c and M_d indicate the contact and applied external moments acting on the particle, respectively.

2.2 Fluid Drag Force Model

For particles with no exposed area, the corresponding fluid drag force is zero, whereas for the exposed particles [13], the fluid drag force F_D is

$$F_D = m\frac{V_r}{T_p}f(\mathrm{Re}_p) \tag{3}$$

where $T_p = \rho_p d_p^2/(18\rho_f \nu) \sim 1.0e - 2$ is the relaxation time [10], ρ_f and ν stand for the density and kinetic viscosity of air, respectively, $V_r = v_f - v_p$ means the relative velocity of the fluid and particle, $f(\mathrm{Re}_p)$ indicates a function related to the Reynolds number calculated as follows [14]

$$f(\mathrm{Re}_p) = \begin{cases} 1 & (\mathrm{Re}_p < 1) \\ 1 + 0.15\mathrm{Re}_p^{0.687} & (\mathrm{Re}_p \geq 1) \end{cases} \tag{4}$$

where $\mathrm{Re}_p = d_p V_r/\nu$ is the Reynolds number.

2.3 Dimensionless Parameters of Crust Strength and Thickness

The crust in the model was consistent with the structural crust described by Valentin and Bresson [6]. Compared to the biological crust, the structural crust has a simpler structure. Due to the lack of consistency in measuring crust properties, the following dimensionless parameters were proposed to characterize the crust strength and thickness

$$CS = F_\phi/G \tag{5}$$

$$CT = H/\mathrm{D} \tag{6}$$

where CS is the dimensionless crust strength, F_ϕ means the cohesions between crust particles, G stands for the gravitational force of the crust particle, CT is the dimensionless crust thickness, H indicates the vertical thickness of the crust, D means the diameter of the crust particle.

Field observations demonstrated that the thickness of the structural crust ranged from 1000 to 3000 μm, and wind tunnel experiments conducted by Sharratt and Vaddella revealed that some crusts thicker than two millimeters did not reach the threshold friction velocities [15]. Therefore, in this study, it was assumed that the crust thickness corresponded to two to ten times the diameter of the crust particles, i.e., a range of 400 to 2,000 μm under realistic conditions.

2.4 Discrete-Continuum Coupled Model for Sand-Crust Particles

A. The parallel bonding model for characterizing the crust

In this study, the cohesion force of the crust particles was characterized by the parallel bonding model [16], which was also commonly used in other cohesive particles such as ice [17] and snow [18]. The cohesion force was described with the following equations

$$\sigma = -\frac{F_n}{A} + |T|\frac{r_b}{L} \tag{7}$$

$$\tau = -\frac{F_s}{A} \tag{8}$$

where F_n and F_s are normal force and shear force, respectively, r_b means the bond radius, $A = \pi r_b^2$ indicates the bond area, T stands for the bending moment, $L = \pi r_b^4/4$ denotes the moment of inertia. If the normal or tensile stress exceeds the normal or tensile strength, the bond breaks and both F_n and F_s are set to zero, but the contact forces are not altered.

B. The spring damping model for characterizing sand particles

The sand particles were characterized by the spring damping model [12], with the following equations

$$F_n = \begin{cases} k_n \times \Delta\delta_n - \beta_n \times v & \Delta\delta_n \geq 0 \\ 0 & \Delta\delta_n < 0 \end{cases} \tag{9}$$

$$F_s = \begin{cases} k_s \times \Delta\delta_s - \beta_s \times r & F_s \leq \mu F_n \\ \mu \times F_n & F_s > \mu F_n \end{cases} \tag{10}$$

where F_n and F_s are normal force and shear force, $\Delta\delta_n$ and $\Delta\delta_s$ stand for normal and shear overlap of particles, β_n and β_s indicate normal and shear viscous damping ratio, respectively, v and r denote the normal relative velocity and shear relative deformation rate, respectively, μ means the frictional coefficient.

2.5 Wind Profile

The wind profile (Fig. 1) is applied through a standard logarithmic law [3]

$$u_x = \frac{u_*}{\kappa}\ln\left(\frac{z}{z_0}\right) \tag{11}$$

where u_* is the fluid shear velocity, $\kappa = 0.4$ is the Karman constant, z stands for the altitude above the bed surface, $z_0 = d_p/30$ means the aerodynamic roughness [1].

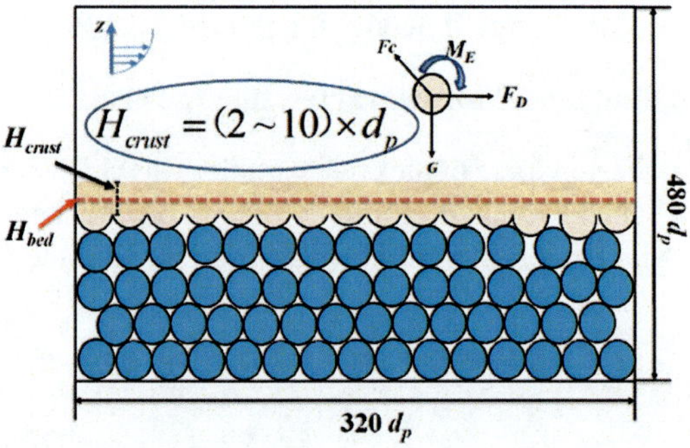

Fig. 1. Schematic diagram for determination of the bed height, crust thickness and forces applied on the airborne crust particle ($d_p = 200$ μm)

2.6 Simulation Procedures and Model Settings

The numerical simulation domain was set as $L_x \times L_y = 320\,d_p \times 480\,d_p$. The top and bottom boundaries were set reflective conditions, and the left and right boundaries were set periodic conditions. Simulation procedures are as follows.

1. A amount of 8000 particles were randomly generated within the domain to form a sediment bed. The criterion for the calm bed was $V_{p\max} \leq 1.0e - 4$ m/s.
2. The particles above H_0 ($H_0 = 12d_p$) were removed to form a flat sediment bed. Three repetitions of the above step ensured a calm and steady state. The porosity of the bed was approximately 0.5. The bed height $H_{bed} = H_0 - pd_p$ at which the mean wind velocity reduced to zero was determined (Fig. 1).

After generation, particles were divided into two groups including crust and sand, and different physical properties and contact models (Table 1), were set to characterize the continuum and the discrete type. The force chain network of the coupled model was shown in Fig. 2.

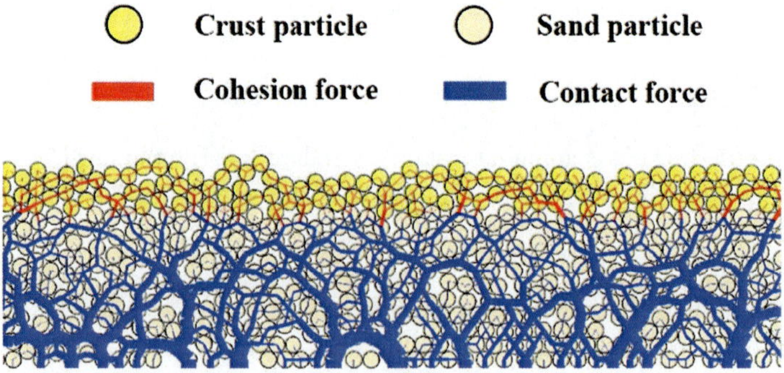

Fig. 2. Distributions of cohesion forces and contact forces in force chain network formed by the discrete-continuum coupled model

Table 1. Parameters used in DEM simulation

Parameter	Description	Value	Unit
k_n	Normal stiffness	2e6	Nm^{-1}
k_s	Shear stiffness	2e6	Nm^{-1}
μ	Frictional coefficient	0.30	–
σ_c	Normal bonding strength	4e4–5e4	Pa
τ_c	Shear bonding strength	4e4–5e4	Pa
ρ_p	Particle density	2650	kgm^{-3}
d_p	Particle diameter	200	μm
ρ_f	Air density	1.225	kgm^{-3}
v	Kinematic viscosity of air	1.5e-5	ms^{-2}
βn	Normal viscous damping ratio	0.2	–
βs	Shear viscous damping ratio	0.2	–
E	Young's modulus	1e6	Nm^{-2}

3 Results and Discussion

3.1 Model Validation

The threshold friction velocity determines the subsequent movement of particles. As shown in Fig. 3a, the linear scaling law between the particle diameter and the threshold friction velocity were verified, compared to the existing simulation result. In addition, three groups of field experiments under crusted conditions were included.

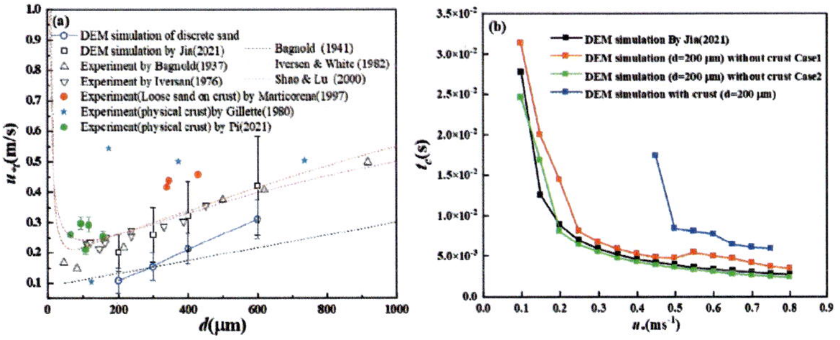

Fig. 3. **a.** Comparisons of simulated and measured threshold velocities u_{*t} as a function of particle diameter d; and **b** validations of response time t_c as a function of threshold velocities u_{*t} under sand and crusted conditions compared to Jia [10]

The response time of the first lifting particle was also investigated under crusted conditions. As shown in Fig. 3b, with the inter-particle cohesions, the exponential decline law between response time and friction velocity under crusted conditions was consistent with those in sand cases.

3.2 Influences of Crust Strength and Crust Thickness on the Threshold Aerodynamic Friction Velocity

Under the condition of no inter-particle cohesions, the threshold friction velocity of sand particles ($d_p = 200$ μm) was approximately 0.1 m/s, as shown in Fig. 4. The threshold friction velocity increased by approximately two to seven times under crusted conditions, and the threshold friction velocity increased linearly with crust strength and thickness. However, once the crust thickness reached the critical value of 5 D, the threshold friction velocity nearly remained constant as the crust thickness increased.

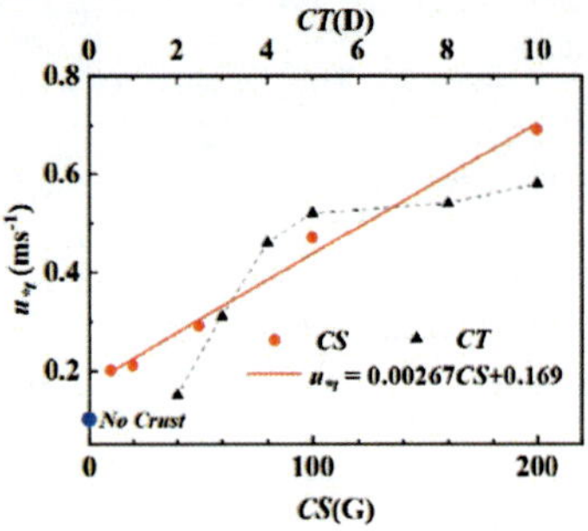

Fig. 4. Crust strength (CS) and Crust thickness (CT) as functions of threshold velocities u_{*t}. The red line is the linear regression equation of crust strength F_Φ versus threshold velocities u_{*t}, and the dash line with scatter for crust thickness H versus threshold velocities u_{*t}

Since the fluid drag force was related to the friction velocity, the linear increase between threshold friction velocity and crust strength could be explained by the larger fluid drag force required to brake the parallel bonds.

This phenomenon occurred because the distance between crust particles was constrained by the inter-particle cohesion force. Within a certain range, crust particles from the lower layer had a "delaying" effect on the aerodynamic entrainment of crust particles from the upper layer. However, beyond a certain range, this "delaying" effect was only manifested by the influence of particles in the upper layer and was independent of crust thickness.

3.3 Distributions of Take-Off Velocity and Angle Under Different Crust Strength and Crust Thickness

Here, the particle take-off velocity is defined as $\mu_p = \sqrt{\mu_u^2 + \mu_w^2}$, where μ_p is the particle velocity, decomposed into the streamwise velocity (μ_u) and vertical velocity (μ_w) along the X-axis and Y-axis, respectively. The particle velocity was obtained directly from the program output.

The streamwise and vertical velocities of particle take-off exhibited a consistent linear scaling law in the absence of cohesions. As shown in Fig. 5a and 5b, the linear scaling law of take-off velocity with the friction velocity still remained when the crust strength reached 100 G and 200 G, which was consistent with the simulation results

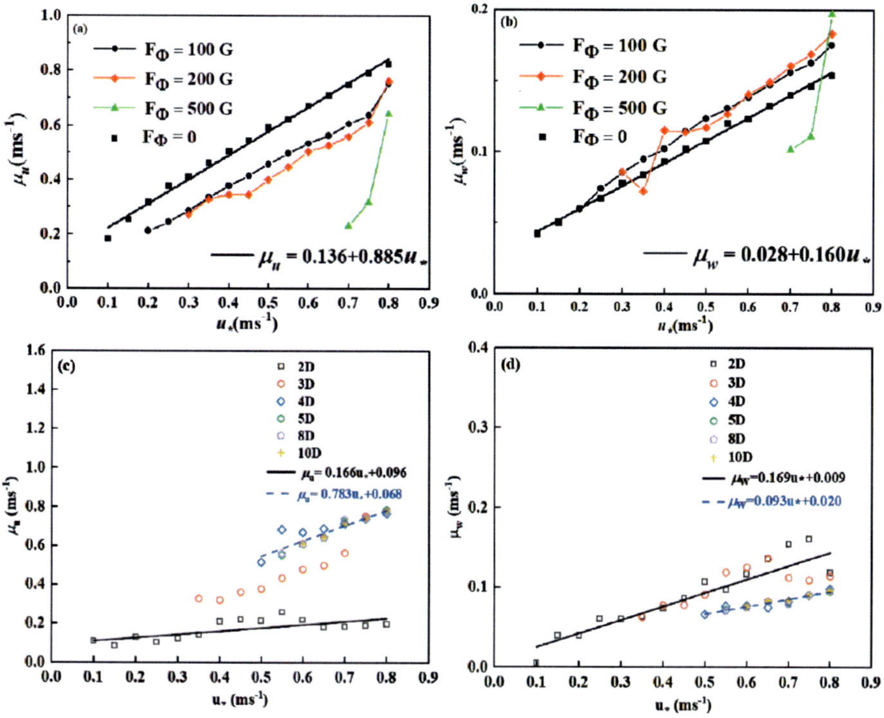

Fig. 5. Evolutions of streamwise velocities μ_u and vertical velocities μ_w versus the friction velocity $u*$ under different crust strengths F_Φ for **a** and **b**, respectively; and crust thicknesses H for **c** and **d**. The data of $H = 2$ D, 5 D, 8 D and 10 D are calculated in linear regression

of Jia and Wang [10]. When the crust strength exceeded 200 G, the take-off velocity changed from a linear scaling law to an exponential scaling law, and this relationship was significant at 500 G.

When the crust thickness exceeded 5 D, the take-off velocity remained almost unchanged with the increase of crust thickness. The critical value of the crust thickness affecting take-off velocity was between 4 D and 5 D. The take-off velocity and angle reflected the first lifting particle kinetic energy, and the aerodynamic entrainment process converted fluid kinetic energy into particle kinetic energy. The inter-particle cohesions just postponed this conversion process without affecting the linear scaling law, whereas the excessive cohesions prevented particles from accumulating energy from sliding and collision processes of the aerodynamic entrainment.

The particle take-off angle is defined as $\alpha_u = \arctan(\mu_w/\mu_u)$, where μ_w is the vertical velocity and μ_u is the streamwise velocity.

Figure 6 presented the take-off angle α_u as a function of friction velocities $u*$ under conditions of sand and crust thicknesses. It can be found that the take-off angle α_u was roughly distributed between 12° and 13° without F_Φ, and the maximum take-off angle at various $u*$ increased from roughly 15° to 26° with the increase of F_Φ.

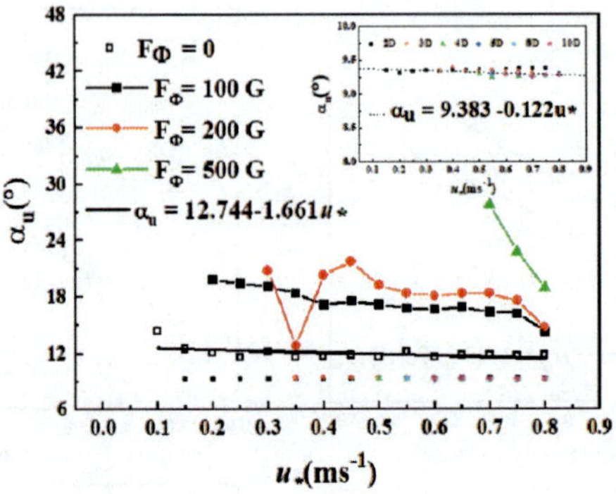

Fig. 6. Take-off angle α_u as a function of friction velocities u_* under conditions of sand and crust thicknesses H

Seen from Fig. 6, the take-off angle α_u remained basically constant with the increase of crust thickness H, roughly distributed in the range of 9.2° to 9.4°. When H exceeded 5 D, the evolution of the α_u with u_* remained relatively stable. In this case, the u_* did not affect the α_u under thicker crusted conditions.

4 Conclusions

Combining the parallel bonding model and the spring damping model, this study presents a coupled continuous-discrete model of crust-sand sediment bed, and investigates the effects of crust strength and thickness on the aerodynamic entrainment process. The study reveals that the crust strength delays the response time and increases the threshold friction velocity of aerodynamic entrainment to about two to seven times compared to non-crusted conditions, but it does not change distribution laws of particle take-off velocity and angle with the friction velocity. A critical value of five times of crust thicknesses is determined and the take-off angle under crusted conditions is approximately fifty percent smaller than that of non-crusted conditions. The fluid drag force, inter-particle cohesions and the gravity determine the aerodynamic entrainment, and the crust exhibits obstructive and delaying characteristics in the process.

This study proposes a comprehensive aerodynamic entrainment scheme for sand particles under crusted conditions, providing more accurate predictions for sandstorms and desertification, and offers an insight into the optimization of desertification control measures.

Acknowledgements. This work was supported by the National Natural Science Foundation of China (11902341, 12102364), the China Postdoctoral Science Foundation (2021M702679), the Young Talent Fund of University Association for Science and Technology in Shaanxi, China (20210503), and the Chinese Universities Scientific Fund (2452021048, 2452020266).

References

1. Bagnold, R.A.: The Physics of Blown Sand and Desert Dunes. Methuen, London (1942)

2. Kok, J.F., Parteli, E.J., Michaels, T.I., Karam, D.B.: The physics of wind-blown sand and dust. Rep. Prog. Phys. Phys. Soc. **75**, 106901 (2012)
3. Shao, Y.: Physics and Modelling of wind Erosion. Springer, Newyork (2008). https://doi.org/10.1007/978-1-4020-8895-7_9
4. Li, G., Zhang, J., Herrmann, H.J., Shao, Y., Huang, N.: Study of aerodynamic grain entrainment in aeolian transport. Geophys. Res. Lett. **47**, e2019GL086574 (2020)
5. Hagen, L., Zobeck, T., Skidmore, E., Elminyawi, I.: WEPS technical documentation: soil submodel. In: SWCS WEPP/WEPS Symposium, Ankeny, IA, pp. 9–11. Soil and Water Conservation Society, Ankeny (1995)
6. Valentin, C., Bresson, L.-M.: Morphology, genesis and classification of surface crusts in loamy and sandy soils. Geoderma **55**, 225–245 (1992)
7. Chepil, W.S.: Soil Conditions that Influence Wind Erosion. USDA-ARS Tech. Bull, Washington, DC (1958)
8. Zobeck, T.M.: Soil properties affecting wind erosion. J. Soil Water Conserv. **46**, 112–118 (1991)
9. Anderson, R.S., Haff, P.K.: Simulation of eolian saltation. Science **241**, 820–823 (1988)
10. Jia, S., Wang, Z.: Simulation of aerodynamic entrainment with interparticle cohesions based on discrete element method. Earth. Surf. Proc. Land. **46**, 1410–1418 (2021)
11. Yeom, S., Sjoblom, K.: Structural soil crust development from raindrop impacts using two-dimensional discrete element method. Comput. Geosci. **97**, 49–57 (2016)
12. Cundall, P.A., Strack, O.D.: A discrete numerical model for granular assemblies. Geotechnique **29**, 47–65 (1979)
13. Dey, S., Papanicolaou, A.: Sediment threshold under stream flow: a state-of-the-art review. KSCE J. Civ. Eng. **12**, 45–60 (2008)
14. Clift, R., Grace, J.R., Weber, M.E.: Bubbles, Drops, and Particles. Dover Publications, Newyork (2005)
15. Sharratt, B., Vaddella, V.: Threshold friction velocity of crusted windblown soils in the Columbia Plateau. Aeol. Res. **15**, 227–234 (2014)
16. André, D., Iordanoff, I., Charles, J.-L., Néauport, J.: Discrete element method to simulate continuous material by using the cohesive beam model. Comput. Methods Appl. Mech. Eng. **213–216**, 113–125 (2012)
17. Long, X., Liu, L., Liu, S., Ji, S.: Discrete element analysis of high-pressure zones of sea ice on vertical structures. J. Mar. Sci. Eng. **9**(3), 348 (2021)
18. Comola, F., Kok, J.F., Gaume, J., Paterna, E., Lehning, M.: Fragmentation of wind-blown snow crystals. Geophys. Res. Lett. **44**(9), 4195–4203 (2017)

Turbulent Thermal Convection Driven by Heat-Releasing Particles

Liangbing Chen, Zimo Liao, Zhenhua Wan, Nansheng Liu(✉), and Xiyun Lu

Department of Modern Mechanics, University of Science and Technology of China, Hefei 230026, People's Republic of China
{chlb,zimoLiao}@mail.ustc.edu.cn, {wanzh,lns,xlu}@ustc.edu.cn

Abstract. Thermal convection driven by heat-releasing particles in a quasi-3D cavity has been investigated through four-way coupled Euler-Lagrange direct numerical simulations. To examine the effects of thermal buoyancy and particle inertia, the Rayleigh-Robert number and density ratio are chosen in the range of $4.97 \times 10^5 \leq Rr \leq 4.97 \times 10^8$ and $1 \leq \rho_p/\rho_f \leq 250$, respectively, and the Prandtl number $Pr = 1$. Compared to the convection driven by uniform internal heating, it is found that the particles of small inertia ($\rho_p/\rho_f = 1$) have a weak effect on enhancing convective heat transfer as Rr increases. The ensemble-averaged temperature is highly dependent on the spatial distribution of particles. For higher inertia ($\rho_p/\rho_f \geq 10$), the particles are expelled away from the vortex core by the centrifugal force originating in the particle-vortex interaction. It results in the preferential accumulation of particles commonly at the vortex edges and, consequently, a highly non-uniform distribution of heat sources. Of interest, the occurrence of preferential accumulation may suppress or promote the heat escaping through the walls, namely, redistributing the heat flux on the top and bottom walls.

Keywords: particle-laden flow · Heat-releasing particles · thermal convection

1 Introduction

Particle-laden thermal flow can be commonly encountered in various natural and engineering applications, such as volcanic ash clouds spreading into the atmosphere [1] and solar receivers that utilize particle-laden molten salt as a medium to absorb solar heat [2], etc. A distinctive feature of these flow systems is the coupling of momentum and heat between the carrier phase and the dispersed phase, which spans multiple scales. In addition, the fluid motions are affected by the heat sources from particles, which brings a flurry of interesting phenomena and daunting challenges.

The physical model that buoyancy convection driven by the radiative heating of particles has gradually gained attention in recent years [3–7]. In this model, the fluid motions are driven by the heat sources from particles, so the temperature fields around particles are completely different from the isothermal particles in thermal flow. Zamansky [8,9] numerically studied turbulent thermal convection driven by particle heat

X. Zheng and S. Balachandar (Eds.): IUTAM 2023, IUTAM Bookseries 41, pp. 188–195, 2024.
https://doi.org/10.1007/978-3-031-47258-9_12

release in a three-period square cavity and analyzed the dependence of velocity and temperature on Stokes Number (St). Pan [3] and Yang [5,6,10] investigated the Rayleigh-Bérnard convection (RBC) with heat-releasing particles and they focused on the effect of different particle sizes on the boundary layer thickness and Nusselt number (Nu). The most recent work of Du [4] conducted the wall-bounded thermal turbulent convection driven by heat-releasing inertial particles with two-way thermal coupling and one-way momentum coupling. The mechanism of particle concentration at the wall is discovered and scaled.

The work mentioned above is mainly conducted by two-way or even one-way coupling, which is valid for dilute ($10^{-6} \leq \Phi_v \leq 10^{-3}$) or dusty ($\Phi_v \leq 10^{-6}$) flow, where Φ_v is the volume fraction of particles. However, the two-way coupled simulation of particle-laden flow initially lying in the dilute dispersion regime is likely to fail due to the extreme local accumulation of particles. When the dispersed phase is dense as $\Phi_v \geq 10^{-3}$, the interaction between particles, including collision, heat conduction, agglomeration, and breakup, is important, and only four-way coupling can characterize these flows properly.

In this work, thermal convection driven by heat-releasing particles in a quasi-3D cavity has been investigated through four-way coupled Euler-Lagrangian direct numerical simulations. It is found that flow motions and heat transfer processes are highly dependent on the spatial distribution of particles. Moreover, the centrifugal force originating in the inertia and particle-vortex interaction results in the preferential accumulation of particles and, consequently, a highly non-uniform distribution of heat sources.

2 Problem Set-Up and Governing Equations

As Fig. 1 illustrates, a square cavity with a height of H and width of W is placed vertically, in which a layer of particle-laden fluid is filled. The width of the square cavity is small compared to the height and large compared to the particle diameter, namely $d \ll W \ll H$, where d is the particle diameter. In this model, we assume that the velocity of fluid in z-direction is small so that can be neglected. The flow characteristic tends to be 2D and the particle feature is still 3D.

No-slip and no-penetration boundary conditions are adopted for all four walls. The side walls are thermal insulation, while the top and bottom walls keep a constant temperature T_0. In the initial state, the temperature of the particles and fluid is T_0, and particles are randomly distributed in the square cavity. Particles raise their temperature by absorbing external radiation. With the Oberbeck-Boussinesq approximation, the fluid density is simply assumed to be linearly dependent on temperature, namely $\rho = \rho_f[1 - \beta(T - T_0)]$, where ρ_f and β are the fluid density and thermal expansion coefficient at reference temperature T_0, respectively. The particle diameter is assumed to be smaller than the Kolmogorov scale. The dimensionless governing equations can be readily obtained as

$$\nabla \cdot \boldsymbol{u} = 0 \tag{1a}$$

$$\partial_t \boldsymbol{u} + \nabla \cdot (\boldsymbol{u}\boldsymbol{u}) + \nabla p = Pr^{1/2} Rr^{-1/2} \nabla \cdot (\nabla \boldsymbol{u}) + T\boldsymbol{e}_y + \boldsymbol{F}_p \tag{1b}$$

$$\partial_t T + \nabla(\boldsymbol{u}T) = Pr^{-1/2} Rr^{-1/2} \nabla \cdot (\nabla T) + Q_p + Q_f^o \tag{1c}$$

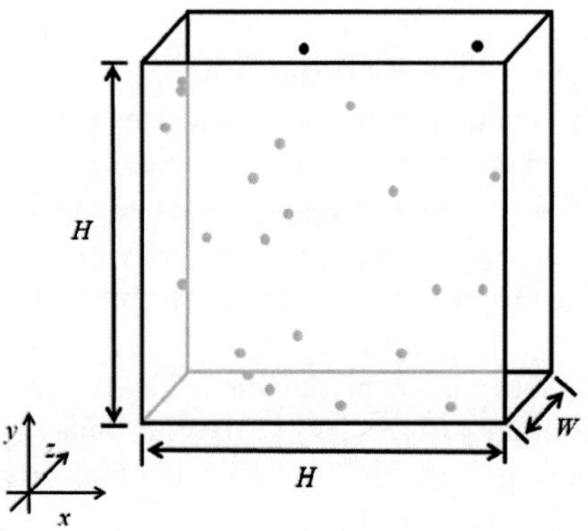

Fig. 1. The configuration of the convection driven by heat-releasing particles.

$$\frac{\mathrm{d}\boldsymbol{v}_i}{\mathrm{d}t} = \sum_j \boldsymbol{F}_{ij}^c + \frac{C_d Re_p}{St_p}(\boldsymbol{u}_i - \boldsymbol{v}_i) \tag{1d}$$

$$\frac{\mathrm{d}T_i}{\mathrm{d}t} = \sum_j Q_{ij}^c + \frac{Nu_p}{St_\theta}(T_{fi} - T_i) + Q_i^o \tag{1e}$$

which are nondimensionalized by the characteristic scales

$$T_{ref}^* = \frac{H^{*2} Q_f^{o^*}}{8\kappa\rho_0 C_{pf}^*} \tag{2a}$$

$$\boldsymbol{u}_{ref}^* = \sqrt{g^* \beta T_{ref}^* H^*} \tag{2b}$$

$$\tau_{ref}^* = \frac{H^*}{\sqrt{g^* \beta T_{ref}^* H^*}} \tag{2c}$$

$$H_{ref}^* = H^* \tag{2d}$$

Superscript $(\cdot)^*$ emphasize that the physical quantities and operators are dimensional. $\boldsymbol{u}$, p, T, ν, κ are the velocity vector, pressure, temperature, kinematic viscosity, and thermal diffusivity of the carrier phase, respectively. $\boldsymbol{g}$ is the gravitational acceleration vector that points in the negative direction of the y-axis. $\boldsymbol{F}_p$ and Q_p represent the momentum and heat exchange rate between fluid and particles. Q_f^o is the heat flux that the fluid absorbs from external sources. $\boldsymbol{v}_i$, and T_i are the velocity vector and temperature of the i-th particle, $\boldsymbol{F}_{ij}^c$ and Q_{ij}^c are the contact force and heat conduction flux between the i-th and j-th particle or j-th particle and wall when they come into contact. Q_i^o is the power of external energy absorbed by the i-th particle. For simplification, we assume that all particles have the same physical properties, including Q_i^o. The particle Reynolds number Re_p is defined as $Re_p = \rho_f d_p |\boldsymbol{u}_i - \boldsymbol{v}_i| / \mu_f$.

The key parameters include the Prandtl number Pr, the Rayleigh-Robert number Rr, the momentum Stokes number St_p, the thermal Stokes number St_θ, which are defined respectively as

$$Pr = \frac{\nu}{\kappa}, \quad Rr = \frac{g\beta T_r e f H^3}{\nu\kappa}, \quad St_p = \frac{\tau_p}{\tau_f}, \quad St_\theta = \frac{\tau_\theta}{\tau_f}, \quad n_p = \frac{NH}{W} \tag{3}$$

where $\tau_p = \rho_p d^2/18\rho_0\nu$ is the momentum relaxation time and $\tau_\theta = d^2\rho_p C_{pp}/\kappa\rho_0 C_{pf}$ the thermal relaxation time. St_p and St_θ are the ratio of particle momentum and thermal relaxation time to flow characteristic time $\tau_f = \tau^*_{ref}$, respectively. Due to the influence of particles on τ_f, here we use density ratio ρ_p/ρ_f to represent the strength of inertial effect of particle. n_p indicates the number of particles per unit volume. The definition of Rr has the same form as that in RBC, and represents the strength of buoyancy.

The discrete element method (DEM) is adopted to calculate the motion and temperature of each individual particle separately. The soft-particle approach is adopted here to model the complex interactions between particles, namely $\boldsymbol{F}^c_{ij}$ and Q^c_{ij} [11]. In such an approach, the particles are permitted to suffer minute deformations, and the elastic, plastic, and frictional forces, as well as the heat flux between particles, can be modeled. The drag coefficient C_d and Nusselt number Nu_p are calculated by semi-theoretical and semi-empirical model [12,13].

The CFD-DEM method we adopted, as well as the consideration of four-way coupling between particles and fluid, is able to guarantee the accuracy of numerical results at higher particle volume fractions ($\Phi_v \leq 0.1$). Therefore, it is more reliable for the simulation of particle-laden flows with high local aggregation.

3 Results Discussion

The parameters covered by the present simulations are shown in Fig. 2, the Prandtl number is fixed at $Pr = 0.7$, and the thermal Stokes number is $St_\theta \ll 1$. All the above simulations have reached the statistically steady state, and the flow duration exceeds $1260\tau_f$ and increase to approximately $9960\tau_f$ as Rr increase.

According to the study of Wang [14], the $\langle T\rangle$ exhibit a $-1/5$ power relationship to Rr, where $\langle\cdot\rangle$ is the ensemble average, namely the temporal and spatial average. Figure 3(*a*) show the effect of n_p and Rr on $\langle T\rangle$ under the same density ratio $\rho_p/\rho_f = 1$. And $n_p = 0$ represents the reference case that convection is driven by uniform internal

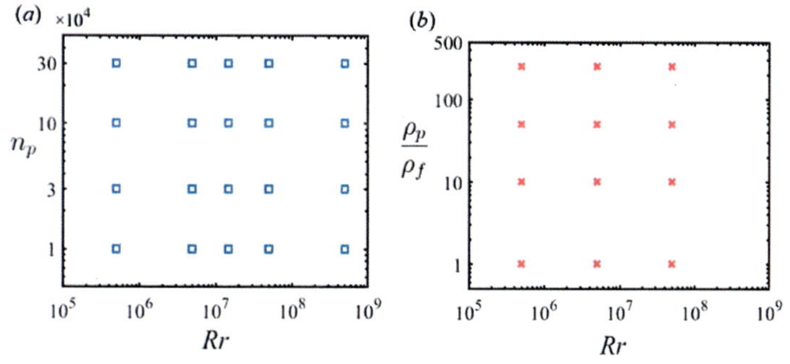

Fig. 2. The parameter space in the $Rr - n_p$ plane (a) and $Rr - \rho_p/\rho_f$ plane (b).

heating that heat sources are from the fluid directly. It is found that the power relationship does not change in particle-driven thermal convection, which indicates the inherent unity between these flows. The particles with small inertia ($\rho_p/\rho_f = 1$) have a weak effect on the global average temperature $\langle T \rangle$. However, as the inertia of particles increases, the average temperature $\langle T \rangle$ decrease and the decrease becomes more significant at higher Rr.

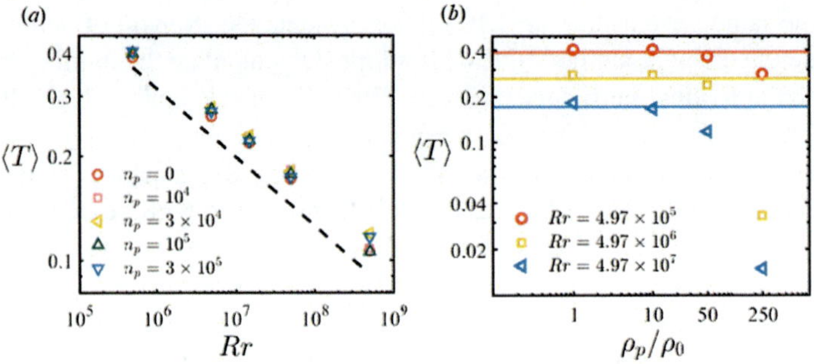

Fig. 3. The global average temperature with $\rho_p/\rho_f = 1$, different (n_p, Rr) (a) and with $n_p = 10^4$, different $(\rho_p/\rho_f, Rr)$ (b). The dashed line has a slope of $-1/5$, and the solid line represents the value of internal heating convection at the corresponding Rr. The internal heating convection can be regarded as convection induced by a large number of heat-releasing particles that are uniformly distributed and have no momentum coupling with the fluid

For the current particle-laden flow, the heat generated by the particles will eventually escape through the top and bottom walls. To separate the contributions of different mechanisms to heat transfer, we take a temporal and spatial average of the Eq. (1c) at the statistically steady state, and then we have

$$\langle wT \rangle = -Pr^{-1/2} Rr^{1/2} \left. \frac{\mathrm{d}\overline{T}}{\mathrm{d}z} \right|_{z=0} + \int_0^1 \int_0^z \overline{Q_p} dz' dz \tag{4}$$

It can be considered that the heat transport due to convection $q_{conv} = \langle wT \rangle$ and the nonuniform distribution of particles $q_{part} = 1 - \int_0^1 \int_0^z \overline{Q_p} dz' dz$ determines the heat flux $q_b = Pr^{-1/2} Rr^{1/2} \left. \frac{\mathrm{d}\overline{T}}{\mathrm{d}z} \right|_{z=0}$ on the bottom wall. So we can rewrite the Eq. (4) as $q_b = 1 - q_{part} - q_{conv}$. Note that $q_{part} = 1/2$ when particles are uniformly distributed in the cavity, and local accumulation of particles may cause q_b to deviate from this equilibrium value $1/2$. Thus q_{part} can reflect the spatial distribution tendency of particles.

As Fig. 4(a) illustrates, the particles of small inertia ($\rho_p/\rho_f = 1$) enhance convective heat transfer as Rr increases, which is consistent with the findings of Kuerten [15], while the enhancement becomes weaker as n_p increases. This enhancement originates from the temperature gradient around the heat-releasing particles, which effectively promotes the mixing of heat. The number density n_p may play a key role in the transfer of the particles, which consequently affects heat source distribution. As shown in Fig. 4(b),

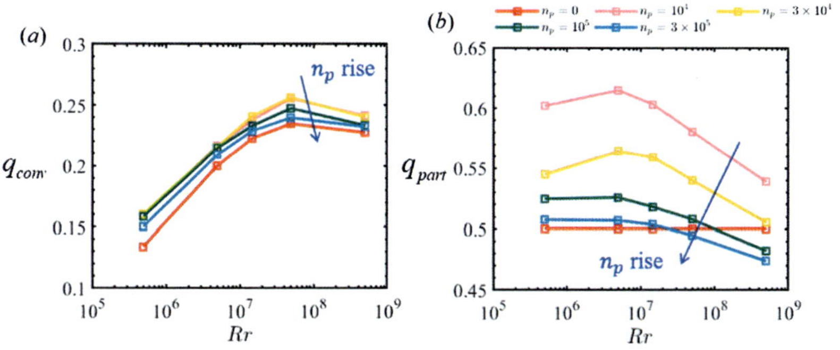

Fig. 4. The variations of the q_{conv} (a) and q_{part} (b) with increasing Rr, n_p for fixed $\rho_p/\rho_f = 1$

when the Rr increases, q_{part} drops more slowly at higher n_p. q_{part} decreases to less than $1/2$ implying that the region where particles tend to stay is shifted from $z > 1/2$ to $z > 1/2$.

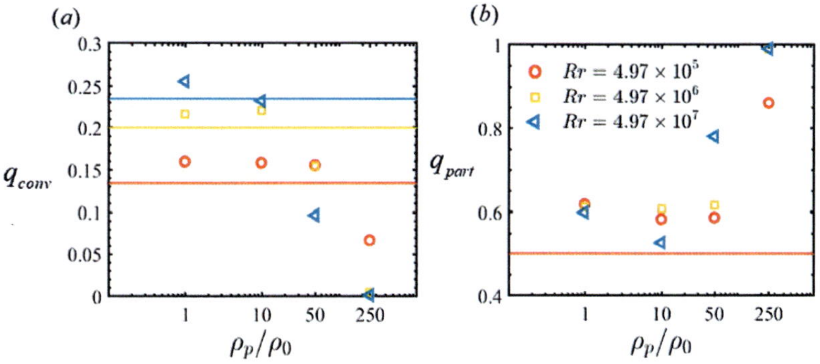

Fig. 5. The variations of the q_{conv} (a) and q_{part} (b) with increasing Rr, ρ_p/ρ_f for fixed $n_p = 10^4$. The solid line represents the value of single-phase internal heating convection at the corresponding Rr

However, as the density ratio ρ_p/ρ_f increases, q_{conv} decreases rapidly until it approaches 0, q_{part} increases dramatically and closes to 1 at $\rho_p/\rho_f = 250$ as Fig. 5(a) and (b) shows. It can be seen that the inertia of the particles plays a crucial role in the current particle-laden flow. The fundamental causes lie in the coupling of momentum and heat between the two phases. Thus, the spatial distribution characteristics of the particles have a drastic effect on the flow. The inertia of the particle, meanwhile, dominates the particle transfer process. As Fig. 6 illustrates, the particles with a small density have no significant preference to be distributed in the square cavity. However, as the density ratio increases, the particles are expelled away from the vortex core by the centrifugal force originating in the particle-vortex interaction, which leads to almost all particles accumulating near the wall. Once most of the heat is released near the walls, it is difficult for the heat-releasing particles to drive convection.

(a) (b)

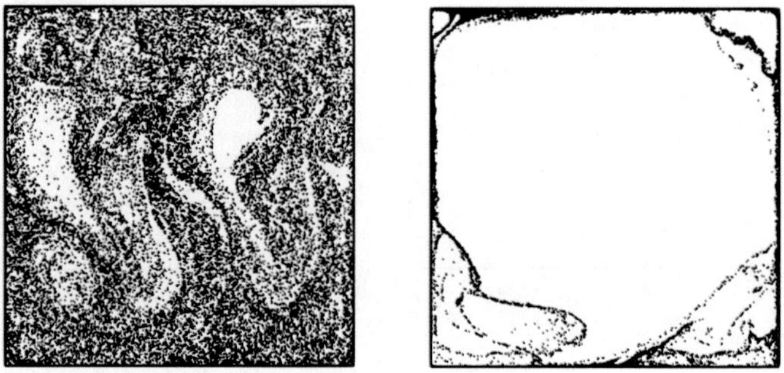

Fig. 6. Instantaneous distribution of particles after reaching a statistically steady state with $Rr = 4.97 \times 10^8$, $n_p = 10^5$ and $(a) : \rho_p/\rho_f = 1$, $(b) : \rho_p/\rho_f = 250$

4 Conclusion

In this paper, four-way coupled Euler-Lagrange direct numerical simulations have been conducted for thermal convection driven by heat-releasing particles in a quasi-3D cavity. The effects of density ratio ρ_p/ρ_f and number density n_p on the heat transfer process at different *Rr* are focused on. The results show that the heat-releasing particle-driven convection is highly similar to the uniform internal heating convection at small density ratios $\rho_p/\rho_f = 1$, and particles enhance convective heat transfer to a certain extent. However, at high-density ratios $\rho_p/\rho_f = 250$, the flow motion is critically dependent on the distribution properties of the particles. As a result of the inertial mechanism, the heavy particles will concentrate near the wall, eventually leading to extremely weak or even no convection.

References

1. Lube, G., Breard, E.C., Esposti-Ongaro, T., Dufek, J., Brand, B.: Multiphase flow behaviour and hazard prediction of pyroclastic density currents. Nat. Rev. Earth Environ. **1**(7), 348–365 (2020)
2. Pouransari, H., Mani, A.: Effects of preferential concentration on heat transfer in particle-based solar receivers. J. Solar Energy Eng. **139**(2) (2017)
3. Pan, M., Dong, Y., Zhou, Q., Shen, L.: Flow modulation and heat transport of radiatively heated particles settling in Rayleigh-bénard convection. Comput. Fluids **241**, 105454 (2022)
4. Du, Y., Yang, Y.: Wall-bounded thermal turbulent convection driven by heat-releasing point particles. J. Fluid Mech. **953**, A41 (2022)
5. Yang, W., Zhang, Y.-Z., Wang, B.-F., Dong, Y., Zhou, Q.: Dynamic coupling between carrier and dispersed phases in Rayleigh-bénard convection laden with inertial isothermal particles. J. Fluid Mech. **930**, A24 (2022)
6. Yang, W., Wan, Z.-H., Zhou, Q., Dong, Y.: On the energy transport and heat transfer efficiency in radiatively heated particle-laden Rayleigh-bénard convection. J. Fluid Mech. **953**, A35 (2022)

7. Rahmani, M., Geraci, G., Iaccarino, G., Mani, A.: Effects of particle polydispersity on radiative heat transfer in particle-laden turbulent flows. Int. J. Multiph. Flow **104**, 42–59 (2018)
8. Zamansky, R., Coletti, F., Massot, M., Mani, A.: Radiation induces turbulence in particle-laden fluids. Phys. Fluids **26**(7), 071701 (2014)
9. Zamansky, R., Coletti, F., Massot, M., Mani, A.: Turbulent thermal convection driven by heated inertial particles. J. Fluid Mech. **809**, 390–437 (2016)
10. Yang, W., Wang, B.-F., Tang, S., Zhou, Q., Dong, Y.: Transport modes of inertial particles and their effects on flow structures and heat transfer in Rayleigh-bénard convection. Phys. Fluids **34**(4), 043309 (2022)
11. Di Renzo, A., Di Maio, F.P.: Comparison of contact-force models for the simulation of collisions in DEM-based granular flow codes. Chem. Eng. Sci. **59**(3), 525–541 (2004)
12. Li, J., Mason, D.: A computational investigation of transient heat transfer in pneumatic transport of granular particles. Powder Technol. **112**(3), 273–282 (2000)
13. Zhu, H., Zhou, Z.Y., Yang, R., Yu, A.: Discrete particle simulation of particulate systems: theoretical developments. Chem. Eng. Sci. **62**(13), 3378–3396 (2007)
14. Wang, Q., Lohse, D., Shishkina, O.: Scaling in internally heated convection: a unifying theory. Geophys. Res. Lett. **48**(4), e2020GL091198 (2021)
15. Kuerten, J.G., Van der Geld, C., Geurts, B.J.: Turbulence modification and heat transfer enhancement by inertial particles in turbulent channel flow. Phys. Fluids **23**(12), 123301 (2011)

Check for updates

Reynolds Number Effects in Wall-Bounded Turbulent Flows

Wei Zhu, Xiaocang Ji, and Guohua Wang(✉)

Center for Particle-Laden Turbulence, Lanzhou University, Lanzhou 730000, China
ghwang@lzu.edu.cn

Abstract. Experimental measurements of the turbulent boundary layer flows with moderate and low Reynolds numbers in a wind tunnel are combined with measurements of the high Reynolds number wall turbulent flows in the atmospheric surface layer (ASL) at the Qingtu Lake Observation Array (QLOA) to reveal the Reynolds number effect on some basic turbulence statistics and the turbulent structures. The results show that the Karman constants of the ASL flows with $Re_\tau \sim O(10^6)$ are larger than those of the low- and moderate-Reynolds-number boundary layer flows, and the Karman constants in sand-laden flows are greater than that in clean airflows. Taking the amplitude modulation coefficient between turbulent structures equal to 0 as the criterion to identify the midpoint of the logarithmic layer, the inner-scaled height of the midpoint and the lower bound of the logarithmic layer both increase linearly with the Reynolds number; The streamwise scales of the turbulent structures in the outer layer firstly increase and then tend to stabilize with the Reynolds number. The critical Reynolds number above which the scales of the very large scale motions (VLSMs) do not increase is about $Re_\tau = 1200$, and the critical Reynolds number at various heights has the same values. The energy contribution of the VLSMs in the outer layer increases with the Reynolds number in a log-linear way, and the maximum contributions of the VLSMs in ASL can exceed 65%.

Keywords: wall-bounded turbulent flows · Reynolds number effects · logarithimc layer · turbulent structures

1 Introduction

Since Kim and Adrian firstly identified the existence of one new type of turbulent coherent structures with very large streamwise length scales in the outer region of the pipe flow at $Re_\tau = 3175$, i.e. the very large scale motions (VLSMs) [1], many careful measurements and simulations have attempted to reveal the specific features of the VLSMs [2] from the perspective of comparing with the coherent structures in the wall turbulence at lower Reynolds numbers, such as large-scale motions (LSMs), near-wall streaky structures etc. The consensus reached includes: in the turbulent boundary layer flow, the streamwise scales of the VLSMs reach 6δ, while those in pipe and channel flows can exceed 10δ

X. Zheng and S. Balachandar (Eds.): IUTAM 2023, IUTAM Bookseries 41, pp. 196–211, 2024.
https://doi.org/10.1007/978-3-031-47258-9_13

(δ represents the thickness of the boundary layer, the half-height of the channel or the radius of the pipe); the inclination angles of the VLSMs are suggested to be about $10°–20°$ which decrease with the friction wind velocity [3]. In addition, it was found that as the Reynolds number increasing one second peak of the turbulent kinetic energy (TKE) gradually appears in the middle of the logarithmic layer which has close relationship with the VLSMs [?]. Since these new phenomena were all captured in high Reynolds number turbulent flows, more and more attentions have been attracted to the Reynolds number effect on the turbulence statistics and turbulent structures in wall-bounded turbulent flows, especially the application scope of the logarithmic scaling law for the overlap region, the variations of the TKE and the scales and energies of the coherent structures with Reynolds number.

One important aspect of the Reynolds number effect is the turbulence statistics including the application scope of the logarithmic mean profile that has been widely accepted, its parameters such as the von Karman constant, and the representative model for the TKE profile. In the classic theory, the region above the buffer layer ($z^+ > 30$) is considered as the logarithmic region [5], but Smits and Zagarola [6] believed that the logarithmic region of the mean profile in the pipe flow appears only at $Re_\tau > 9000$, and the wall-normal range for the log layer is $600 < z^+ < 0.07Re_\tau$. Further, McKeon et al. [7] suggested that the log layer appears at $Re_\tau > 5000$, and within $600 < z^+ < 0.12Re_\tau$. Marusic et al. [8] compared the data in the turbulent boundary layer, pipe flow and the atmospheric surface layer (ASL) and suggested that the lower bound of the log layer is related to the Reynolds number . The range of the log layer is about $30 < z^+ < 0.15Re_\tau$.The turbulent boundary layer experiment by Vallikivi et al. [9] pointed out that when $Re_\tau > 2.0 \times 10^4$ a universal logarithmic scaling law appears in the range of $400 < z^+ < 0.15Re_\tau$ which was supported by the DNS simulations given by Lee and Moser [10] with $Re_\tau = 5200$. Willert et al. [11] believed that the lower bound of the log layer is $z^+ \sim 200$ based on their experiments in the pipe flow with $2383 < Re_\tau < 40000$. However, with the similar pipe flow experiments Furuichi et al. [12] presented support to Marusic et al. [8] on the range of the log layer, that is, the lower bound of the log layer is related to the Reynolds number, but the upper bound of the log layer in the pipe flow locates at about $0.2Re_\tau$. In summary, the range of the log layer is generally believed to be related to the Reynolds number, but its quantitative relationship with the Reynolds number has not reached consensus yet. Besides, in the classical logarithmic law the von Karman constant (κ) is considered as a universal constant that does not depend on the flow type and the Reynolds number. In classic literatures and textbooks (for example, [13]), it has been widely accepted that $\kappa = 0.40 - 0.41$. However, a large amount of recent evidences show that the Karman constant of the logarithmic scaling law has the Reynolds number effect, and the value of κ tends to be related to the flow type when the Reynolds number approaches high enough [14,15]. For example, for the pipe flow $\kappa = 0.41 \pm 0.02$ [16], while for the turbulent boundary layer and channel flows, $0.384 < \kappa < 0.389$ [17–19]. Given the considerable controversies on the specific value of the Karman constant [4], the classic viewpoint that κ is a universal constant is questioned. In addition, the logarithmic scaling law of the TKE distribution along with the height becomes gradually clearer as the Reynolds number increases [8], and in different flow types including pipe, channel and turbulent boundary layer flows, as well as the ASL flow the streamwise TKE also satisfies a logarithmic decreasing law with the height, i.e., $u'^{2+} = B_1 - A_1 ln(z/\delta)$. Although it is revealed that the scale coefficients A_1 and B_1 have different values for various flow types and suggested that the logarithmic scaling law of the streamwise TKE holds only at high-enough Reynolds number [8], it does not reflect the Reynolds number effect. If the coefficients in the logarithmic scaling law of the TKE are all constants, then the TKE is only related to the height, or saying, at the same outer-scaled height the TKE does not vary with the Reynolds number. Actually, studies have shown that with the increase of the Reynolds number the TKE in the outer region still shows a trend of logarithmically linear increase [20]. Therefore, it is necessary to clarify the change of the TKE with the Reynolds number through experiments and measurements at higher Reynolds numbers or to find out how high the Reynolds number is enough above which the TKE profile is independent with the Reynolds number.

Another aspect of the Reynolds number effect is on the scales and energy contributions of the coherent structures in wall turbulence. Ascending to the 1950s, Grant discovered the existence of the large-scale turbulent structures through detecting the long tail of the autocorrelation of the streamwise wind fluctuations [21]. Discovered the existence of the large-scale turbulent structures through detecting the long tail of the autocorrelation of the streamwise wind fluctuations. Tritton carried out experiments in the turbulent boundary layer in wind tunnel with $Re_\tau \approx 880$, the one-dimensional two-point correlation of the streamwise wind fluctuations reveal the turbulent structures appearing at the top of the log layer have streamwise scales larger than 5δ [22]. Moin and Kim performed the large eddy simulations (LES) on the channel flow with $Re_\tau \approx 640$, the two-point correlation shows the structures at the top of the log layer have streamwise scales of about 6δ which is much larger than the spanwise scales [23]. Hutchins and Marusic reported the streamwise scales of the LSMs in the turbulent boundary layer at $Re_\tau = 1120, 7610, 1438$ and 19960 respectively with the high Reynolds number boundary layer wind tunnel (HRNBLWT), and found that the streamwise scales of the LSMs at $Re_\tau = 1120$ are about 4δ the top of the log layer, which are slightly smaller than those at the other three Reynolds number conditions. It implies that the scales of the VLSMs might have a Reynolds number dependence. However, higher Reynolds number measurements in the ASL with $Re_\tau \sim O(10^5 - 10^6)$, such as Hutchins & Marusic [24], Wang & Zheng [25] present a consistent conclusion that the streamwise scales of the VLSMs in the high Reynolds number wall turbulence do not display Reynolds number dependence comparing with the low and moderate Reynolds number experiments. As for the energy contributions of the VLSMs, Balakumar & Adrian conducted experiments in the turbulent boundary layer with $61476 < Re_\tau < 2395$, and suggested that the energy contribution of the VLSMs to the total TKE is about 50%

[26]. Subjected to the relatively narrow range of the Reynolds number, it failed to reveal the variations of the energy fraction of the VLSMs with the Reynolds number. Due to the lack of experimental data of high Reynolds number wall turbulence, it is still an open issue that whether and how the three-dimensional feature of the VLSMs varies with the Reynolds number. For this purpose, Liu et al. integrated a large number of high Reynolds number data obtained in the ASL through the QLOA [27]. Through combining with the low and moderate Reynolds number TBL data given by previous studies, they analyzed the variations of the three-dimensional scales of the VLSMs with the Reynolds number and suggested that the scales of the VLSMs have no obvious Reynolds number dependence within the range of $Re_\tau \sim (10^3 - 10^6)$. Assuming this conclusion is true, the following question is when the VLSMs appear and when their scales no longer increase with the Reynolds number. It is an open issue that has not been answered, even little attention been paid to. Also, it is still unclear how the energy contribution of the VLSMs varies with the Reynolds number in the wider range of Reynolds number.

In summary, the researches on the turbulence statistics and turbulent structures of wall turbulence have not systematically revealed the Reynolds number effect up to date. It is still an open question that the range of Reynolds number across multiple orders of magnitude indicates new phenomena or scaling law in the turbulence statistics and the structural characteristics of wall turbulence or not. In this study, the Reynolds number effects on the turbulence statistics, the scales and energy of the turbulent structures were investigated by means of low- and moderate-Reynolds-number turbulent boundary layer experiments with $Re_\tau \sim (10^2 - 10^3)$ combined with the high-Reynolds-number ASL measurements. Section 2 presents an introduction to the experimental and field measurements and analysis methods; Sect. 3 gives the discussions on the turbulence statistics; Sect. 4 gives the discussions on the turbulent structures, and finally Sect. 5 presents conclusions.

2 Methods

In this study, a variety of methods and facilities were used including wind tunnel experiments and field observations in the ASL to investigate the wall turbulence over the Reynolds number range spanning 5 orders of magnitude, i.e. $Re_\tau \sim O(10^2 - 10^6)$. Through comparing the results of the wind field data obtained at various Reynolds numbers, the Reynolds number effect in wall turbulence was investigated. This section presents an introduction to the research methods and experimental setup of this study.

2.1 Field Observation in the ASL

Field observations on the high Reynolds number wall turbulence in the ASL were conducted at the QLOA site. The QLOA site as shown in Fig. 1 is located on the dry lake bed of the Qingtu Lake between the Badain Jaran Desert and the Tengger Desert in western China (E:103°40′03", N:39°12′27"). The introductions of QLOA site and observations refer to [25] and [27]. The clean-air flow data adopted in this study have also been reported in the previous studies and the corresponding flow conditions are listed in Table 1. Further, the values of the Karman constants in the sand-laden ASL flows were investigated with the sand-laden flow data obtained by the QLOA. The corresponding flow-condition parameters of the sand-laden flows and the corresponding mass flux measured in the ASL are listed in Table 2, in which the friction velocity u_τ was estimated by the Reynold stress measured at 2.4 m: $u_\tau =< -uw >^{-1/2}$, and the thickness of the ASL was taken the same value as that in clean-air flows, that is, $\delta = 150$ m.

Fig. 1. The panoramic photo of the Qingtu Lake Observation Array (QLOA).

Table 1. The flow conditions of the clean-air flow data in the ASL

No.	year	time and date	$Re_\tau(\times 10^6)$	$u_\tau(ms^{-1})$	$\nu(10^{-5}m^2s^{-1})$	z/L	$\delta(m)$
1	2013	23 May 17–18	3.2	0.36	1.78	–0.01	158
2	2014	18 April 04–05	2.7	0.31	1.65	0.007	144
3	2014	1 May 06–07	2.7	0.29	1.63	0.01	151
4	2015	1 April 18–19	2.4	0.27	1.63	–0.02	145
5	2015	10 April 18–19	3.1	0.34	1.7	–0.006	155
6	2015	14 April 18–19	2.7	0.31	1.78	0.01	155

2.2 Wind Tunnel Experiments

The tubulent boundary layer experiments at moderate and low Reynolds number were conducted in the wind tunnel of Lanzhou University. The wind tunnel is a blown-type wind tunnel with a section of 1.4 m (width) × 1.3 m (height). The working section is 20.0 m long, and the axial wind velocity is adjustable from

Table 2. The flow conditions of the particle-laden flow data in the ASL

No.	year	time and date	$Re_\tau(\times 10^6)$	$u_\tau(ms^{-1})$	$\nu(10^{-5}m^2s^{-1})$	z/L	$Q(gm^{-2}s^{-1})$
1	2016	11 April 09–10	4.24	0.48	1.70	–0.014	1.30
2	2016	14 April 19–20	4.89	0.55	1.69	0.010	5.11
3	2016	14 April 20–21	4.59	0.53	1.73	0.014	4.22
4	2016	14 April 22–23	5.82	0.66	1.70	–0.007	2.64
5	2016	14 April 23–24	5.53	0.63	1.71	0.006	7.26
6	2016	16 April 18–19	4.16	0.46	1.66	–0.006	3.77

3 ms^{-1} to 40 ms^{-1}. In this study, the instruments for measuring the wind velocity is the Dantec streamline hot wire anemometers (54N81). In order to ensure the size of the hot-wire probe not affect the spatial accuracy of the turbulence measurement, the corresponding dimensionless length scale, length and diameter ratio etc. of the probe all satisfy the requirement proposed by Hutchins et al. [29]. In the wind tunnel experiment, the sample frequency of the hot-wire anemometer is 10 kHz. The low Reynolds number experiment ($Re_\tau < 1000$) was carried out on a flat thin plate above the boundary layer of the wind tunnel and parallel to the bottom of the wind tunnel. The thin plate is 3.6 m long and 1.2 m wide, 0.3 m away from the bottom wall. The lower part of the front edge of the plate was chamfered to avoid flow separation caused by the leading edge effect of a bluff body. The plate was located 8.0 m away from the entrance of the working section, and the measurements were conducted at 0.2 m–3 m away from the upstream edge of the plate. With the boundary layer flow developing on the plate, the measured friction velocity u_τ and the thickness of the boundary layer δ were varying along the incoming flow direction, so that the variable Reynolds number conditions were achieved through the experimental setup.

In the moderate Reynolds number experiments, the measurements were conducted directly on the bottom wall of the tunnel, and 16.0 m away from the entrance of the working section of the wind tunnel, as shown in Fig. 2. It should be noted that the thin plate was removed in the moderate-Reynolds-number experiment. The flow conditions and measurement parameters of the low- and moderate-Reynolds-number wind tunnel experiments are listed in Table 3. In which the friction velocity was estimated by the Coles law (see [28]).

3 On the Turbulent Statistics

To validate the experimental data, we firstly compared the turbulence statistics including the mean velocity and the turbulent kinetic energy (TKE) profiles with the previous experimental and theoretical results. Figure 3 shows the mean velocity profiles measured at various Reynolds numbers, in which the hollow points represent the results obtained in the wind tunnel at moderate and low Reynolds numbers, and the black solid points are the measured results in the ASL by the QLOA. It can be seen that the mean velocity profiles at moderate and low

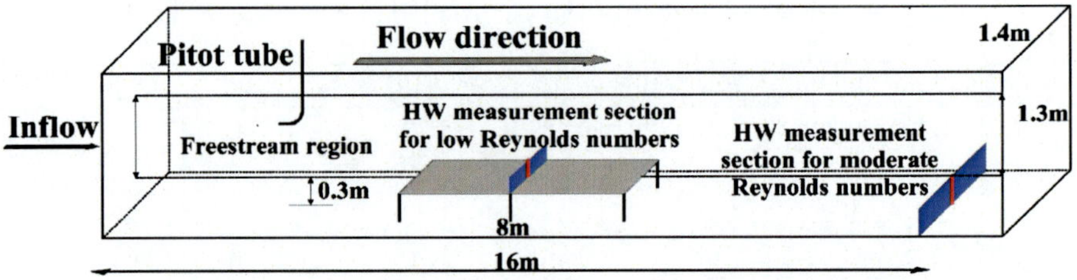

Fig. 2. Schematic view of the wind tunnel experiment.

Table 3. Flow conditions and parameters of low- and moderate-Reynolds-number wind tunnel experiments.

Re_τ	$u_\tau(ms^{-1})$	$\delta(m)$	$\nu(10^{-5}m^2s^{-1})$	U_∞
267	0.1676	0.0266	1.7	3.31
387	0.1985	0.033	1.7	4.001
501	0.2761	0.0308	1.7	5.5597
685	0.2568	0.0448	1.7	5.6115
840	0.2121	0.0671	1.7	4.9198
923	0.2012	0.0771	1.7	5.0048
971	0.2114	0.0774	1.7	5.0499
1152	0.1295	0.1506	1.7	3.0917
1383	0.2767	0.0848	1.7	6.5593
1664	0.171	0.1648	1.7	4.1391
1876	0.186	0.1713	1.7	4.7885
2860	0.2959	0.1636	1.7	7.5965
3613	0.3661	0.1672	1.7	9.8203
4613	0.443	0.1766	1.7	12.1416
5184	0.5751	0.1533	1.7	16.332
5490	0.5203	0.1788	1.7	14.4605
6438	0.5842	0.1869	1.7	16.3113
7015	0.6011	0.1982	1.7	16.388
8652	0.7836	0.1877	1.7	21.6797

Reynolds numbers agree well with the classic logarithmic law, while the profiles measured in the ASL with high Reynolds numbers are generally consistent with the logarithmic law with downward shifts from the logarithmic relationship of the smooth-walled turbulent boundary layer flows due to the surface roughness at the QLOA [25].

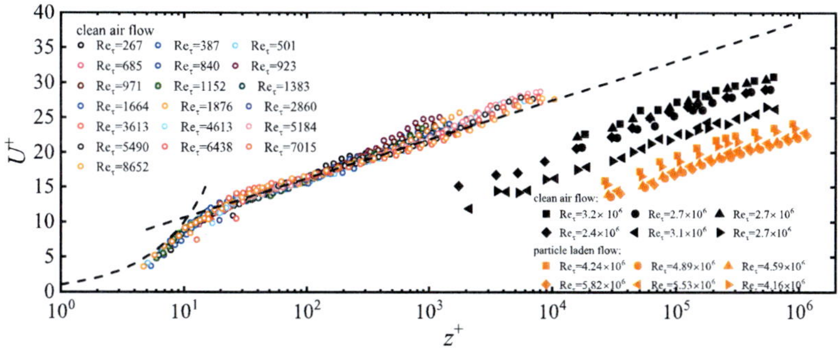

Fig. 3. The mean velocity profiles measured in the wind tunnel (hollow points) and the ASL (black solid points).

Based on the measured mean velocities at different heights, it is possible to derive some key parameters by fitting the logarithmic relationship, such as the Kárman constant κ. Marusic et al. investigated the logarithmic region in wall shear turbulence, and proposed that $\kappa = 0.41 \pm 0.028$ for the ASL flows at the SLTEST site in Utah [8]. In order to obtain the Kárman constant κ in the clean-air and sand-laden flows, streamwise velocity data at the 11 heights of the QLOA are fitted by the least square method with the standard logarithmic relationship:

$$U^{+} = \frac{1}{\kappa} log(z^{+}) + B$$

Here, $U^{+} = u/u_{\tau}$, where u_{τ} is the friction velocity, $z^{+} = zu_{\tau}/\nu$, where ν is the kinematic viscosity, κ is the von Kárman constant, and B is a parameter that depends on the roughness of the surface, and is assumed to be a constant for smooth-walled flows, Fig. 4 shows the estimated κ values of the ASL data in this study and a compilation of the κ values from turbulent boundary layers (see [30]) and the SLTEST [31], where black and yellow stars represent the fitted κ values in the clean-air and sand-laden flows respectively at the QLOA. The κ values vary from 0.38 to 0.42 in the clean-air flows with an average value of 0.41, which is consistent with the conclusion by Marusic et al. [8] based on the measurements in the SLTEST. However, comparing with the values of κ extracted by Nagib and Chauhan [30] for turbulent boundary layers, the results in the ASL reveal slight variations of the Karman constant with the Reynolds number, i.e. κ increases with the increase of Reynolds number. In addition, the average value of the Karman constant in the sand-laden flows by the QLOA is around 0.43 which is larger than that in the clean-air flows. It may come from the enhance of the local Reynolds stress by the particle phase and thus the corresponding increase of the Karman constant based on the logarithmic law.

In order to analyze the Reynolds number dependence of the TKE, the streamwise TKE at one given height in the clean-air flows at various Reynolds numbers conditions are calculated and compared with the existing experimental results. Figure 5 shows the variations of the streamwise TKE with the Reynolds num-

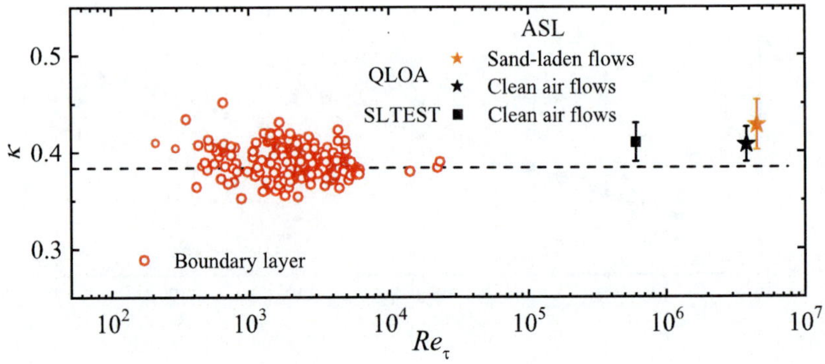

Fig. 4. The estimated κ values of the ASL data in this study and a compilation of the values from Nagib and Chauhan [30] (hollow points) and the SLTEST [31] (solid points).

ber at $z/\delta = 0.036$ in this study, compared with the existing results. It can be seen that although the ASL data by QLOA have great discreteness, the streamwise TKE still shows a clear trend of increasing with the Reynolds number. It means that even if the logarithmic scaling law of the TKE with height has been approved by different types of turbulent boundary layer flows, the influence of Reynolds number can not be ruled out up to $Re_\tau \sim (10^6)$, and further the scaling parameters also have Reynolds number dependence.

In the study of wall turbulence with high Reynolds number, it is suggested that the VLSMs have significant amplitude modulation effect on the small-scale turbulences [32]. For the amplitude modulation, it has been found that the modulation effect decreases with height and the modulation coefficient approaches to zero at a certain height below which the positive fluctuations of the VLSMs enhance the small-scale turbulent motions while above this height the former weaken the latter. Mathis et al. [33] and Marusic et al. [34] revealed that the height where the amplitude modulation coefficient (denoted as RAM) is zero corresponds to the middle of the log layer. It inspires the following investigations

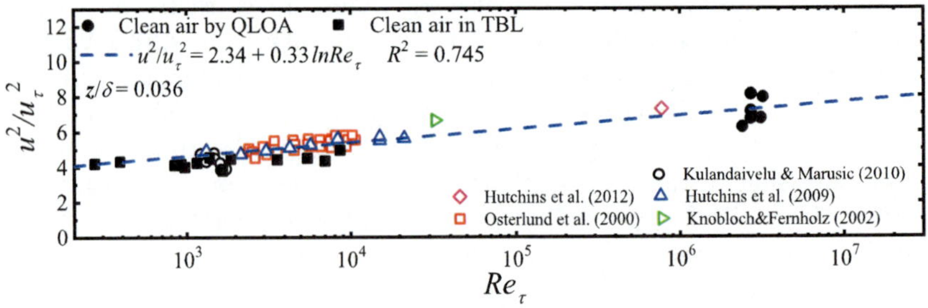

Fig. 5. The variations of the streamwise TKE with Re_τ at $z/\delta = 0.036$ in this study (solid points), comparing with the existing results (hollow points).

to analyze the log layer with the amplitude modulation coefficients. In order to obtain the height at which the amplitude modulation coefficient approaches to zero, we employed the same definition and method as [33] for single-point amplitude modulation coefficient (denoted as RAM), and investigate the modulation effect of the VLSMs onto small-scale turbulent motions at various Reynolds numbers.

According to the previous studies on the amplitude modulation (see [14, 32], and [33]), the cutoff wavelength used in this study was taken as $\lambda_x = \delta$ to decompose the turbulence signals into large- and small-scale motions. The variations of the modulation coefficients with height and the Reynolds number are analyzed. After calculating the amplitude modulation coefficients (RAM) of different types of wall turbulence at different Reynolds numbers, it is possible to extract the position z_0^+ where RAM $= 0$ (that is the middle of the log layer) for different Reynolds number flows, and further figure out the range of the log layer of the ASL. There have been suggested that the range of the log layer can be scaled with two forms, corresponding to two laws of the variations of the inner-scaled height of the middle of the log layer with the Reynolds number, that is, z_0^+ increases with the Reynolds number following a 1/2 power or a 3/4 power law, respectively [33].

This study investigated the variations of the positions where RAM $= 0$ with the Reynolds number and compared with the existing scaling law. Figure 6 shows the variations of the position z_0^+ normalized by the inner-scale with the Reynolds number given by this study, in compilation of the results by Mathis et al. [33] with $Re_\tau = 2800$, 3900, 13600 and 19000 respectively, and the ASL data obtained by the SLTEST with $Re_\tau = 6.5 \times 10^6$. The gray and black dotted lines represent the fitted laws for the variation of the middle of the log layer (z_0^+) with Re_τ based on Mathis et al. [33] respectively, that is, $z_0^+ = 3.9Re^{1/2}$ and $z_0^+ = 0.42Re^{3/4}$. The solid squares and circles are the extracted z_0^+ respectively for the turbulent boundary layer flows in wind tunnel and the ASL flows by the QLOA of this study. It can be found that z_0^+ increases with the Reynolds number which is consistent with the existing results in quality. In quantity, the results of this study display significant deviations from both the fitting curves of the 1/2 and the 3/4 power laws. The goodness of fit (R^2) for the two laws are both about 0.6. Comparisons show that the two power laws both underestimate the growth rate of the position corresponding to the middle of the log layer with the Reynolds number. Furthermore, Fig. 6 shows that z_0^+ increases linearly with the Reynolds number. Fitting the experimental data with a linear function as shown by the blue line in Fig. 6, it can be seen that linear fitting is much better than the 1/2 or the 3/4 power fitting, and the goodness of fit for linear fitting is $R^2 = 0.952$. Therefore, it is debatable whether the inner-scaled height of the middle of the log layer increases with the Reynolds number in a power or a linear law. The results of this study support the latter which can also be verified through the variation of the outer-scaled height of the middle of the log layer with the Reynolds number.

Figure 7 shows the variations of the outer-scaled height (z_0^+) where $RAM = 0$ with the Reynolds number. It is found that the outer-scaled positions representing the middle of the log layer generally do not vary with the Reynolds number which approach to a constant $z_0/\delta = 0.04$ (or $z_0^+ = 0.04Re_\tau$). In the similar way, the fitting curves of the inner-scaled positions based on the results by Mathis et al. [33] are also included as comparing objects, as shown by the gray and black lines respectively in Fig. 7, which both display decreasing trends of the outer-scaled height of the middle of the log layer with the Reynolds number. The discrepancy between Mathis et al. [33] and this study may be caused by the lack of enough Reynolds number involved in the study of Mathis et al. [33] which makes the fitting relationships only work within the limited Reynolds number range and are different from the law governing the whole Reynolds number range.

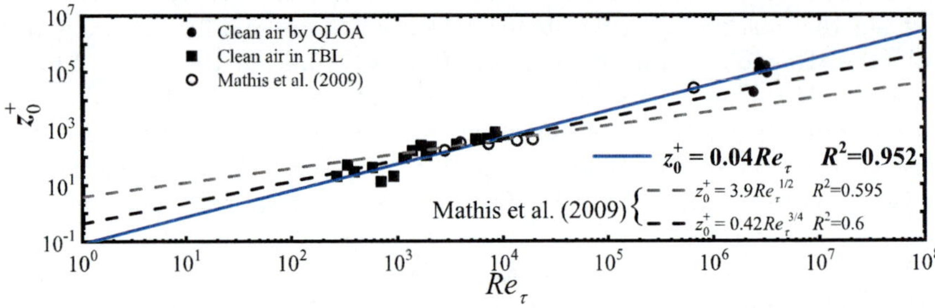

Fig. 6. The variations of the inner-scaled height of the middle of the log layer (z_0^+) with Re_τ.

Given the variation law of the inner- and outer-scaled height representing the middle of the log layer, it is reasonable to estimate the lower and upper bound of the log layer. Till now, it has been generally believed that the upper bound of the log layer is a fixed value, i.e.$z_u = 0.15\delta$ with the outer scale [8] or $z_u^+ = 0.15Re_\tau$ with the inner scale. With the relationship defining the middle position in terms of the upper and the lower bound of the log layer, that is, $lnz_0^+ = (lnz_b^+ + lnz_u^+)/2$, the lower bound of the log layer z_b^+ can be deduced as $z_b^+ \sim (z_0^+)^2/z_u^+$. Substituting the linear law of the middle of the log layer varying

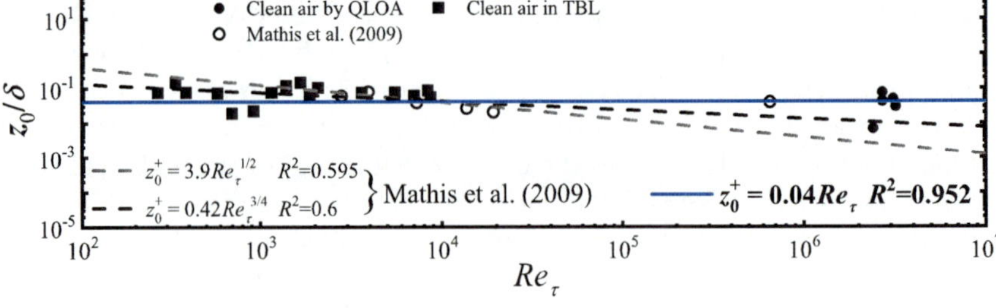

Fig. 7. The variations of the outer-scaled height of the middle of the log layer (z_0/δ) with Re_τ.

with the Reynolds number given in this study, we can obtain the variation of the lower bound of the log layer with the Reynolds number which also display a linear increasing trend as $z_b^+ \sim 0.01Re_\tau$.

4 On the Turbulent Structures in the Outer Layer

Up to date, it has been accepted that VLSMs in the outer layer are coherent structures with streamwise scales larger than 3δ (δ is the thickness of the boundary layer). Based on low-pass filtering (with a filter scale of 3δ), we can distinguish the flow signals of VLSMs and LSMs and further investigate the variations of the energy fractions of VLSMs/LSMs with height and Reynolds number. Figure 8 presents the variation of the energy fraction carried by the VLSMs in the clean-air flow fields with height at different Reynolds numbers. It can be seen that in the near-wall region of the ASL, VLSMs account for less than 20% of the total streamwise TKE, but the energy fraction of the VLSMs increases linearly with height which agrees well with the results reported by Wang & Zheng [25]. At $z = 0.2\delta$, the contribution of the VLSMs to the total TKE is close to or even more than 65%.

The variations of the energy fractions of the VLSMs with the Reynolds number are illustrated in Fig. 8. It should be noted that the absolute variations of the energy fraction contributed by the turbulent motions larger than 3δ are relatively small, and the experimental data are sparse to a certain extent over the whole Reynolds number range. In order to clarify the varying trend of the energy fraction of the VLSMs with the Reynolds number, the experimental data in the similar Reynolds number range were averaged to show in Fig. 8. It can be seen that in the flows at low Reynolds number ($Re_\tau < 10^3$), the VLSMs are not fully developed, and their contributions to the total energy are relatively small. With the increase of the Reynolds number, the energy fraction of the VLSMs increases. In the high Reynolds number ASL with $Re_\tau \sim O(10^6)$ the average energy fraction of the VLSMs achieves more than 50% and the maximum contribution reaches up to 60%, which is significantly higher than those in low-Reynolds-number turbulent boundary layer flows. At the meanwhile, as

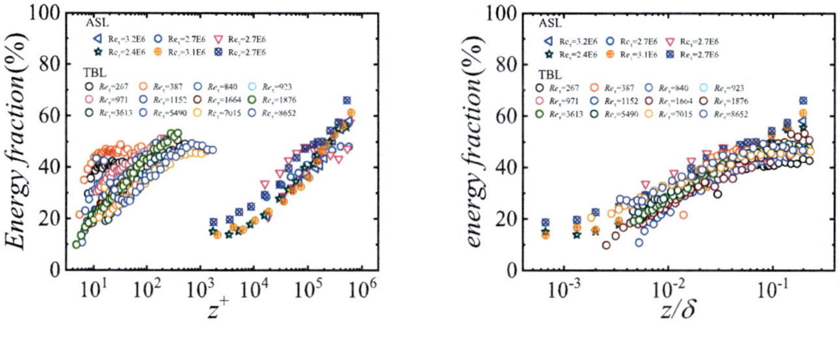

Fig. 8. The energy fractions of the VLSMs in the clean-air flow fields varying with (a) the inner-scaled height and (b) the outer-scaled height.

shown in Fig. 9, the energy fraction of the VLSMs whose streamwise scales larger than 3δ demonstrate log-linear increase trends with the Reynolds number over the whole Re_τ range in this study. Such increase does not show an approximation trend within the range of $Re_\tau \sim (10^2 - 10^6)$.

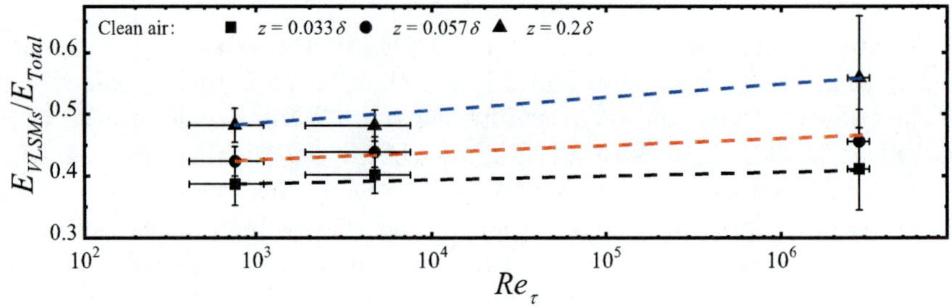

Fig. 9. The variations of the energy fractions of the VLSMs withRe_τ.

Furthermore, the variations of the streamwise scales of the turbulent structures with the Reynolds number were analyzed. Through calculating the autocorrelation function of the streamwise fluctuations at a reference positions, the streamwise scale (L_x) of a turbulent structure can be determined by taking the correlation coefficient 0.05 as a judge for the boundary of a coherent structure. Figure 10 presents the variations of the streamwise scales of the turbulent structures based on autocorrelation with the Reynolds number at different heights. The purple solid triangles are the experimental results obtained in the turbulent boundary layer in the wind tunnel, and the black solid dots are date in the ASL by the QLOA. As shown in this figure, the variation of the streamwise scales of the turbulent structures at $z = 0.02\delta$ with the Reynolds number all demonstrate approximately the same trends. That is, when $Re_\tau < 1200$ the streamwise scales of the structures increase with the Reynolds number, while for $Re_\tau > 1200$ the scales of the turbulent structures normalized by the boundary layer thickness δ no longer vary with the Reynolds number which are generally larger than 3δ in average at different heights and therefore can be regarded as the VLSMs. In another word, when $Re_\tau > 1200$ the streamwise scales of the VLSMs approximately converge to a constant characteristic scale which has no longer the Reynolds-number dependence. This conclusion based on the scale analysis is not the same as the trend given by the energy fraction analysis which demonstrates an increase with the Reynolds number. It means that when the Reynolds number is large enough, the continuous increase of the energy of the VLSMs does not have a significant effect on the growth of theirs scales. One possible reason may be that in the high-Reynolds-number turbulent flows the energy of the largest energy-contained coherent structure in the flow corresponding to the peak of the turbulent energy spectra in the outer layer increases but its scale keeps invariant with the Reynolds number [25,35], and [36], which leads to the correlation scale dominated by the largest energy-contained coherent structure do not vary with the Reynolds number.

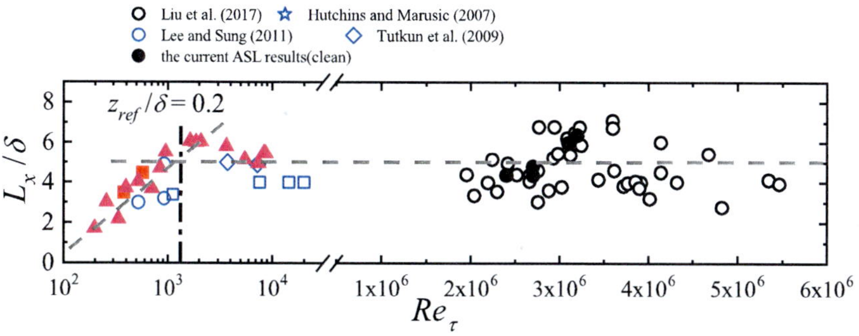

Fig. 10. The variations of the streamwise scales of the turbulent structures based on autocorrelation with Re_τ.

5 Conclusions

In this study, experimental measurements were carried out in a wind tunnel to investigate the Reynolds number effect on the turbulent statistics and coherent structures in the turbulent boundary layer over Reynolds number range of $Re_\tau \sim O(10^2 - 10^3)$. Combing with the high-quality near-neutral data on the high Reynolds number ASL obtained from the QLOA with $Re_\tau \sim O(10^6)$, we examined the variations of the basic turbulence statistics in the boundary layer flow with Reynolds number including the von Karman constant κ in the logarithmic scaling law, the middle and the lower bound of the log-layer, the streamwise TKE as well as the statistical characteristics of the turbulent structures in the outer layer including their streamwise scales and contributions to the turbulent kinetic energy. It is suggested that κ is around 0.41 for the ASL flows with $Re_\tau \sim O(10^6)$ which is consistent with the results obtained in the ASL at SLTEST with $Re_\tau \sim O(10^5)$ by Morris et al. [31], but higher than that in the moderate- and low-Reynolds-number turbulent boundary layer flows. Analyses on the amplitude modulation effect of the LSM in the outer layer onto the small-scale turbulent motions show that the inner-scaled height of the middle position of the log-layer corresponding to the position where the amplitude modulation coefficient is zero increases linearly with the Reynolds number. In addition, the estimated lower bound of the log-layer based on the middle position also increases linearly with the Reynolds number. Through analyzing the streamwise TKE at the same height, it is found that the streamwise TKE increases approximately linearly with the increase of Reynolds number, and such Reynolds-number-dependent trend is still obvious even $Re_\tau \sim O(10^6)$. On the statistical characteristics of the turbulent structures in the outer layer, it is suggested that the streamwise scales of the turbulent structures increase at first and then achieve saturation with the Reynolds number, the critical Reynolds number of the scale invariance is about $Re_\tau = 1200$. Considering that their streamwise scales are larger than 3 times the thickness of the boundary layer, our results imply that the critical Reynolds number of the scale invariance of the VLSMs is about $Re_\tau = 1200$. Further analysis on the TKE of the VLSMs shows that the contribution of the VLSMs to the total TKE increase with the Reynolds number in a log-linear way, and the maximum fraction reach up to 65%.

The results of this study may be helpful to deepen the current understanding of the wall turbulence and the Reynolds number effect. For a long time, there has been a vision to find a Reynolds number above which the turbulent statistics and the turbulent structures have no Reynolds-number-dependent variations, so that reliable turbulence modeling can be achieved. However, the results of this study show that some statistical properties of wall turbulence still demonstrate Reynolds number dependence at $Re_\tau \sim O(10^6)$, which means that more or even higher Reynolds number experimental studies are required to accurately reveal the Reynolds number effect of wall turbulence.

Acknowledgements. This research was supported by the National Natural Science Foundation of China (Grants Nos.12072138 and 92052202).

References

1. Kim, K.C., Adrian, R.J.: Very large-scale motion in the outer layer. Phys. Fluids **11**, 417–422 (1999)
2. Smits, A.J., Mckeon, B.J., Marusic, I.: High-Reynolds number wall turbulence. Ann. Rev. Fluid Mech. **43**, 353–375 (2011)
3. Liu, H., Bo, T., Liang, Y.: The variation of large-scale structure inclination angles in high Reynolds number atmospheric surface layers. Phys. Fluids **29**(3), 035104 (2017)
4. Marusic, I., Mathis, R., Hutchins, N.: High Reynolds number effects in wall turbulence. Int. J. Heat Fluid Flow **31**, 418–428 (2010)
5. Tennekes, H.: A First Course in Turbulence. MIT Press, Cambridge (1972)
6. Smits, A.J., Zagarola, M.V.: Response to "Scaling of the intermediate region in wall-bounded turbulence: the power law.". Phys. Fluids **10**(4), 1045–1046 (1998)
7. Mckeon, B.J., Li, J., Jiang, W., Morrison, J., Smits, A.J.: Further observations on the mean velocity distribution in fully developed pipe flow. J. Fluid Mech. **501**, 135–147 (2004)
8. Marusic, I., Monty, J.P., Hultmark, M., Smits, A.J.: On the logarithmic region in wall turbulence. J. Fluid Mech. **716**, R3 (2013)
9. Vallikivi, M., Hultmark, M., Smits, A.J.: Turbulent boundary layer statistics at very high Reynolds number. J. Fluid Mech. **779**, 371–389 (2015)
10. Lee, M., Moser, R.D.: Direct numerical simulation of turbulent channel flow up to $Re_\tau \sim 5200$. J. Fluid Mech. **774**, 395–415 (2015)
11. Willert, C.E., Soria, J., Stanislas, M., Klinner, J., Amili, O., Eisfelder, M.: Near-wall statistics of a turbulent pipe flow at shear reynolds numbers up to 40000. J. Fluid Mech. **826**, R5 (2017)
12. Furuichi, N., Terao, Y., Wada, Y., Tsuji, Y.: Further experiments for mean velocity profile of pipe flow at high Reynolds number. Phys. Fluids **30**, 7 (2018)
13. Schlichting, H., Gersten, K.: Boundary-Layer Theory. McGraw-Hill, New York (1979)
14. Marusic, I., Mathis, R., Hutchins, N.: Predictive model for wall-bounded turbulent flow. Science **329**(5988), 193–196 (2010)

15. Klewicki, J.C.: Reynolds number dependence, scaling, & dynamics of turbulent boundary layers. J. Fluids Eng.-Trans. ASME **132**, 094001 (2010)
16. Bailey, S.C.C., Vallikivi, M., Hultmark, M., Smits, A.J.: Estimating the value of von Karman's constant in turbulent pipe flow. J. Fluid Mech. **749**, 79–98 (2014)
17. Chauhan, K.A.: Study of canonical wall-bounded turbulent flows. Thesis (Ph.D.), Illinois Institute of Technology, p. 367 (2007)
18. Zanoun, E.S., Durst, F., Nagib, H.: Evaluating the law of the wall in two-dimensional fully developed turbulent channel flows. Phys. Fluids **15**, 3079 (2003)
19. Monty, J.P.: Developments in smooth wall turbulent duct flows. PhD thesis, Univ. Melbourne (2005)
20. Hutchins, N., Marusic, I., Monty, J., Klewicki, J.: Towards reconciling the large-scale structure of turbulent boundary layers in the atmosphere & laboratory. Bound.-Layer Meteorol. **145**, 273–306 (2012)
21. Grant, H.: The large eddies of turbulent motion. J. Fluid Mech. **4**(2), 149–190 (1958)
22. Tritton, D.J.: Some new correlation measurements in a turbulent boundary layer. J. Fluid Mech. **28**(3), 439–462 (1967)
23. Moin, P., Kim, J.: Numerical investigation of turbulent channel flow. J. Fluid Mech. **118**, 341–77 (1982)
24. Hutchins, N., Marusic, I.: Evidence of very long meandering features in the logarithmic region of turbulent boundary layers. J. Fluid Mech. **579**, 1–28 (2007)
25. Wang, G., Zheng, X.: Very large scale motions in the atmospheric surface layer: a field investigation. J. Fluid Mech. **802**, 464–489 (2016)
26. Balakumar, B.J., Adrian, R.J.: Large- & very-large-scale motions in channel & boundary-layer flows. Phil. Trans. Royal Soc. A Math. Phys. Eng. Sci. **365**, 665–681 (2007)
27. Liu, H., Wang, G., Zheng, X.: Spatial length scales of large-scale structures in atmospheric surface layers. Phys. Rev. Fluids **2**(6), 064606 (2017)
28. Coles, D.E., Hirst, E.A.: Compiled data. In: Proceedings of Computation of Turbulent Boundary Layers, AFOSR-IFP Stanford Conference 1968, vol. II (1969)
29. Hutchins, N., Nickels, T.B., Marusic, I., Chong, M.S.: Hot-wire spatial resolution issues in wall-bounded turbulence. J. Fluid Mech. **635**, 103–136 (2009)
30. Nagib, H.M., Chauhan, K.A.: Variations of von kármán coefficient in canonical flows. Phys. Fluids **20**(10), 159–163 (2008)
31. Morris, S.C., Stolpa, S.R., Slaboch, P.E., Klewicki, J.C.: Near-surface particle image velocimetry measurements in a transitionally rough-wall atmospheric boundary layer. J. Fluid Mech. **580**, 319–338 (2007)
32. Hutchins, N., Marusic, I.: Large-scale influences in near-wall turbulence. Phil. Trans. Roy. Soc. A: Math. Phys. Eng. Sci. **365**, 647–664 (2007)
33. Mathis, R., Hutchins, N., Marusic, I.: Large-scale amplitude modulation of the small-scale structures in turbulent boundary layers. J. Fluid Mech. **628**, 311–337 (2009)
34. Marusic, I., Mathis, R., Hutchins, N.: Reynolds number dependence of the amplitude modulated near-wall cycle. In: Progress in Wall Turbulence: Understanding and Modeling. Springer, Netherlands (2011). https://doi.org/10.1007/978-90-481-9603-6_11
35. Vallikivi, M., Ganapathisubramani, B., Smits, A.J.: Spectral scaling in boundary layers & pipes at very high Reynolds numbers. J. Fluid Mech. **771**, 303–326 (2015)
36. Liu, H., Wang, G., Zheng, X.: Amplitude modulation between multi-scale turbulent motions in high-Reynolds-number atmospheric surface layers. J. Fluid Mech. **861**, 585–607 (2019)

Effect of Shear-Induced Lift on Particle Motion and Turbulence Modulation in Fully Developed Compressible Turbulent Channel Flow

Yucang Ruan[1] and Zuoli Xiao[1,2(✉)]

[1] State Key Laboratory for Turbulence and Complex Systems, College of Engineering, Peking University, Beijing 100871, People's Republic of China
z.xiao@pku.edu.cn

[2] HEDPS and Center for Applied Physics and Technology, College of Engineering, Peking University, Beijing 100871, People's Republic of China

Abstract. Particle-laden flows exist widely in natural phenomena and engineering applications. Particle-laden turbulent channel flow is essential for understanding the interaction between particles and wall-bounded turbulence, which sheds light on various problems such as the design of turbo-machine and the evolution of sandstorm. However, the influence of shear-induced lift on particle-turbulence interaction remains unclear. Therefore, the two-way coupling particle-laden compressible gravity-free turbulent channel flow is simulated by point-particle direct numerical simulation (PP-DNS). We examine first the influence of Saffman lift on particle velocity, turbophoresis and turbulence modulation. It is noticed that the Saffman lift will 1) augment near-wall particle velocity, 2) reduce turbophoresis and 3) increase turbulence attenuation. Then, we consider the effect of different feedback methods, i.e., Particle-in-Cell (PIC) and projection methods. It turns out that both methods show few influences on fluid statistics, but the projection method gives a stronger turbophoresis and a smaller particle stream-wise velocity within buffer region.

Keywords: particle-laden flow · compressible turbulent channel flow · two-way coupling · Saffman lift · inertial particles

1 Introduction

Wall-bounded turbulent flows laden with small inertial particles exist widely in environmental phenomena and engineering applications. Such flow is very complex because it is not only chaotic and non-homogeneous but also presents multi-scale effect and strong particle-turbulence interaction [1]. When the particle diameter d is much smaller than the characteristic length of turbulence, i.e., the Kolmogorov length η, the volume of particles can be ignored and the particles can be considered as point-particles. Point-particle direct numerical simulation

X. Zheng and S. Balachandar (Eds.): IUTAM 2023, IUTAM Bookseries 41, pp. 212–228, 2024.
https://doi.org/10.1007/978-3-031-47258-9_14

(PP-DNS) is widely used to study particle-turbulence interaction [2], including particle motions and turbulence modulation, owing to its efficiency and capacity of simulating large number of particles.

The particle-turbulence interaction of PP-DNS can be generally classified into three categories based on particle volume fraction ϕ_v and particle mass fraction ϕ_m [3]: (1) one-way coupling, where the influence of particles on fluid and particle-particle collision are neglected because of low particle volume and mass fractions; (2) two-way coupling, where the large mass fraction makes particle-fluid momentum exchange no longer negligible but the particle-particle collision remains unimportant and (3) four-way coupling, where the particle-fluid momentum exchange and particle-particle collision must be considered due to large particle volume and mass fractions. The particle volume and mass fractions of interest in this paper lead to a two-way coupling particle-fluid interaction.

Special cares are given to the feedback terms of particle on fluid in two-way coupling PP-DNS. Crowe et al. [4] first propose the particle-source-in cell (PIC) method, which allocates the feedback terms to the grid cell where particle is located. Squires and Eaton [5] distribute the feedback terms to the neighbouring eight grids around a particle, i.e., the so-called projection method. Boivin et al. [6] demonstrate that projection method is better at capturing energy spectrum of incompressible homogenous isotropic turbulence than PIC method. However, very little research focus on the difference of these two methods on the particle-laden wall-bounded flows.

In PP-DNS, the carrier phase is usually solved under Eulerian framework and the dispersed phase is simulated by Lagrangian particle-tracking. Therefore, the prediction of particle motion depends strongly on the mechanical model. Stokes [7] is among the first to derive the formula for drag force on a rigid sphere in uniform flow. Maxey and Riley [8] obtain the well-known equation of motion for a small rigid sphere in non-uniform flow. Besides, there exists shear-induced lift in wall-bounded flows [9,10]. McLaughlin [11] introduces the Saffman force to particle motion equation in the simulation of incompressible turbulent channel flow and shows that the Saffman lift is non-negligible compared with the Stokes drag near the wall regions. Zeng et al. [12] perform direct numerical simulation to evaluate the force on an isolated particle at different wall distances in incompressible turbulent channel flow. The results suggest that the standard Stokes drag could almost capture the instantaneous force on the particle. Recently, Costa et al. [13] compare the results of one-way coupling PP-DNS and particle-resolved DNS (PR-DNS) in particle-laden incompressible turbulent channel flow, which manifest that the Saffman lift influences significantly the near-wall particle behaviour such as particle velocity and turbophoresis. Thus, the effect of Saffman force on particle motion and turbulence modulation remains an open question. To this end, the present paper is intended to investigate the influence of Saffman lift using two-way coupling PP-DNS.

Mostly, the research focus has been placed on particle-laden incompressible turbulent flow. Recently, there exist obvious interests in particle-turbulence interaction in compressible turbulent flow. The turbulence modulation [14], preferential concentration of inertial particles [15] and time irreversibility of turbulence [16] are studied in compressible homogenous isotropic turbulence. Xiao et al. [17] consider the particle accumulation in compressible turbulent boundary layer. Dai et al. [18] analyse the influence of particles on characteristic scales of turbulence in mixing layer. The compressibility effect on particle response and turbulence modulation in turbulent channel flow is still unanswered. Therefore, the Mach number effect on the particle behaviour and turbulence modulation in turbulent channel flow is also examined in this study.

The remaining structure of this paper is organized as follows. Presented in Sect. 2 are the numerical methods employed for two-way coupling PP-DNS. In Sect. 3, we give the simulation details of compressible channel flow and the corresponding particle parameters. The numerical results and the corresponding analysis are shown in Sect. 4. Specifically, we first address the influence of Saffman lift on particle-laden turbulence flow. Then, the effect of different two-way coupling feedback methods on the particle motion and turbulence modulation is examined. Furthermore, we explore the Mach number effect on the small particle-laden turbulent channel flow. Conclusions and discussions are given in Sect. 5.

2 Numerical Methods

2.1 Governing Equations for Carrier Phase

In this study, the carrier phase is assumed to be air, a typical compressible Newtonian fluid. The governing equations are the three-dimensional unsteady compressible Navier-Stokes equations [17,19].

The Navier-Stokes equations are non-dimensionlized with the following reference values: $\rho_0 = \rho_{\text{air}}$ for density, $\mu_0 = \mu_{\text{air}}$ for dynamic viscosity, $T_0 = 288.15\,\text{K}$ for temperature, $c_0 = \sqrt{\gamma r T_0}$ for sound speed (where $\gamma = C_p/C_v$ is the ratio of specific heat at constant pressure C_p to that at constant volume C_v, and $r = C_p - C_v$ is the specific gas constant of air), $\kappa_0 = C_p\mu_0/Pr$ for conductivity, $u_0 = c_0 Ma$ for velocity, $L_0 = Re\mu_0/(\rho_0 u_0)$ for length, $t_0 = L_0/u_0$ for time, $p_0 = \rho_0 c_0^2/\gamma$ for pressure, and $E_0 = \rho_0 u_0^2$ for energy per unit volume. The dimensionless parameters are Reynolds number Re, Mach number Ma and Prandtl number Pr. Therefore, the non-dimensional equations of mass, momentum and energy can be written as

$$\frac{\partial \rho}{\partial t} + \frac{\partial \rho u_j}{\partial x_j} = 0, \tag{1}$$

$$\frac{\partial \rho u_i}{\partial t} + \frac{\partial}{\partial x_j}\left(\rho u_i u_j + \frac{p}{\gamma Ma^2}\delta_{ij}\right) = \frac{1}{Re}\frac{\partial \sigma_{ij}}{\partial x_j} + \varphi_{ui}, \tag{2}$$

$$\frac{\partial E}{\partial t} + \frac{\partial}{\partial x_j}\left[\left(E + \frac{p}{\gamma Ma^2}\right)u_j\right] = \frac{1}{(\gamma-1)\,RePrMa^2}\frac{\partial}{\partial x_j}\left(\kappa\frac{\partial T}{\partial x_j}\right) + \frac{1}{Re}\frac{\partial \sigma_{ij}u_i}{\partial x_j} + \varphi_e, \tag{3}$$

with equation of state for ideal gas

$$p = \rho T. \tag{4}$$

Here ρ, u_i, T, p are dimensionless density, velocity in the ith direction, temperature, and pressure of fluid, respectively. δ_{ij} is the Kronecker delta. The dimensionless total energy per unit volume E is defined as

$$E = \rho\left(C_v T + u_i^2/2\right) = \frac{p}{\gamma(\gamma - 1)} + \frac{1}{2}\rho u_i^2. \tag{5}$$

The viscous shear stress tensor for Newtonian fluid is

$$\sigma_{ij} = \mu\left(\frac{\partial u_i}{\partial x_j} + \frac{\partial u_j}{\partial x_i}\right) - \frac{2}{3}\mu\frac{\partial u_k}{\partial x_k}\delta_{ij}. \tag{6}$$

The dynamic viscosity and the thermal conductivity follow Sutherland's law:

$$\mu = \kappa = \frac{(1 + T_{sb}/T_0)T^{3/2}}{T + T_{sb}/T_0} \tag{7}$$

with $T_{sb} = 110.4$ K. The feedback force and energy source of particles are denoted as φ_{ui} and φ_e, respectively. The exact expressions of these feedback terms are given in Sect. 2.3.

2.2 Governing Equations for Particle Motion

The solid particles are supposed to be identical spheres of dimensionless diameter d and mass density $\rho_p = 1000$. The particle diameter is much smaller than the Kolmogorov scale of background turbulence, i.e., $d \ll \eta_K$. Therefore, the Eulerian-Lagrangian point particle method is employed to track particles [1,2]. Meanwhile, we neglect the particle collision, particle rotation and heat transfer between particle and fluid.

The Maxey-Riley equation [8] is employed to describe individual particle motion. Owing to the large mass density of particles, the Stokes drag is dominant in Maxey-Riley equation [20]. In addition, the Saffman lift force is applied because it is comparable to the normal component of Stokes drag in near-wall regions [11]. The gravity is neglected to avoid high stream-wise velocity in vertical channel and asymmetric deposition of particles in horizontal channel. Based on the same characteristic quantities that are used for the fluid equations, the non-dimensional equations for particle motions are written as

$$\frac{dx_i^p}{dt} = v_i^p, \tag{8}$$

$$\frac{dv_i^p}{dt} = f_i^D + f_i^L, \tag{9}$$

with the Stokes drag [7]

$$f_i^D = \frac{f_1}{\tau_p}\left[u_i\left(\mathbf{x}^p\right) - v_i^p\right], \tag{10}$$

and the Saffman lift [9,10]

$$f_i^L = \frac{6.46}{\tau_p}\frac{d}{12\pi}\frac{\partial u_1/\partial x_2}{|\partial u_1/\partial x_2|}\left(Re\frac{\rho}{\mu}\left|\frac{\partial u_1}{\partial x_2}\right|\right)^{1/2}\left[u_1\left(\mathbf{x}^p\right)-v_1^p\right]\delta_{i2}. \tag{11}$$

The particle relaxation time is given as $\tau_p = Re\rho_p d^2/(18\mu)$. The finite particle Reynolds number effect is considered through the empirical correction $f_1 = 1 + 0.15Re_p^{0.685}$ with particle Reynolds number defined as $Re_p = Re\rho|\mathbf{u}\left(\mathbf{x}^p\right) - \mathbf{v}|d/\mu$.

2.3 Two-Way Coupling Feedback Terms

The two-way coupling feedback source terms φ_{ui} in momentum equation and φ_e in energy equation can be obtained by two different methods. One is the so-called Particle-in-Cell (PIC) method [4], which is to simply distribute the feedback force onto the nearest grid to the particle. The other method projects the force exerted by each particle to eight neighbouring grids around the particle [5]. Both the PIC method and projection method can be formulated as

$$\varphi_{ui} = -\frac{\pi d^3}{6}\rho_p \sum_{n \text{ in } V} H\left(\mathbf{x}-\mathbf{x}^{p,n}\right)a_i^{p,n}, \tag{12}$$

$$\varphi_e = -\frac{\pi d^3}{6}\rho_p \sum_{n \text{ in } V} H\left(\mathbf{x}-\mathbf{x}^{p,n}\right)a_i^{p,n}u_i, \tag{13}$$

with different weight functions H. a_i^p denotes the particle acceleration in the ith direction. For the PIC method, the weight function is

$$H(\mathbf{x}-\mathbf{y}) = \begin{cases} 1/\Delta V & \text{if } |x_i - y_i| < \Delta x_i/2,\ i = 1,2,3, \\ 0 & \text{otherwise.} \end{cases} \tag{14}$$

For the projection method, however, the weight function takes the form

$$H(\mathbf{x}-\mathbf{y}) = \begin{cases} (1-r_1)(1-r_2)(1-r_3)/\Delta V & \text{if } |x_i - y_i| < \Delta x_i,\ i = 1,2,3, \\ 0 & \text{otherwise,} \end{cases} \tag{15}$$

with $r_i = |x_i - y_i|/\Delta x_i$.

2.4 Numerical Schemes

In this study, finite difference method is employed for the fluid phase. The Steger-Warming method [21] is used to split the flux. The seventh-order upwind scheme and seventh-order weighted essentially non-oscillatory (WENO) scheme [22] are applied to discretizing convection terms for low-Mach number case and high-Mach number case, respectively. The diffusion terms are discretized by eighth-order accurate central difference scheme. The third-order total variation diminishing (TVD) Runge-Kutta multistage method [23] is used for time integration. For the disperse phase, the particles are tracked by explicit second-order Adams-Bashworth scheme. The third-order Lagrange interpolation scheme is employed to compute the fluid velocity at the particle positions [24].

3 Simulation Details

3.1 Channel Configuration

In this study, turbulent channel flows are considered in both weakly compressible (or approximately incompressible) ($Ma = 0.2$) and compressible ($Ma = 1.5$) regimes. The computational domain is $L_x \times L_y \times L_z = 4\pi\delta \times 2\delta \times 4\pi\delta/3$ with $\delta = 1$ being the half-width of channel and x, y, and z respectively the stream-wise, normal and span-wise directions of the channel. The non-slip boundary conditions are applied at the walls. The periodic conditions are applied in the stream-wise and span-wise directions. The isothermal condition $T = 1$ is applied at two walls [25]. The flow is driven by a body force to maintain a constant mass flux. A laminar parabolic Poiseuille velocity profile is used for the initial condition.

Table 1. Parameters for approximately incompressible and compressible channel flows.

Re	Ma	Re_τ	N_x	N_y	N_z	Δx^+	Δz^+	$\tau_{k,cp}$
3000	0.2	192	240	129	140	10.05	5.74	1.76
3000	1.5	216	240	129	140	11.31	6.46	2.13

The same wall units are used as done by Huang et al. [26]. The friction velocity is defined as $u_\tau = \sqrt{\tau_w/\rho_w}$ where τ_w is the total shear stress at the wall and $\rho_w = \langle\rho\rangle_w$ is the Reynolds-averaged density at the wall. Therefore, the wall unit is given by $y^+ = \rho_w y u_\tau/\mu_w$ with $\mu_w = \langle\mu\rangle_w$ and the friction Reynolds number is defined as $Re_\tau = \rho_w u_\tau \delta/\mu_w$. The simulation parameters, including grid numbers in the stream-wise direction (N_x), wall-normal direction (N_y) and span-wise direction (N_z), for two different channels are depicted in Table 1. The grid number for approximately incompressible channel flow takes the same criteria as those in [27] such that $\Delta x^+ \approx 10$ and $\Delta z^+ \approx 6$. The grid number for compressible channel flow keeps the same as that for approximately incompressible one of equal Reynolds number.

3.2 Particle Parameters

Here, we consider identical heavy spherical particles. The Stokes number is defined as the ratio of particle relaxation time to fluid characteristic time. However, the possible choice of fluid characteristic time is endless, including the Kolmogorov time scale at any layer and friction time scale, due to the inhomogeneity of wall-bounded flows. The friction time τ^+, Kolmogorov time at centre-plane $\tau_{k,cp}$ and Kolmogorov time at the walls $\tau_{k,w}$ are the three most frequently used fluid characteristic time scales. In this study, $\tau_{k,cp}$ is used to describe particle inertia relative to turbulence. The values of $\tau_{k,cp}$ for different cases are given in Table 1.

The particle Stokes number St, particle density ρ_p and Kolmogorov time at centre-plane are fixed for each simulation. The particle diameter needs to be calculated by the definition of particle relaxation time, i.e.,

$$d = \sqrt{\frac{18\mu St \tau_{k,cp}}{Re \rho_p}}, \tag{16}$$

which indicates that $d \sim \sqrt{St}$ and the point-particle method can not simulate particles of too large Stokes number because their diameter will exceed the Kolmogorov scale. Therefore, one must pay attention to the Stokes number of particles under consideration. The number of particles N_p can be obtained by the mass loading rate ϕ_m as

$$N_p = \frac{6 L_x L_y L_z \phi_m}{\pi \rho_p d^3}. \tag{17}$$

The volume loading rate ϕ_v is related to the mass loading rate in the form $\phi_v = \phi_m / \rho_p$.

Table 2. Parameters for approximately incompressible and compressible channel flows.

Case number	Ma	particle motion model	two-way coupling method
A	0.2	Stokes drag	PIC
B	0.2	Stokes drag + Saffman lift	PIC
C	0.2	Stokes drag + Saffman lift	Projection
D	1.5	Stokes drag + Saffman lift	Projection

For the present study, the particle Stokes number is $St = 3$, and the particle mass loading is $\phi_m = 10\%$, which leads to a volume loading rate of 1×10^{-4}, satisfying the two-way coupling condition [3]. Specifically, four different cases (see Table 2) are simulated using two-way coupling PP-DNS method to investigate (a) influence of Saffman lift (case A versus case B), (b) influence of two-way coupling method (case B versus case C), and (c) influence of Mach number (case C versus case D).

4 Results and Discussions

4.1 Influence of Saffman Lift

We first consider the stream-wise mean velocities of carrier phase U^+ and dispersed particles U_p^+, which are normalized by the friction velocity u_τ. It is observed in Fig. 1 that the Saffman lift has little effect on the stream-wise mean velocity of carrier phase, which is very close to the result of clean channel. This

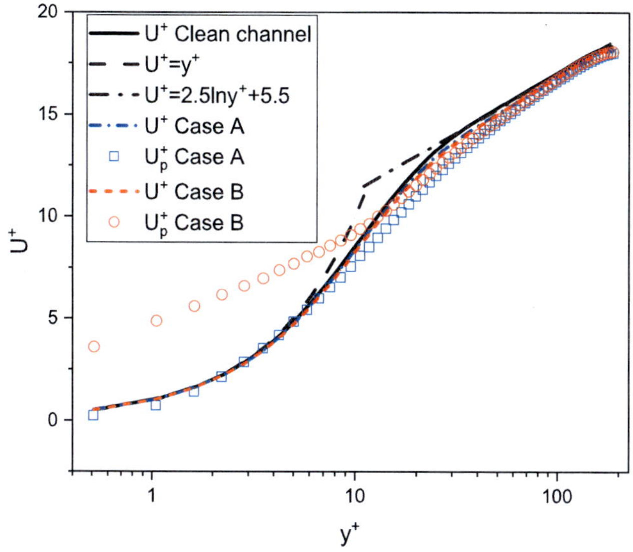

Fig. 1. Stream-wise mean velocity of carrier fluid (lines) and dispersed particles (scatters). Results for the case without Saffman lift are in blue and those with Saffman lift are in red. The black lines are the results for clean channel.

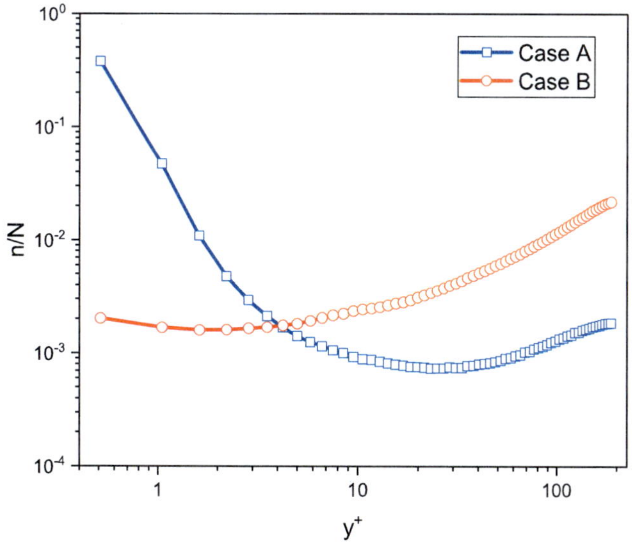

Fig. 2. Distribution of particle number normalized by total particle number at different positions from the wall: case A in blue line with squares; case B in red line with circles.

result confirms again that the effect of particles with diameter less than the Kolmogorov length scale on mean flow field is negligible [28]. However, it is clear that the Saffman lift will augment significantly the mean particle velocity in the stream-wise direction near the wall region ($y^+ < 20$). This is because the Saffman lift is no longer negligible compared with Stokes drag under the buffer region [11].

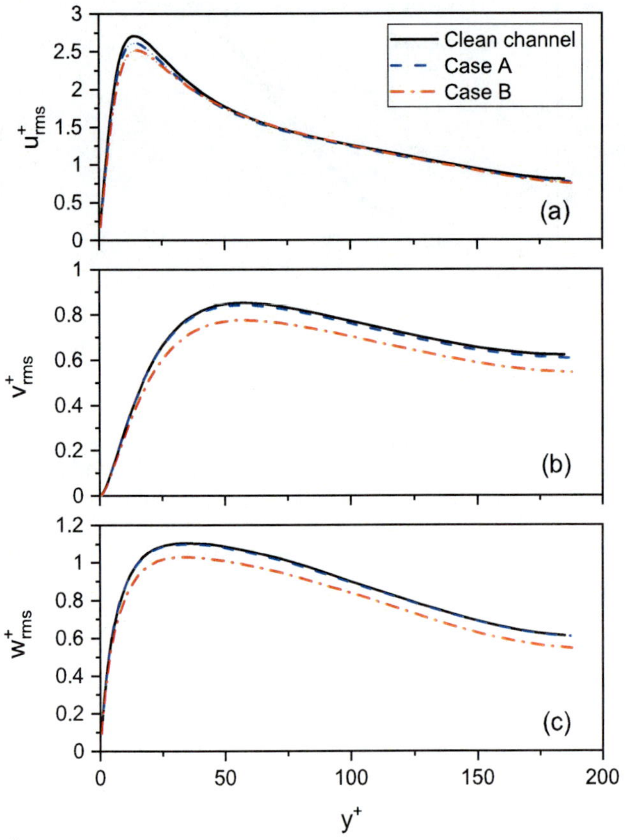

Fig. 3. Profiles of turbulence intensities in the stream-wise (a), wall-normal (b) and span-wise (c) directions: clean channel in black solid line, case A in blue dashed line, and case B in red dash-dotted line.

Then, we display the distribution of particle number at different positions from the wall in Fig. 2. As the particles are initially uniformly distributed in channel, the distribution of particle number at steady stage could represent the strength of turbophoresis. As can be observed, the Saffman lift can reduce significantly the strength of turbophoresis. The majority of particles are transported to the wall when the Saffman lift is absent. However, when Saffman lift is considered in particle motion equation, the turbophoresis becomes very weak. This can be explained by the fact that the Saffman lift will reduce the duration of a particle captured by viscous sublayer [13] as the Saffman lift will give particles enough kinetic energy to escape from viscous sublayer.

Figure 3 shows the turbulence intensities in stream-wise, wall-normal and span-wise directions, respectively. The present numerical results reveal that the Saffman lift increase the turbulence attenuation in all the three directions, confirming the turbulence attenuation by small particles observed in experiments [28]. When Saffman force is not included in particle motion equation, the turbulence intensities in the wall-normal direction and span-wise direction are very close to those for clean channel flow because the particles concentrate in the near-wall region and the momentum exchange in wall-normal direction is very weak. As for stream-wise turbulence intensity, we point out that the Saffman lift has great influence in the buffer region, where the turbulent motion is very strong and active. However, the Saffman lift shows little effect on the stream-wise turbulence intensity at the centre of channel.

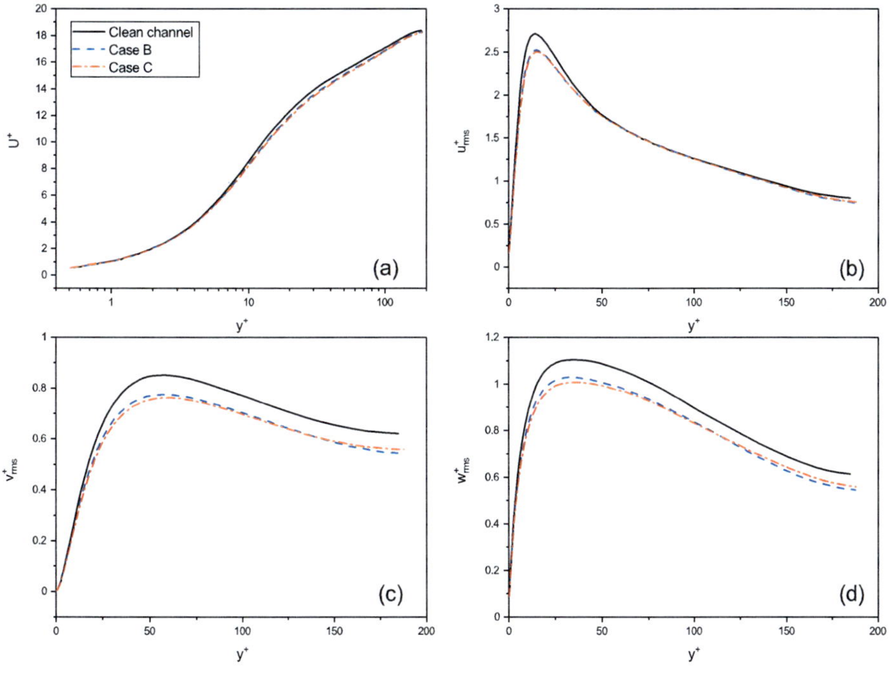

Fig. 4. Profiles of mean velocity (a), and turbulence intensities in the stream-wise (b), wall-normal (c) and span-wise (d) directions: clean channel in black solid line, case B in blue dashed line and case C in red dash-dotted line.

4.2 Influence of Different Two-Way Coupling Methods

In this subsection, we discuss the influence of different two-way coupling feedback terms on the first-order and second-order statistics of fluid velocities and averaged information of particles.

Depicted in Fig. 4 are the profile of mean stream-wise velocity and root mean squares of fluid velocity fluctuations in the stream-wise, wall-normal and span-wise directions, respectively. The numerical results imply that the PIC method and projection method give very similar fluid statistics. In other words, the difference of fluid statistics caused by two-way coupling feedback method is very limited. Therefore, we conclude that the influence of two-way coupling terms on fluid statistics is negligible when the total particle number in channel is large enough.

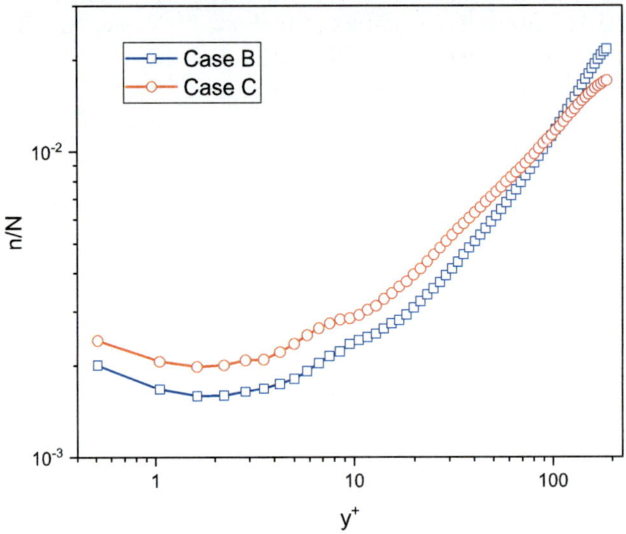

Fig. 5. Distribution of particle number normalized by total particle number at different positions from the wall: case B in blue line with squares and case C in red line with circles.

However, we observe in Fig. 5 that the two-way coupling method will change the particle distribution along the wall-normal direction, i.e., change the turbophoresis strength. The projection method leads to a stronger turbophoresis than the PIC method, which may be explained by the fact than projection method has a larger influence domain in fluid that PIC method. Besides, some visible difference exists in mean particle stream-wise velocity near the wall region ($y^+ < 20$) as depicted in Fig. 6. Nevertheless, such difference is very small, and seems not to influence the qualitative results.

Therefore, we suggest that both methods can be employed to study the particle-laden flows when there are enough particles in the carrier fluid. The quantitative results of particle velocity and fluid statistics are not influenced by the choice of feedback method. Special cares need to be paid when the turbophoresis is studied. The feedback force will cause some non-negligible difference, but the qualitative conclusion should not be changed. Last but not least,

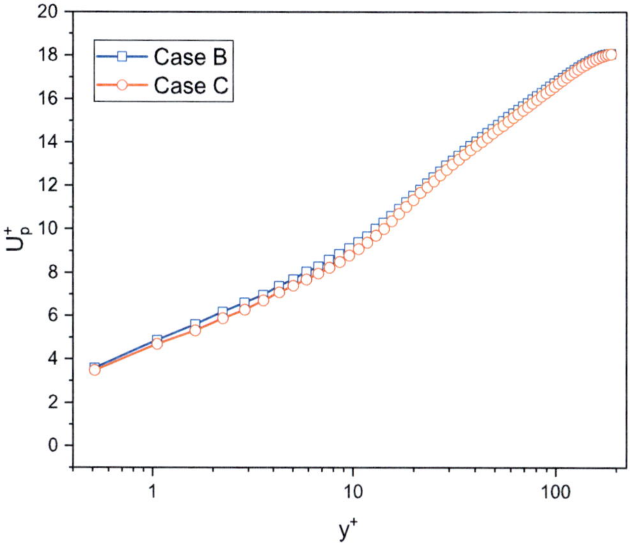

Fig. 6. Profile of mean particle stream-wise velocity in the wall-normal direction: case B in blue line with squares and case C in red line with circles.

the projection method might be more reasonable than the PIC method if one considers the nature of feedback force as a body force.

4.3 Mach Number Effect

In this subsection, we examine the Mach number effect on the particle-laden channel flows. According to Huang et al. [26], the dilatation dissipation and the mean product of pressure and dilatation fluctuations are very small in compressible turbulent channel flows. Therefore, the compressibility effect is mainly due to the mean density variation of fluid. The Van Driest transformation is frequently used to align the compressible mean stream-wise velocity with the incompressible case. In this study, however, the Van Driest transformation is not employed because we would like to compare the particle velocity with fluid velocity and it seems to make no sense to weight particle velocity by fluid density. Thus, all stream-wise velocities are simply normalized by the wall friction velocity in this paper.

Shown in Fig. 7 are the mean stream-wise velocities of carrier fluid and dispersed particles at $Ma = 0.2$ and $Ma = 1.5$, respectively. The numerical results suggest that the particle effect on mean stream-wise velocities of carrier phase is very small as the mean stream-wise velocity profiles of particle-laden cases are very close to those of single-phase channel flows. The mean stream-wise particle velocity in log-law region and core region of channel almost collapses onto the fluid velocity. The main deviation of particle velocity profile from that of carrier phase occurs in sublayer and buffer region, where the Saffman force is important.

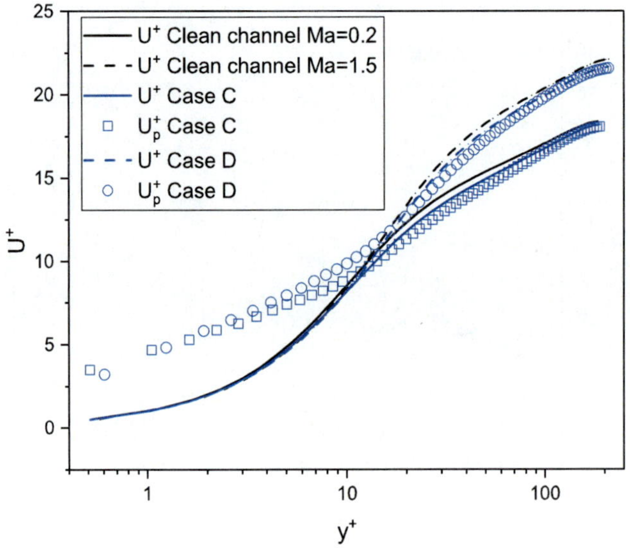

Fig. 7. Profiles of mean particle/fluid stream-wise velocity in the wall-normal direction: case C in solid line (fluid) with squares (particles), and case D in dashed line (fluid) with circles (particles).

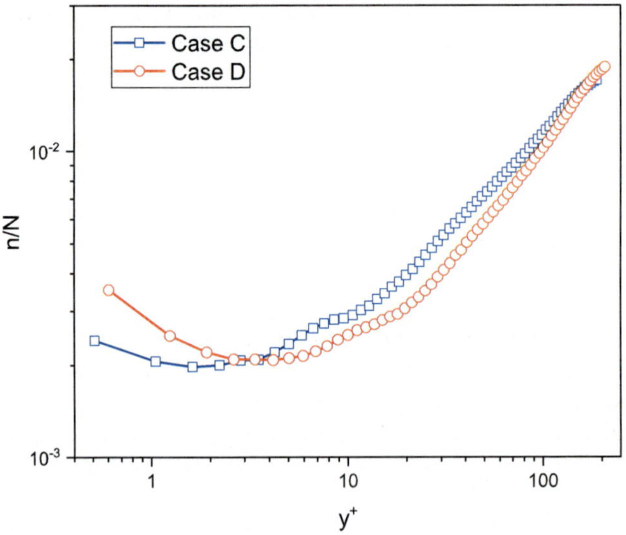

Fig. 8. Distribution of particle number normalized by total particle number at different positions from the wall: case C in blue line with squares, and case D in red line with circles.

However, the Mach number seems not to change the normalized particle velocity near the wall. Therefore, the Mach number effect on mean stream-wise particle velocity lies in two aspects. (1) In the near-wall region, the Mach number effect tends to alter the wall friction velocity and the normalized velocity is independent of Mach number. (2) Far from the wall, the Mach number effect appears as the mean density variation of fluid and the particle velocity follows the fluid velocity.

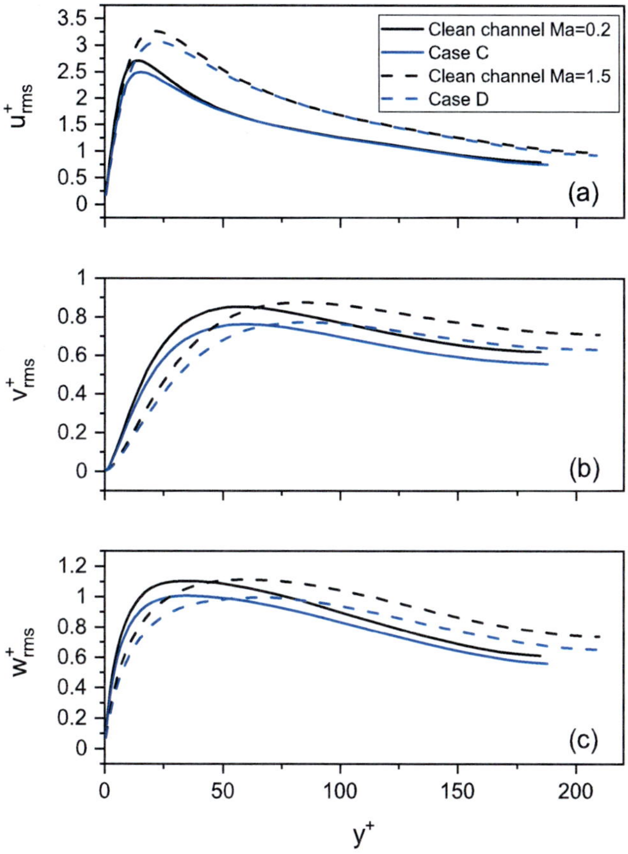

Fig. 9. Profiles of turbulence intensities in the stream-wise (a), wall-normal (b) and span-wise (c) directions for particle-laden approximately incompressible and compressible channel flows.

Furthermore, we evaluate the Mach number effect on the particle turbophoresis. It need to be pointed out that the particle diameters and total particle numbers in approximately incompressible case and compressible case are different due to the change of Kolmogorov time scale at the centre plane of channel. Thus, it is necessary to normalize the regional particle number by the total particle number in the two cases as shown in Fig. 8. The compressibility effect will enhance the turbophoresis as more particles are transported to the wall region and the centre region of the channel, where the turbulence activity is relatively weak compared with that in the buffer region.

In addition, the Mach number effect on turbulence fluctuations is studied. Turbulence intensities in the stream-wise, wall-normal and span-wise directions are plotted in Fig. 9. The basic fact is that small particles will cause turbulence attenuation, which is dependent on Mach number. Higher Mach number tends to increase the normalized turbulence intensity in all directions and the peak of turbulence intensity occurs at a larger y^+ than approximately incompressible case. The increase in Mach number will increase the turbulence attenuation in the wall-normal (about 10% attenuation for approximately incompressible case and 11% attenuation for compressible case in log-law layer and core region) and span-wise (about 9% attenuation for approximately incompressible case and 12% attenuation for compressible case in log-law layer and core region) directions. However, such increased turbulence attenuation can hardly be observed in the stream-wise direction. The insufficiently large Mach number might be a possible reason. One may also observe the increase of turbulence attenuation in the stream-wise direction when Mach number becomes much larger. Further research and numerical simulations with large variety of fluid/particle parameters are required to deeply understand the Mach number effect on fluid-particle interaction. Only some very preliminary observations are presented in this paper.

5 Conclusions

In this paper, the particle-fluid interaction is investigated based on two-way coupling point-particle direct numerical simulation of turbulent channel flow laden with heavy particles at the Stokes number $St = 3$.

First of all, we evaluate the influence of Saffman lift on particle motion and turbulence modulation. The numerical results suggest that the Saffman lift will augment significantly the particle stream-wise velocity near the wall region and reduce greatly the turbophoresis. It also turns out that the Saffman force results in a stronger turbulence attenuation compared with the cases where only the Stokes drag is included in the particle motion equation. In general, the Saffman lift is recommended when the wall constraint is present in particle-laden turbulent flow.

Furthermore, the influence of different two-way coupling methods is considered. The numerical results indicate that the difference in fluid statistics caused by the PIC method and the projection method is negligible. However, the two-way coupling method has some influence on particle behaviour, especially the turbophoresis. The deviation of particle velocity between these two methods is very limited. As a result, both the PIC and projection methods can be employed to qualitatively and quantitatively study turbulence modulation and particle response (except turbophoresis). When the turbophoresis is of interest, it is essential to use the same two-way coupling method to qualitatively compare the results for different cases. It should be mentioned that any quantitative study of turbophoresis by numerical simulation is not convincing enough unless better two-way coupling method is developed or particle-resolved simulations are performed.

Finally, we investigate the Mach number effect on fluid-particle interaction. It is found that the normalized stream-wise velocity of particles in the near-wall region in compressible turbulent channel flow almost collapses onto that in approximately incompressible case. The Mach number will influence the normalized stream-wise velocity of particles in log-law and centre regions of the channel due to the density-weighted fluid velocity. The increase in Mach number will also increase the level of turbulence attenuation in both the wall-normal and span-wise directions.

Acknowledgments. Numerical simulations were finished on the Polaris computing platform of Peking University in Beijing, China. The authors acknowledge the financial supports provided by National Natural Science Foundation of China (Grants No. 92152202 and No. 11988102).

References

1. Balachandar, S., Eaton, J.K.: Turbulent dispersed multiphase flow. Annu. Rev. Fluid Mech. **42**, 111–133 (2010). https://doi.org/10.1146/annurev.fluid.010908.165243
2. Kuerten, J.G.M.: Point-particle DNS and LES of particle-laden turbulent flow - a state-of-the-art review. Flow Turbul. Combust. **97**(3), 689–713 (2016). https://doi.org/10.1007/s10494-016-9765-y
3. Elghobashi, S.: On predicting particle-laden turbulent flows. Appl. Sci. Res. **52**, 309–329 (1994). https://doi.org/10.1007/BF00936835
4. Crowe, C.T., Sharama, M.P., Stock, D.E.: The particle-source-in cell (PSI-CELL) model for gas-droplet flows. J. Fluids Eng. **99**(2), 325–332 (1977). https://doi.org/10.1115/1.3448756
5. Squires, K.D., Eaton, J.K.: Particle response and turbulence modification in isotropic turbulence. Phys. Fluid A **2**(7), 1191–1203 (1990). https://doi.org/10.1063/1.857620
6. Boivin, M., Simonin, O., Squires, K.D.: Direct numerical simulation of turbulence modulation by particles in isotropic turbulence. J. Fluid Mech. **375**, 235–263 (1998). https://doi.org/10.1017/S0022112098002821
7. Stokes, G.G.: On the effect of the internal friction of fluids on the motion of pendulums. Trans. Cambridge Philos. Soc. 6, [8] (1850). https://doi.org/10.1017/CBO9780511702266
8. Maxey, M.R., Riley, J.J.: Equation of motion for a small rigid sphere in a nonuniform flow. Phys. Fluids **26**(4), 883–889 (1983). https://doi.org/10.1063/1.864230
9. Saffman, P.G.: The lift on a small sphere in a slow shear flow. J. Fluid Mech. **22**(2), 385–400 (1965). https://doi.org/10.1017/S0022112065000824
10. Saffman, P.G.: The lift on a small sphere in a slow shear flow - Corrigendum. J. Fluid Mech. **31**(3), 624 (1968). https://doi.org/10.1017/S0022112068999990
11. McLaughlin, J.B.: Aerosol particle deposition in numerically simulated channel flow. Phys. Fluids A **1**(7), 1211–1224 (1989). https://doi.org/10.1063/1.857344
12. Zeng, L., Balachandar, S., Fischer, P., Najjar, F.: Interactions of a stationary finite-sized particle with wall turbulence. J. Fluid Mech. **594**, 271–305 (2008). https://doi.org/10.1017/S0022112007009056

13. Costa, P., Brandt, L., Picano, F.: Interface-resolved simulations of small inertial particles in turbulent channel flow. J. Fluid Mech. **883**, A54 (2020). https://doi.org/10.1017/jfm.2019.918
14. Xia, Z., Shi, Y., Zhang, Q., Chen, S.: Modulation to compressible homogenous turbulence by heavy point particles. I. Effect of particles' density. Phys. Fluids **28**(1), 016103 (2016). https://doi.org/10.1063/1.4939794
15. Zhang, Q., Liu, H., Ma, Z., Xiao, Z.: Preferential concentration of heavy particles in compressible isotropic turbulence. Phys. Fluids **28**(5), 055104 (2016). https://doi.org/10.1063/1.4948810
16. Zhang, Q., Xiao, Z.: On the time irreversibility of compressible turbulence reflected by particles of various inertias. Phys. Fluids **33**(3), 036113 (2021). https://doi.org/10.1063/5.0041210
17. Xiao, W., Jin, T., Luo, K., Dai, Q., Fan, J.: Eulerian-Lagrangian direct numerical simulation of preferential accumulation of inertial particles in a compressible turbulent boundary layer. J. Fluid Mech. **903**, A19 (2020). https://doi.org/10.1017/jfm.2020.601
18. Dai, Q., Jin, T., Luo, K., Fan, J.: Direct numerical simulation of a three-dimensional spatially evolving compressible mixing layer laden with particles. I. Turbulent structures and asymmetric properties. Phys. Fluids **31**(8), 083302 (2019). https://doi.org/10.1063/1.5099469
19. Dai, Q., Luo, K., Jin, T., Fan, J.: Direct numerical simulation of turbulence modulation by particles in compressible isotropic turbulence. J. Fluid Mech. **832**, 438–482 (2017). https://doi.org/10.1017/jfm.2017.672
20. Armenio, V., Fiorotto, V.: The importance of the forces acting on particles in turbulent flows. Phys. Fluids **13**(8), 2437–2440 (2001). https://doi.org/10.1063/1.1385390
21. Steger, J.L., Warming, R.F.: Flux vector splitting of the inviscid gas dynamic equations with application to finite-difference methods. J. Comput. Phys. **40**(2), 263–293 (1981). https://doi.org/10.1016/0021-9991(81)90210-2
22. Jiang, G.-S., Shu, C.-W.: Efficient implementation of weighted ENO schemes. J. Comput. Phys. **126**(1), 202–228 (1996). https://doi.org/10.1006/jcph.1996.0130
23. Shu, C.-W., Osher, S.: Efficient implementation of essentially non-oscillatory shock-capturing schemes. J. Comput. Phys. **77**(2), 439–471 (1988). https://doi.org/10.1016/0021-9991(88)90177-5
24. Yeung, P.K., Pope, S.B.: An algorithm for tracking fluid particles in numerical simulations of homogeneous turbulence. J. Comput. Phys. **79**(2), 373–416 (1988). https://doi.org/10.1016/0021-9991(88)90022-8
25. Coleman, G.N., Kim, J., Moser, R.D.: A numerical study of turbulent supersonic isothermal-wall channel flow. J. Fluid Mech. **305**, 159–183 (1995). https://doi.org/10.1017/S0022112095004587
26. Huang, P.G., Coleman, G.N., Bradshaw, P.: Compressible turbulent channel flows: DNS results and modelling. J. Fluid Mech. **305**, 185–218 (1995). https://doi.org/10.1017/S0022112095004599
27. Bernardini, M., Pirozzoli, S., Orlandi, P.: Velocity statistics in turbulent channel flow up to $Re_\tau = 4000$. J. Fluid Mech. **742**, 171–191 (2014). https://doi.org/10.1017/jfm.2013.674
28. Kulick, J.D., Fessler, J.R., Eaton, J.K.: Particle response and turbulence modification in fully developed channel flow. J. Fluid Mech. **277**, 109–134 (1994). https://doi.org/10.1017/S0022112094002703

Flow-Induced Reconfiguration of and Force on Elastic Cantilevers

Jie Wei[1,2] and Jianjun Wu[1,2](✉)

[1] Key Laboratory of Mechanics on Disaster and Environment in Western China, Attached to the Ministry of Education of China, Lanzhou University, Lanzhou 730000, Gansu, People's Republic of China
wujjun@lzu.edu.cn

[2] College of Civil Engineering and Mechanics, Lanzhou University, Lanzhou 730000, Gansu, People's Republic of China

Abstract. The reconfiguration of elastic beam-like structures (e.g. the foliage) that subjected to cross-flow is widely encountered in nature and engineering applications. Based on a fully coupled numerical algorithm, the fluid–structure interaction between elastic cantilever and cross-flow is investigated in this paper. The immersed boundary-lattice Boltzmann method (IB-LBM) is used to solve the flow involving moving boundaries. The elastic cantilevers are modeled as inextensible Euler-Bernoulli beams and their dynamics are governed by geometrically nonlinear equations. The results suggest that the flow-induced dynamics and fluid forces are sensitive to Cauchy number (Cy). In essence, the flow-induced reconfiguration of elastic cantilevers discussed in this paper is the superposition of large deformation and sustained oscillation. The amplitude as well as the periodicity of drag and lift force are significantly modulated by structure motion. A greater Cy is found to yield a lift force which is comparable with the drag force. The Cauchy-Reconfiguration number curve overall fits the empirical values, in addition to the underestimated values obtained by the present simulations, we attribute it to the streamlined morphology and fluid force in transverse.

Keywords: Reconfiguration · Fluid–structure interactions · Immersed boundary-lattice Boltzmann method · Cauchy number

1 Introduction

The dynamic interactions between fluid flows and flexible structures (FSI) are associate with a broad range of physical process. In nature, the terrestrial/aquatic plants provide important ecosystem services. Due to the flexibility attribute, fluid-structure interaction is ubiquitous for various plant communities and plays a critical role in agriculture and costal protection, involving the lodging of crops and wave attenuation by submerged vegetation [1–3]. For a vegetation canopy, a wave-like motion, which is referred to as *Honami/Monami* for terrestrial/aquatic vegetation flow, would be observed when the blades or stalks are forced to bend and vibrate by aero/hydrodynamics loads [4, 5].

X. Zheng and S. Balachandar (Eds.): IUTAM 2023, IUTAM Bookseries 41, pp. 229–249, 2024.
https://doi.org/10.1007/978-3-031-47258-9_15

Such coherent motion is induced by the flexibility attributes of plants and the mixing-layer instability associated with the inflection point of the wind velocity profile [6]. This strong coupling between flow and flexible plants has been confirmed to be correlated with several key scalar transport process, e.g. the dispersion of seeds, spores, and pollens [7, 8]. The fluid-structure interactions are also observed in engineering areas with the increasing broad application of flexible structures. For the sake of flow control, passive slender structures are adopted to augment the wake of bluff bodies [9]. Also, the energy harvesting devices are applied to extract energy from the natural renewable resources (e.g. currents and wind flow) by the special designed flexible oscillating foils [10, 11]. Therefore, with the increasing demands in making precise prediction for both structural dynamics and resistance of the flexible cantilevered structures over a wide range of physical properties and flow conditions, such physical interaction has attracted growing attentions by researchers in the past decade. Moreover, the strong coupling between flexible structures (e.g. flat blades, elastic flags, filaments) and surrounding flows has become a hot and frontier topic in recent interdisciplinary research. Since the relevance in various areas, the related theoretical achievements are crucial to help us comprehend such interactions systematically and design various engineering structures, which take advantage the FSI; on the other hand, predict conditions for structure failure.

Among these FSI cases, the elastic cantilevered structures (e.g. flat plates and filaments) that exposed to cross-flow are widely seen, and the flexibility plays a vital role. With the bending stiffness being overcome by fluid loads, these structures would experience a set of complex dynamics responses, including significant deformations accompanied by large amplitude vibrations [12]. In another way, the flow-driven reconfiguration modulates the surrounding flow and yield a mitigated drag for a structure. Due to this, the risks of failure in uprooting and fracture by surrounding flow are reduced for natural plants. The early efforts, which tried to quantitatively reveal the relation between reconfiguration and drag reduction for vegetation elements, could be dated to Vogel et al. [13, 14]. As is known that, the resistance of rigid bluff bodies varies with stream velocity and follows the classical quadratic scaling law, namely, $F_d \propto U^2$. Yet the variation in resistance versus velocity was found to go more gently for the flexible ones and the Vogel exponent v was introduced to quantify it, which reads as $F_d \propto U^{2+v}$. Since then, the characterization and quantification of the drag scaling law as well as the Vogel exponent have become an open problem. A more comprehensively accepted scope for the Vogel exponent is $-1.3 < v < 0$ and v varies around -0.7, which is summarized from abundant experimental and numerical results, and also coincides with the $v = -2/3$ recovered by dimensional analysis [15, 16]. The dimensional analysis also suggests that Cauchy number (Cy) is the primary parameters in characterizing the coupling system. Remarkably, both the reconfiguration and drag reduction exhibit similar behaviors for those elastic cantilevered specimens with different morphologies, dimensions and flexural rigidity [17, 18]. This suggests that the relation between reconfiguration and drag reduction of these structures with different morphology is similar and controlled by the dimensionless Cy. Following this framework, Cy has been used in the model of flow-flexible vegetation interactions to characterize the impact of reconfiguration on the resistance of an individual vegetation element [19, 20].

Since the recent decade, the interactions between elastic cantilevered structures and flows have been explored with abundant useful insights being gained by studies, especially in establishing the relation between reconfiguration and drag reduction. Nonetheless, further works should be carried out on this issue. On one hand, plenty of experiments have been performed, which verified the results derived from theoretical analysis and also revealed more physical mechanisms. While, in terms of experimental measurements for such FSI system, it is still restrained by the fewer choice of adjustable physical parameters. Some variables, which are associated with the flow conditions and also the structural attribution, are even hard to control. On the other hand, remarkably, the earlier numerical models on flow-induced reconfiguration were simplified. In which, structural motion was confined to static deformation forced by drag and the background flow was reduced to steady state. Thus, some crucial dynamic effects, e.g. the instantaneous, non-uniform distribution structural loads induced by vortex shedding as well as the flow field perturbation from structural vibration were all ignored. However, with the rapid developments in numerical techniques. The FSI numerical methods e.g. Arbitrary Lagrangian-Eulerian Method (ALE), immersed boundary method (IBM) have been confirmed to be capable in fully capturing the coupling between fluid flows and various deformable structures. So far, few numerical attempts have been performed on the elastic cantilevered structure interacts with cross-flow. Therefore, the immersed boundary method is adopted here within the framework of fluid-structure interaction, the present work would focus on the effects of Cauchy number on dynamics behaviours and time-dependent fluid forces of slender elastic cantilevers.

2 Physical Model Statement

In order to emphasize the basic mechanisms of flow-induced dynamics for elastic cantilevers, we consider the structure of length L and is subjected to a uniform incoming flow (see Fig. 1). Available experimental observations suggested that the dynamics of such slender structures were mainly two-dimensional [21]. Thus, following assumptions are made for simplification: (i) The elastic cantilever is inextensible with constant physical attribute of density (ρ_f), elastic modulus (E), width ($w = 0.125L$), thickness ($t = 0.012w$) along the length. (ii) The cantilever only bend in x-y plane without torsion and any out-of-plane deformation. (iii) The 2D background flow is viscous and incompressible. (iv) Gravity and buoyant are disregarded due to structural morphology.

The schematic diagram of physical model is illustrated by Fig. 1. The computational domain is $[-11L, 31L] \times [-11L, 11L]$. The origin of Euler coordinate is fixed at the cantilever's mid-point of initial configuration. The x-axis and y-axis are respectively parallel and normal to the incoming flow.

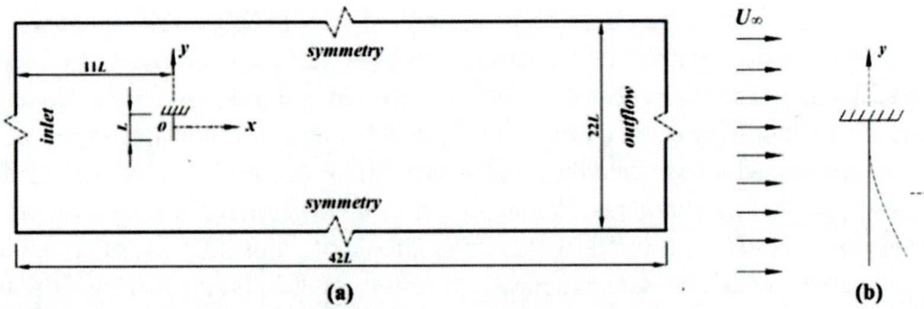

Fig. 1. Schematic of physical model (a). Computational domain; (b). Partial enlarged detail

3 Governing Equations and Numerical Algorithm

Within the framework of IB-LBM, the flow would be solved on fixed Cartesian mesh, while a set of discretized Lagrange grids are used to represent the 1-D elastic cantilever. The effects of structure reconfiguration on the surrounding flow is modeled by a forcing term incorporated into the momentum equations. Thus, the present study implement the FSI simulation by solving the fluid field and structure motion sequentially. Such coupling scheme has been confirmed to be accurate, efficient and robust in papers that involve the similar issue of moving boundary [22–24].

3.1 Mathematical Formulation of Lattice Boltzmann Method for Fluid Flow

As an alternative solver, LBM has been widely adopted by recent papers to solve the flows that governed by N-S equations, especially for the issues involving deformable solid boundaries. Here, the evolution of particle distribution function for the IB-LBM method in single-relaxation form is expressed as follows [25]

$$f_\alpha(\boldsymbol{r}+\boldsymbol{e}_\alpha\delta_t, t+\delta_t) - f_\alpha(\boldsymbol{r}, t) = -\frac{1}{\tau}[f_\alpha(\boldsymbol{r}, t) - f_\alpha^{eq}(\boldsymbol{r}, t)] + \delta t F_\alpha(\boldsymbol{r}, t) \tag{1}$$

$$F_\alpha = (1-\frac{1}{2\tau})\omega_\alpha[\frac{(\boldsymbol{e}_\alpha - \boldsymbol{u})}{c_s^2} + \frac{(\boldsymbol{e}_\alpha \cdot \boldsymbol{u})}{c_s^4}\boldsymbol{e}_\alpha] \cdot \mathbf{f} \tag{2}$$

Equations (1)–(2) include the external forcing term. Through the Chapman–Enskog multi-scale expansion, they can be recovered back to N-S equations. Here, $\boldsymbol{u}$ is the macroscopic velocity, $\mathbf{f}$ is the forcing term. α is the index of discretized velocity space. $\boldsymbol{e}_\alpha$ is the local particle velocity. f_α and f_α^{eq} are the density distribution function and the equilibrium distribution function, respectively. F_α is the forcing term in discretized velocity space. τ is the dimensionless relaxation time and defined as follows

$$\tau = 0.5 + \upsilon / c_s^2 \delta t \tag{3}$$

An incompressible D2Q9 LBGK model [26] is adopted here. It is also denoted as **D2G9**, whose particle velocity set is same with that of the **D2Q9** and the correspondent equilibrium distribution function are given by

$$f_\alpha^{eq} = \begin{cases} -4\sigma \frac{p}{c^2} + s_0(\boldsymbol{u}); \alpha = 0 \\ \lambda \frac{p}{c^2} + s_\alpha(\boldsymbol{u}); \alpha = 1 \sim 4 \\ \gamma \frac{p}{c^2} + s_\alpha(\boldsymbol{u}); \alpha = 5 \sim 8 \end{cases} \tag{4}$$

where, $s_\alpha(\boldsymbol{u}) = \omega_\alpha[\frac{\boldsymbol{e}_\alpha \cdot \boldsymbol{u}}{c_s^2} + \frac{(\boldsymbol{e}_\alpha \cdot \boldsymbol{u})^2}{2c_s^4} - \frac{|\boldsymbol{u}|^2}{2c_s^2}]$, the sound speed $c_s = 1/\sqrt{3}$, the weight coefficients are $\omega_0 = 4/9$, $\omega_{1-4} = 1/9$, $\omega_{5-8} = 1/36$ and $\delta = 5/12$, $\lambda = 1/3$, $\gamma = 1/12$. Furthermore, the velocity $\boldsymbol{u}$ and pressure p are computed by following formulas

$$\boldsymbol{u} = (\sum_\alpha f_\alpha \boldsymbol{e}_\alpha)/\rho_f + \delta t \boldsymbol{f}/2 \tag{5}$$

$$p = c_s^2[\sum_{\alpha=1}^{8} f_\alpha + s_0(\boldsymbol{u})]/(1 - \omega_0) \tag{6}$$

3.2 Equations for Structure Dynamics

Within large-deflection regime, the elastic cantilevers are modelled as inextensible Euler-Bernoulli beams, the geometrically nonlinear dynamics equations [27], which is defined in curvilinear coordinate system, and given by

$$\frac{\partial^2 \boldsymbol{X}}{\partial t^2} = \frac{\partial}{\partial s}(T\frac{\partial \boldsymbol{X}}{\partial s}) + \frac{\partial^2}{\partial s^2}(-EI\frac{\partial^2 \boldsymbol{X}}{\partial s^2}) + \boldsymbol{F} \tag{7}$$

the Poisson equation for tension T, which is derived from the inextensibility condition ($\partial \boldsymbol{X}/\partial$ s) ($\partial \boldsymbol{X}/\partial$ s) = 1, is expressed as

$$\frac{\partial \boldsymbol{X}}{\partial s} \cdot \frac{\partial}{\partial s}(\frac{\partial}{\partial s}(T\frac{\partial \boldsymbol{X}}{\partial s})) = \frac{1}{2}\frac{\partial^2}{\partial t^2}(\frac{\partial \boldsymbol{X}}{\partial s} \cdot \frac{\partial \boldsymbol{X}}{\partial s}) - \frac{\partial^2 \boldsymbol{X}}{\partial t \partial s} \cdot \frac{\partial^2 \boldsymbol{X}}{\partial t \partial s} - \frac{\partial \boldsymbol{X}}{\partial s} \cdot \frac{\partial}{\partial s}(\boldsymbol{F}_b + \boldsymbol{F}) \tag{8}$$

in which, s denotes local coordinates of the Lagrangian points along the length, $\boldsymbol{X} = (X, Y)$ is the Eulerian coordinates of the Lagrange points; $\boldsymbol{F}$ is the external force exerted by the surrounding flow; $\boldsymbol{F}_b$, i.e. the second term on the right-hand side of Eq. (7), is elastic bending force. Equations (7) and (8) are discretized by finite-difference method (FDM) and calculated on staggered grids. More detailed discretization of the governing equations and time marching scheme are described in [28], they would not be covered here.

3.3 Immersed Boundary Method

In this study, the implicit velocity correction-based immersed boundary-lattice Boltzmann method proposed by Wu and Shu (2009) [29] is adopted to solve the flow. As compared to the conventional IB treatments, e.g. the penalty function method [30] and the direct forcing method [31], the forcing terms of this IB scheme are calculated implicitly along with the merits of high computational efficiency and easy implementation. In addition, the non-slip boundary condition could be satisfied precisely without any user-specified parameter. In which, the force density $\mathbf{f}$ is taken as unknown, the velocity $\boldsymbol{u}$ is compartmentalized into the intermediate velocity $\boldsymbol{u}^*$ and the velocity correction $\boldsymbol{\delta u}$, i.e.

$$\boldsymbol{u} = \boldsymbol{u}^* + \boldsymbol{\delta u} \tag{9}$$

According to Eq. (5), they are defined as

$$\boldsymbol{u}^* = (\sum_\alpha f_\alpha \boldsymbol{e}_\alpha)/\rho_f \tag{10}$$

$$\boldsymbol{\delta u} = (\delta \mathbf{t}/2\rho_f)\mathbf{f} \tag{11}$$

For a given boundary point $\mathbf{X}_B(s_k)$, the unknown velocity correction $\delta \boldsymbol{u}_B$ is introduced here, $k = 0, 1, \ldots \mathrm{N}$. The velocity correction $\boldsymbol{\delta u}$ could be interpolated by

$$\delta \boldsymbol{u}(\boldsymbol{r},t) = \int_\Gamma \delta \boldsymbol{u}_B\left(\mathbf{X}_B^k,t\right)\delta\left(\boldsymbol{r} - \mathbf{X}_B^k(s,t)\right)ds \tag{12}$$

$$\delta\left(\boldsymbol{r} - \mathbf{X}_B^k(s,t)\right) = D_{i,j}^k\left(\boldsymbol{r}_{i,j} - \mathbf{X}_B^k\right) = \frac{1}{\Delta x \Delta y}\delta\left(\frac{x_{i,j} - X_B^k}{\Delta x}\right)\delta\left(\frac{y_{i,j} - Y_B^k}{\Delta y}\right) \tag{13}$$

in which, D is Dirac delta function, and the kernel distribution function as

$$\delta(r) = \begin{cases} (1 + \cos(\pi|r|/2))/4; & |r| < 2 \\ 0; & |r| \geq 2 \end{cases} \tag{14}$$

The following expression of $\boldsymbol{\delta u}$ is obtained by substituting Eq. (13) into Eq. (12)

$$\begin{aligned} \delta \boldsymbol{u}(\boldsymbol{r}_{i,j},t) &= \sum_k \delta \boldsymbol{u}_B(\mathbf{X}_B^k, t) D_{i,j}^k(\boldsymbol{r}_{i,j} - \mathbf{X}_B^k)\Delta s_k \\ &= \sum_k \frac{1}{\Delta x \Delta y}\delta \boldsymbol{u}_B(\mathbf{X}_B^k, t)\delta(\frac{x_{i,j} - X_B^k}{\Delta x})\delta(\frac{y_{i,j} - Y_B^k}{\Delta y})\Delta s_k \end{aligned} \tag{15}$$

where, velocity of the Lagrangian points $\mathbf{U}_B$, is interpolated by Dirac delta function

$$\mathbf{U}_B(\mathbf{X}_B^k, t) = \sum_{i,j} \boldsymbol{u}(\boldsymbol{r}_{i,j}, t) D_{i,j}^k(\boldsymbol{r}_{i,j} - \mathbf{X}_B^k)\Delta x \Delta y \tag{16}$$

it can be further re-written by substituting Eq. (9) and Eq. (15) into Eq. (16) as follows

$$\begin{aligned}\mathbf{U}_B(\mathbf{X}_B^k, t) &= \sum_{i,j} (\boldsymbol{u}^*(\boldsymbol{r}_{i,j}, t) + \delta\boldsymbol{u}(\boldsymbol{r}_{i,j}, t))D_{i,j}^k(\boldsymbol{r}_{i,j} - \mathbf{X}_B^k)\Delta x \Delta y \\ &= \sum_{i,j} \boldsymbol{u}^*(\boldsymbol{r}_{i,j}, t)D_{i,j}^k(\boldsymbol{r}_{i,j} - \mathbf{X}_B^k)\Delta x \Delta y \\ &+ \sum_{i,j}\sum_{m} \delta\boldsymbol{u}_B(\mathbf{X}_B^m, t)D_{i,j}^m(\boldsymbol{r}_{i,j} - \mathbf{X}_B^m)\Delta s_m D_{i,j}^k(\boldsymbol{r}_{i,j} - \mathbf{X}_B^k)\Delta x \Delta y\end{aligned} \tag{17}$$

Apparently, Eq. (17) could be written in the form of $[\mathbf{A}]\{\delta\boldsymbol{u}_B\} = \{\mathbf{B}\}$, in which the unknown $\delta\mathbf{u}_B = \{\delta\mathbf{u}_0^B, \delta\mathbf{u}_B^1, \ldots, \delta\mathbf{u}_B^m, \ \ldots \delta\mathbf{u}_B^N\}$. Thus, the velocity correction $\boldsymbol{\delta u}$ could be obtained by substituting the solved $\delta\boldsymbol{u}_B$ into Eq. (12) and the force density $\mathbf{f}$ is also determined by Eq. (11). Furthermore, the fluid load acting on the solid boundary, is

$$\mathrm{f}' = -2\rho_f \delta\boldsymbol{u}_B/\delta t \tag{18}$$

The drag and lift force could be calculated by

$$(F_D, F_L) = \int_\Gamma (f_x', f_y')ds \tag{19}$$

therefore, the drag coefficient and lift coefficient are defined as

$$(Cd, CL) = \frac{(F_D, F_L)}{(1/2)\rho U_\infty^2 L} \tag{20}$$

3.4 Boundary Conditions and Discretization Schemes

For the fluid domain, the uniform velocity incoming flow is imposed on the boundary of $x = -11L$, i.e.

$$u(y) = U_\infty;\ v = 0 \tag{21}$$

The outflow condition is specified at $x = 31L$, i.e.

$$\partial(u, v, p)/\partial x = 0 \tag{22}$$

On the top and bottom boundary, $y = \pm 11L$, the symmetry condition is specified, i.e.

$$\partial(u, p)/\partial y = 0 \tag{23}$$

The elastic cantilever is treated as moving no-slip wall boundary with its upper point clamped at $\boldsymbol{X}_0(0, 0.5L)$, i.e.

$$\boldsymbol{X} = \boldsymbol{X}_0,\ \partial\boldsymbol{X}/\partial\boldsymbol{s} = (0, -1) \tag{24}$$

For the free end of a cantilever, the boundary condition is given by

$$\partial^3 \boldsymbol{X}/\partial^3 s = \boldsymbol{0},\ \ \partial^2 \boldsymbol{X}/\partial^2 s = \boldsymbol{0}, T = 0 \tag{25}$$

For the sake of reducing the amount of discretized lattice points and achieving a higher computational efficiency, a non-uniform spatial discretization scheme is employed by this work and the Taylor series least square-based LBM (TLLBM) is adopted to update the particle distribution function. A detailed description for TLLBM could be referred to [32], here it would not be repeat. For the following FSI simulations, uniform grids are generated in the vicinity of the cantilever (a square domain of $[-2L, 2L] \times [-2L, 2L]$) with the minimum mesh size of the computation domain, i.e. $\Delta x_0 = \Delta y_0 = L/80$. While, the mesh size in both x- and y-direction is design to increase gradually in the region away from the cantilever, the maximum mesh size is specified as $\Delta x_{\max} = \Delta y_{\max} = 3.0\Delta x_0$. While, the cantilevers are discretized by uniformly-spaced Lagrangian points along the length, and the structure mesh size is $\Delta s = L/50$.

4 Numerical Results and Discussion

Based on the numerical algorithm described above, the present study implements the FSI simulation by solving Eqs. (1), (2), Eqs. (7), (8) and Eq. (17) sequentially. The FSI simulation code is written by C++ programming language. From Eq. (1), (2), the algorism is discovered to be featured by a good parallelism property. The OpenMP parallelism scheme, therefore, is introduced to decrease the CPU time costs.

4.1 Verification of the Present Numerical Method

Both the treatment on immersed boundary and the numerical solution for structure dynamics that governed by Eqs. (7), (8) are verified as bellow.

For the purpose of giving an assessment for the IB-LBM code on the calculation of flow field, the numerical simulation of flow past a fixed cylinder is performed. It is a classical research in CFD and abundant results are accessible in literatures. The diameter of the circular cylinder is L, the minimum mesh size is set to be $\Delta x_0 = \Delta y_0 = L/80$. Numerical simulations for both the state and unsteady state flow are performed.

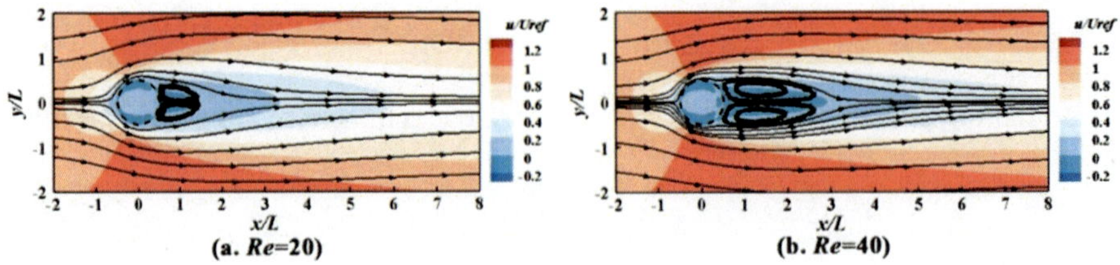

Fig. 2. Velocity Contour and streamlines for flow past a circular cylinder

From Fig. 2, the flow reach the final steady state at $Re = 20$ and 40. A pair of Fopple vortices developed in the wake of the cylinder forming the recirculating region, whose length increases with *Re*. While, the flow becomes unstable at $Re = 100$. The Fopple vortices shed from the cylinder alternately, forming the kármán vortex street in the wake. The non-dimensional vortex shedding frequency, which is also referred to as the Strouhal number, is $St = f_v L/U$. The time-dependent drag and lift coefficient of the cylinder are defined as Eq. (20) and shown in Fig. 3.

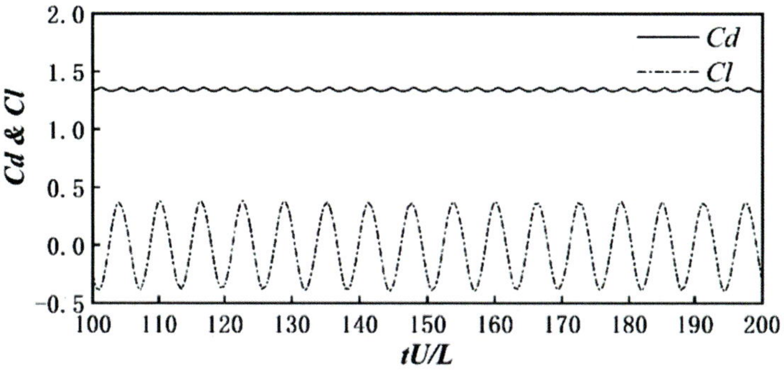

Fig. 3. Evolution of drag and lift coefficient for flow past a circular cylinder at $Re = 100$

Table 1 shows a good coincidence of our data with the data given by the previous literatures. Based on the above analysis, we can conclude that the present numerical algorithm and code are accurate enough to solve flow over bluff body and capable of calculating the fluid loads exerted on the solid-boundary.

Table 1. Drag coefficient, Strouhal number for flow past a cylinder

Case	Reference	*Cd*	*St*
$Re = 20$	Hu et al. [33]	2.21	–
	Niu et al. [34]	2.14	–
	Cui et al. [35]	2.11	–
	Present	2.21	–
$Re = 40$	Kim et al. [36]	1.51	–
	Wu and Shu [29]	1.56	–
	Cui et al. [35]	1.56	–
	Present	1.51	–
$Re = 100$ (time-averaged)	Peng et al. [31]	1.18	0.167
	Wu and Shu [29]	1.33	0.163
	Cui et al. [35]	1.36	0.167
	Present	1.30	0.162

Similarly, the following simulation of a rope-pendulum like motion is carried out to assess the performance of the structural dynamics code segment in the calculation of a forced-vibrating slender beam. The governing equations and boundary conditions are described by Eqs. (7), (8) and Eqs. (24), (25), respectively. The structure is initialized by

$$\boldsymbol{X}(s,0) = (L-s)(\cos\theta, \sin\theta),\ \boldsymbol{X}'(s,0) = (0,0) \tag{26}$$

It would start to swing due to a time-independent force $\boldsymbol{F} = (0, Fy)$. Its length is $L = 1$, the initial angle is $\theta = \pi/10$, the bending rigidity $\gamma = 0.01$ and $Fy = 10$. All of these parameters are identical to that adopted in the study of [28]. The time history of y-displacement of the free-end is plotted in Fig. 4 and agree well with the previous result.

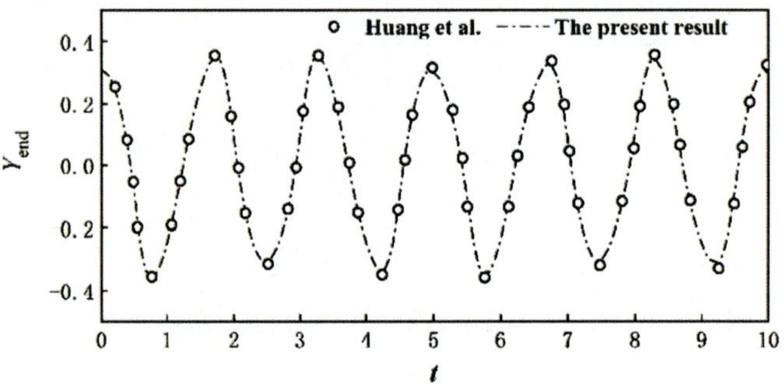

Fig. 4. Time history of y-displacement of free-end

One can conclude from the above facts that both the numerical method and code adopted here are feasible for the further studies in dealing with the flow-structure interactions issue.

4.2 Flow-Induced Reconfiguration and Fluid Forces of Elastic Cantilevers

To systematically explore the interactions between elastic cantilevers and fluid flows, a non-dimensional banding stiffness of flexible structures is introduced, which is defined by Eq. (27) and referred to as Cauchy number (Cy) [2, 16].

$$Cy = Cd_0 \rho_f U^2 A_0 / 2EI \tag{27}$$

in which, Cd_0 and A_0 are the drag coefficient and reference area of an equivalent rigid structure at the same Reynolds number, respectively. EI is the flexural rigidity of an elastic cantilever. In this study, Cy varies within $0.65 \leq Cy \leq 7.20$, while both the Reynolds number and mass ratio are kept and fixed to be $Re = U_\infty L/\nu = 200$, $m^* = 735$. According to the associated literatures, the overall attributes of the coupling system, including the structure vibration and vortex dynamics, are fundamentally similar within a wide range of Re. Therefore, they are confirmed to be relative insensitive to the variation of *Re*, especially for $Re > 100$ [37]. Thus, the parameters adopted here are considered to qualify for the target of capturing the rich dynamics features of such FSI system, while at the same time to provide a limitation in computation expense. In following sections, we will focus on the effects of *Cy* on dynamics behaviors of elastic cantilevers and fluid forces exerted on them.

Effects of Cauchy Number on Flow-Induced Vibration

First, the flow-induced dynamics of elastic cantilevers is investigated in this section. All the FSI calculations are terminated at dimensionless time $t^* = tU/L = 1400$, the time span is sufficient for the coupled fluid-structure system to get the fully developed state.

The time history of non-dimensional displacement component $X*$ free-end, which is defined as $X*$ free-end $= X_{\text{free-end}}/L$, is shown in Fig. 5 for 6 selected cases with different *Cy*.

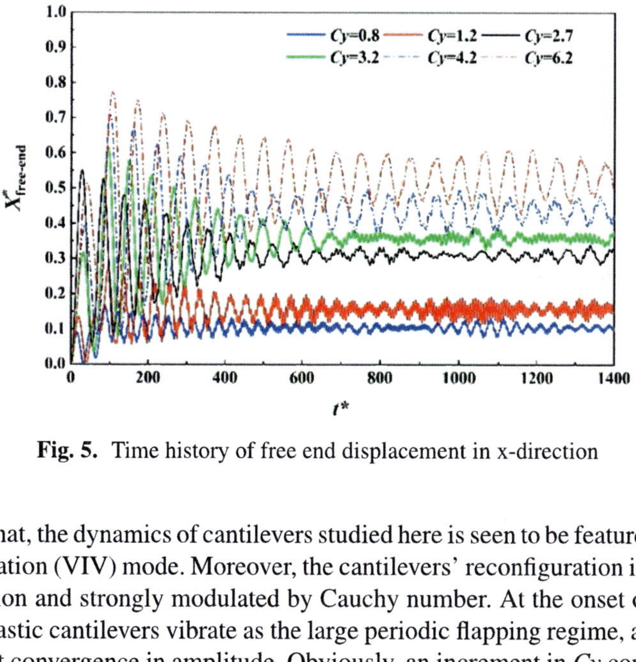

Fig. 5. Time history of free end displacement in x-direction

Clearly that, the dynamics of cantilevers studied here is seen to be featured by vortex-induced vibration (VIV) mode. Moreover, the cantilevers' reconfiguration is featured by large deflection and strongly modulated by Cauchy number. At the onset of reconfiguration, the elastic cantilevers vibrate as the large periodic flapping regime, accompanied by an evident convergence in amplitude. Obviously, an increment in *Cy* consequent in a increment in vibration period for this regime. For the cases of $Cy < 4.2$, a transition from flapping regime to flutter regime is seen over time, and then the structures vibrate within the small amplitude-high frequency mode. For the flutter regime, the mean deflection decrease for a reduced Cauchy number, also a smaller *Cy* tends to suppress the amplitude. While, it is interesting to note that the beat vibration is discovered at $Cy = 1.2$. We attribute it to resonance, as the vortex shedding frequency is found to approach the 2-th order structural natural frequency. A detailed spectrum analysis of displacement will be addressed in following sections. In order to give a better visualization for the main signatures of structural motion, the instantaneous configuration of two cantilevers are plotted in Fig. 6.

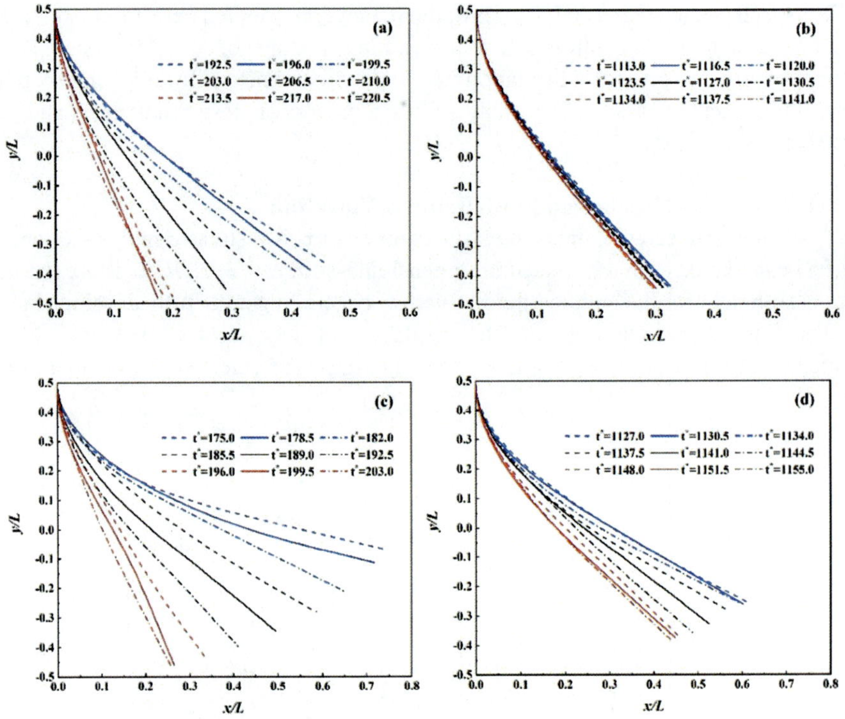

Fig. 6. Instantaneous configuration of the elastic cantilever: (a, b). $Cy = 2.7$; (c, d). $Cy = 6.2$

With the flexural rigidity being overcome by the fluid loads, the elastic cantilevers get aligned with flow through bending and deform to the streamlined morphology, which further result in a reduced frontal area for structures, see Fig. 6. Moreover, the flow-induced reconfiguration of the immersed structures is essentially the superposition of a mean deflection and a sustained oscillation around it. This general trend of structural motion is qualitatively similar to that given by [38, 39]. Figure 6(a, b) present the structural motion within flapping regime and flutter regime during half period of vibration. As the amplitude is far less than the mean deformation, the cantilever is analogous to 'motionless' in Fig. 6(b). While, as shown in Fig. 5 and Fig. 6(c, d), the transition of flapping regime to flutter regime is not observed for $Cy > 4.2$ and the vibration becomes stable with an approximate constant amplitude. Here, we suggest that a threshold value of Cy exists for the coupling system, and if Cy exceeds it, the structural displacement would be fully controlled by flapping. For the present FSI system with $Re = 200$, $m^* = 735$ and aspect ratio $L/t = 2000/3$, the critical Cy is within the range of $3.5 < Cy < 4.2$ and around 4.0. One can also conclude that, for a flow-elastic cantilever system, a greater Cauchy number tends to sustain the flapping regime.

The dynamics of the elastic cantilevers is further explored from the perspective of frequency domain property, the normalized power spectrum density (PSD) of $X^*_{\text{free-end}}(t^*)$ are illustrated by Fig. 7.

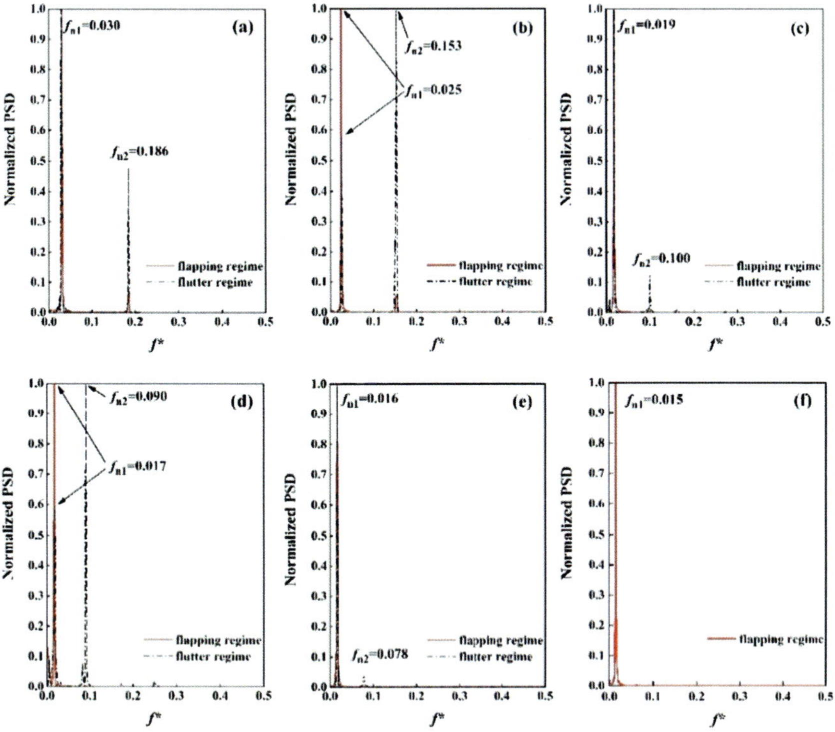

Fig. 7. Normalized power spectrum density of free end displacement in x-direction: (a). $Cy = 0.8$; (b). $Cy = 1.2$; (c). $Cy = 2.7$; (d). $Cy = 3.2$; (e). $Cy = 4.2$; (f). $Cy = 6.2$

At the onset of reconfiguration, the dynamics is featured by flapping and dominated only by the 1-th order natural frequency (f_{n1}). While, the motion is dominated jointly by 1-th and 2-th order natural frequency (f_{n2}) for the subsequent flutter regime. Thus, high frequency oscillation that featured by quasi-periodic is apparent in Fig. 5. It can be inferred from the PSD results that, within large deformation regime, the immersed elastic cantilevers vibrate at the well-defined natural frequency. But it doesn't mean that the effects of the cross-flow, which acts as the excitation source, on dynamics of the elastic cantilevers is absent. As is seen from Fig. 7 (b, d), the dominant frequency is f_{n2} for cantilevers under flutter regime. Here, the ratios of Strouhal number (*St*) to the 2-th order natural frequency are $St/f_{n2} = 1.03, 1.92$ at $Cy = 1.2$ and 3.2. They are approach to the integer times of f_{n2}, indicating that the impact of f_{n2} on displacement response is amplified by resonance in these cases.

Effects of Cauchy Number on Lift Force and Vortex Shedding

With the increase in *Cy*, the streamlining becomes more pronounced in cantilevers' morphology. The fluid load component in transverse is thus observed and further forms the lift force. The transverse loads, however, were underemphasized in previous studies. The vortex shedding frequency (f_v), could also be calculated by applying fast Fourier transform (FFT) on the time series of lift force. Figures 8 and 9 show the evolution of lift coefficient and the corresponding frequency spectrum for the selected cantilevers.

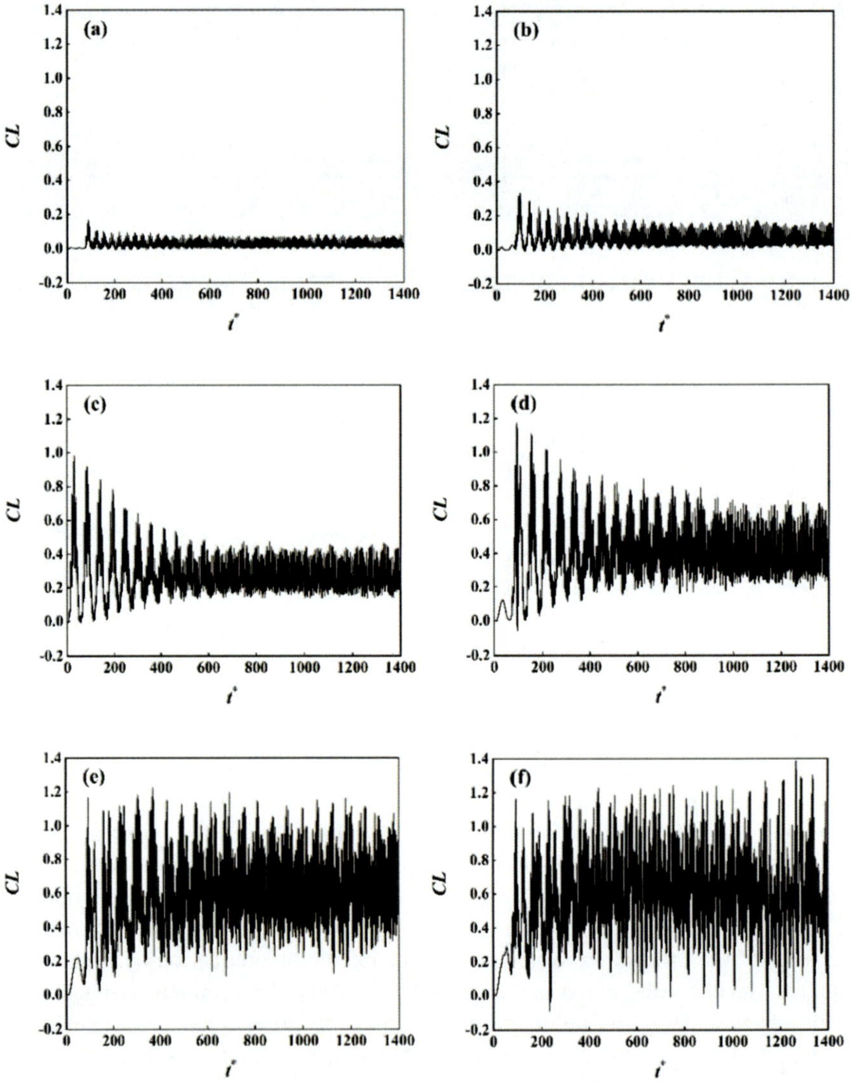

Fig. 8. Time history of lift coefficient for elastic cantilevers with different *Cy*: (a). $Cy = 0.65$; (b). $Cy = 1.2$; (c). $Cy = 2.7$; (d). $Cy = 3.7$; (e). $Cy = 6.2$; (f). $Cy = 7.2$

From Fig. 8, the flow-induced reconfiguration of the examined elastic cantilevers yielded a significant increment in lift coefficient. In terms of amplitude, the transition of structural motion from flapping to flutter is seen to induce an obvious convergence in amplitude of $CL(t^*)$, see Fig. 8(a, d). Note that, at high values of *Cy*, the lift force could be comparable with the stream-wise fluid force, which would be presented in following section. Furthermore, the time series are characterized by random if *Cy* exceeds the threshold, see Fig. 8(e, f). Also the isolated and brief snapping events are discovered to be evident, which are associated with the large amplitude flapping motion.

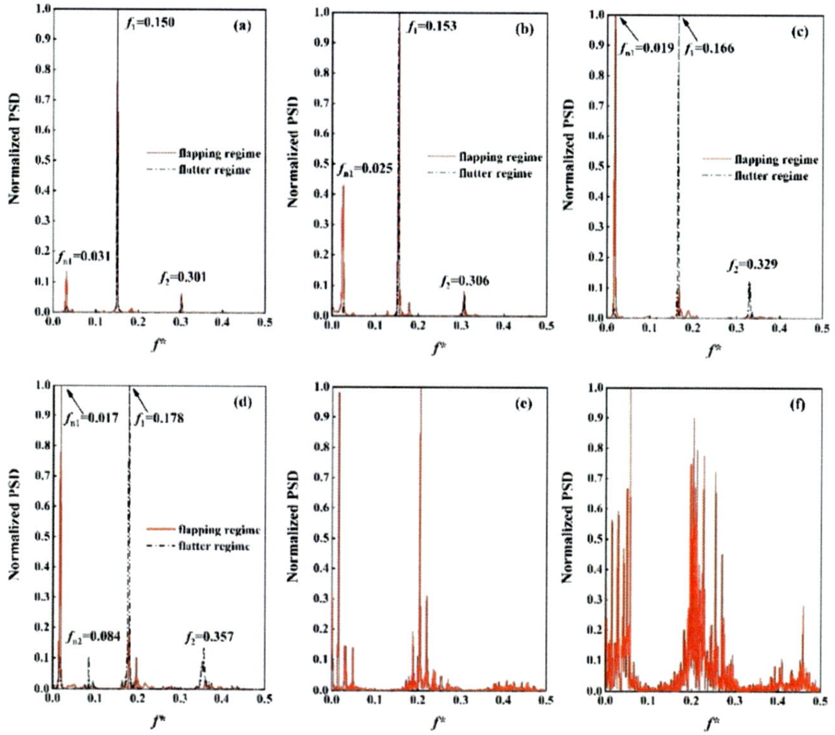

Fig. 9. Normalized power spectrum density of lift coefficient: (a). $Cy = 0.65$; (b). $Cy = 1.2$; (c). $Cy = 2.7$; (d). $Cy = 3.7$; (e). $Cy = 6.2$; (f). $Cy = 7.2$

As is shown in Fig. 9(a, d), four frequency components are evident in the two frequency spectrum plots that are featured by multiple peaks. They are natural frequency f_{n1} and f_{n2}; fundamental frequency f_1 and its harmonic frequency f_2, both of them are associated with Reynolds number and structural geometry. The dimensionless vortex shedding frequency is determined by the dominant frequency, which is treated in the same way to the case of flow past a rigid bluff body. Here, f_1 is identified as the Strouhal number. At $Cy = 0.65$ and 1.2, the dominant frequency is f_1 for both the flapping and flutter regime. With *Cy* increases, the modulation of frequency f_{n1} on time series $CL(t^*)$, also the vortex shedding, becomes more significantly until f_{n1} supersede f_1 as the dominant frequency within flapping stage, see Fig. 9 (d). While, for flutter regime of structural motion, vibration amplitude is far less than mean deflection; the resulting structural morphology is approximately fixed and the flow is essentially similar to that past a rigid bluff body. Thus, the vortex shedding is dominated by f_1. Furthermore, an evident increment in *St* for flutter regime could be also seen from Fig. 9(a–d) with the increase in *Cy*. At relatively high values of *Cy* (e.g. $Cy = 6.2$ and 7.2), the vibration mode is maintained to flapping. In addition, the frequency spectrum of time series $CL(t^*)$ manifests as continuous spectrum and presents the typical chaotic feature. Therefore, no definitive peak could be chosen as the dominant frequency.

Effects of Cauchy Number on Drag Force

The effects of Cauchy number on drag force is explored in this section. For the sake of comparison, the drag coefficient (*Cd*) evolution of the rigid structure is illustrated by Fig. 10(a). At $Cy = 0$, the time series of *Cd* features by high-frequency fluctuation with the time-averaged value $Cd_{\text{mean}} = 3.20$ when the flow get fully developed.

From Fig. 10, the elastic cantilevers' resistance is also found to be sensitive to *Cy*. Obviously, the amplitude of drag force as well as the lift force is seen to be strongly correlated to the periodicity and envelope of the structural displacement, especially for flapping regime at onset of the structural reconfiguration, see Fig. 10(c–f). The reconfiguration results in an evident reduction in frontal area. Furthermore, both the streamlined morphology and the resulting reduced frontal area are confirmed to be the two key mechanisms which would result in drag reduction. More visibly in Fig. 10, The mean value of $Cd(t^*)$ gradually decrease with *Cy*. Here, the Reconfiguration number (*R*) is introduced to assess the drag reduction quantitatively and defined as

$$R = \frac{F_d}{Cd_0 \rho_f U^2 Lb/2} \tag{28}$$

in which, F_d is the time-averaged drag force exerted on the elastic cantilever, Cd_0 is the drag coefficient of the equivalent rigid specimen [18]. For a rigid body, $R = 1$.

The *Cy*-*R* curve is plotted in logarithmic coordinate and illustrated by Fig. 11. It is found that the reconfiguration numbers calculated by this work overall agree with the values given in [15]. All the reconfiguration numbers of the simulated cantilevers are less than 1.0. At small values of *Cy*, the mean deflection of the given elastic cantilevers is indistinctive, thus the reconfiguration curve is seen to decrease mildly within the range $Cy < 1.0$. In terms of structural resistance, their performance close to the rigid ones. While, for $Cy > 1.0$, the curve starts to decline evidently accompanied by a great slope. The corresponding Vogel exponent is $v = 2\text{d}(\log R)/\text{d}(\log Cy) \approx -0.86$, which falls near the theoretical value of −2/3 and within the common accepted range of $v = (-1.2, -0.2)$ [40]. Although the overall tendency of *R* versus *Cy* reasonably fits the results that derived from theoretical model, a deviation also exists. Several factors may be responsible for the underestimation of our drag values. (i), it has been shown that the elastic cantilevers become more and more streamlined as Cauchy increases. With the help of the bi-directional coupling mode, the cantilevers' time-dependent configuration could be solved precisely by this work. For the theoretical model proposed in [15], however, the streamlining effects of structures on flow were not introduced. (ii), at greater values of *Cy*, e.g. $Cy > 1$, as is seen from above analysis, the fluid force component in transverse is of the same order of the magnitude as drag force. So, it also promote to a more streamlined morphology for cantilevers. This contribution from the transverse fluid loads for deformation, however, was neglected by the theoretical prediction. (iii), the energy exchange between structures and the surrounding flows is omitted by the theoretical model. Through the sustained storage-release of mechanical energy by structures, a

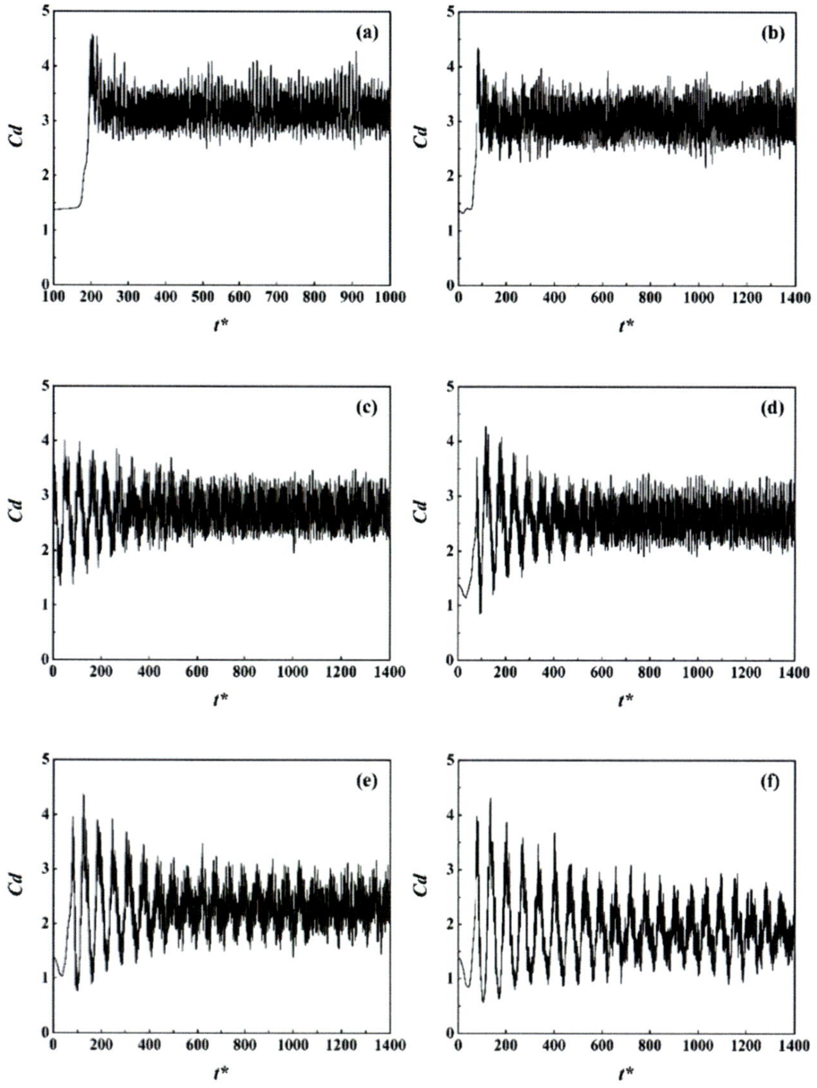

Fig. 10. Time history of drag coefficient for elastic cantilevers with different *Cy*: (a). $Cy = 0$; (b). $Cy = 1.2$; (c). $Cy = 2.7$; (d). $Cy = 3.7$; (e). $Cy = 6.2$; (f). $Cy = 7.2$

smaller drag is needed to maintain the mean deformation. These facts are likely to be responsible for our underestimated value of *R*, especially at $Cy > 1$.

The effects of *Cy* on elastic cantilevers' drag in terms of the frequency domain property are illustrated by Fig. 12. Also, the PSD result of the rigid case is given.

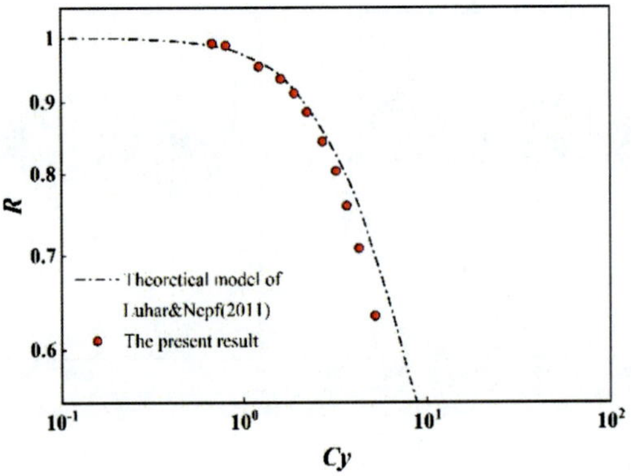

Fig. 11. Reconfiguration number R versus Cauchy number of elastic cantilevers

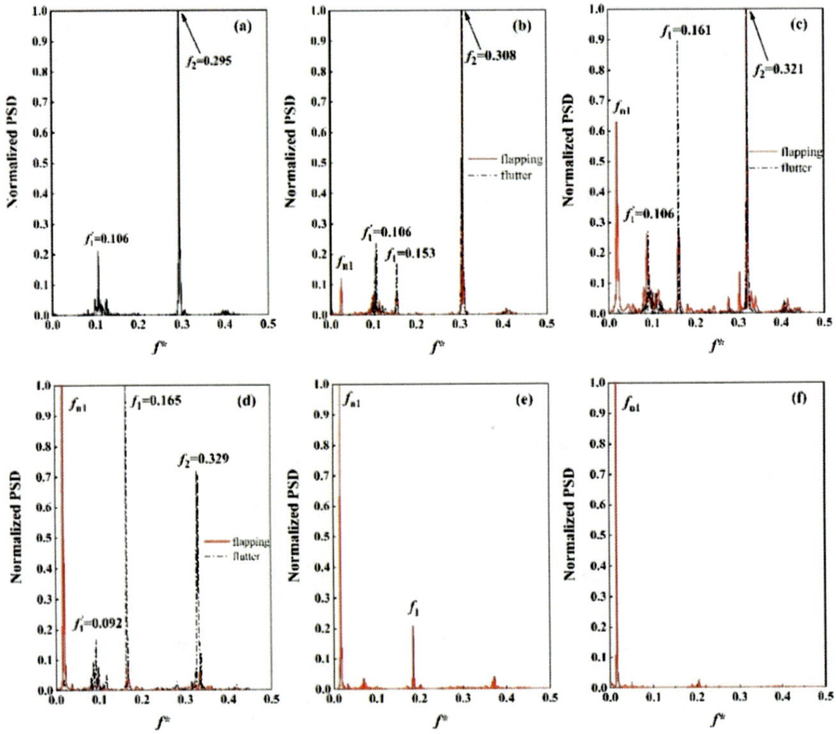

Fig. 12. Normalized power spectrum density of drag coefficient: (a). $Cy = 0$; (b). $Cy = 1.2$; (c). $Cy = 2.2$; (d). $Cy = 2.7$; (e). $Cy = 4.2$; (f). $Cy = 6.2$

In Fig. 12(a), two spectral peaks are apparent, i.e. $f'_1 = 0.106, f_2 = 0.295$. The ratio of them is an irrational number, indicating that the time series has a typical quasi-periodic feature. Usually, we have the dominant frequency $f_2 = 2f_1 = 2St$ for flow past a bluff body. For elastic cantilevers studied in this paper, multiple peaks appear in the frequency spectrum of time series $Cd(t^*)$. In addition to f' **1**and f_2, the frequency f_{n1} and f_1 are also observed. At $Cy = 1.2$ and 2.2, f_2 dominates the fluctuation of $Cd(t^*)$ under both the flapping and flutter regime. The natural frequency f_{n1} shows a growing modulation effect on $Cd(t^*)$ in flapping regime as Cy increases while it is less than the critical value. In Fig. 12(d), f_{n1} supersedes f_2 as the dominant frequency for flapping regime. For flutter regime, on the other hand, as the cantilevers become more streamlined with Cy, a more pronounced modulation effect of f_1 on time series $Cd(t^*)$ is observed until f_1 displace f_2 as the dominant frequency, see Fig. 12(b–d). Moreover, as Cy exceeds the threshold, the structural dynamics would be fully controlled by flapping mode, thus f_{n1} is seen to dominant the fluctuation of time series $Cd(t^*)$, which is in common with the frequency spectra of structure motion.

5 Conclusions

In the framework of the immersed boundary-lattice Boltzmann method, this study focused on the effects of Cauchy number on dynamics behaviors and fluid force exerted on slender elastic cantilevers within $0.65 \leq Cy \leq 7.2$.

In terms of the displacement response, the dynamic reconfiguration of the elastic cantilevers discussed in this paper was mainly characterized by VIV and sensitive to Cy. Two different vibration regimes, namely, flapping and flutter were identified based on the periodicity of vibration. The flow-induced reconfiguration was essentially the superposition of large deformation and sustained vibration. While, a greater Cy yielded an increase in mean deflection for the structures and was found to sustain the flapping regime. In addition, the PSD results revealed that the immersed elastic cantilevers vibrate at the well-defined natural frequency under the large deformation regime.

The strong coupling between dynamic reconfiguration and fluid flow also induced multiple frequency features in fluid forces on the structures. With the increase in Cy, the modulation of natural frequency on both the fluid forces and the vortex shedding process become more significant. Consequently, the amplitude of drag and lift force, as well as the vortex shedding frequency is found to be strongly correlated to the periodicity and the envelope of structural displacement, especially for the flapping regime at the onset of reconfiguration. In transverse, a greater Cy also yielded fluid force component which could not be negligible and its value could be even comparable with the drag force in magnitude. In terms of the drag reduction performance, the time-averaged drag simulated by this study overall agree with the previous results deduced from theoretical model, except for the deviation of underestimation. The streamlined morphology of the structures as well as the transverse fluid load were supposed to be responsible for the deviation.

References

1. Joseph, G., Mohammadi, M., Sterling, M., et al.: Determination of crop dynamic and aerodynamic parameters for lodging prediction. J. Wind Eng. Ind. Aerodyn. **202**, 104169 (2020)

2. Luhar, M., Infantes, E., Nepf, H.: Seagrass blade motion under waves and its impact on wave decay. J. Geophys. Res. Oceans **122**, 3736–3752 (2017)
3. Lei, J., Nepf, H.: Wave damping by flexible vegetation: connecting individual blade dynamics to the meadow scale. Coast. Eng. **147**, 138–148 (2019)
4. Langre, E.D.: Plant vibrations at all scales: a review. J. Exp. Bot. **70**(15), 3521–3531 (2019)
5. Tschisgale, S., Löhrer, B., Meller, R., et al.: Large eddy simulation of the fluid–structure interaction in an abstracted aquatic canopy consisting of flexible blades. J. Fluid Mech. **916**, A43 (2021)
6. Brunet, Y.: Turbulent flow in plant canopies: historical perspective and overview. Bound.-Layer Meteorol. **177**(2–3), 315–364 (2020)
7. Finnigan, J.: Turbulence in plant canopies. Annu. Rev. Fluid Mech. **32**(1), 519–571 (2000)
8. De Langre, E.: Effects of wind on plants. Annu. Rev. Fluid Mech. **40**, 141–168 (2008)
9. O' Connor, J., Revell, A.: Dynamic interactions of multiple wall-mounted flexible flaps. J. Fluid Mech. **870**, 189–216 (2019)
10. Fernandez-Feria, R.: Flutter stability analysis of an elastically supported flexible foil. Application to the energy harvesting of a fully-passive flexible flapping-foil of small amplitude. J. Fluids Struct. **109**, 103454 (2022)
11. Fernandez-Feria, R., Sanmiguel-Rojas, E.: On the feasibility of a flexible foil with passive heave to extract energy from low wind speeds. J. Fluids Struct. **114**, 103751 (2022)
12. Luhar, M., Nepf, H.M.: Wave-induced dynamics of flexible blades. J. Fluids Struct. **61**, 20–41 (2016)
13. Vogel, S.: Drag and flexibility in sessile organisms. Am. Zool. **24**(1), 37–44 (1984)
14. Vogel, S.: Drag and reconfiguration of broad leaves in high winds. J. Exp. Bot. **40**(8), 941–948 (1989)
15. Luhar, M., Nepf, H.M.: Flow-induced reconfiguration of buoyant and flexible aquatic vegetation. Limnol. Oceanogr. **56**(6), 2003–2017 (2011)
16. Leclercq, T., de Langre, E.: Drag reduction by elastic reconfiguration of non-uniform beams in non-uniform flows. J. Fluids Struct. **60**, 114–129 (2016)
17. Gosselin, F., De Langre, E., Machado-Almeida, B.A.: Drag reduction of flexible plates by reconfiguration. J. Fluid Mech. **650**, 319–341 (2010)
18. Gosselin, F.P., De Langre, E.: Drag reduction by reconfiguration of a poroelastic system. J. Fluids Struct. **27**(7), 1111–1123 (2011)
19. Beudin, A., Kalra, T.S., Ganju, N.K., et al.: Development of a coupled wave-flow-vegetation interaction model. Comput. Geosci. **100**, 76–86 (2017)
20. Lu, J., Dai, H.C.: Numerical modeling of pollution transport in flexible vegetation. Appl. Math. Model. **64**, 93–105 (2018)
21. Schouveiler, L., Eloy, C., Le Gal, P.: Flow-induced vibrations of high mass ratio flexible filaments freely hanging in a flow. Phys. Fluids **17**(4), 047104 (2005)
22. Revell, A., Mandal, P., Day, P.: Application of a lattice Boltzmann-immersed boundary method for fluid-filament dynamics and flow sensing. J. Biomech. **49**(11), 2143–2151 (2016)
23. Kanchan, M., Maniyeri, R.: Numerical simulation of buckling and asymmetric behavior of flexible filament using temporal second-order immersed boundary method. Int. J. Numer. Meth. Heat Fluid Flow **30**(3), 1047–1095 (2020)
24. Wu, J., Cheng, Y., Zhang, C., et al.: Simulating vortex induced vibration of an impulsively started flexible filament by an implicit IB–LB coupling scheme. Comput. Math. Appl. **79**(1), 159–173 (2020)
25. Guo, Z., Zheng, C., Shi, B.: Discrete lattice effects on the forcing term in the lattice Boltzmann method. Phys. Rev. E **65**(4), 046308 (2002)
26. Guo, Z., Shi, B., Wang, N.: Lattice BGK model for incompressible Navier-Stokes equation. J. Comput. Phys. **165**(1), 288–306 (2000)

27. He, S., Liu, H., Shen, L.: Simulation-based study of turbulent aquatic canopy flows with flexible stems. J. Fluid Mech. **947**, A33 (2022)
28. Huang, W.X., Shin, S.J., Sung, H.J.: Simulation of flexible filaments in a uniform flow by the immersed boundary method. J. Comput. Phys. **226**(2), 2206–2228 (2007)
29. Wu, J., Shu, C.: Implicit velocity correction-based immersed boundary-lattice Boltzmann method and its applications. J. Comput. Phys. **228**, 1963–1979 (2009)
30. Feng, Z.G., Michaelides, E.E.: The immersed boundary-lattice Boltzmann method for solving fluid-particles interaction problems. J. Comput. Phys. **195**, 602–628 (2004)
31. Fadlun, E.A., Verzicco, R., Orlandi, P., et al.: Combined immersed-boundary finite-difference methods for three dimensional complex flow simulations. J. Comput. Phys. **161**, 35–60 (2000)
32. Shu, C., Niu, X.D., Chew, Y.T.: Taylor-series expansion and least-squares-based lattice Boltzmann method: two-dimensional formulation and its applications. Int. J. Mod. Phys. C **14**(07), 925–944 (2003)
33. Hu, Y., Yuan, H., Shu, S., et al.: An improved momentum exchanged-based immersed boundary–lattice Boltzmann method by using an iterative technique. Comput. Math. Appl. **68**, 140–155 (2014)
34. Niu, X.D., Shu, C., Chew, Y.T., et al.: A momentum exchange-based immersed boundary-lattice Boltzmann method for simulating incompressible viscous flows. Phys. Lett. A **354**, 173–182 (2006)
35. Cui, X.W., Yao, X.L., Wang, Z.K., et al.: A hybrid wavelet-based adaptive immersed boundary finite-difference lattice Boltzmann method for two-dimensional fluid–structure interaction. J. Comput. Phys. **333**, 24–48 (2017)
36. Kim, J., Kim, D., Choi, H.: An immersed-boundary finite-volume method for simulations of flow in complex geometries. J. Comput. Phys. **171**, 132–150 (2001)
37. Mazharmanesh, S., Young, J., Tian, F.B., et al.: Energy harvesting of inverted piezoelectric flags in an oscillating flow. J. Fluids Struct. **115**, 103762 (2022)
38. Zhang, X., He, G., Zhang, X.: Fluid–structure interactions of single and dual wall-mounted 2D flexible filaments in a laminar boundary layer. J. Fluids Struct. **92**, 102787 (2020)
39. Silva-Leon, J., Cioncolini, A., Filippone, A., et al.: Flow-induced motions of flexible filaments hanging in cross-flow. Exp. Thermal Fluid Sci. **97**, 254–269 (2018)
40. Henriquez, S., Barrero-Gil, A.: Reconfiguration of flexible plates in sheared flow. Mech. Res. Commun. **62**(1), 1–4 (2014)

Check for updates

Volumetric Flow Measurement on a Displacement Piston Model for a β-Type Stirling Engine

Yun Liu(✉), Yasmeen Jojo-Cunningham, and Nuri Zeytinoglu

Purdue University Northwest, Hammond, IN, USA
liu739@pnw.edu

Abstract. Although many thermodynamic analyses and CFD simulations have been conducted on Stirling engines, the internal flow physics of Stirling engines is still unclear from experimental observations. Particularly, there is a significant lack of knowledge about the unsteady/oscillating three-dimensional internal flow field, which is critical to the understanding of the fundamental heat transfer process and performance of Stirling engines. Therefore, in this study, a Particle-Tracking-Velocimetry system: the Shake-The-Box system is implemented on a vertically oscillating disk inside a confined cylindrical container, filled with Sodium Iodide solution, to resolve the three-dimensional flow field around an experimentally simulated displacement piston inside a β type Stirling engine. Using the refractive index matching method, the image distortion due to the curved cylindrical wall was greatly suppressed and clear particle images were obtained. For the first time, the volumetric flow field below the displacement piston model is resolved and investigated experimentally under three different oscillating frequencies when it moves to the Top Dead Center. The flow measurement result suggests that a circular high velocity jet is created through the gap between the displacement piston model and the cylindrical wall of the container. As this high velocity jet is responsible for the heat transfer of convection at the bottom surface of the container, therefore can directly impact the performance of the Stirling engine. With the velocity of the jet estimated from the volumetric flow measurement data, the average heat convective coefficient $\overline{h}$ value is calculated and estimated at different oscillating frequencies.

Keywords: Stirling engine · Displacement piston · Shake-The-Box · Volumetric flow field

1 Introduction

The Stirling engine, invented by Robert Stirling in 1816, is quite different from conventional internal combustion engines and can be simply powered by the heat transfer between high and low temperature sources [1]. While conventional internal combustion engines, like diesel engine, vent the waste gas out of the engine cylinders, Stirling engines have concealed cylinders with the working fluid been permanently contained

X. Zheng and S. Balachandar (Eds.): IUTAM 2023, IUTAM Bookseries 41, pp. 250–264, 2024.
https://doi.org/10.1007/978-3-031-47258-9_16

inside the engine. Inside the concealed cylinder of a typical Stirling engine, like beta type Stirling engine, a power piston will output work under the pressure alternation between the high-pressure field with the heat transferred into the cylinder and the low-pressure field with the heat transferred out of the cylinder [1]. To realize this pressure alternation, a displacer piston will be moving between the bottom and top of the cylinder to transport the working fluid between a high temperature heat source and a low temperature heat sink. An ideal Stirling engine cycle consists of four basic stages [1]: During the first stage of isothermal compression process, driven by the flywheel's inertia and the low pressure of the working fluid, the power piston will be moving toward the bottom of the cylinder and compressed the working fluid. At the second stage of constant-volume heating process, the power piston stays at the Bottom Dead Center (lowers line) while the displacer moves and transports most working fluid to the heat source side of the engine to rise the working fluid pressure. During the third stage of isothermal expansion process, the high pressure working fluid expands and pushes the power piston away from the Bottom Dead Center to output work and drive the flywheel. During the last stage of constant volume cooling, the power piston will stays at the Top Dead Center while the displacer moves and transports most working fluid to the heat sink side of the engine to lower the working fluid temperature and pressure for the next cycle. Normally, to further improve the performance of the Stirling engines, heat regenerators (porous media to store the heat) will be added to the Stirling engines at the transportation of the working fluid to restore and reuse some of the heat, transporting between the heat source and sink [2]. As the pressure alternation of the working fluid is directly depending on the transportation of the working fluid between the heat source and sink, the working fluid transportation can directly impact the performance of the Stirling engine [3]. However, due to the complex and opaque engine structure, it is quite difficult to perceive and investigate the flow field inside the concealed Stirling engine experimentally. Instead, the CFD simulations have been implemented extensively on Stirling engine to study the internal flow field [4–9]. Using the OpenFoam, Mohammad and Ali investigated the oscillating flow on two β type stirling engines, including a TEMPO and a GPU-3 Stirling engine, with Dynamic Mesh and Volume Averaging technique to simulate the complex turbulent flow inside the engine [4]. Abuelyamen and Mansor studied the heat transfer rate and total pressure of α, β, γ type Stirling engines using a commercial CFD software of ANSYS to examine the performances of three different types of Stirling engines [5]. Using a two-dimensional axisymmetric CFD model in ANSYS-FLUENT, Chi *et al.* studied a β type free piston Stirling engine. Different thermal specification is examined to explore the operation cycle of the engine with detailed flow field simulated [6]. On a similar Stirling engine testing rig, CFD simulation work was conducted by Ibrahim to compare the detailed CFD simulation with experimental results, revealing a reasonable agreement on the transient velocities [7].

Although these above mentioned CFD simulations have revealed the detailed unsteady/oscillating flow field which is essential for comprehending the fundamental mechanics and performance of Stirling engines, most CFD numerical studies on Stirling engines are limited to two-dimensional flow simulation whereas the actual flow inside the Stirling engines is highly three-dimensional and turbulent. Therefore, to have a better understanding on the internal flow field of a Stirling engine, it is inevitable to exam the three-dimensional flow field inside the engines. Cheng and Chen made the first attempt to implement a 3D-CFD model to study the thermofluidic phenomena on a 1-kW β type Stirling engine [8]. The effect of regenerator porosity, initial charged pressure and heater temperature on engine power output is explored systematically. On a low temperature differential γ type Stiring engine, a 3D - CFD simulation is also realized to investigate the overall transient heat transfer rate and temperature field [9]. Meanwhile, the velocity field is examined under different gap widths between the displacer and displacer wall. However, it is quite challenging to resolve the 3-D internal flow on a Stirling engine/model experimentally. The difficulties of such experimental work mainly lays on following aspects. First, the opaque metal structure of the engine forbid conventional imaging based measurement. Second, even with a transparent cylindrical engine cylinder wall, the curved wall create significant image distortion due to refraction, therefore preventing 2-D conventional image based flow measurement inside the engine not to mention the 3-D images based measurement. On a curved transparent wall, as the light entering and exiting to different media, the light refraction happens and distorts the particle images. Therefore, to achieve the 3-D flow measurement inside a Stirling engine/model, a transparent cylinder wall and method of image distortion elimination have to be implemented. Therefore, in this study, the cylinder of a β-type Stirling engine is simulated with a cylindrical wall and the unsteady flow around a displacer model is investigated. The Shake-The-Box system, a Particle Tracking Velocimetry system, is utilized to resolve the three-dimensional flow field around the displacer model. To eliminate and suppress the image distortion due to the curved transparent wall, the method of refractive index matching is implemented with an outer hexagon container and sodium Iodide (NaI) solution.

2 Experimental Setup

2.1 Displacer Model and Refractive Index Matching

A displacer piston model was built and driven by a servo motor to move vertically inside a cylindrical container that has a cap on the top to simulate the concealed space of a β type Stirling engine cylinder (Fig. 1 *c*). The cylindrical container has an inner diameter of 70 mms while the displacer piston model has a diameter of 64 mms with a thickness of 10 mms so the displacer can freely move inside the cylindrical container. The servo motor is controlled by an Arduino program and can drive the displacer model to oscillate vertically at three frequencies of 0.2, 0.3, 0.6 Hz with an amplitude of 22 mms. The parameters of the experiment are summarized in Table 1. Yet, conventional flow field measurement methods, such as Particle Image Velocimetry and Particle Tracking Velocimetry, cannot be directly implemented on the flow field because the curved surface of the cylindrical container will introduce significant particle image distortion due to refraction. Therefore, the refractive index matching method is implemented in this study to eliminate the image distortion. Moreover, as a four-camera particle tracking velocimetry system is utilized to film the particle images from four different perspectives (camera views have to be independent and intersecting), an outer hexagon tank is utilized to accommodate the angles of imagining (cameras filming the particles perpendicular to the flat wall) when the cameras are filming the seeding particles inside the inner cylindrical tank, installed at the center of the hexagon tank (Fig. 1 *a, b*). As the entire tank is made of plexiglass, a Sodium Iodide solution (density: 1789 kg/m^3, kinematic viscosity: 1.069×10^{-6} m^2/s) is prepared with its refractive equals/close to the one of plexiglass (refractive index of plexiglass is at 1.49) and filled in the inner cylindrical tank as well as the outer hexagon tank. Consequently, the scattered light from the seeding particles can pass through the curved plexi-glass wall and solution with little light refraction, eliminating the particle image distortion when implementing the Particle Tracking Velocimetry.

2.2 Particle Tracking Velocimetry System

The state-of-the-art: Shake-the-Box system (Lavision, Gottingen) is implemented on a displacer model inside a cylindrical tank to measure the volumetric velocity field where the flow is seeded with 8~12 microns hollow glass spheres (TSI, Minnesota). A high repetition rate Nd: YLF single cavity laser (Photonics DM-30-527) and four high-speed cameras (Phantom VEO 640), mounted on two sides, were utilized to illuminate the seeding particles in the flow as well as to record the particle images. The camera lenses (Tokina Macro) have a focal length of 100 mms and are set at aperture size of f – 16 and f - 22 during the experiments (different aperture sizes are used to accommodate the scattered light intensity). Cylindrical optical lenses are added on the laser head to generate a 10 mm thick laser light that penetrates and illuminates the inner cylindrical tank from the side (Fig. 1 *a*). To measure and resolve the oscillating flow field on the displacer, moving at three different frequencies of 0.2, 0.3, 0.6 Hz, the sampling frequency of the images/laser was set at 300 Hz with an image resolution of 2560 $\times$ 1600 pixels on all four high-speed cameras. Due to the laser reflections on the displacer and connecting rod surfaces, only the flow field beneath the displacer is analyzed when the displacer moves to the Top Dead Center (Highest limit), resolving the flow field within a volume of 8.4 $\times$ 69.8 $\times$ 8.4 mm (26 $\times$ 4 $\times$ 4 Voxel, allowed triangulation error: 1 Voxels). The calibration, data collection and velocity field constructions were realized using the software Davis 10 (LaVision, Gottingen). A 55 $\times$ 55 mm calibration target was utilized and placed in the field of view (with seal cap and displacer model removed) and Sodium Iodide solution for taking the images by the high-speed cameras. Then the calibration was performed from the target images in the Davis 10. Finally, the volumetric flow measurement was achieved on the oscillating displacer piston model that is driven by the servo motor (Fig. 1 *a*). To realize the volumetric measurement with the system and improve sub-pixel accuracy, the volume self-calibration is required by the Shake-the-Box system. And volumetric self-calibration is carried out with the particle images taken during the experiments. Afterward, the particle reconstruction/tracking was carried out with the Shake-the-Box algorithm (LaVision, Gottingen) through shaking the particle position by 1 voxels during the iterations. During the post-processing within the Davis 10, the instantaneous volumetric velocity field was constructed from binning the particle tracking results, resulting three 26 $\times$ 4 $\times$ 4 matrices for the volumetric velocity in the three dimensions with a spatial resolution of 2.79 mm/velocity vector. For details about the algorithm of Shake-The-Box, please refer to Schanz et al. [10].

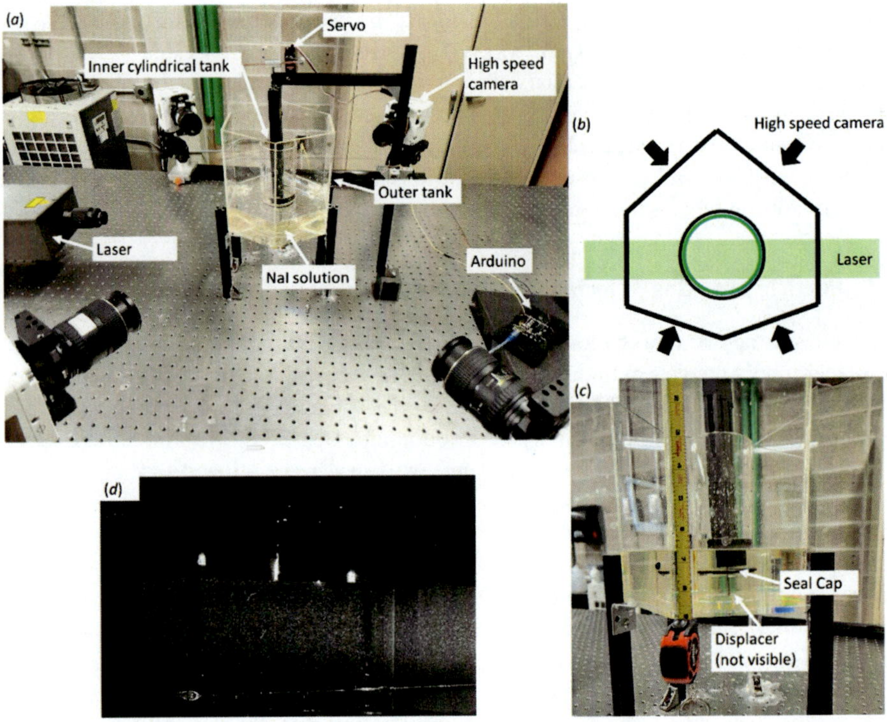

Fig. 1. Experimental setup. *a*) Shake-The-Box PTV system with the displacer model. *b*) The top-view of the experimental setup sketch. *c*) The displacer model in the NaI solution (not visible due to refractive index match). *d*) Sample particle images from the experiment and the flow field under the displacer is studied when the displacer moves to the Top Dead Center.

3 Experimental Results and Impinging JET Model

Although the implemented method of refractive index matching can significantly minimize the particle image distortions and enable the three-dimensional particle tracking, the light reflection on the solid surface is inevitable and can saturate the images. Particularly, the light reflection from the metal rod connection on the displacer model, when the displacer moves to the Bottom Dead Center, creates significant image saturation and poor particle images. Consequently, in this work, only the flow field below the displacer model when the displacer moves to the Top Dead Center is analyzed and discussed under the three different oscillating frequencies of 0.2, 0.3 and 0.6 Hz. Figures 2, 3, 4 present the averaged volumetric flow field below the displacer model during the first 1/18 period when the displacer reaches the Top Dead Center. The volumetric velocity vector field and three-dimensional streamlines under the three different displacer oscillating frequencies are plotted in the part *a* and *d* of the figures while the iso-surfaces of the velocity magnitude and angular velocity magnitude are presented in the part *b* and *c* of the figures.

In Fig. 2, with the displacer oscillating at 0.2 Hz, high flow velocity at 1m/s can be observed at the left and right ends of the flow field close to the cylindrical wall (y = −

30 mm and 30 mm). Yet, the flow velocity is significantly lower at the center (y = 0 mm) of the measurement domain compared to the ends. Moreover, the three-dimensional streamline plot suggests the flow field is quite different between the center and end regions with the streamlines spreading at the center but concentrating at the ends. The iso-surface plot of the angular velocity magnitude also presents a high vorticity field at the ends but low vorticity field at the center which is expected due the high velocity gradient/different close to the cylindrical wall.

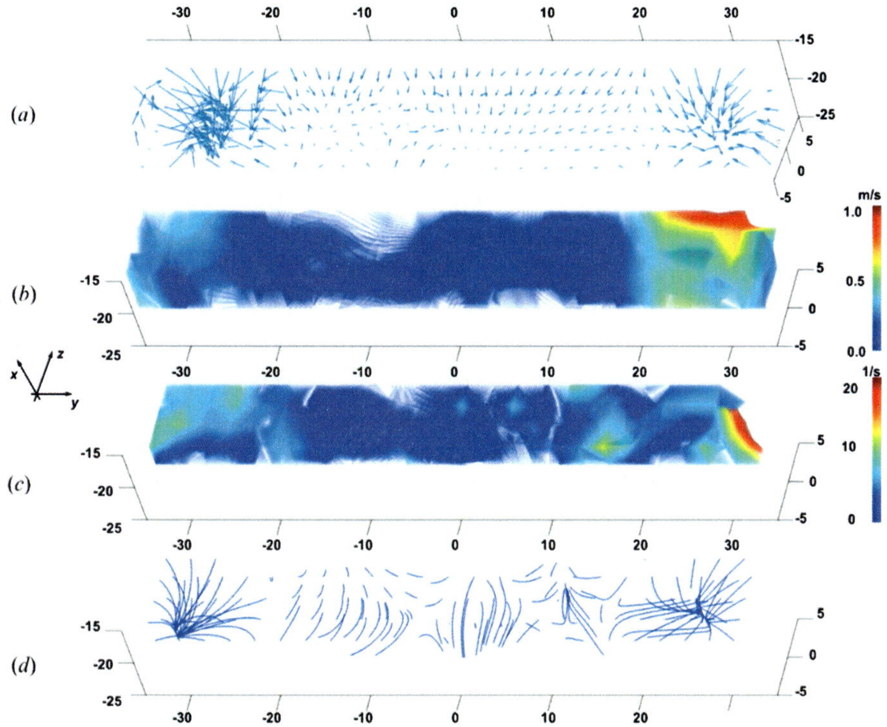

Fig. 2. The flow measurement results at a displacer oscillating frequency of 0.2 Hz. *a*) The volumetric velocity vector field. *b*) The velocity magnitude plot with transparent iso-surfaces. *c*) The constant angular velocity magnitude plot with transparent iso-surfaces. *d*) 3D streamlines.

At an oscillating frequency of 0.3 Hz, in Fig. 3, high flow velocity concentrates at the ends of the measurement domain while the low flow velocity, close to stagnant flow, spreads at the center of the measurement domain. However, comparing to the flow field at 0.2 Hz oscillating frequency, the three-dimensional streamlines, and iso-surfaces of angular velocity magnitude plots at 0.3 Hz oscillating frequency show different behaviour. At the oscillating frequency of 0.3 Hz, the three-dimensional streamlines concentrate only at the right side (y = 30 mm) of the measurement domain while the streamlines intersecting at the center of the measurement domain (the asymmetry of the flow field will be addressed later in the conclusion and discussion). Besides, high vorticity flow is observed not only close to the cylindrical wall but also at the center of the flow measurement domain. When the oscillating frequency of the displacer is increased to 0.6 Hz, the flow velocity at the ends of the flow measurement domain is significantly elevated. In Fig. 4, the flow velocity close to the cylindrical wall can reach up to 4 m/s when the flow at the center stays at a low velocity which is similar to the other two cases at lower oscillating frequencies. Likewise, the high vorticity/rotational flow is observed at the two ends of the measurement domain where a high velocity gradient is expected. The three-dimensional streamlines at 0.6 Hz is similar to the one at 0.2 Hz where the streamlines concentrates at the ends but spread at the center of the flow measurement domain.

Therefore, from Figs. 2, 3, and 4 a general overview of the volumetric flow field below the displacer model can be achieved. As the flow field is concealed within a cylindrical container with a height of 25 mm, when the displacer is driven (the displacer is driven by a servo motor in this study and driven by flying wheel and power piston on a real Stirling engine) to move vertically up and down, the displacer will not only displace the fluid but also will push and squeeze the fluid in the confined space, therefore creating high-speed downward (or upward) flow through the gap between the displacer and the cylindrical wall. This flow phenomenon can be further observed and investigated in the 2D plots from Figs. 5 and 6.

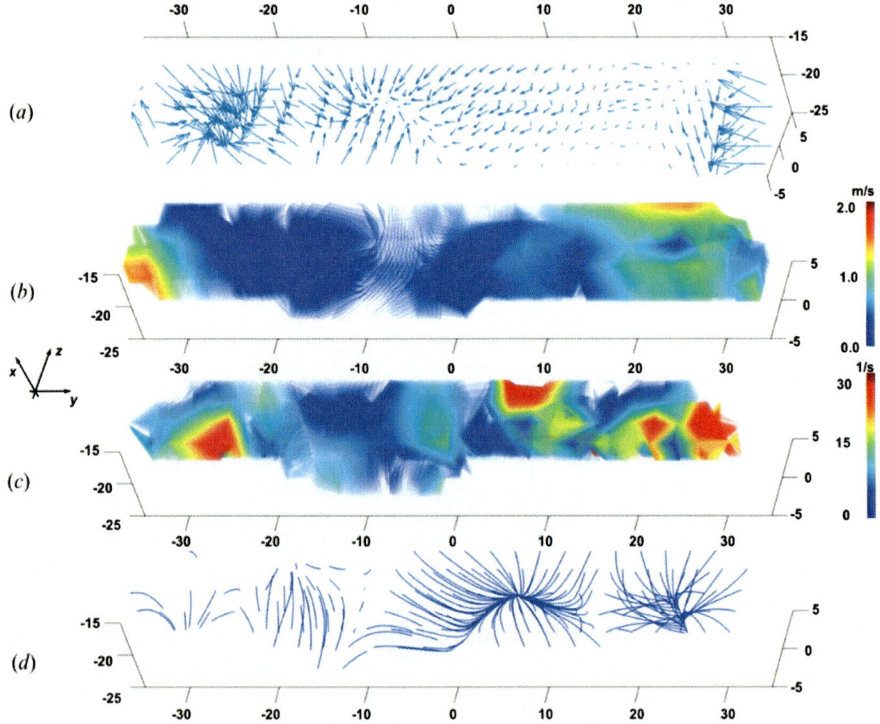

Fig. 3. The flow measurement results at a displacer oscillating frequency of 0.3 Hz. *a*) The volumetric velocity vector field. *b*) The velocity magnitude plot with transparent iso-surfaces. *c*) The constant angular velocity magnitude plot with transparent iso-surfaces. *d*) 3D streamlines.

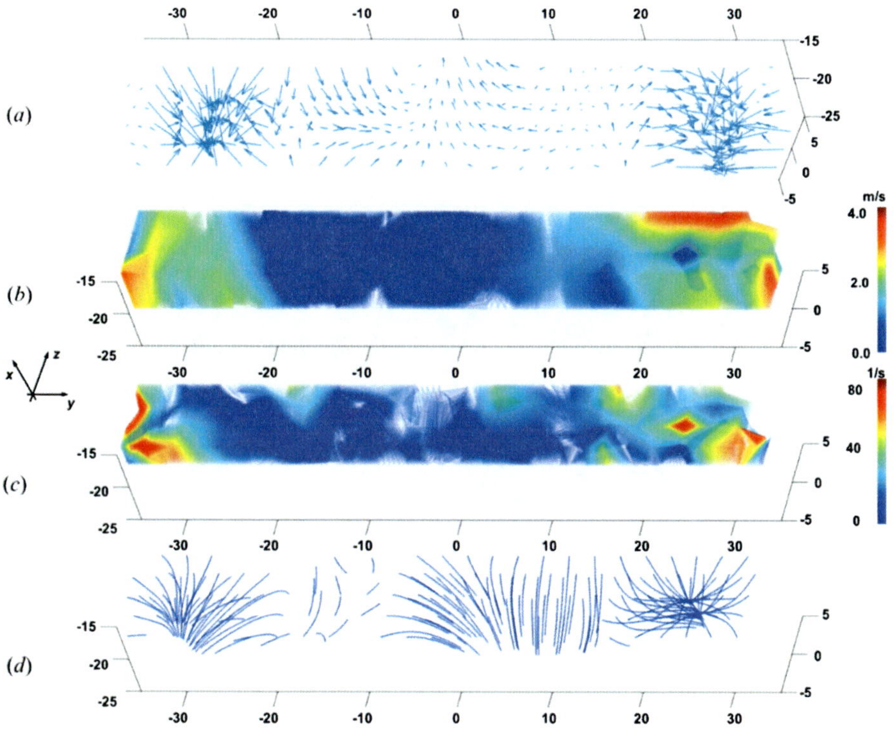

Fig. 4. The flow measurement results at a displacer oscillating frequency of 0.6 Hz. *a*) The volumetric velocity vector field. *b*) The velocity magnitude plot with transparent iso-surfaces. *c*) The constant angular velocity magnitude plot with transparent iso-surfaces. *d*) 3D streamlines.

Figure 5 presents the contour plot of the upward velocity (velocity in the x direction) close to the central plane of the measurement domain ($z = -1.8$ mm) as well as the velocity vector field on the plane under the three different displacer oscillating frequencies. As the displacer moves toward the Top Dead Center, the displacer will push and squeeze the fluid above the displacer, therefore creating high speed downward jet flow through the circular gap between the displacer and the cylindrical wall. Figure 5 clearly presents the two high-speed jet flow close to the end of the measurement domain ($y = -30$ and 30 mm) where the circular gap is located. In Fig. 5 *a*), at $y = -30$ mm, a downward jet flow with a magnitude of 0.5~1.0 m/s is created while a downward jet flow with a magnitude of 0.1~0.4 m/s is created at $y = 30$ mm. Likewise, in Fig. 5 *b*), two downward jets with a magnitude of 0.8~1.7 m/s and 0.1 m/s are created at $y = -30$ mm and 30 mm respectively. And 1.8 ~ 3.8 m/s and 0.6~2.0 m/s are observed at $y = -30$ mm and 30 mm in Fig. 5 *c*) with an oscillating frequency of 0.6 Hz. Clearly, as the displacer oscillating frequency increases, the downward jet flow is significantly intensified. Moreover, the downward jet flow influencing region (the blue contour in the plots) extends and considerable velocity can be observed at the center region ($y = 0$ mm). Notably, a counter upward flow can be seen in the flow field at $y = -30$ mm but no clear evidence of such counter upward flow is observed at y = 30 mm, suggesting a considerable asymmetry of the flow field.

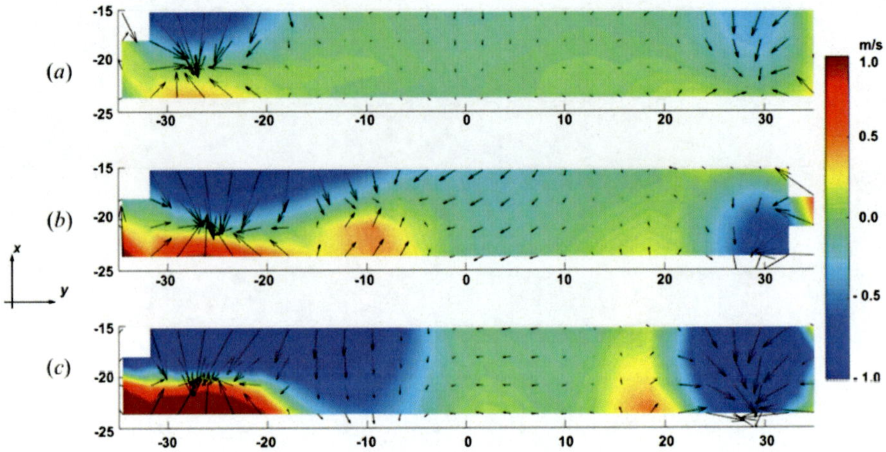

Fig. 5. *a*) Vertical velocity contour plots at z = −1.8 mm with a displacer oscillating frequency of 0.2 Hz *b*) Vertical velocity contour plots at z = −1.8 mm with a displacer oscillating frequency of 0.3 Hz. *c*) Vertical velocity contour plots at z = −1.8 mm with a displacer oscillating frequency of 0.6 Hz.

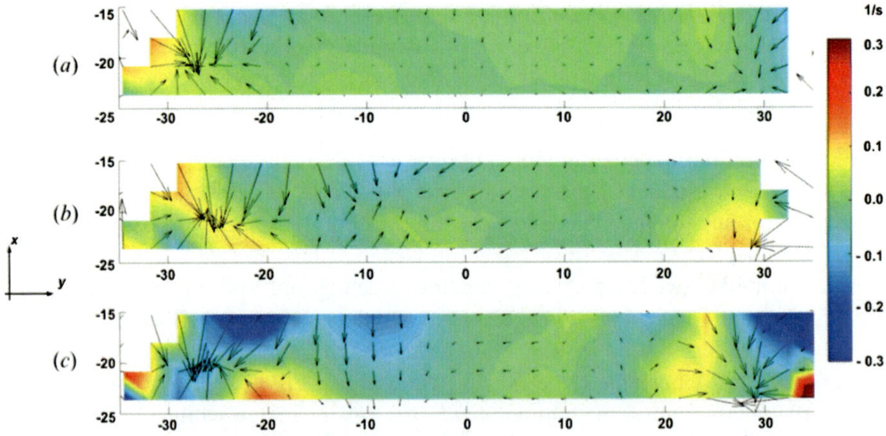

Fig. 6. *a*) Out-of-plane vorticity contour plot at z = −1.8 mm with a displacer oscillating frequency of 0.2 Hz *b*) Out-of-plane vorticity contour plot at z = −1.8 mm with a displacer oscillating frequency of 0.3 Hz. *c*) Out-of-plane vorticity contour plot at z = −1.8 mm with a displacer oscillating frequency of 0.6 Hz.

To further investigate and understand the dynamics of the flow field, the contour plot of in plane vorticity (in the Z direction) is plotted in Fig. 6. As the displacer oscillating frequency increases, high vorticity/rotational flow can be observed. Particularly, at the ends of flow domain, close to the cylindrical wall, strong rotational flow appears owing to the high velocity gradient in the region. On a moving disk with its direction perpendicular to the disk, if the disk is moving in a quiescent environment, a vortex ring is expected under the disk [11]. However, in this study with the displacer moving inside a confined domain, no evidence of counter-rotating vortex pair/vortex ring is observed in the flow field, differentiating the flow field from the one on a moving disk. Instead, two high velocity jets are formed through the gaps between the displacer and the cylindrical wall and impinged on the bottom surface of the tank. A flow diagram is sketched in Fig. 7 to illustrate the flow field around the displacer when it moves to the Top Dead Center. Inside the cylinder of a β type Stirling engine, the top and bottom plates of the cylinder will experience two different temperatures to generate the heat transfer between the internal fluid and the plates. When the displacer moves to the top, inside the cylinder, a high-speed jet flow will be created through the gap between the edge of the displacer and the cylindrical wall. In three-dimensional space, this high-speed jet flow will be formed in the circular gap around the edge of the displacer and impinged at the bottom wall of the cylinder. As the heat transfer mainly takes place at the inner surface of the cylinder, particular on the inner surface of the top and bottom plates, the heat transfer between the internal fluid flow and solid surface can be characterized as heat convention at the bottom surface. And the internal flow can be treated as the impinging jets through two slots nozzles when the displacer moves to the Top Dead Center (Fig. 7b).

As a result, to quantify the heat convection process, the Nusselt number can be estimated using an equation derived on an array of slot nozzles [12]. Equation 1 and 2 presents the calculation of Nusselt number on an array of slot nozzles that produce impinging jets. Where Re is the Reynolds number which depends on the impinging jet velocity at the nozzle V_e (gap in this study) and width of the nozzle W (with of the gap in this study). A_r is the ratio of the nozzle width W (width of the gap) and adjacent nozzles distance S (the distance between the gaps). H is the height of the impinging jet or the distance between the gap and the bottom surface of the wall. Besides the jet velocity V_e at the nozzle/gap, most quantities in the equation can be derived directly from the dimensions of the experimental setup and fluid properties. To derive the jet velocity V_e, the volumetric flow field data can be utilized.

$$\frac{\overline{Nu}}{Pr^{0.42}} = \frac{2}{3}A_{r,o}^{3/4}\left(\frac{2Re}{A_r/A_{r,o} + A_{r,o}/A_r}\right)^{2/3} \quad (1)$$

$$A_{r,o} = \left[60 + 4\left(\frac{H}{2W} - 2\right)^2\right]^{-1/2} \quad (2)$$

Figure 8 presents the contour plot of the downward velocity at the top surface of the measured flow domain under three different displacer oscillating frequencies. Similar to the observation made in Fig. 5, significant downward velocity are observed in the region at y = −30 mm and 30 mm where the impinging jets should be created. As the plots in Fig. 8 are generated on top of the flow measurement domain which is close to the gap between the displacer and cylindrical wall, therefore, the downward velocity in the regions can be considered as the impinging jet speed V_e. To estimate V_e for all the three cases of different displacer oscillating frequencies, the mean values of downward velocity in the regions: −34 mm < y < −26 mm and 26 mm < y < 34 mm are calculated to derive $V_{e,Left}$ and $V_{e,Right}$ respectively. Then, due to the asymmetry of the flow field,

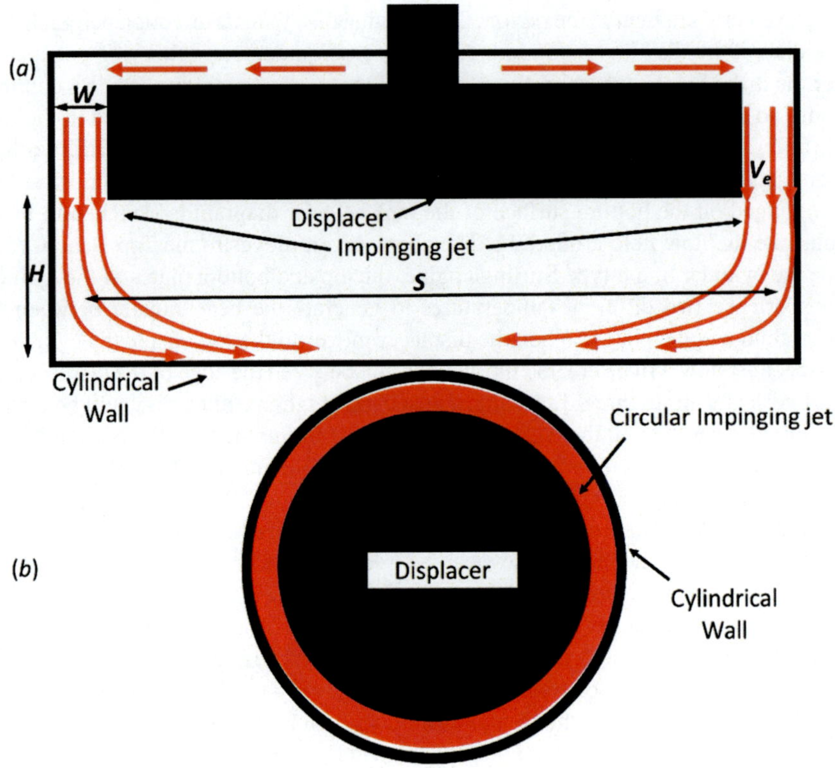

Fig. 7. Impinging jet model. *a*) A sketch of impinging jet flow inside a β type Stirling engine when the displacer moves to the Top Dead Center. *b*) A side view sketch of impinging jet flow inside a β type Stirling engine when the displacer moves to the Top Dead Center.

the average value of the $V_{e,Left}$ and $V_{e,Right}$ is calculated and considered as the input value for impinging jet speed $V_{e.}$ Subsequently, the V_e is estimated to be 0.5 m/s, 0.6 m/s and 2.0 m/s at displacer oscillating frequencies of 0.2 Hz, 0.3 Hz and 0.6 Hz respectively. Using the estimated velocity values of V_e, experimental setup dimensions and fluid properties (fluid property of water at room temperature) as the input for the Eq. 1 and 2, the Nusselt number are estimated. Table 1 summarizes the calculated values of main parameters formulated in the Eq. 1 and 2. As expected, the heat convective coefficient value h increases with the frequency, as higher impinging jet velocity V_e is generated at higher displacer oscillating frequency and strongerbulk motion of the fluid creates more heat convection. The averaged heat convective coefficient $\overline{h}$ is estimated to be 190 $W/m^2{\cdot}K$, 220 $W/m^2{\cdot}K$ and 490 $W/m^2{\cdot}K$ at 0.2 Hz, 0.3 Hz and 0.6 Hz respectively.

4 Conclusions and Discussion

In this work, a volumetric flow measurement within an 8.4 × 69.8 × 8.4 mm volume is conducted on an oscillating disk inside a confined cylindrical container to simulate the complex flow field around a displacer model inside a β type Stirling engine, utilizing the

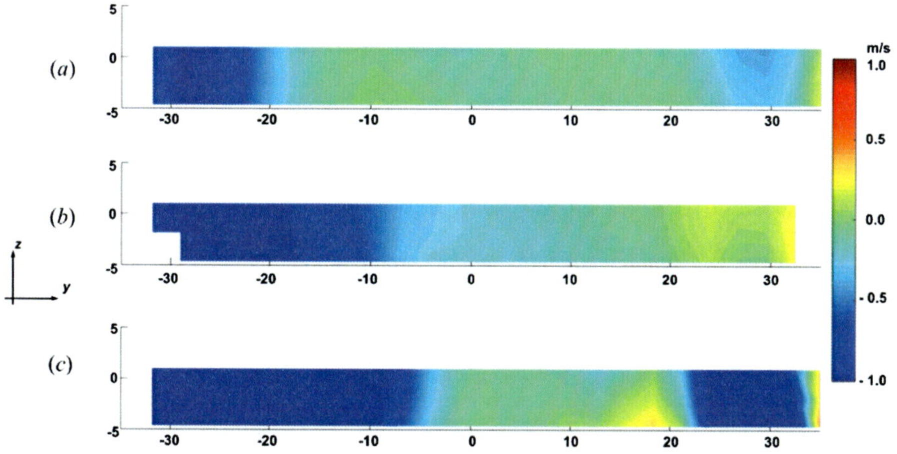

Fig. 8. Contour plot of vertical velocity at the top of the flow measurement domain (X = −15.24 mm) to determine the value of V_e. *a*) Contour plot of vertical velocity result with a displacer oscillating frequency of 0.2 Hz. *b*) Contour plot of vertical velocity result with a displacer oscillating frequency of 0.3 Hz. *c*) Contour plot of vertical velocity result with a displacer oscillating frequency of 0.6 Hz.

Table 1. Calculation of Nusselt Numbers and h value using Eq. 1 and 2.

	Frequency (Hz)	H (mm)	W (mm)	S (mm)	V_e (m/s)	$\overline{Nu}$	Re	$\overline{h}$ (W/m^{2}K)
Case I	0.2	24	10	263	0.5	42	2800	192
Case II	0.3	24	10	395	0.6	48	3400	220
Case III	0.6	24	10	790	2.0	107	11200	489

Shake-The-Box system and the refractive index matching method. For the first time, the volumetric flow field on a displacer model with three different oscillating frequencies is resolved and investigated when the displacer model moves to the Top Dead Center. The volumetric flow field result suggests that high speed jet flow is created through the gap between the displacer and the cylindrical wall and impinges at the bottom surface of the cylindrical container. As the heat transfer on a β type Stirling engine is primarily happening between the internal fluid and the top/bottom inner surfaces of the engine, thus this impinging jet flow is largely responsible for the heat transfer of convection between the working fluid and the engine wall. Using an empirical equation about the impinging jet through nozzle slots, we estimated the average Nusselt number $\overline{Nu}$ and heat convective coefficient $\overline{h}$ value with the impinging jet velocity V_e obtained from the volumetric flow measurement, suggesting the $\overline{h}$ is in the range of 200~500 under current study when the Na I solution is used as the fluid medium.

On the calculations of $\overline{h}$ at different displacer oscillating frequencies, the impinging jet velocity V_e is the primary variable that can be attribute to the displacer frequency

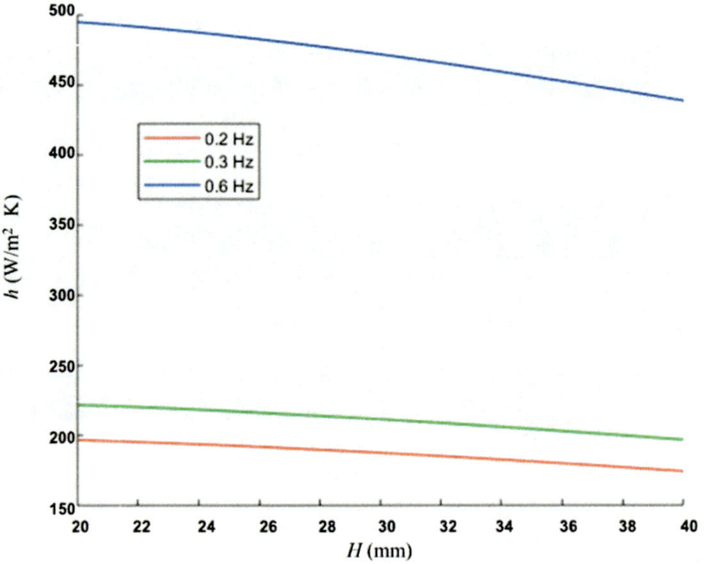

Fig. 9. Estimated h value vs different height of the impinging jet

variation. Yet, in this study, we didn't change the dimensions of the displacer or the cylindrical container that could affect the outcome of the $\overline{h}$. If the diameter of the displacer is changed, it can directly impact the width of the gap between the displacer edge and the cylindrical wall, because the impinging jet is created by pushing the fluids through the gap. Therefore, the only dimensional variation that will not affect V_e substantially is the height of the jet H. Here a brief estimation of $\overline{h}$ value and its dependency on the height of the jet H are conducted and plotted with all the parameters kept the same as the experiment but with only H varying from 20 to 40 mm. Figure 9 presents the $\overline{h}$ value with H changing from 20 to 40 mm under the three different displacer oscillating frequencies. As expected, with height H increases, the heat convective coefficient h decreases. Moreover, the rate of decreasing is growing with higher H value. Therefore, to maintain an effective heat transfer, the height of the impinging jet or the height of the cylindrical container should remain at a relative low value. However, the strength of the impinging jet also depends on the height of the container. For example, if the cylindrical container has a too small height, the displacer may have very little room to move vertically therefore restraining the velocity of the displacer so as the impinging jet velocity. Therefore, to accurately determine the $\overline{h}$ and understand its correlation with the dimensions of the setup, a more systematic experiment is needed.

Furthermore, a significant out of plane velocity (Z direction velocity in Figs. 2, 3, 4) field is captured and resolved inside the cylindrical container, suggesting a strong three-dimensional effect of the flow field. However, most CFD simulations on Stirling engine rely on simplified models of two-dimensional flow which considers the flow field inside the Stirling engine as either symmetric or axial symmetric. Consequently, with CFD simulation, to achieve an accurate and comprehensive understanding about the Striling engine, the flow model is preferable to be three dimensional to include the out of plane velocity. Meanwhile, a significant asymmetry of flow field is observed in Figs. 2, 3, 4 and the impinging jet flow at y = −30 mm is clearly stronger than the jet flow at y = 30 mm (Figs. 5, 6, and 8). This flow asymmetry could be created due to the asymmetry of the experimental setup. Particularly, during the experiment when the displacer is oscillating vertically, the displacer and the cylindrical wall could be slightly eccentric and create non-axisymmetric setup, therefore generating asymmetric flow field in the measurement domain. Besides, it is realized that the displacer model was not on a perfect horizontal plane but slightly titled with a small angle which could also cause a complex flow field with asymmetry. It should be reminded that the above discussed flow field result is the mean velocity field over the first 1/18 period and the flow around the displacer is inherently unsteady and periodic. From the entire particle tracking velocimetry data (not shown in this paper), after the first 1/18 cycle, the flow field under the displacer gradually settled down and momentum dissipated before the displacer moves again. This could explain the observation of counter upward flow in the Fig. 5 where upward flow is observed against the impinging jet flow and the downward jet flow does not reach the bottom of the container. Due to the unsteadiness of the flow field, the jet flow could impinge the bottom of the container intermittently and the averaged velocity field result may neglect that intermittency. Therefore, to derive an accurate V_e value, the downward velocity at the top of the flow domain is used for a robust estimation of $\overline{h}$ and $\overline{Nu}$ for the beginning of the flow development when the displacer moves to the Top Dead Center.

Acknowledgment. The acquisition of the Shake-The-Box system in this work is supported by the National Science Foundation under Grant No. CMMI-1919726.

References

1. Gheith, R., Hachem, H., Aloui, F., Nasrallah. S.: Stirling Engines. In: Comprehensive Energy Systems Energy conversion, vol. 4, Elsevier, pp. 169–208 (2018)
2. Xiao, G., et al.: Design optimization with computational fluid dynamics analysis of b - type Stirling engine. Appl. Therm. Eng. **113**, 87–102 (2017)
3. Timoumi, Y., Tlili, I., Nasrallah, S.: Performance optimization of Stirling engines. Renew. Energy **33**, 2134–2144 (2008)
4. Mohammadi, M., Jafarian, A.: CFD simulation to investigate hydrodynamics of oscillating flow in a beta-type Stirling engine. Energy **153**, 287–300 (2018)
5. Abuelyamen, A., Ben-Mansour, R.: Energy efficiency comparison of Stirling engine types (α, β, and γ) using detailed CFD modeling. Int. J. Thermal Sci. **132**, 411–423 (2018). https://doi.org/10.1016/j.ijthermalsci.2018.06.026
6. Chi, C., Mou, J., Lin, M., Hone, G.: CFD simulation and investigation on the operating mechanism of a beta-type free piston Stirling engine. Appl. Therm. Eng. **166**, 114751 (2020)
7. Ibrahim, M.: 2D CFD simulation of the heat transfer and fluid dynamics in an experimental model of the Hot end of a stirling engine. In: IEEE Energytech Conference, May 29–31th, Cleveland, OH, USA (2012)
8. Cheng, C.-H., Chen, Y.-F.: Numerical simulation of thermal and flow fields inside a 1-kW beta-type Stirling engine. Appl. Thermal Eng. **121**, 554–561 (2017). https://doi.org/10.1016/j.applthermaleng.2017.04.105

9. Chen, W.: A study on the effects of geometric parameters in a low-temperature-differential γ-type Stirling engine using CFD. Int. J. Heat Mass Transf. **107**, 1002–1013 (2017)
10. Schanz, D., Gesemann, S., Schröder, A.: Shake-the-box: Lagrangian particle tracking at high particle image densities. Exp. Fluids **57**, 70 (2016)
11. Tian, X., Xiao, L., Zhang, X., Yang, J., Tao, L., Yang, D.: Flow around an oscillating circular disk at low to moderate Reynolds numbers. J. Fluid Mech. **812**, 1119–1145 (2017)
12. Bergman, T.L., Lavine, A.S., Incropera, F.P., DeWitt, D.P.: Fundamentals of Heat and Mass Transfer. 7th Edn. John Wiley, (2011)

The Lift Force on the Finite-Sized Particle Along Particle Trajectory in the Wall-Turbulent Flow over the Sediment Bed

Yinghaonan Lei and Ping Wang(✉)

Key Laboratory of Mechanics on Disaster and Environment in Western China, The Ministry of Education of China, Department of Mechanics, Lanzhou University, Lanzhou 730000, People's Republic of China
wping@lzu.edu.cn

Abstract. The data from a particle-resolved direct numerical simulation (PR-DNS) of two-phase over the sediment bed are used to examine the existing lift force models, including flow shear, wall effect and particle rotation. It is found that, in the current case, the rotation-induced lift force is the dominant during particle saltating, and the classical lift models could not predict the resolved lift force alone in particle motion. On this basis, a new lift force correction is proposed. By evaluation, it can reasonably forecast the lift force by PR-DNS along the particle trajectory.

Keywords: multiphase flow · turbulence simulation · turbulent boundary layers

1 Introduction

The forces on particles are vital to characterize the dynamic of turbulent multiphase flow common in the natural environment [1–3], industrial applications [4,5] and biomedicine [6,7], etc. Although the lift force on the particle is usually not as large as the drag force, and the transverse motion in the wall-normal direction is often weaker than the streamwise motion, it can play an important role in particle deposition, entrainment and transport [8–10]. For example, the lift force will transfer kinetic energy to particles migrating towards the wall in the viscous sublayer and influence the particle deposition by varying their vertical velocity [11,12]. This effect was further quantified by Marchioli, Picciotto & Soldati [13] through the direct numerical simulation (DNS) of particle-laden turbulent upward/downward flow in a vertical channel, which shows that the particle deposition rate induced by lift force differs for different particle Stokes number St. Amiri, Hannani & Mashayek [14] showed that the lift force exerts a considerable impact on the mean streamwise velocity of particles in turbulent channel flow, and this effect will increase with the increase of particle Stokes

X. Zheng and S. Balachandar (Eds.): IUTAM 2023, IUTAM Bookseries 41, pp. 265–284, 2024.
https://doi.org/10.1007/978-3-031-47258-9_17

number St and particle Reynolds number Re_p. For two-phase flow with a low particle-to-fluid density ratio, the instantaneous lift force may be even larger than the particle submerged weight near a certain height of the effective bed surface [9], which might cause the direct entrainment of bed particles. In addition, Mazzitelli, Lohse & Toschi [15] found that the lift force enhances the accumulation of bubbles on the downward flow by comparing the simulation results with and without the lift force in a microbubble-laden homogeneous and isotropic turbulent flow.

Generally, the lift force on a particle is mainly affected by three factors: ambient shear or fluid vorticity around particles, the wall and particle rotation [16]. Saffman [17,18] first showed a functional dependence of a unidirectional lift force perpendicular to the streamwise direction on a sphere moving in a simple unbounded linear shear flow using matched asymptotic expansions. This force eventually became known as the Saffman lift force expression and is given as

$$F_L = 6.46\nu\rho_f\left(\frac{D_p}{2}\right)^2 U_c\left(\frac{\gamma}{\nu}\right)^2 - \left(\frac{11}{8}\right)\rho_f U_c\gamma\left(\frac{D_p}{3}\right)^3 + \pi\rho_f U_c\omega_p\left(\frac{D_p}{2}\right)^3 \quad (1)$$

where ρ_f and ν are the density and kinematic viscosity of the fluid, U_c is the streamwise fluid velocity at the center of the sphere, ω_p is spanwise component of the freely rotational speed of particle, D_p is the diameter of the sphere, and γ is the fluid shear rate. This expression is strictly limited to the conditions of $\mathrm{Re}_p \ll 1$, $\mathrm{Re}_\gamma \ll 1$, ${\mathrm{Re}_\gamma}^{1/2} \gg \mathrm{Re}_p$ and $\mathrm{Re}_{\omega_p} \ll 1$, where Re_p is the particle Reynolds number defined as $\mathrm{Re}_p = |\boldsymbol{u}_{\mathrm{slip}}|D_p/\nu$, $\mathrm{Re}_\gamma = |\gamma|D_p^2/\nu$ is the shear Reynolds number and $\mathrm{Re}_{\omega_p} = |\boldsymbol{\omega}_p|D_p^2/\nu$ is the particle rotational Reynolds number. Note that $\boldsymbol{u}_{\mathrm{slip}}$ is the slip velocity vector between the fluid and the particle, and $\boldsymbol{\omega}_p$ is the particle angular rotation vector in a more general sense. In the practical application of this expression (e.g. [12,19]), the second and third terms are usually neglected since, as Saffman [17] illustrated, the lift force component caused by self-induced free rotation is less than the shear-induced lift force by an order of magnitude for the condition of $\mathrm{Re}_\gamma \ll 1$. The second term is less important than the first when ${\mathrm{Re}_\gamma}^{1/2} \gg \mathrm{Re}_p$. McLaughlin [20] and Asmolov [21] relaxed the limit of ${\mathrm{Re}_\gamma}^{1/2} \gg \mathrm{Re}_p$ by multiplying a correction factor $J(\epsilon)$ on the first term of the Saffman lift force, where $\epsilon = {\mathrm{Re}_\gamma}^{1/2}/\mathrm{Re}_p$. Since then, many researchers have tried to expand the limit of the Saffman lift expression through theoretical analysis [22], experimental measurement [23] and numerical simulation [24].

The force on the particle can be strongly affected by the wall. It is known that the drag force of particles will significantly increase when the particle is close to the wall [8,25,26] compared to the unbounded situation. A similar increase in the lift force due to the wall has also long been an important issue. The presence of the wall creates an additional wallward force due to the pressure in the gap between the particle and the wall. This wall-induced lift force, which may be directed either towards the wall or away from it, depends on the flow conditions [27]. In the theoretical analysis of wall-induced lift force, the wall can often reside either in the inner (Stokes) or the outer (Oseen) region of the particle. The inner region models require the particle to be close to the wall where the viscous effects are more significant than the inertial effects, and the outer models solve the Oseen-type equation via an asymptotic expansion [28]. In the pioneering work of Cox & Brenner [29], they proposed an implicit expression for the inner-region lift force model, which only valid in $L/D_p \gg 1$. Subsequently, this restriction was gradually relaxed for non-rotating [23] or freely rotating [30] particles. The Saffman lift force expression can be regarded as an outer-region model when $L/D_p \rightarrow \infty$. Zeng et al. [31] and Ekanayake, Berry & Harvie [32] also proposed the lift force models accounting for the distance L/D_p (from nearly touching to sufficiently far from the wall $0.5 \leq L/D_p \leq 4$) at finite particle Reynolds number ($2 \leq \mathrm{Re}_p \leq 250$) and small Reynolds numbers ($10^{-3} \leq \mathrm{Re}_p \leq 10^{-1}$), respectively.

In addition to shear-induced and wall-induced lift force, particles will encounter a lift force component caused by particle rotation and orthogonal to its direction of motion even in a uniform unbounded flow. Reviewing the history of the rotation-induced lift force, Robins [33] and Magnus [34] confirmed the existence of a transverse force on a rotating sphere by the qualitative experiment, which is well known as the Robins effect or Magnus effect [35]. This force was first quantitatively described by Rubinow & Keller [36] with an asymptotic expansion on a spinning sphere moving in the viscous fluid at a low Reynolds number. Note here, although this rotation-induced lift force in Eq. 1 for a freely rotating particle due to the flow shear is believed to be small, it could be as large as the drag force in magnitude when the particle rotates with high angular velocity or $Re_{\omega_p} \gg 1$ which is common in multiphase flow due to particle-particle and particle-wall collision. Barkla & Auchterlonie [35] utilized a conical pendulum method to determine the lift and drag coefficient of a rotating sphere suspended with $1500 \leq \mathrm{Re}_p \leq 3000$. Tsuji, Morikawa & Mizuno [37] studied the trajectories of spheres bouncing on an inclined plate and proposed a new relationship between the lift force and particle rotations for particle Reynolds numbers in the range of $550 \leq \mathrm{Re}_p \leq 1600$. Oesterlé & Dinh [38] estimated the lift coefficient for particle with $10 \leq \mathrm{Re}_p \leq 140$ and $2 \leq {\omega_p}^* \leq 12$, where ${\omega_p}^* = \omega_p D_p / |\boldsymbol{u}_{\mathrm{slip}}|$ is the dimensionless angular velocity. The lift force on a rotating sphere in a simply unbounded linear shear flow was given by Kurose & Komori [39] with the parameters in the range of $1 \leq \mathrm{Re}_p \leq 500$, $0 \leq \gamma^* \leq 0.8$, $0 \leq {\omega_p}^* \leq 0.5$, where $\gamma^* = \gamma D_p / |\boldsymbol{u}_{\mathrm{slip}}|$ is the dimensionless flow shear rate.

Indeed, one of the research areas on the lift and drag model at the present stage is to seek a generalized model valid for arbitrary particle Reynolds numbers, particle-wall separation distances, the incoming flow shear rate and the rotation velocity of the particle. In a low Reynolds number regime, $\mathrm{Re}_p \ll 1$, $\mathrm{Re}_\gamma \ll 1$ and $\mathrm{Re}_{\omega_p} \ll 1$, these models are typically the binary interactions of fluid shear, particle rotation and the wall [40, 41]. However, for the finite Reynolds numbers, a complex non-linear relationship among fluid shear, particle rotation (freely or forced) and the wall should be expected. The accuracy of these empirical and semi-empirical correlations is not guaranteed if they are used outside of their proposed range of validity and conditions. For instance, the lift force is also influenced by turbulent fluctuation and neighboring particles, to name a few. Bagchi & Balachandar [42] discovered that as particle size increased, the standard drag and lift correlations became less accurate in predicting instantaneous force in an isotropic turbulent flow. Kim & Balachandar [43] suggested that it is crucial to include both deterministic and stochastic components in the force model to accurately predict the movement of a particle in turbulence, and the modelling of lift force should also take into account the stochastic contribution that arises from self-induced vortex shedding. Zeng et al. [26] simulated flow around an isolated fixed particle with varying particle-wall distances and particle sizes in a turbulent channel flow. They found that the lift force is very significant compared with the drag force in viscous sublayer and further emphasized the relationship between a sharp increased lift force and turbulent structures, which is also observed in the Li et al. [44] over a rough bed. It has already explored that the drag force is significantly influenced by surrounding particles (the volume fraction) [45–47]. In addition, Akiki & Balachandar [48] found that the presence of neighbouring particles causes a significant variation of the lift force on each particle in a randomly arranged particle array while the mean lift force on particles remains consistent.

Although many lift force models with complex forms and combined effects have been proposed, their validity in real multiphase flow has yet to be discovered. Referring to the latest research on drag force, the impact of complex factors such as particle velocity fluctuations [49,50] and particle volume fraction gradients [51] on the lift model still needs to be clarified.

This paper investigates the lift forces acting on a spherical particle saltating in a wall-bounded flow via particle-resolved (PR) DNS. Note that this study is an extension of our previous work [3], which investigates the drag force on particles saltating over the sediment bed. In Sect. 2, we briefly introduce the methodology of DNS. The method of calculating the local variables around the particle are presented in Sect. 3. Finally, we analysis the results of lift force models in Sect. 4 and draw the conclusions in Sect. 5.

2 Numerical Methods and Data Description

The same data from our previous work [3] are used in the present study. In the previous work, a PR-DNS [52] was performed to simulate a two-phase turbulent flow over an erodible sediment bed. We only provide a brief description of the calculation method here. The three-dimensional compressible continuity equation in 2 and Navier-Stokes Eqs. 3 governing the Newtonian carrier phase are solved with a second-order central difference scheme for fluid spatial discretization and a second-order Runge-Kutta iteration for fluid time advancement [53–55].

$$\nabla \cdot \boldsymbol{u}_f = 0 \tag{2}$$

$$\frac{\boldsymbol{u}_f}{\partial t} + \nabla \cdot (\boldsymbol{u}_f \boldsymbol{u}_f) = -\frac{1}{\rho_f}\nabla p + \nu \nabla^2 \boldsymbol{u}_f + \boldsymbol{f} \tag{3}$$

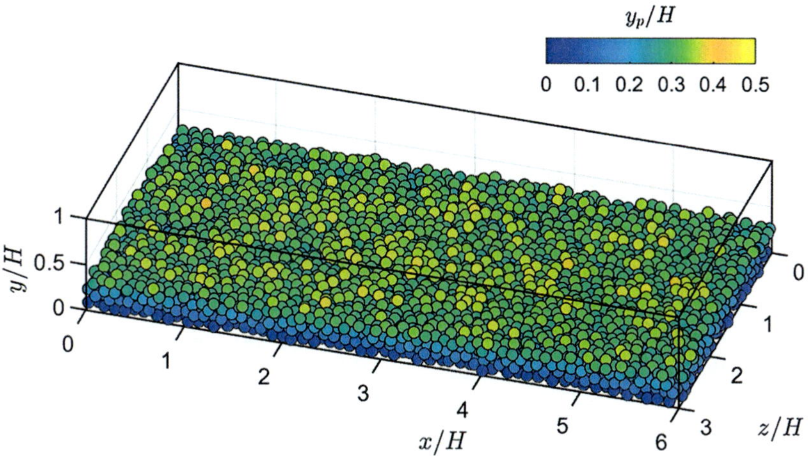

Fig. 1. The sediment bed. All particles are colored by the vertical height of their center y_p.

In Eq. 2 and 3, $\boldsymbol{u}_f$ is the fluid velocity, t is the time, p is the dynamic fluid pressure, f is the volume force. The motion of finite-sized particles is obtained by solving the equations governing particle position $\boldsymbol{x}_p$, translational velocity $\boldsymbol{u}_p$ and angular velocity $\boldsymbol{\omega}_p$ as

$$\frac{\mathrm{d}\boldsymbol{x}_p}{\mathrm{d}t} = \boldsymbol{u}_p \tag{4}$$

$$\rho_p V_p \frac{\mathrm{d}\boldsymbol{u}_p}{\mathrm{d}t} = \boldsymbol{F}_h + (\rho_p - \rho_f) V_p \boldsymbol{g} + \boldsymbol{F}_c = \oint_{\partial V} \boldsymbol{\tau} \cdot \boldsymbol{n}_p \, \mathrm{d}A + (\rho_p - \rho_f) V_p \boldsymbol{g} + \boldsymbol{F}_c \tag{5}$$

$$I_p \frac{\mathrm{d}\boldsymbol{\omega}_p}{\mathrm{d}t} = \boldsymbol{T}_h + \boldsymbol{T}_c = \oint_{\partial V} \boldsymbol{r} \times (\boldsymbol{\tau} \cdot \boldsymbol{n}_p) \, \mathrm{d}A + \boldsymbol{T}_c \tag{6}$$

where ρ_p is the particle density, V_p is the volume of the particle, I_p is the moment of inertia of the spheroidal particle, $\boldsymbol{F}_h$ and $\boldsymbol{T}_h$ are the hydrodynamic force and torque, $\boldsymbol{g} = [0, G, 0]^T$ is the gravitational acceleration, the superscript T stands for transpose, $\boldsymbol{F}_c$ and $\boldsymbol{T}_c$ are the collision force and torque acting on the particle, $\boldsymbol{\tau}$ is the outward unit normal vector at the particle surface denoted by ∂V, and $\boldsymbol{r}$ is the position vector of the particle surface relative to the particle center, respectively. The interaction between flow field and particle is solved by the immersed boundary method used in Uhlmann [56] and Breugem [57]. The lubrication model of Costa et al. [58] is adopted for lubrication force and we use the adaptive collision time model (ACTM) [59,60] to account for the collision force.

The simulation domain is $[L_x \times L_y \times L_z]/D_p = 60 \times 10 \times 30$ and is discretized on a uniform Cartesian grid with $N_x \times N_y \times N_z = 540 \times 90 \times 270$, where H is the vertical height of the entire computing domain, the subscripts x, y, and z denote the streamwise, vertical, and spanwise directions, respectively. Periodic conditions are enforced in both the streamwise and spanwise directions. No-slip boundary conditions are employed on both the bottom wall and sphere surfaces. A free-slip condition is applied at the top of the simulation domain. The particles are randomly released within the simulation domain. These particles will deposit onto the wall due to the gravity (without the hydrodynamic force) and collision, and eventually form a sediment bed when the total kinetic energy of the particle decays to close to zero, as shown in Fig. 1. The sediment bed is composed of four layers of monodisperse spheres in a resting state and the total number of the particles N_p is 7200. From this moment on, the sediment bed is added into the already fully developed open channel turbulence, and the two-way coupled solver is activated. Under the action of the hydrodynamic force, particle motion (rolling, slipping and saltating) will be triggered, and the two-phase flow will reach a statistically steady state after a rather long period due to the particle-bed-turbulence interaction.

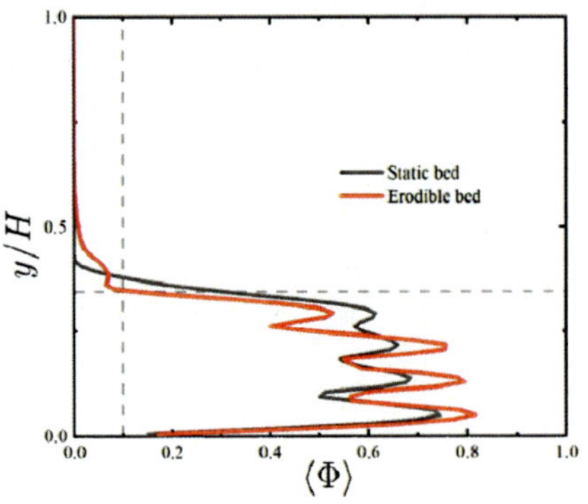

Fig. 2. The profiles of mean solid indicator fraction $\langle\Phi\rangle$ over static and erodible bed.

At this point, the mean solid indicator fraction $\langle\Phi\rangle$ decays with height, where $\langle\Phi\rangle$ is calculated by the indicator function of particles defined in [3,61,62]. The effective height H_b of erodible particle bed is determined as $H_b = 0.34$ when $\langle\Phi\rangle$ equals to 0.1 following Kidanemariam & Uhlmann (2014) and Kidanemariam & Uhlmann (2017). The frictional Reynolds number Re_τ^b, defined by the height $H - H_b$ and the frictional fluid velocity u_τ at H_b, equals 91. Figure 2 presents the profiles of $\langle\Phi\rangle$ on static and erodible beds. It can be seen that the particles in the uppermost layer are delaminated due to the turbulent flow near the wall. Meanwhile, the arrangement of particles at the bottom layer will become tighter.

3 Local Variables on the Finite-Sized Particles

Before discussing the force, it is necessary first to calculate the undisturbed fluid velocity, the slip velocity, the undisturbed fluid shear rate and the undisturbed fluid rotation angular velocity based on the flow field information around the particles [63–66] in order to use and validate the classic force models.

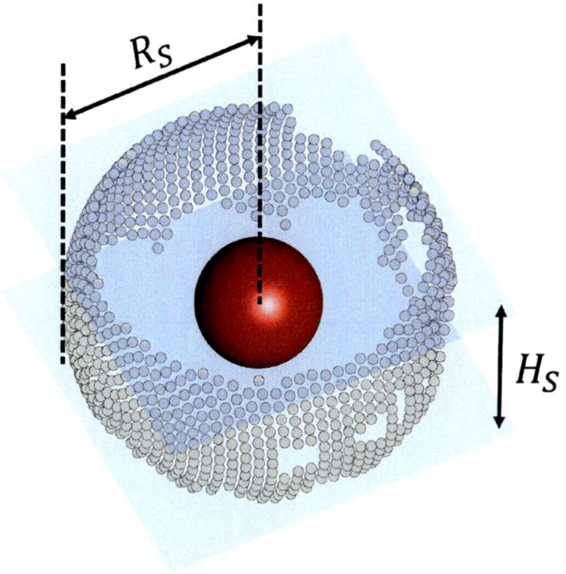

Fig. 3. A schematic diagram of the spherical shell enclosing particles for the calculation of the slip velocity, the undisturbed fluid shear rate and the undisturbed fluid rotation, where the red sphere in the centre of the figure represents the particle, the surrounding grey points represent Lagrangian points on the shell with radius R_S. The two parallel planes separated by $H_S < R_S$ intersect this shell to form a spherical segment, as did in Kidanemariam et al. [67].

As before, we define the undisturbed fluid velocity and the slip velocity using the method proposed by Cisse, Homann & Bec [68]. That is, a spherical shell centre at the particle location and a radius of R_s is created by the Lagrangian grid generation scheme [57]. The unit norm vector $\boldsymbol{e}_{\text{slip}}$ on the direction of the slip velocity is calculated as

$$\boldsymbol{e}_{\text{slip}} = \frac{\left| \oiint_{\partial V_S} \phi_f(\boldsymbol{x}) \left[\left(\boldsymbol{u}_f^S - \boldsymbol{u}_p \right) \cdot \boldsymbol{n}_S \right] \boldsymbol{n}_S \, \mathrm{d}A \right|}{\oiint_{\partial V_S} \phi_f(\boldsymbol{x}) \left[\left(\boldsymbol{u}_f^S - \boldsymbol{u}_p \right) \cdot \boldsymbol{n}_S \right] \boldsymbol{n}_S \, \mathrm{d}A} \tag{7}$$

where $\boldsymbol{n}_S$ is the outward unit normal vector at the spherical shell denoted by ∂V_S, $\boldsymbol{u}_f^S$ is the fluid velocity on the spherical shell in the microelement area $\mathrm{d}A$. Ω_f is the fluid domain, and $\phi_f(\boldsymbol{x})$ is an indicator function [67].

$$\phi_f(\boldsymbol{x}) = \begin{cases} 1 & \text{if } x \in \Omega_f \\ 0 & \text{else} \end{cases} \tag{8}$$

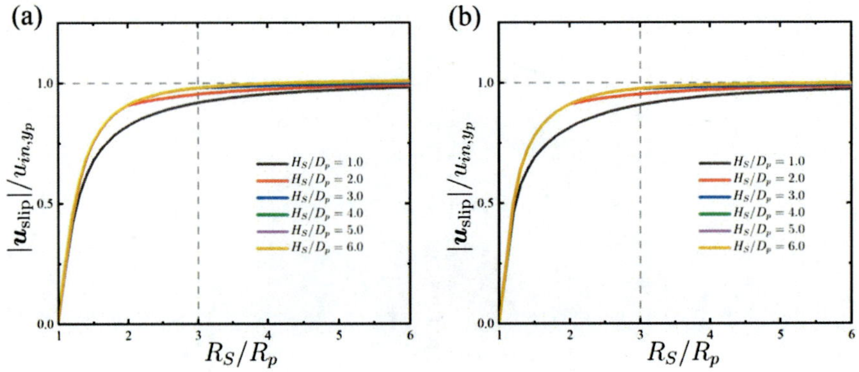

Fig. 4. The slip velocity $\boldsymbol{u}_{\text{slip}}$ for (a) uniform inflow case and (b) the linear shear inflow case. The horizontal gray dotted line indicates $\boldsymbol{u}_{\text{slip}}/u_{in,y_p} = 1.0$, and the vertical gray dotted line indicates $R_S/R_p = 3$.

Once $\boldsymbol{e}_{\text{slip}}$ is obtained, the slip velocity $\boldsymbol{u}_{\text{slip}}$ can be calculated by

$$\boldsymbol{u}_{\text{slip}} = \frac{\oiint_{\partial V_S} \phi_f(\boldsymbol{x}) \left[\left(\boldsymbol{u}_f^S - \boldsymbol{u}_p \right) \cdot \boldsymbol{e}_{\text{slip}} \right] \mathrm{d}A}{\oiint_{\partial V_S} \phi_f(\boldsymbol{x}) \mathrm{d}A} \boldsymbol{e}_{\text{slip}} \tag{9}$$

This approach was also adopted by Luo [5,69]. Usually, R_S is set to be several times the particle radius R_p. Kidanemariam et al. [67] suggested that the averaging spherical surface should be trimmed by two wall-parallel planes (at a distance of H_S) to avoid sampling bias due to the inhomogeneity in channel flow. We also additionally examine the effects of spherical segment with H_S [67] on $\boldsymbol{u}_{\text{slip}}$. The schematic diagram of the spherical shell enclosing particles for calculating the slip velocity is shown in Fig. 3.

To further discuss this method, we perform auxiliary simulations of the fluid field around a fixed spherical particle in uniform incoming flow and linear shear flow by PR-DNS, respectively. The computational domain is $[L_x \times L_y \times L_z]/D_p = 30 \times 12 \times 12$ and the governing equations are resolved at uniform gird of $N_x \times N_y \times N_z = 600 \times 240 \times 240$. The particle is placed at the centre of the domain. The symmetric conditions are imposed at the side boundaries, and the outflow boundary condition $\partial u_{f,x}/\partial x = 0$ is specified at the outlet of the simulation domain. In the case of uniform inflow, the free-slip but no-penetration conditions are imposed at the top and bottom surfaces of the domain. For linear shear incoming flow, the inlet flow shear rate is forced to be $\boldsymbol{u}_f(x = 0, y, z) = [\gamma y, 0, 0]^T$. The boundary condition on the top surface of the computational domain is $\boldsymbol{u}_f(x, y = L_y, z) = [\gamma L_y, 0, 0]^T$ and a no-slip condition is imposed on the bottom boundary. The particle Reynolds numbers $\mathrm{Re}_p^{in} = u_{in,y_p} D_p/\nu$ based on u_{in,y_p} for uniform incoming flow case and linear shear flow case are 50 and 150, respectively, where u_{in,y_p} is the incoming fluid velocity at the height of particle centre y_p, which can be regarded as the undisturbed fluid velocity. The dimensionless linear shear rates γ^* for the uniform incoming flow and linear shear flow are 0 and 1/6, respectively.

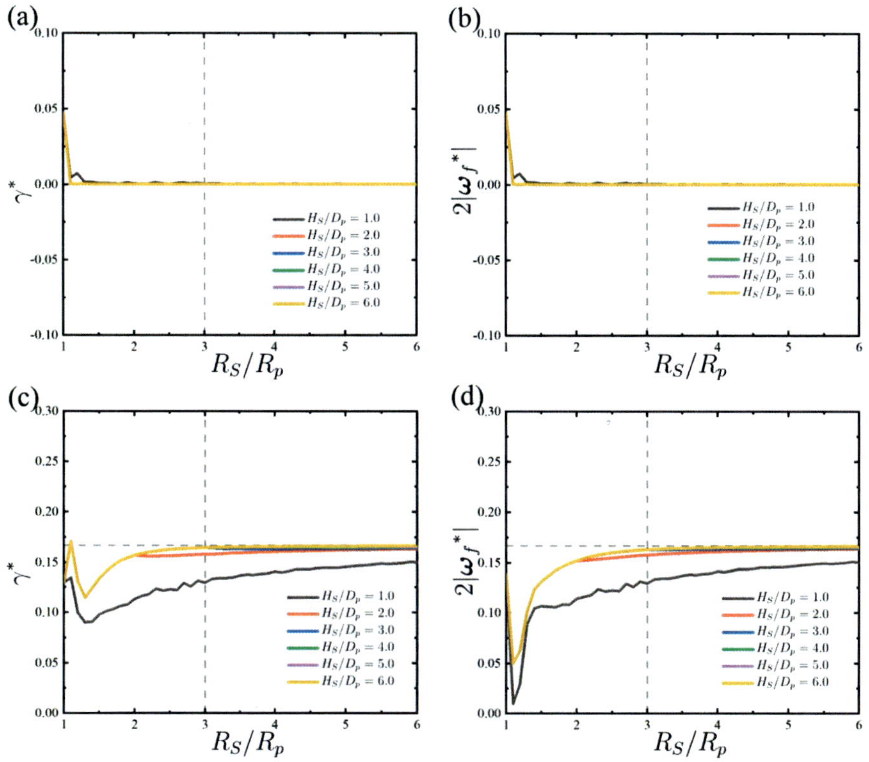

Fig. 5. The calculated dimensionless undisturbed fluid shear rate γ^* and the dimensionless undisturbed the fluid rotation $|\boldsymbol{\omega}_f{}^*|$ around the particle calculated by Eq. 10 and 11 in (a-b) uniform flow case and (c-d) the linear shear flow case. The horizontal gray dotted line indicates (a-b) $\gamma^* = 0.0$ and (c-d) $\gamma^* = 1/6$, and the vertical gray dotted line indicates $R_S/R_p = 3$.

Figure 4 shows the calculated slip velocity for the uniform inflow and linear shear incoming flow cases. We get a similar conclusion as Kidanemariam et al. [67], Cisse et al. [68] and Luo et al. [69], that is, the slip velocity $\boldsymbol{u}_{\text{slip}}$ can reach 98.02% of the incoming flow velocity u_{in,y_p} at $R_S = 3R_p$ for the former case, as seen in Fig. 4(a), regardless of H_S. It is also found that $\boldsymbol{u}_{\text{slip}}$ can reach 97.56% in latter case, as seen in Fig. 4(b). Therefore, $R_S = 3R_p$ is enough to obtain the convergent results. A spherical segment of H_S is unnecessary for calculating the slip velocity.

Further, the undisturbed fluid shear rate γ and the undisturbed fluid rotation angular velocity $\boldsymbol{\omega}_f$ around the particle are similarly defined as

$$\gamma = \frac{\oiint_{\partial V_S} \phi_f(\boldsymbol{x})\,(\partial u_{f,x}/\partial y)\,\mathrm{d}A}{\oiint_{\partial V_S} \phi_f(\boldsymbol{x})\mathrm{d}A} \tag{10}$$

$$\boldsymbol{\omega}_f = \frac{\oiint_{\partial V_S} \phi_f(\boldsymbol{x})\,(1/2\nabla \times \boldsymbol{u}_f)\,\mathrm{d}A}{\oiint_{\partial V_S} \phi_f(\boldsymbol{x})\mathrm{d}A} \tag{11}$$

The results are shown in Fig. 5, where $\boldsymbol{\omega}_f{}^* = \boldsymbol{\omega}_f D_p/|\boldsymbol{u}_{\mathrm{slip}}|$ is the dimensionless undisturbed fluid rotation angular velocity around the particle.

We observe the same phenomenon that all curves converge at $R_S = 3R_p$ in Fig. 5, regardless of whether the shell is cut or not. In the uniform inflow case, the dimensionless undisturbed fluid shear rate γ^* and the dimensionless undisturbed fluid angular velocity $|\boldsymbol{\omega}_f{}^*|$ around the particle calculated by current method are -1.75×10^{-6} and 2.33×10^{-6} respectively, which can be considered almost equal to zero. In the linear shear flow, γ^* is 1.6437×10^{-1} and $2|\boldsymbol{\omega}_f{}^*|$ is 1.6284×10^{-1}, which are very close with the dimensionless fluid shear $\gamma^* = 1/6$ at inlet surface. Therefore, we believe that the method of calculating the undisturbed fluid variables around particles through spherical shells with the radius $R_S = 3R_p$ is suitable. Therefore, this method is used for the subsequent analysis.

4 Results

It is also curial to convincingly define the lift and drag force according to the hydrodynamic force (obtained by integrating the pressure and viscous stresses on the resolved particle surface). In our previous study [3], the drag force is defined as the hydrodynamic force component in the direction of the slip velocity and, accordingly, the lift force is here defined as the hydrodynamic force component in the direction perpendicular to the slip velocity. Therefore, the drag force $\boldsymbol{F}_D$ and the lift force $\boldsymbol{F}_L$ are calculated by

$$\cos\alpha = \frac{\boldsymbol{F}_h \cdot \boldsymbol{u}_{\mathrm{slip}}}{|\boldsymbol{F}_h|\,|\boldsymbol{u}_{\mathrm{slip}}|} \tag{12}$$

$$\boldsymbol{F}_D = |\boldsymbol{F}_h| \cos\alpha \frac{\boldsymbol{u}_{\mathrm{slip}}}{|\boldsymbol{u}_{\mathrm{slip}}|} \tag{13}$$

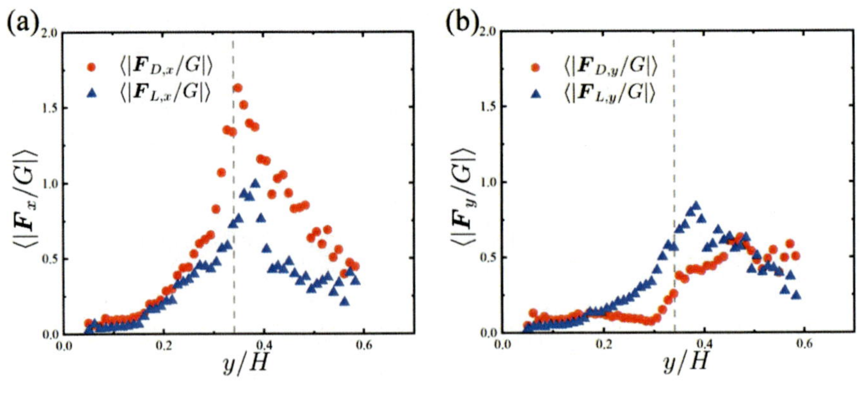

Fig. 6. The profiles of the mean drag force $\boldsymbol{F}_D$ and lift force $\boldsymbol{F}_L$ in (a) streamwise and (b) wall-normal directions. The dotted lines indicate the effective bed height H_b.

$$\boldsymbol{F}_L = \boldsymbol{F}_h - \boldsymbol{F}_D \tag{14}$$

where α is an angle between the slip velocity $\boldsymbol{u}_{\mathrm{slip}}$ and the total hydrodynamic force $\boldsymbol{F}_h$. Note that this definition for freely moving particles in wall turbulence differs from that for a fixed sphere [8,26,28,32,70].

Figure 6 shows the profiles of the streamwise and the vertical components of the drag force $\boldsymbol{F}_D$ and lift force $\boldsymbol{F}_L$, respectively. From Fig. 6(a), it is seen that both the drag force and the lift force in the streamwise reach the peak near the bed and the drag force plays a major role. The drag force is always greater than the lift force in the streamwise and the maximum of the average lift force is close to the particle submerged gravity in magnitude. Figure 6(b) shows that the mean vertical lift force on particles reaches a maximum at about one particle diameter above the effective bed while the drag force continues to grow along the height. Near the bed, the lift force is greater than the drag force in the wall-normal direction. While the vertical component of the lift force is smaller than that of drag force away from the bed where the shear rate becomes low. Ji et al. [9] found that the instantaneous lift force can be larger than the submerged weight of the particle. Therefore, the lift force on the particle cannot be ignored, especially near the bed.

As before, we are concerned about the force on saltating particles in this work. We filter the data to test the lift force models and determine the parameter ranges that saltating particles experienced during their trajectories. Figure 7 shows the probability density distribution function (P.D.F) of the particle Reynolds number Re_p, the fluid rotation Reynolds numbers Re_{ω_f}, particle rotation Reynolds numbers Re_{ω_p}, and the Reynolds numbers Re_{ω_r} based on relative angular velocity $\boldsymbol{\omega}_r = \boldsymbol{\omega}_f - \boldsymbol{\omega}_p$. It is seen that $0 \leq \mathrm{Re}_p \leq 70$, $1 \leq \mathrm{Re}_{\omega_f} \leq 40$, $0 \leq \mathrm{Re}_{\omega_p} \leq 80$ and $0 \leq \mathrm{Re}_{\omega_r} \leq 70$. The Reynolds numbers corresponding to the curve peaks are on the same order of magnitude. Based on these parameter ranges, the lift force models proposed by Mei [71], Zeng et al. [31] and a combination lift formula by Rubinow & Keller [36] and Oesterlé & Dinh [38] in Li et al. [72] are chosen to be compared with PR-DNS results of the two-phase flow. For the convenience of subsequent representation, we will call these models as the Mei, Zeng, and RKOD models in the following text. These models are listed in Table 1.

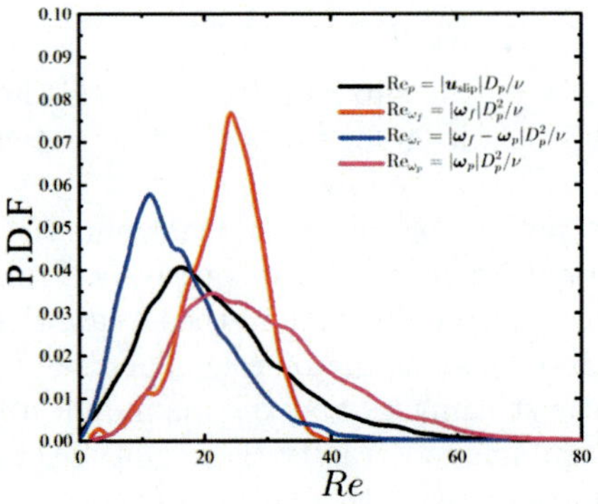

Fig. 7. The probability density distribution (P.D.F) of various Reynolds numbers.

A typical saltating particle (the same as that for the drag force discussion [3]) is chosen to display the related variables along its parabolic trajectory, seen Fig. 8, where the abscissa t is non-dimensionalized by the inner scale ν/u_τ^2. Figure 8(a) shows the wall-normal height of the particle center above the effective bed surface during two successive jump processes. The four Reynolds numbers of the particle are presented in Fig. 8(b) and the PR-DNS lift force it experienced is depicted in Fig. 8(c)–(d), together with the lift forces predicted by various models using the local variables (the calculation method is introduced and verified in section Sect. 3).

The first phenomenon of concern is the extreme lift force just after the collision at $t^+ = 2.05$, $t^+ = 14.52$, $t^+ = 72.01$ and $t^+ = 114.10$. Note that the vertical coordinate range of Fig. 8(c) and 8(d) is only $[-3, 3]$ to show the model predictions clearly. Meanwhile, the lift force on the particle increases rapidly during the ascending phase and is higher than that in the descending phase. This is likely because the particle has stronger rotational velocity and hence high Re_{ω_p} after the collision, as seen in Fig. 8(b). The predicted rotational lift is much higher than the shear- and wall-induced lift. The rotation-induced lift model of RKOD obviously better captures the variation characteristics of lift force on this tracked saltating particle after the collision, while the models of Zeng and Mei all have very poor performance. The lift forces calculated by the latter two models are much lower in magnitude than PR-DNS results. In addition, the small-scale fluctuations are more pronounced in the lift force curves than in the drag curves presented in Wang et al. [3], indicating that the lift force might be more sensitive to turbulence. Nevertheless, the force models cannot well predict the lift force acting on the particle, regardless of whether wall and rotation corrections are included. It is necessary to improve the lift model.

Our previous work on the drag force [3] proposed a correction for the drag force with respect to the vertical particle velocity, in addition to particle slip Reynolds number, through a simple theoretical analysis of the kinetic energy, potential energy and the work done by the hydrodynamic force along particle trajectory. We do a similar treatment for the lift force here. We store all instan-

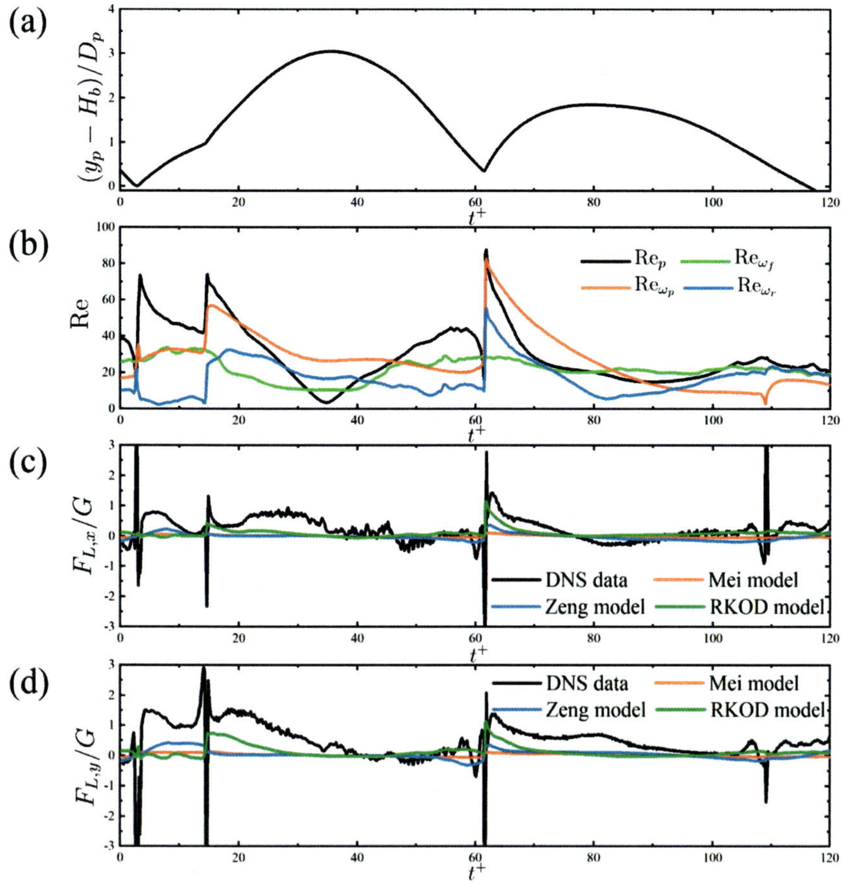

Fig. 8. The dynamic response of a typical saltating particle within two successive saltating processes. (a)The vertical position of the particle center y_p. (b)The time history of the Re_p, Re_{ω_f}, Re_{ω_p} and Re_{ω_r}. (c) The time history of the streamwise component of the lift force $F_{L,x}$ and (d) the vertical component of particle lift force $F_{L,y}$.

taneous lift force $|\boldsymbol{F}_L|$ of saltating particles into a dataset $F(t, x)$ and divide the data into different bins of particle Reynolds number Re_p and the dimensionless vertical particle velocity $u_{p,y}/u_{p,0}$, where $u_{p,0}$ is the initial vertical liftoff velocity of saltating particle just leaves the surface bed, $u_{p,y}$ is the vertical particle velocity. Figure 9 shows the ratios of the modulus of the PR-DNS lift force $|\boldsymbol{F}_L|$ to the lift force predicted by various models $|\boldsymbol{F}_L^{model}|$ for various Re_p and $u_{p,y}/u_{p,0}$ bins. In order to highlight the effects of different factors, $|\boldsymbol{F}_L^{model}|$ is taken to be $\boldsymbol{F}_L^{model} = \boldsymbol{F}_L^{Mei} + \boldsymbol{F}_L^{RKOD}$ and $\boldsymbol{F}_L^{model} = \boldsymbol{F}_L^{Zeng} + \boldsymbol{F}_L^{RKOD}$, respectively. It is seen from Fig. 9 that the former combination predicts better results since the ratio of $(|\boldsymbol{F}_L|)/(|\boldsymbol{F}_L^{model})$ is smaller, indicating that the lift force in the simulated two-phase flow over sediment bed is dominated by rotation and particle-wall separation. The interesting phenomenon that both combinations capture the $(|\boldsymbol{F}_L|)/(|\boldsymbol{F}_L^{model}) \sim 1/\mathrm{Re}_p$ and $(|\boldsymbol{F}_L|)/(|\boldsymbol{F}_L^{model}) \sim u_{p,y}/u_{p,0}$ relationships implies that there is certain underlying similarity in the variation of lift and drag force along the trajectory. Therefore, we fit the data and propose a correction model of the lift force on saltating particles as

$$\boldsymbol{F}_L^{Fit} = C_R(\boldsymbol{F}_L^{Zeng} + \boldsymbol{F}_L^{RKOD}) \tag{15}$$

We can obtain the correction factor C_R in Eq. 15 as

$$C_R = \frac{22\left(\frac{u_{p,y}}{u_{p,0}}\right)^2 + 57}{\mathrm{Re}_p} \tag{16}$$

Similarly, we can also use the friction fluid velocity u_τ to nondimensionalize the vertical particle velocity $u_{p,y}$ and obtain the correction as

$$C_R = \frac{64\left(\frac{u_{p,y}}{u_\tau}\right)^2 + 10\frac{u_{p,y}}{u_{p,0}} + 60}{\mathrm{Re}_p} \tag{17}$$

In Eq. 17, the form of the fitted formula is slightly more complex than Eq. 16.

Since the instantaneous lift prediction for each saltating particle is impossible, we evaluate this correction model according to the mean lift force along particle trajectory. In fact, there are many ways to consider the mean characteristics of forces on particles in Fig. 10. The time averaging method is often used in the case of fixed particles [26,39] seen the sketch in Fig. 10(a). Moreover, the box average is commonly used for moving particles or the particle array [5,47,50], seen the sketch in Fig. 10(b). While in Wang et al. [3], the average operation was performed based on particle saltating angle β, seen the sketch in Fig. 10(c). Beyond question, this angle average method is much better for saltating particles due to the inhomogeneity in the particle distribution and the time evolution of particle motion.

Finally, Fig. 11 presents the mean lift force $\widetilde{\boldsymbol{F}}_L$ of more than 500 parabolic particle trajectories, where the superscript $\sim$ denotes the average in Fig. 10(c). The predictions $\widetilde{\boldsymbol{F}}_L^{model}$ of the selected lift drag models and the combining model are also shown in Fig. 11. It is seen that the correction models significantly improve the prediction results as compared with Mei model, Zeng model and RKOD model. Equation 16, as function of $u_{p,y}/u_{p,0}$, and Eq. 17, as function of $u_{p,y}/u_\tau$, agree well except for $\beta > 20^o$, where particles move near the bed during their ascending phase.

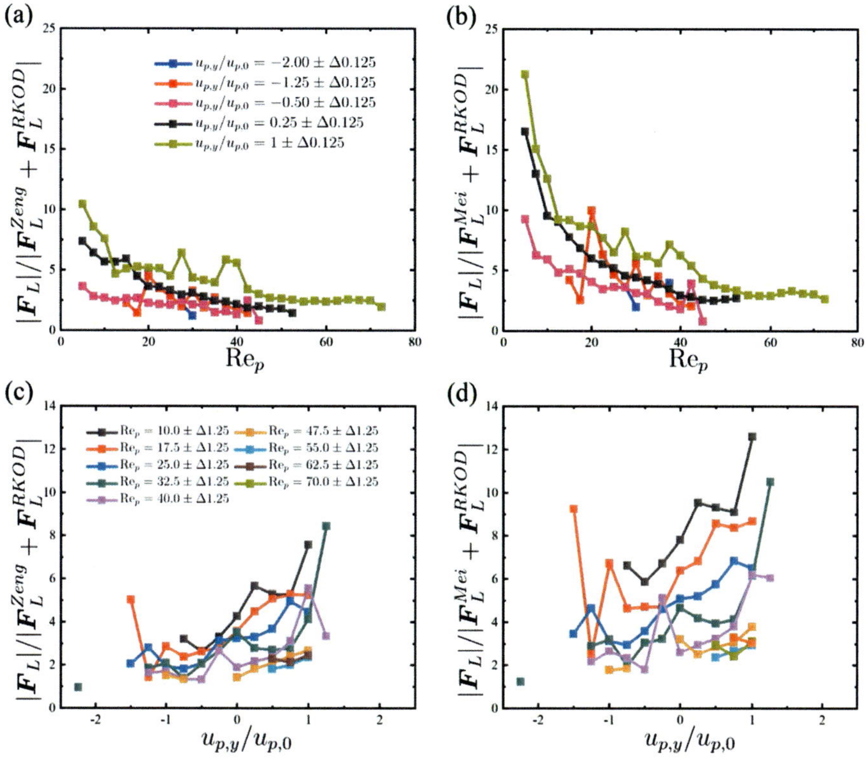

Fig. 9. A comparison between the lift force obtained by PR-DNS and lift models as a function of Re_p and $u_{p,y}/u_\tau$. (a) and (c) are a combination of Zeng model and RKOD model. (b) and (d) are the same as (a) and (c) but for a combination of Mei model and RKOD model.

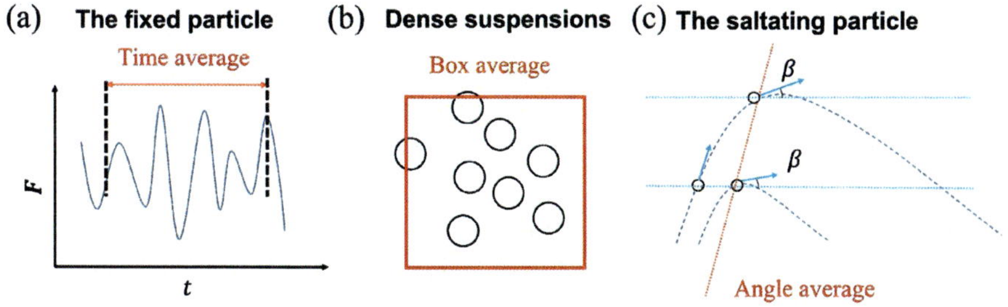

Fig. 10. Several typical calculation methods for the average force on particles. (a) The time average method, (b) the box average method and (c) the angle average method.

Table 1. The expression of the different lift force models.

Study	Lift models	Type
Mei model [71]	$\boldsymbol{F}_L = \frac{\pi}{8}\rho_f D_p^3 \frac{4.1126}{\mathrm{Re}_{\omega_f}^{0.5}} C_{LS}\left[(\boldsymbol{u}_f - \boldsymbol{u}_p) \times \boldsymbol{\omega}_f\right]$ $C_{LS} = \begin{cases} \left(1 - 0.3314\beta^{\frac{1}{2}}\right)\exp\left(-\frac{\mathrm{Re}_p}{10}\right) + 0.3314\beta^{1/2}, \mathrm{Re}_p \le 40 \\ 0.0524\left(\beta \mathrm{Re}_p\right)^{1/2}, \mathrm{Re}_p > 40 \end{cases}$ $\beta = \frac{1}{2}\mathrm{Re}_{\omega_f}\mathrm{Re}_p$	shear-induced lift model
Zeng model [31]	$\boldsymbol{F}_L = \frac{\pi}{8} C_{LS} D_p^2 \left(L + \frac{D_p}{2}\right)\left[(\boldsymbol{u}_f - \boldsymbol{u}_p) \times \boldsymbol{\Omega}_f\right]$ $\boldsymbol{\Omega}_f = \nabla \times \boldsymbol{u}_f$ $C_{LS} = \frac{3.663}{\left(\mathrm{Re}_p^2 + 0.1173\right)^{0.22}} \exp\left[-0.5\left(L/D_p\right)\left(\frac{\mathrm{Re}_p}{250}\right)^{\frac{4}{3}}\right] \times \left\{\exp\left[\alpha_L\left(\frac{L}{D_p}\right)^{\beta_L}\right] - \lambda_L\right\}$ $\alpha_L = -\exp\left(-0.3 + 0.025\mathrm{Re}_p\right)$ $\beta_L = 0.8 + 0.01\mathrm{Re}_p$ $\lambda_L = \left[1 - \exp\left(-L/D_p\right)\right]\left(\mathrm{Re}_p/250\right)^{5/2}$	shear-induced lift model with wall correction
RKOD model [36,38]	$\boldsymbol{F}_L = \frac{\pi}{8}\rho_f D_p^2 C_{LR}\left\|\boldsymbol{u}_f - \boldsymbol{u}_p\right\| \frac{\boldsymbol{\omega}_r \times (\boldsymbol{u}_f - \boldsymbol{u}_p)}{\left\|\boldsymbol{\omega}_r\right\|}$ $C_{LR} = \begin{cases} \mathrm{Re}_{\omega_r}/\mathrm{Re}_p, & \mathrm{Re}_p \le 1 \\ 0.45 + \left(\mathrm{Re}_{\omega_r}/\mathrm{Re}_{\omega_p} - 0.45\right)\exp\left(-0.05684\mathrm{Re}_{\omega_r}^{0.4}\mathrm{Re}_p^{0.3}\right), & 1 < \mathrm{Re}_p < 140 \end{cases}$	rotation-induced lift model

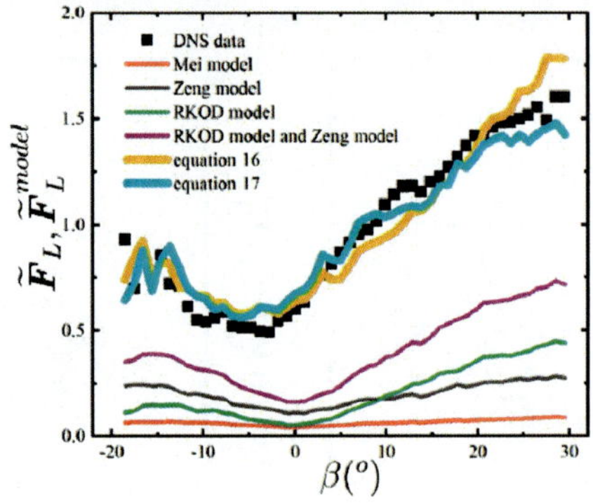

Fig. 11. The mean lift force along particle trajectories.

5 Conclusions

The lift force plays an important role in particle deposition, entrainment and transport. A great number of reports have demonstrated that the lift force is induced by rotation and shear rate and can be strongly affected by the wall and turbulence, etc. Although many lift models have been proposed under simple, ideal conditions at finite Reynolds number, their validity in real two-phase flow is still an open question.

We analyzed the lift force on saltating particles using the PR-DNS data of turbulent two-phase flow over sediment bed and reveal that the lift force acting on finite size saltating particle in this flow is even as important as the drag force, especially near the bed. Again, previously proposed lift force model cannot well predict the lift force on saltating particle. Finally, we present a new lift force correlation in terms of particle Reynolds number and particle vertical velocity. The results confirm that, the corrected lift force model can qualitatively and quantitatively captures the variation of the lift force, on average, along particle trajectory.

Acknowledgements. This research was supported by the National Natural Science Foundation of China (Grants Nos.12072138 and 92052202) and the National Numerical Wind tunnel project of China.

References

1. Sullivan, R., et al.: Wind-driven particle mobility on Mars: insights from mars exploration rover observations at "El Dorado" and surroundings at Gusev Crater. J. Geophys. Res.: Planets **113**(E6) (2008)
2. Zheng, X., Feng, S., Wang, P.: Modulation of turbulence by saltating particles on erodible bed surface. J. Fluid Mech. **918** (2021)
3. Wang, P., Lei, Y., Zhu, Z., Zheng, X.: Drag model of finite-sized particle in turbulent wall-bound flow over sediment bed. J. Fluid Mech. **964**, A9 (2023)
4. Luo, M., Koh, C.G., Bai, W., Gao, M.: A particle method for two-phase flows with compressible air pocket. Int. J. Numer. Meth. Eng. **108**(7), 695–721 (2016)
5. Luo, K., et al.: Analysis and development of novel data-driven drag models based on direct numerical simulations of fluidized beds. Chem. Eng. Sci. **231**, 116245 (2021)
6. Lee, M.G., Shin, J.H., Choi, S., Park, J.K.: Enhanced blood plasma separation by modulation of inertial lift force. Sens. Actuators B Chem. **190**, 311–317 (2014)
7. Fogelson, A.L., Neeves, K.B.: Fluid mechanics of blood clot formation. Annu. Rev. Fluid Mech. **47**(1), 377–403 (2015)
8. Zeng, L., Balachandar, S., Fischer, P.: Wall-induced forces on a rigid sphere at finite Reynolds number. J. Fluid Mech. **536**, 1–25 (2005)
9. Ji, C., Munjiza, A., Avital, E., Ma, J., Williams, J.J.R.: Direct numerical simulation of sediment entrainment in turbulent channel flow. Phys. Fluids **25**(5), 056601 (2013)
10. Ji, C., Munjiza, A., Avital, E., Xu, D., Williams, J.: Saltation of particles in turbulent channel flow. Phys. Rev. E **89**(5), 052202 (2014)
11. McLaughlin, J.B.: Aerosol particle deposition in numerically simulated channel flow. Phys. Fluids A Fluid Dyn. **1**(7), 1211–1224 (1989)
12. Wang, Q., Squires, K.D.: Large eddy simulation of particle deposition in a vertical turbulent channel flow. Int. J. Multiph. Flow **22**(4), 667–683 (1996)
13. Marchioli, C., Picciotto, M., Soldati, A.: Influence of gravity and lift on particle velocity statistics and transfer rates in turbulent vertical channel flow. Int. J. Multiph. Flow **33**(3), 227–251 (2007)
14. Amiri, A.E., Hannani, S.K., Mashayek, F.: Large-eddy simulation of heavy-particle transport in turbulent channel flow. Numer. Heat Transfer Part B Fundam. **50**(4), 285–313 (2006)
15. Mazzitelli, I.M., Lohse, D., Toschi, F.: On the relevance of the lift force in bubbly turbulence. J. Fluid Mech. **488**, 283–313 (2003)
16. Guan, L., Salinas, J., Zgheib, N., Balachandar, S.: Force and torque model sensitivity and coarse graining for bedload-dominated sediment transport. Eur. J. Mech. B. Fluids **90**, 137–151 (2021)
17. Saffman, P.G.: The lift on a small sphere in a slow shear flow. J. Fluid Mech. **22**(2), 385–400 (1965)
18. Saffman, P.G.: The lift on a small sphere in a slow shear flow-corrigendum. J. Fluid Mech. **31**(3), 624 (1968)
19. Wiberg, P.L., Dungan Smith, J.: Model for calculating bed load transport of sediment. J. Hydraul. Eng. **115**(1), 101–123 (1989)

20. McLaughlin, J.B.: Inertial migration of a small sphere in linear shear flows. J. Fluid Mech. **224**, 261–274 (1991)
21. Asmolov, E.: Dynamics of a spherical particle in a laminar boundary layer. Fluid Dyn. **25**(6), 886–890 (1990)
22. McLaughlin, J.B.: The lift on a small sphere in wall-bounded linear shear flows. J. Fluid Mech. **246**, 249–265 (1993)
23. Cherukat, P., McLaughlin, J.B., Graham, A.L.: The inertial lift on a rigid sphere translating in a linear shear flow field. Int. J. Multiph. Flow **20**(2), 339–353 (1994)
24. Legendre, D., Magnaudet, J.: The lift force on a spherical bubble in a viscous linear shear flow. J. Fluid Mech. **368**, 81–126 (1998)
25. Goldman, A.J., Cox, R.G., Brenner, H.: Slow viscous motion of a sphere parallel to a plane wall—I Motion through a quiescent fluid. Chem. Eng. Sci. **22**(4), 637–651 (1967)
26. Zeng, L., Balachandar, S., Fischer, P., Najjar, F.: Interactions of a stationary finite-sized particle with wall turbulence. J. Fluid Mech. **594**, 271–305 (2008)
27. Shi, P., Rzehak, R.: Lift forces on solid spherical particles in unbounded flows. Chem. Eng. Sci. **208**, 115145 (2019)
28. Ekanayake, N.I.K., et al.: Lift and drag forces acting on a particle moving with zero slip in a linear shear flow near a wall. J. Fluid Mech. **904**, A6 (2020)
29. Cox, R.G., Brenner, H.: The lateral migration of solid particles in Poiseuille flow—I theory. Chem. Eng. Sci. **23**(2), 147–173 (1968)
30. Magnaudet, J., Takagi, S., Legendre, D.: Drag, deformation and lateral migration of a buoyant drop moving near a wall. J. Fluid Mech. **476**, 115–157 (2003)
31. Zeng, L., Najjar, F., Balachandar, S., Fischer, P.: Forces on a finite-sized particle located close to a wall in a linear shear flow. Phys. Fluids **21**(3), 033302 (2009)
32. Ekanayake, N.I., Berry, J.D., Harvie, D.J.: Lift and drag forces acting on a particle moving in the presence of slip and shear near a wall. J. Fluid Mech. **915**, A103 (2021)
33. Robins, B.: New Principles of Gunnery: Containing the Determination of the Force of Gunpowder, and Investigation of the Difference in the Resisting Power of the Air to Swift Ad Slow Motions.... By Benjamin Robins,... with an Account of His Life and Writings, by James Wilson, M. D. F. Wingrave, in the Strand (1805)
34. Magnus, G.: Ueber die Abweichung der Geschosse, und: Ueber eine auffallende Erscheinung bei rotirenden Körpern. Ann. Phys. **164**(1), 1–29 (1853)
35. Barkla, H.M., Auchterlonie, L.J.: The Magnus or Robins effect on rotating spheres. J. Fluid Mech. **47**(3), 437–447 (1971)
36. Rubinow, S.I., Keller, J.B.: The transverse force on a spinning sphere moving in a viscous fluid. J. Fluid Mech. **11**(3), 447–459 (1961)
37. Tsuji, Y., Morikawa, Y., Mizuno, O.: Experimental measurement of the magnus force on a rotating sphere at low Reynolds numbers. J. Fluids Eng. **107**(4), 484–488 (1985)
38. Oesterlé, B., Dinh, T.B.: Experiments on the lift of a spinning sphere in a range of intermediate Reynolds numbers. Exp. Fluids **25**(1), 16–22 (1998)
39. Kurose, R., Komori, S.: Drag and lift forces on a rotating sphere in a linear shear flow. J. Fluid Mech. **384**, 183–206 (1999)
40. Krishnan, G.P., Leighton, D.T.: Inertial lift on a moving sphere in contact with a plane wall in a shear flow. Phys. Fluids **7**(11), 2538–2545 (1995)
41. Lee, H., Balachandar, S.: Drag and lift forces on a spherical particle moving on a wall in a shear flow at finite Re. J. Fluid Mech. **657**, 89–125 (2010)
42. Bagchi, P., Balachandar, S.: Effect of turbulence on the drag and lift of a particle. Phys. Fluids **15**(11), 3496–3513 (2003)

43. Kim, J., Balachandar, S.: Mean and fluctuating components of drag and lift forces on an isolated finite-sized particle in turbulence. Theor. Comput. Fluid Dyn. **26**(1), 185–204 (2012)
44. Li, X., Balachandar, S., Lee, H., Bai, B.: Fully resolved simulations of a stationary finite-sized particle in wall turbulence over a rough bed. Phys. Rev. Fluids **4**(9), 094302 (2019)
45. Hill, R.J., Koch, D.L., Ladd, A.J.C.: Moderate-Reynolds-number flows in ordered and random arrays of spheres. J. Fluid Mech. **448**, 243–278 (2001)
46. Beetstra, R., van der Hoef, M.A., Kuipers, J.A.M.: Drag force of intermediate Reynolds number flow past mono- and bidisperse arrays of spheres. AIChE J. **53**(2), 489–501 (2007)
47. Tenneti, S., Garg, R., Subramaniam, S.: Drag law for monodisperse gas–solid systems using particle-resolved direct numerical simulation of flow past fixed assemblies of spheres. Int. J. Multiph. Flow **37**(9), 1072–1092 (2011)
48. Akiki, G., Balachandar, S.: Shear-induced lift force on spheres in a viscous linear shear flow at finite volume fractions. Phys. Fluids **32**(11), 113306 (2020)
49. Zhou, Q., Fan, L.S.: A second-order accurate immersed boundary-lattice Boltzmann method for particle-laden flows. J. Comput. Phys. **268**, 269–301 (2014)
50. Tang, Y., Peters, E.A.J.F., Kuipers, J.A.M.: Direct numerical simulations of dynamic gas-solid suspensions. AIChE J. **62**(6), 1958–1969 (2016)
51. Chen, X., Song, N., Jiang, M., Ma, T., Zhou, Q.: A microscopic gas-solid drag model considering the effect of interface between dilute and dense phases. Int. J. Multiph. Flow **128**, 103266 (2020)
52. Zhu, Z., Hu, R., Lei, Y., Shen, L., Zheng, X.: Particle resolved simulation of sediment transport by a hybrid parallel approach. Int. J. Multiph. Flow **152**, 104072 (2022)
53. Yang, Z., Lu, X.H., Guo, X., Liu, Y., Shen, L.: Numerical simulation of sediment suspension and transport under plunging breaking waves. Comput. Fluids **158**, 57–71 (2017)
54. Cui, Z., Yang, Z., Jiang, H.Z., Huang, W.X., Shen, L.: A sharp-interface immersed boundary method for simulating incompressible flows with arbitrarily deforming smooth boundaries. Int. J. Comput. Methods **15**(01), 1750080 (2018)
55. Yang, Z., Deng, B.Q., Shen, L.: Direct numerical simulation of wind turbulence over breaking waves. J. Fluid Mech. **850**, 120–155 (2018)
56. Uhlmann, M.: An immersed boundary method with direct forcing for the simulation of particulate flows. J. Comput. Phys. **209**(2), 448–476 (2005)
57. Breugem, W.P.: A second-order accurate immersed boundary method for fully resolved simulations of particle-laden flows. J. Comput. Phys. **231**(13), 4469–4498 (2012)
58. Costa, P., Boersma, B.J., Westerweel, J., Breugem, W.P.: Collision model for fully resolved simulations of flows laden with finite-size particles. Phys. Rev. E **92**(5), 053012 (2015)
59. Kempe, T., Fröhlich, J.: Collision modelling for the interface-resolved simulation of spherical particles in viscous fluids. J. Fluid Mech. **709**, 445–489 (2012)
60. Biegert, E., Vowinckel, B., Meiburg, E.: A collision model for grain-resolving simulations of flows over dense, mobile, polydisperse granular sediment beds. J. Comput. Phys. **340**, 105–127 (2017)
61. Kidanemariam, A.G., Uhlmann, M.: Direct numerical simulation of pattern formation in subaqueous sediment. J. Fluid Mech. **750**, R2 (2014)

62. Kidanemariam, A.G., Uhlmann, M.: Formation of sediment patterns in channel flow: minimal unstable systems and their temporal evolution. J. Fluid Mech. **818**, 716–743 (2017)
63. Gualtieri, P., Picano, F., Sardina, G., Casciola, C.M.: Exact regularized point particle method for multiphase flows in the two-way coupling regime. J. Fluid Mech. **773**, 520–561 (2015)
64. Horwitz, J., Mani, A.: Accurate calculation of Stokes drag for point–particle tracking in two-way coupled flows. J. Comput. Phys. **318**, 85–109 (2016)
65. Esmaily, M., Horwitz, J.: A correction scheme for two-way coupled point-particle simulations on anisotropic grids. J. Comput. Phys. **375**, 960–982 (2018)
66. Balachandar, S., Liu, K., Lakhote, M.: Self-induced velocity correction for improved drag estimation in Euler-Lagrange point-particle simulations. J. Comput. Phys. **376**, 160–185 (2019)
67. Kidanemariam, A.G., Chan-Braun, C., Doychev, T., Uhlmann, M.: Direct numerical simulation of horizontal open channel flow with finite-size, heavy particles at low solid volume fraction. New J. Phys. **15**(2), 025031 (2013)
68. Cisse, M., Homann, H., Bec, J.: Slipping motion of large neutrally buoyant particles in turbulence. J. Fluid Mech. **735**, R1 (2013)
69. Luo, K., Wang, Z., Tan, J., Fan, J.: An improved direct-forcing immersed boundary method with inward retraction of Lagrangian points for simulation of particle-laden flows. J. Comput. Phys. **376**, 210–227 (2019)
70. Akiki, G., Jackson, T.L., Balachandar, S.: Force variation within arrays of monodisperse spherical particles. Phys. Rev. Fluids **1**(4), 044202 (2016)
71. Mei, R.: An approximate expression for the shear lift force on a spherical particle at finite Reynolds number. Int. J. Multiph. Flow **18**(1), 145–147 (1992)
72. Li, D., Wei, A., Luo, K., Fan, J.: Direct numerical simulation of a particle-laden flow in a flat plate boundary layer. Int. J. Multiph. Flow **79**, 124–143 (2016)

Resolving the Volumetric Flow Field Inside a Cylindrical Water Ladle Model with Gas Stirring

Yasmeen Jojo-Cunningham, Xipeng Guo, Chenn Zhou, and Yun Liu(✉)

Purdue University Northwest, Hammond, IN 46323, USA
liu739@pnw.edu

Abstract. Although ladle metallurgy plays a critical role in steelmaking by stirring the molten steel to gain active control over the steel's composition and properties, the physical processes of molten steel flow inside a ladle lack clarity due to its inherent difficulties of perceiving under the extreme conditions. Thus, improving our understanding of the flow processes, particularly inclusion formation and removal mechanisms in steel refining through the stirred flow field, would allow potential enhancement in steelmaking's efficiency and quality. In this project, an advanced Particle-Tracking-Velocimetry system of Shake-the-Box was implemented on a cylindrical water ladle model, made of plexi-glass with compressed air injections through two circular plugs at the bottom, to resolve the three-dimensional flow field inside a $6 \times 6 \times 2$ cm domain. To suppress the flow image distortion due to the curved plexi-glass wall, the method of refractive matching was utilized with an outer polygon tank and sodium iodide solution. Totally, eight conditions of compressed air injection were investigated revealing a highly unsteady and three-dimensional flow field inside the water ladle model.

Keywords: Ladle Metallurgy · Water Ladle model · Volumetric Flow Field · Shake-the-Box system

1 Introduction

In the secondary steelmaking process, the refining of the molten metal from the primary process, gas stirring is a critical operation that involves injecting an inert gas into the molten metal to achieve uniform mixing. It facilitates the homogenization of the chemical composition of alloy elements and the removal of inclusions from the molten steel. The use of stirring aids enhances the quality and purity of the steel, improves its mechanical properties, and minimizes defects in the final product. Furthermore, gas stirring increases the efficiency of alloying and deoxidation, which leads to substantial cost saving and process improvements. The injection of inert gas from the bottom of the ladle into the molten steel creates a slag eye, which is a depression in the slag layer caused by gas bubbles rising from the bottom of the ladle. The behavior of the slag layer is a critical factor in the liquid metal refinement process, as the interaction between the liquid metal and the slag is more prominent in the region surrounding the slag eye compared to

X. Zheng and S. Balachandar (Eds.): IUTAM 2023, IUTAM Bookseries 41, pp. 285–301, 2024.
https://doi.org/10.1007/978-3-031-47258-9_18

other areas in the ladle. The formation of the slag eye facilitates the efficient removal of impurities and inclusions in the molten steel, leading to an enhancement in overall quality and purity. Over the past few decades, researchers have extensively studied the formation of slag eye in a gas stirred ladles. To study this phenomenon, in a practical application, an equivalent scaled simplified water model apparatus has been used, which involves using water at room temperature instead of molten steel. Water at room temperature has the similar kinetic viscosity to molten steel [1], and thus the properly scaled water ladle apparatus is a suitable substitution of steel ladles to be studied for understanding the related flow process.

The use of physical modeling in studying the flow characteristics of molten steel during the secondary steelmaking process has been a topic of interest for researchers for several decades. Szekely et al. [2] first created a simplified water model to study the flow characteristics of a ladle by injecting gas from the bottom, assuming a constant bubble size, and using Spalding's k-ω model to solve the Navier-Stokes equations for velocity and turbulence predictions. The model was developed to simulate the thermal and flow behavior of molten steel during the pouring process using a water-cooled ladle by offering a simple and effective way to predict the behavior of molten steel during pouring by its simple geometry of the ladle and assuming that the thermal and flow behavior could be accurately represented by a single stream of fluid flowing from the ladle spout. Debroy et al. [3] improved Szekely's model by refining the bubble model and accounting for the effects of turbulence and bubble coalescence. These added terms helped to predict the behavior of bubbles more accurately in the liquid metal and their impact on the mixing and refining of the steel. Johansen et al. [4] furthered the research by discovering that bubbles can create turbulence and affect flow velocity in the bubble plume region thus adapting a bottom injection water model. Peranandhantan et al. [5] conducted an experiment to study the behavior of the slag eye in a ladle during steelmaking process by injecting air into a ladle filled with water to simulate the injection of gas into molten steel. The researchers then measured the size and shape of the slag eye using high-speed photography and image analysis techniques. Based on their observation from multiple variables tested including gas flow rate, slag thickness, liquid depth, an empirical expression was developed to formulate the slag eye size in terms of the gas flow rate, the dimensions of the ladle, gravity, surface tension, and the momentum of the gas bubbles on the behavior of the slag eye. Mazumdar et al. [6] reviewed several studies related to the physical modeling and empirical correlation of gas-stirred ladles and highlighted the impact of various variables involved in the gas-stirring process, specifically the impact of plug positions. Several researchers experimented with changing the plug positions in their water ladle model to improve better mixing and wall shear stress distribution [6–8]. It is argued that plug design can change the bubble size distribution close to the plug, but not the average size and distribution in the whole ladle [5, 10–12]. In 2019, Owusu et al. [13] used Particle Image Velocimetry (PIV) to investigate the behaviour of bubbles and their effect on turbulence kinetic energy in the cross-sectional plane of the water ladle. So far, many researchers have used PIV method to study the flow field in the ladle, but the study only resolved the flow field results in two-dimensions. The measurement of velocity and turbulence distribution in three-dimension space is still undeveloped. To further investigate the three-dimensional internal flow field in a water ladle model of a cylindrical container, this study implements a Particle Tracking Velocimetry system with refractive index matching to quantify the unsteady/three-dimensional flow field while minimizing the imaging distortion. Such study cannot only improve our understanding of the complex flow behavior in gas-stirred ladles but also be used to validate the CFD simulation models.

2 Experimental Setup

To achieve the volumetric flow measurement on a cylindrical gas stirred water ladle model, a Particle Tracking Velocimetry system is utilized along with a Refractive Index Matching method. The flow measurements were conducted under eight conditions of gas stirring/injection with a volumetric flow rate ranging from 0.1 LPM to 0.4 LPM. Table 1 summarized the testing conditions of the experiments.

Table 1. Eight test conditions of gas injection flow rate in liters per minute.

Condition	1	2	3	4	5	6	7	8
Plug 1 [LPM]	0.1	0.2	0.3	0.4	0.1	0.2	0.3	0.4
Plug 2 [LPM]	0.1	0.2	0.3	0.4	0	0	0	0

Cylindrical Water Ladle Model and Refractive Index Matching. In this experiment, a cylindrical container with an inner diameter of 2.75 in (70 mm) is implemented to simulate the cylindrical ladle. Due to the curved surface of the cylindrical container that would introduce significant particle image distortion from refraction, the conventional flow field measurement methods, such as Particle Image Velocimetry and Particle Tracking Velocimetry, cannot be directly implemented to measure the flow field. Therefore, this study adopted the refractive index matching method to eliminate image distortion from refraction. Additionally, to film the particle images from four different perspectives, a four-camera particle tracking velocimetry system was utilized on a larger hexagon tank with six flat walls built accordingly to accommodate the angle of imagining. This larger hexagon tank was designed to allow the cameras to film the flow inside the cylindrical tank, which was positioned at the center of the larger hexagon tank, with the cameras filming perpendicular to the flat walls. Both tanks were made of plexi-glasses, and a Sodium Iodide solution was prepared and used to fill the tanks to a solution height of 3 in. As the solution has a refractive index close to plexi-glass, the scattered light from the seeding particles can pass through the curved plexi-glass wall and solution with minimum light refraction, eliminating particle image distortion when the Particle Tracking Velocimetry is employed.

Particle Tracking Velocimetry System. The state-of-the-art: Shake-the-Box system (Lavision, Gottingen) is implemented on the cylindrical water ladle model, which inject compressed air to stir the flow. Hollow glass spheres (diameter of 8–12 μm) are used as seeding particles in the ladle model to track the gas stirred water flow. To capture the seeding particles in the flow field and film the particle images, a high repetition rate laser (Nd: YLF single cavity, Photonics DM-30-527) and four high-speed cameras (Phantom VEO 640), mounted on two sides, were utilized. The imaging system included lenses (Tokina Macro) with a focal length of 100 mms and aperture size of f - 4.5. Cylindrical optical lenses are added to the laser head to generate a 20 mm thick laser light that penetrates and illuminates the inner cylindrical tank from the side. To measure the flow field from the injection plugs, the sampling frequency of the images/laser was set at 100 Hz with an image resolution of 1024 × 1024 pixels on all the four high-speed cameras, resolving the three-dimensional flow field within a volume of 6 × 6 × 2 cm domain. Calibration, data collection and velocity field constructions are realized using the software Davis 10 (LaVision, Gottingen). A 55 × 55 mm calibrate target was utilized and placed in the field of view in the Sodium Iodide solution to capture images for calibration. Calibration was then performed in the Davis 10 using four images of the target. To improve sub-pixel accuracy and realize the volumetric measurement with the system, volume self-calibration was required by the Shake-the-Box system, which was conducted using particle images taken during the experiments. The Shake-the-Box algorithm (LaVision, Gottingen) was then used to carry out particle reconstruction/tracking by shaking the particle position by 1 voxel during the iterations. The instantaneous volumetric velocity field was eventually reconstructed through the post-processing in the Davis 10, resulting in three 15 × 13 × 5 matrices for the volumetric velocity in the three dimensions with a spatial resolution of 4.7 mm/velocity vector. For details about the algorithm of Shake-The-Box, please refer to Schanz et al. [14].

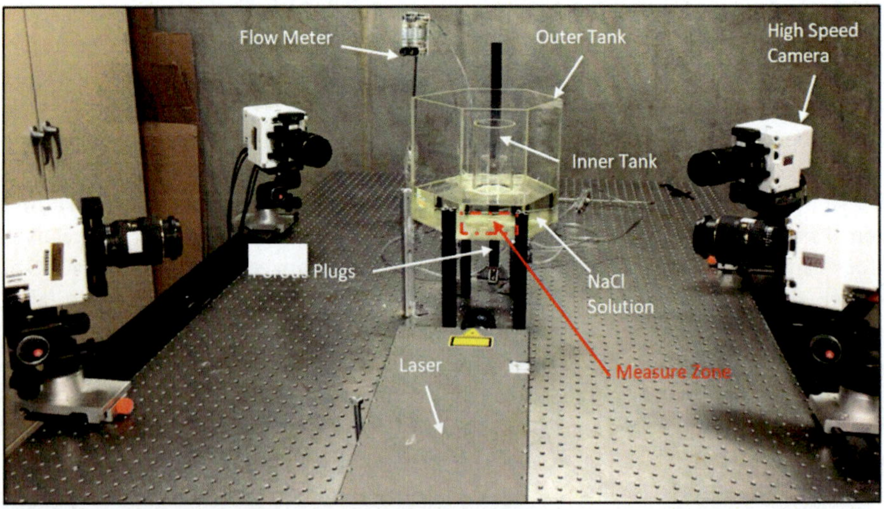

Fig. 1. The Experimental Setup of the Shake-The-Box system

3 Experimental Results

During the experiments, under each testing condition (see Table 1), a total of 2000 data samples (2000 images on each high-speed camera) were taken over a time duration of 20 s, then the seeding particles velocity and trajectories under the 2000-time instants were acquired from the particle tracking process in the Davis 10 software and only the mean velocity field and its characteristics are considered and investigated in this paper.

3.1 Mean Velocity Magnitude

Figure 2 depicts the mean volumetric velocity magnitude field, which exhibits several interesting characteristics of the flow. Specifically, in the case of double gas injection, as the flow rate increasing, the high-velocity region becomes more pronounced, and its location extends from the top towards the bottom of the ladle model. And the high-velocity region is concentrated on the outer edges, specifically on the right-hand side of the ladle model (at the side of $X = 30$ mm), with lower velocity concentrations being observed closed to the bottom. However, in the case of a single gas injection, as the flow rate increasing the high-velocity region mainly concentrated at the corner of the measured flow domain. These observations suggest that the flow behavior under double and single gas injection is markedly different, and further investigation is warranted to better understand its underlying physical process.

A more detailed velocity magnitude analysis is presented in the Figs. 3 and 4, showing the mean volumetric velocity field in the x, y, and z directions for both double and single gas injections.

In the case of double gas injection, the volumetric flow field under the injection rate of 0.1 LPM exhibits a distinct behavior compared to the other conditions. The flow pattern of velocity in the x-direction shows high velocity concentration on diagonal corners that extend along the diagonal. For the y-direction velocity flow field, high velocity concentrations are observed at the center of the top boundary. Moving away from the center, a lower flow velocity surrounds the high velocity concentration, resulting in a concave bowl formation connected to the center line that extends downwards. For the z-direction velocity, the high velocity flow pattern is more concentrated at the center of the x-axis at the top boundary. As the injected gas flow rate increases, the flow pattern becomes more random, lacking a clear structure/pattern. In the case of a single gas injection, as the flow rate increases, the flow contour pattern remains consistent and becomes more distinct and prominent by intensifying the concentrated areas.

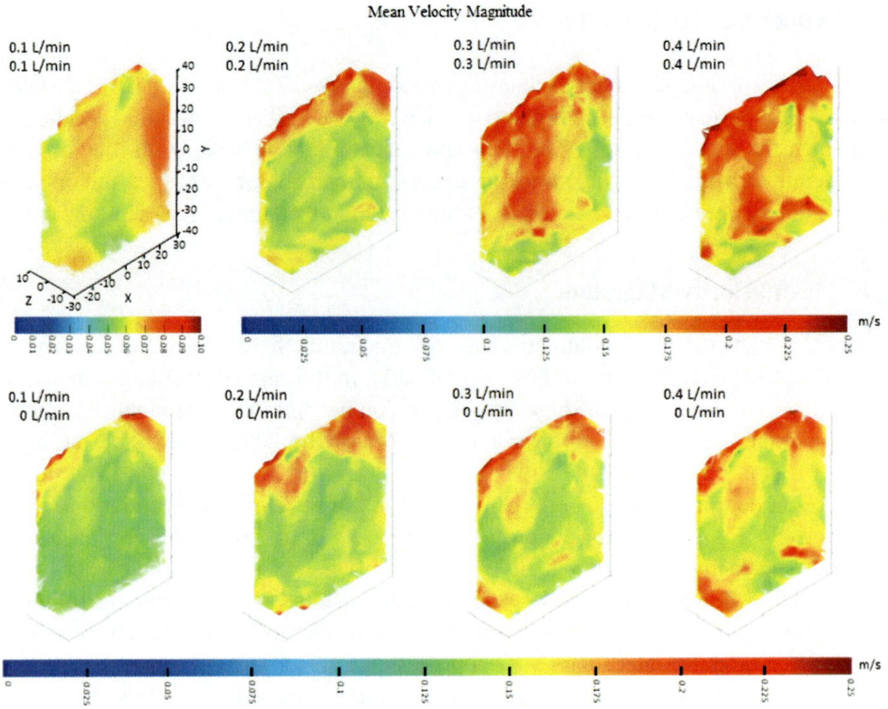

Fig. 2. Volumetric plot of the iso-surfaces of mean velocity magnitude.

3.2 Mean Velocity Streamlines

From the mean velocity field, the three-dimensional streamlines can be obtained and is a helpful tool for visualizing the fluid flow to understand the complex flow behavior. Figure 5 presents the three-dimensional streamlines for double and single gas injections. For double gas injection, at the lower flow rate of 0.1 L per minute, the three-dimensional streamlines are highly structured and organized. The flow follows a well-defined, well-organized pattern in X-Y planes with the streamlines being symmetrically distributed about the centerline in x direction. As the gas injection flow rate increasing, the streamlines become more complex and less organized. On the other hand, the streamline patterns for single gas injection are highly disordered and lack of clear pattern, even at low gas injection rates of 0.1 L per minutes. The streamlines do not follow any particular pattern, and there are groups or clusters in the flow region at any rate of gas injection.

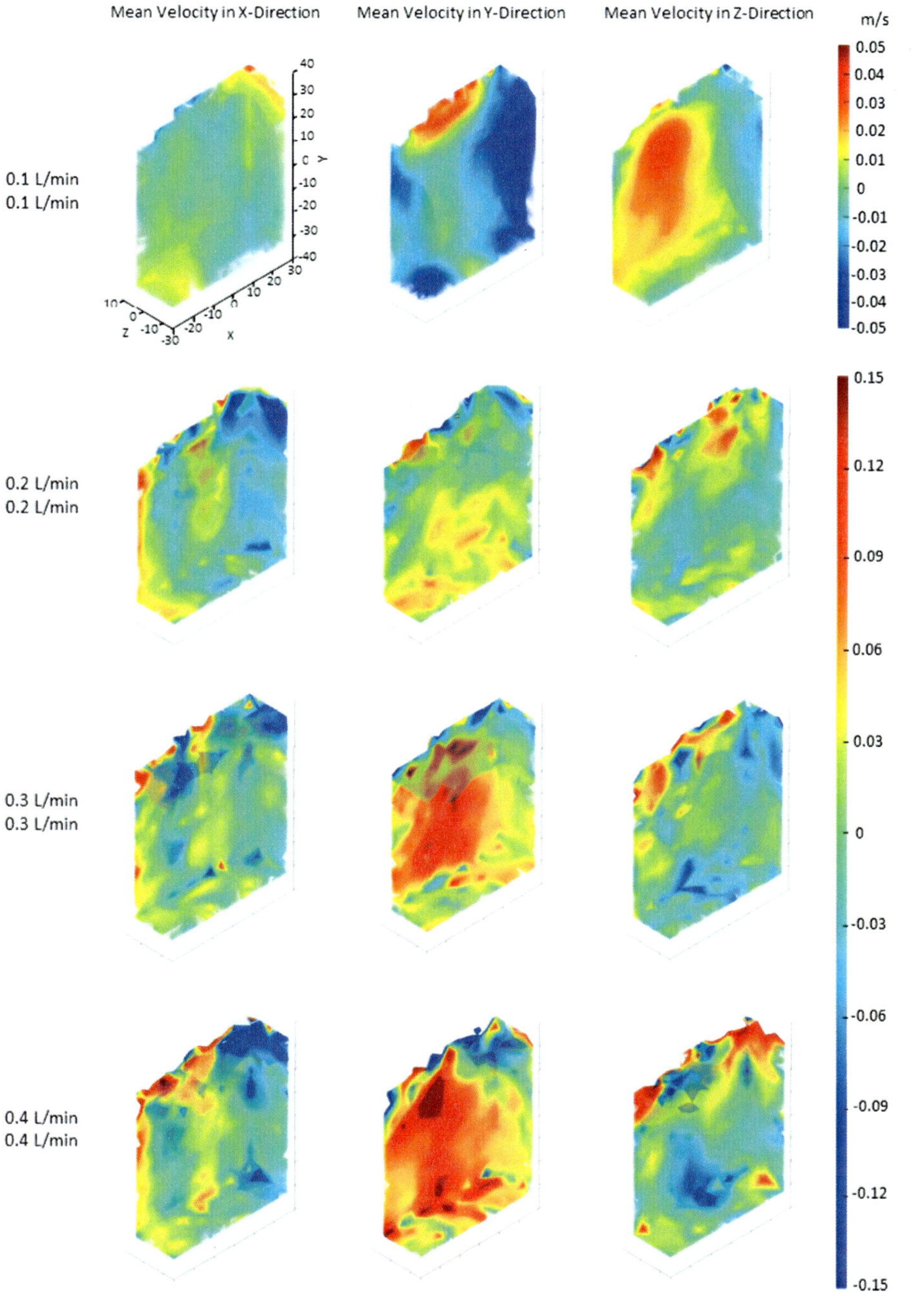

Fig. 3. Mean velocity field in X, Y, Z directions with double gas injections.

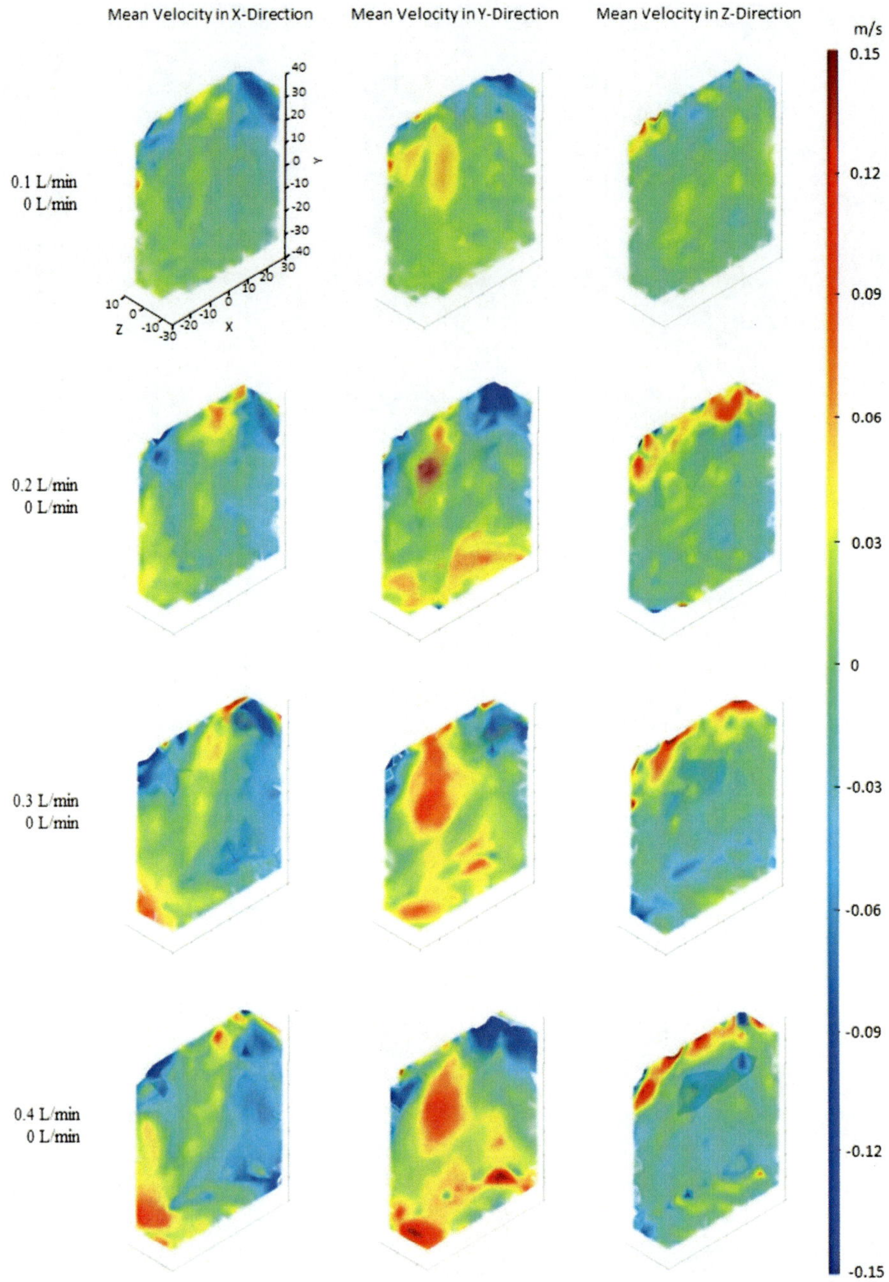

Fig. 4. Mean velocity field in X, Y, Z directions with single gas injection.

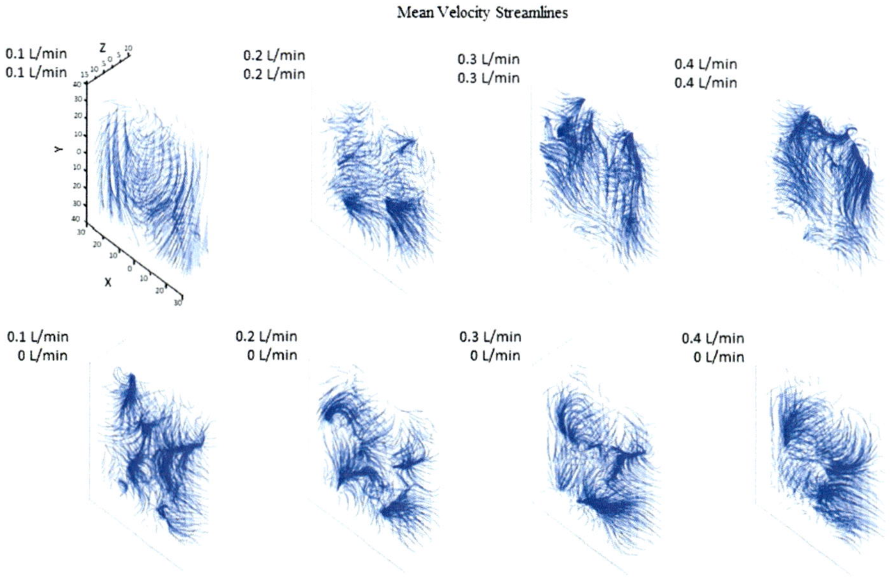

Fig. 5. 3D streamlines for mean velocity field.

When examining streamlines from mean velocity field on two dimensional planes, nodal points of separation and attachment, focus points of separation and attachment, and saddle points can be observed. A nodal point of attachment is a point in the flow field where the velocity lines are directed outward away from the node, while nodal points of separation are directed inward towards the node. On the other hand, a focus differs from a nodal point in that it spirals around the singular point, either from it (a focus of attachment) or into it (a focus of separation). At a saddle point, the streamlines are moving in different directions along two perpendicular planes, forming a point of flow stagnation. As the fluid moves away from the saddle point, it diverges in one direction and converges in the other direction, forming a region of flow separation. These findings are in line with previous studies that have shown the existence of these features in various fluid flow systems [15]. To illustrate these features, in Fig. 6 to Fig. 9, the streamline is plotted on the planes of the x = −5 mm, y = −5 mm and z = −5 mm with and without the velocity magnitude contour. The results showed that these characteristic points were presented in all planes and were clearly influenced by the gas injection flow rate. At lower gas injection flow rate, the streamlines tend to be more organized and follow a more predictable pattern, whereas at higher gas injection flow rate, the streamlines become more chaotic and disordered, making it difficult to discern any specific pattern or feature.

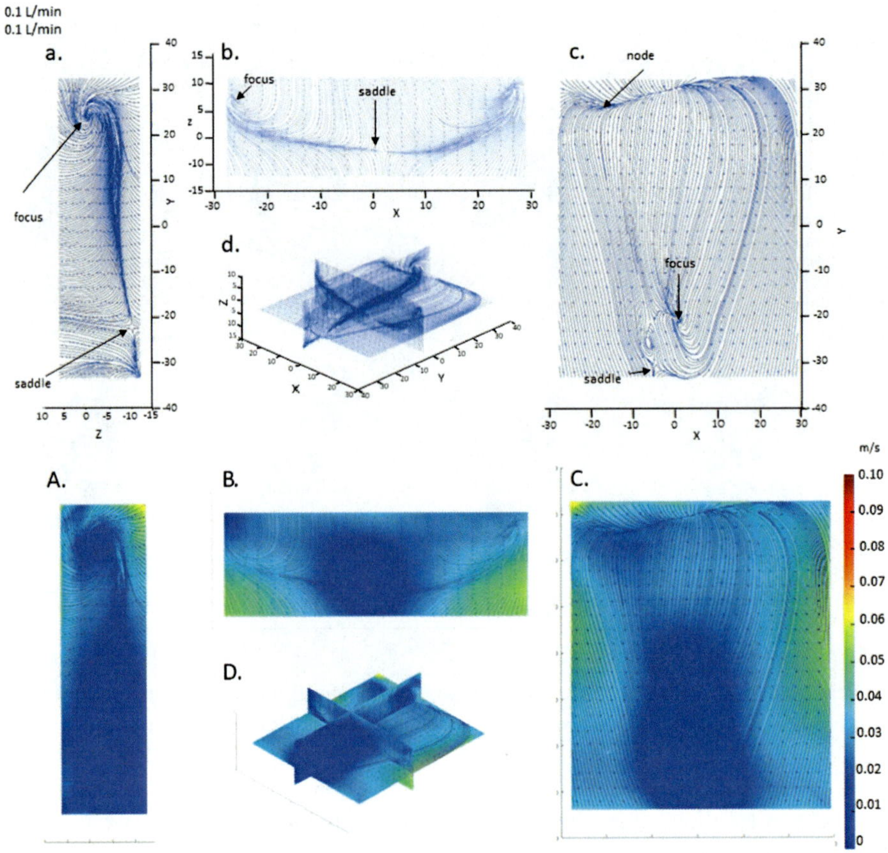

Fig. 6. Streamlines and velocity magnitude on different planes with double gas injection at 0.1 L/Min.

To further elaborate on the observation, Fig. 6 displays a clear and structured flow pattern for the double gas injection at a gas injection rate of 0.1 L per minute. In Fig. 6b, The X-Z plane at $z = -5$ mm exhibits a saddle point near the x-centerline where the flow move into opposite directions, leading to flow stagnation. The foci points are mirrored about this saddle point, which spirals around the singular point. In the X-Y plane (Fig. 6c), the streamlines illustrate an organized flow pattern with a slight deviation from perfect symmetry. Furthermore, downward flow momentum dominates the measured flow field opposing the direction of gas injection, and the low velocity is concentrated throughout the Z-Y plane (Fig. 6a, 6A), near the bottom boundary of the X-Y plane, and the center of X-Z plane (Fig. 6c and 6C). In Fig. 7, when the gas flow rate is increased to 0.3 LPM for the double gas injections, the flow starts to lack organized structure, like previously observed. Notably, the dominate momentum in the flow field changes from downward to upward. Additionally, high velocity is concentrated more throughout each plane and shows a more random pattern of flow compared to Fig. 6. Figure 8 presents the two-dimensional flow patter with single gas injection at a flow rate 0.1 L per minute. The flow field lacks dominated momentum with certain direction and can be characterized by multiple directions present. Low velocity concentration dominates throughout the flow field, with high velocity on the tips of the top boundary, showing no clear pattern of the flow. Finally, Fig. 9 shows the results for a single gas injection at the highest gas injection rate of 0.4 L per minute. The flow is still disrupted and random with no dominated momentum. High velocity concentration is intensified at the outer boundaries with no clear/consistent pattern observed. Overall, 2D streamline visualization provides valuable insights into the fluid flow behavior and flow patterns. The results also suggest that the injection rate and the type of gas injection have a significant impact on the fluid flow behavior and the resulting streamline patterns. At lower flow rates, with double injections, the flow pattern is more structured and organized, whereas at higher flow rates, the flow becomes more random and chaotic. Additionally, the type of gas injection (single or double) can significantly affect the flow pattern.

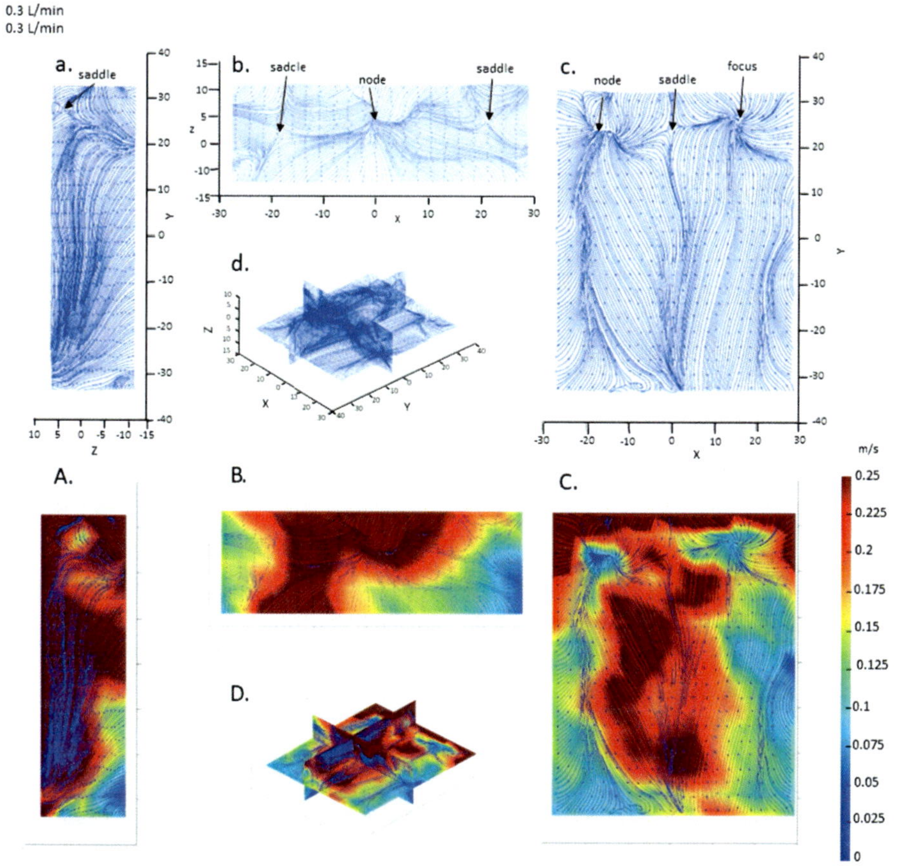

Fig. 7. Streamlines and velocity magnitude on different planes with double gas injection at 0.3 L/Min.

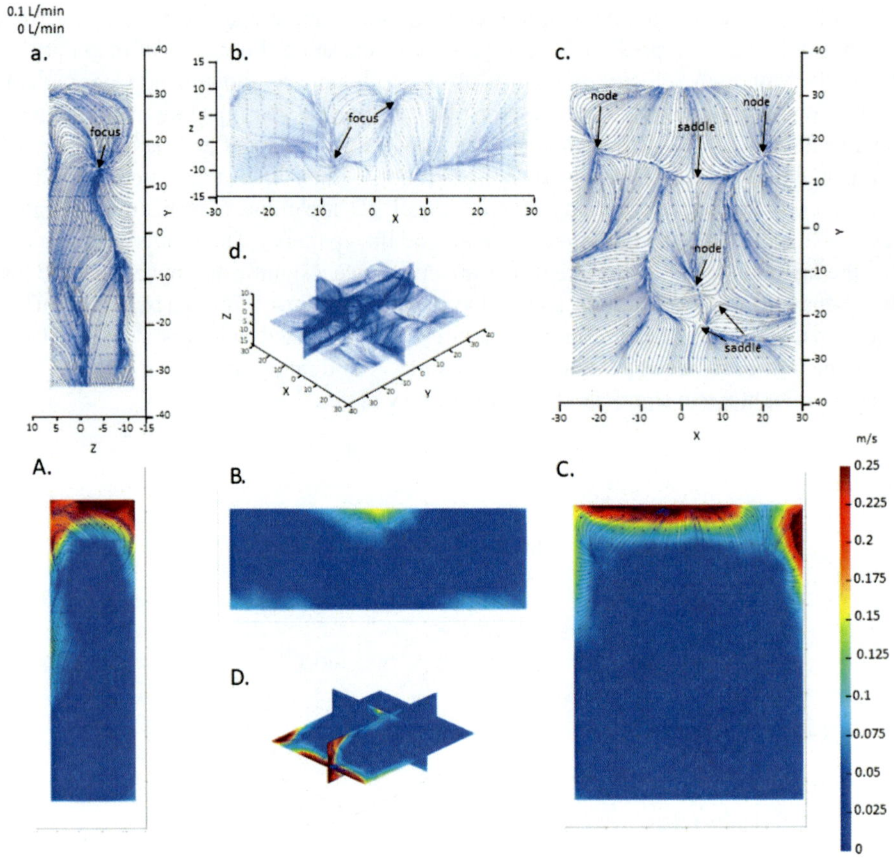

Fig. 8. Streamlines and velocity magnitude on different planes with single as injection at 0.1 L/Min.

4 Discussion

To gain a more comprehensive understanding of the complex process of ladle metallurgy, an analysis was conducted on the velocity distribution within the volumetric flow of a water ladle model, with results presented in Sect. 4.1. In Sect. 4.2, the findings of a Computational Fluid Dynamics (CFD) simulation of the same model are discussed and compared with the experimental data. Finally, a quantitative comparison between the experiment results and the CFD simulation is presented and reviewed.

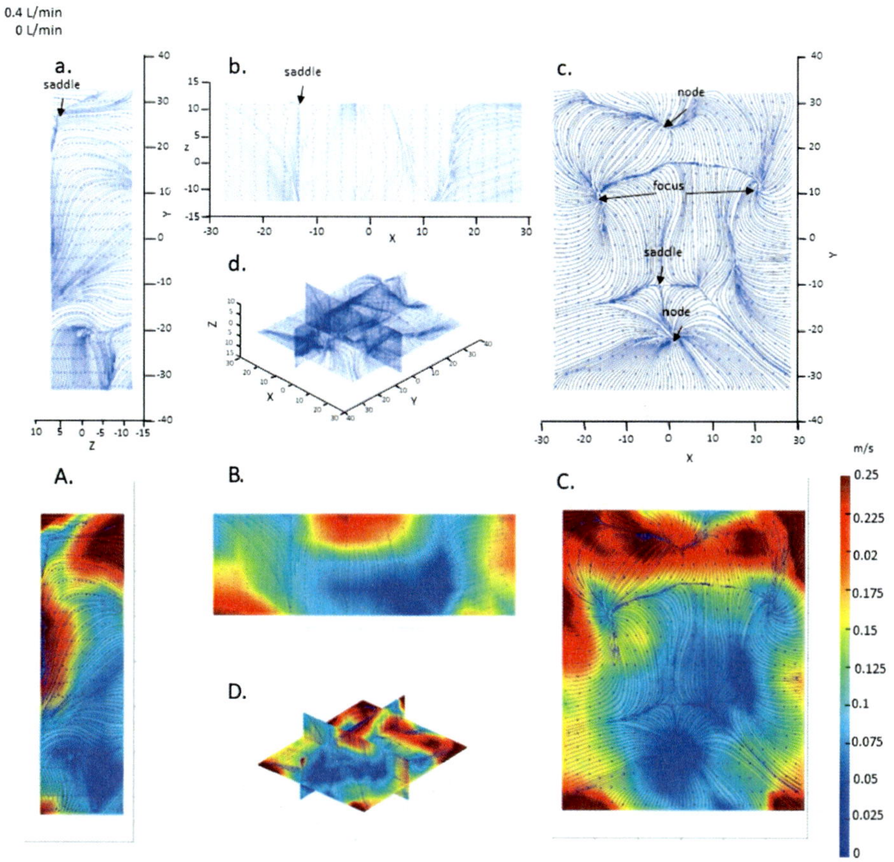

Fig. 9. Streamlines and velocity magnitude on different planes with single as injection at 0.4 L/Min.

4.1 Transition of the flow pattern

Through the velocity field analysis in the volumetric flow inside the water ladle model using mean velocity magnitude (see Sect. 2.1) and streamlines (see Sect. 2.2), consistent trend have been revealed and can be validated across multiple figures. For double gas injection, as the gas flow rate increases, the high velocity concentration extended from top to bottom in the vertical direction, resulting in a more uniform flow with higher fluid velocity magnitude throughout the volumetric region as seen in Fig. 2 and Fig. 3. Further examination of the mean velocity field revealed a clear flow field transition with increased gas injections. With the flow rate of double gas injection increased from 0.1 to 0.2 L per minute, flow field changes from an organized and structured flow pattern to a disorderly, random, and unpredictable flow pattern, as seen in the mean velocity in the y-direction in Fig. 3. As the flow rate increases further, the disorderly flow pattern continues with elevated velocity magnitude throughout the studied region. The trend of increasing gas flow rates leading to irregular and asymmetrical flow patterns is also supported by a particularly clearer visual representation in Fig. 5, the mean 3D velocity streamlines. With the gas injection flow rate increases from 0.1 to 0.2 L per minute, the symmetrical, organized and mostly vertical streamlines transition to an irregular and prominently horizontal streamlines. Not as pronounced but still visual, another transition occurs by increasing the flow rate again to 0.3 L per minute. This can be observed in the changes in the orientations of the streamlines, which become more vertical. Upon closer examination of the mean velocity streamlines at various planes with and without the mean velocity magnitude contour, further verification of the observed trends can be found in Fig. 6 and 7. Particularly, Fig. 6 confirms the highly organized, clear structure, and balanced flow pattern at a low double gas injection of flow rate at 0.1 L per minute. Additionally, it can be seen that downward momentum dominates the flow measurement region, despite the gas injection should introduce an upwards momentum. By investigating the streamlines and mean velocity on the stream-slices, it can be noted that low velocity concentrates around the focus point on the X-Y plane (Fig. 6c). The existence of focus point in the flow field is undesired and problematic as it causes fluid accumulation and prevents the refining process. As the gas flow rate is increased to 0.3 L per minute, in Fig. 7, the flow lacks an orderly structure. In this instance, the dominate flow direction of the flow field changes from downward to upward momentum with less outwards curvature. Additionally, high velocity is more concentrated throughout each plane. Therefore, these trends suggest that higher gas flow rates may be desirable to achieve the desired level of flow uniformity and to reduce the accumulation of stagnant fluid. For single gas injection, unlike double gas injection, increasing the gas flow rate did not lead to changes in the concentration contour or flow pattern. Instead, it enhanced the high velocity concentration and kept the contour pattern consistent, as demonstrated in Fig. 2 and Fig. 4. As the flow rate increases, the contour patterns of the high velocity regions become more pronounced and distinct. This suggests that the flow field is becoming more focused and concentrated as the gas flow rate increases, resulting in a higher concentration of fluid velocity in certain areas of the volume region. Additionally, as seen in Fig. 5, the streamlines continuously lack structure at any flow rate, exhibiting a chaotic, noisy, and irregular flow pattern with multiple directions and strong concentrated clusters. Further verification was obtained by examining the mean velocity streamlines at the planes of $x = -5$ mm, $y = -5$ mm, and $z = -5$ mm with and without the mean velocity magnitude contour, as shown in Fig. 8 and Fig. 9. Figure 8 confirmed the chaotic flow pattern and the absence of a dominant momentum direction, which could be the result of multiple nodal, focus, and saddle points. Moreover, the flow field was dominated by low velocity concentration, which could cause the fluid to be stationary less turbulent within the volume. With the flow rate increased to 0.4 L per minute, as demonstrated in Fig. 9, the flow remained disrupted with no dominate momentum. However, the concentration of high velocity was intensified and enhanced, extending to the outer boundaries and improving the fluid bulk motion. Overall, the analysis of single gas injection showed that increasing the gas injection flow rate can help the enhancement of high velocity concentration but did not significantly change the flow pattern or behavior.

By comparing the double and single gas injection, notable differences can be observed. In Fig. 2, the high velocity concentration of double injection is more evenly distributed on the top, whereas for a single injection, it concentrates on the edges of the top boundary. As the flow rate increases, the high velocity concentration for single injection extends towards the middle of the top boundary, approaching that of double injection. At a flow rate of 0.1 L per minute, the flow was observed to be less organized. Figure 5 shows that double injections become more oriented in the upward/downward direction as the flow rate increases, while single injections continue to have fluid flow in multiple directions. Additionally, single injections have more concentrated groups, while double injections are more uniform throughout the region. In summary, the analysis of the data using various plots has revealed some common trends across different conditions. In the case of double injection at 0.1 LPM, the flow pattern exhibited irregularity phenomenon by having organized, symmetrical, and lowest velocity contour, with a dominant downward momentum in the flow. We suspect that this organized flow pattern is corresponded to the laminar flow. As the flow increased to 0.2 LPM and then 0.3 LPM, flow pattern transition is observed, possibly indicating a transition from laminar to turbulent flow. With the gas injection flow rate increases, high velocity concentration becomes more uniform throughout the flow field, extending from the top to the bottom of the flow field, and the momentum becomes more dominant in the upwards direction. On the other hand, for a single injection, the flow pattern was consistently disorderly and chaotic, with high concentration enhancing as the flow rate increased. This flow behavior is likely due to the unbalanced gas injection which leads to random and unpredictable fluid flow. The node, focus, and saddle points were also observed as the sources of velocity changes across all cases. These finding underscores the significance of careful consideration of injection method and flow rate for industrial applications aiming to achieve desired flow patterns.

4.2 Comparison with CFD Simulation

In recent years, CFD simulation has become the most commonly used method for studying complex fluid flow phenomena such as those encountered in industrial applications. However, due to the inherent complexities of the flow processes, there are still significant uncertainties associated with computational fluid dynamics (CFD) modeling. In order to address these uncertainties and validate the CFD simulation, experiment validation is highly desired. And experiments can provide valuable insights into the limitations and strengths of CFD modeling in capturing the complex flow behavior. In an ongoing research, a CFD simulation is conducted to simulate the complex flow field of the water ladle model that is implemented in the current volumetric flow measurement experiment. Preliminary results from the CFD experiment are shown in Fig. 10 where the simulated mean velocity field in x, y, and z directions with double gas injection at 0.1 L per minute is presented. The simulation has been performed using a VOF-DPM multiphase model to track multiphase interaction and interface behavior, and a Large Eddy Simulation (LES) turbulence model has been employed. In this setup, two injectors are placed at the bottom of the tank, each maintaining a constant flow rate of 0.1 LPM. A qualitative comparison of the velocity distribution in each component direction is presented in the figure. While the velocity distribution pattern is similar, the CFD results demonstrate a more symmetric behavior compared to the experimental results which might not be perfectly symmetric. Specifically, the velocity field in the x-direction has good compatibility while the velocity field in the y-direction is similar in the upper region with higher velocity, but the CFD experiment shows a more intense velocity distribution with an over-predicted plume region. Therefore, these findings highlight the qualitative similarity of the flow field between the experiments and CFD simulations, suggesting the implemented numerical method can derive a reasonable and accurate flow field.

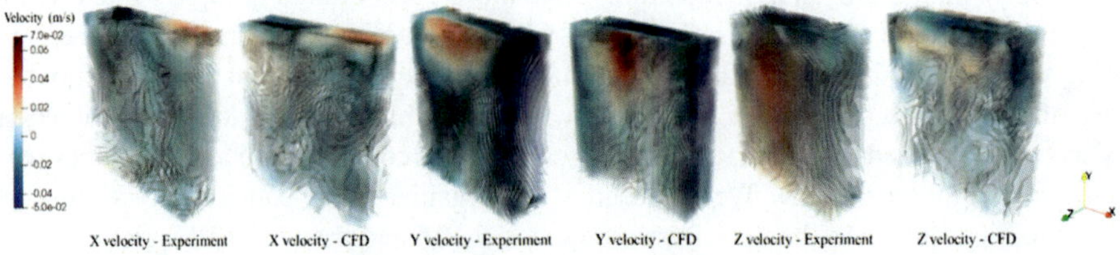

Fig. 10. CFD simulation of mean velocity field in X, Y, Z directions with double gas injections at 0.1 L/Min.

To achieve a quantitative comparison, the experimental and CFD flow field data are reprocessed to extract mean velocity plots that can be compared among the experiment and CFD results. Figure 11 presents the vertical velocity plots estimated on YZ planes and XY planes (mean value of vertical velocity is calculated on each planes along the X and Z directions) under double gas injection at 0.1 L per minute. Although the velocity distribution between the experimental data and the CFD simulation is similar, a noticeable discrepancy is observed. Specifically, the graph shows a vertical shift upward from the experimental data points. Despite this discrepancy, the overall trends match well between the experimental and simulated data, providing promising evidence for the validity of the CFD model. Further analysis is currently being conducted on the other conditions investigated in this project. The aim is to compare the CFD simulation results to the experimental data for all the compressed air injection conditions, to further validate the CFD model and assess its accuracy in predicting the flow behavior in the water ladle model.

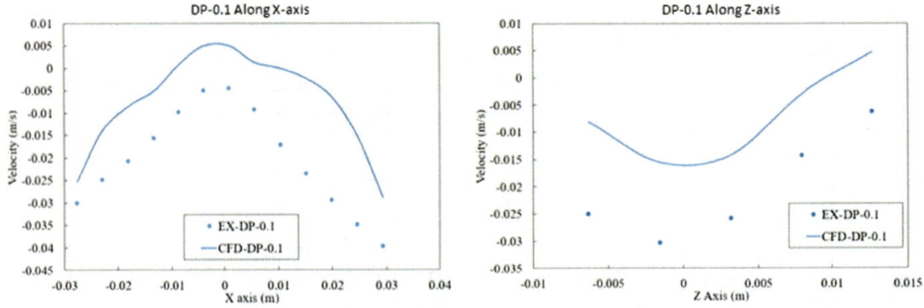

Fig. 11. Quantitative comparison of mean velocity field with double gas injections at 0.1 L/Min.

Acknowledgements. This work was supported in part by the National Science Foundation through its MRI program (CMMI-1919726) and GOALI program (CMMI-2113967). Support from all members of the Steel Manufacturing Simulation and Visualization Consortium (SMSVC), the SMSVC Ladle Project Technical Committee members, and Nucor Steel and the permission to publish this work are greatly acknowledge.

References

1. Mazumdar, D., Guthrie, R.I.L.: The physical and mathematical modelling of gas stirred ladle systems. ISIJ Int. **35**(1), 1–20 (1995)
2. Szekely, J., Asai, S.: The general mathematical statement of turbulent recirculatory flows. Trans. Iron Steel Inst. Jpn. **15**(5), 270–275 (1975)
3. DebRoy, T., Majumdar, A.K., Spalding, D.B.: Appl. Math. Model. **2**, 146 (1978)
4. Johansen, S.T., Robertson, D.G.C., Woje, K., Engh, T.: Metall. Mater. Trans. B **19**, 745 (1988)
5. Peranandhanthan, M., Mazumdar, D.: Modeling of slag eye area in argon stirred ladles. ISIJ Int. **50**(11), 1622–1631 (2010)
6. Mazumdar, D., Guthrie, R.I.L.: Modeling energy dissipation in slag-covered steel baths in steelmaking ladles. Metall. Mater. Trans. B **41**(5), 976–989 (2010)
7. Nunes, R.P., et al.: Visualisation and analysis of the fluid flow structure inside an elliptical steelmaking ladle through image processing techniques. J. Eng. Sci. Technol. **2**(2), 139–150 (2007)
8. Li, L., et al.: Water model and CFD-PBM coupled model of gas-liquid-slag three-phase flow in ladle metallurgy. ISIJ Int. **55**(7), 1337–1346 (2015)
9. Liu, Z., Li, L., Li, B.: Modeling of gas-steel-slag three-phase flow in ladle metallurgy: Part I. Physical modeling. ISIJ Int. **57**(11), 1971–1979 (2017)
10. Vazquez, A., et al.: A look at three measurement techniques for bubble size determination. Exp. Thermal Fluid Sci. **30**(1), 49–57 (2005)
11. Aoki, J., Thomas, B.G., Peter, J., Peaslee, K.D.: Experimental and theoretical investigation of mixing in a bottom gas-stirred ladle. In: AISTech – Iron and Steel Technology Conference Proceedings, vol. 1, pp. 1045–1056 (2004)
12. Fukuji, Y., Mori, Y., Fujita, S.: Sizes and size distributions of bubbles in a bubble column. Chem. Eng. Jpn. **12**(1), 5–9 (2006)
13. Owusu, K.B., et al.: Interaction of injector design, bubble size, flow structure, and turbulence in ladle metallurgy. Steel Res. Int. **90**(2), 1800346 (2019)
14. Schanz, D., Gesemann, S., Schröder, A.: Shake-the-box: Lagrangian particle tracking at high particle image densities. Exp. Fluids **57**(5) (2016). https://doi.org/10.1007/s00348-016-2157-1
15. Tobak, M., Peake, D.J.: Topology of three-dimensional separated flows. Annu. Rev. Fluid Mech. **14**(1), 61–85 (1982). https://doi.org/10.1146/annurev.fl.14.010182.000425

Local Intermittency of Electrified Dust Storms in the Near-Surface Layer

Huan Zhang[1,2,3(✉)] and Yuan Liu[1,2,3]

[1] Center for Particle-laden Turbulence, Lanzhou University, Lanzhou 730000, People's Republic of China
zhanghuan@lzu.edu.cn
[2] Key Laboratory of Mechanics on Disaster and Environment in Western China Attached to the Ministry of Education of China, Lanzhou University, Lanzhou 730000, People's Republic of China
[3] Department of Mechanics and Engineering Science, College of Civil Engineering and Mechanics, Lanzhou University, Lanzhou 730000, People's Republic of China

Abstract. Dust storms are unique dispersed two-phase turbulent flows laden with charged particles, where significant coupling occurs between dust particles and turbulence, as well as between particles and electric fields. Existing works have mainly focused on the intermittent behavior of multiple physical fields in the region far from the surface, where the interphase couplings are weak. As a result, limited attention has been paid to the intermittent behavior of multiple physical fields near the surface. Here, by combining wavelet analysis approaches and the observed data of wind velocity, PM10 dust concentration, and electric field at a height of 0.9 m above the surface during dust storms, we uncover the local intermittent characteristics of multiple physical fields when the interphase couplings are relatively strong. Our findings indicate that PM10 dust concentration and electric field exhibit similar time-averaged intermittent behavior, but with more pronounced intermittency compared to wind velocity. Additionally, the PM10 dust concentration and electric field display relatively sparse local intermittency in the time domain. These results will contribute to our understanding of the multi-field characteristics of complex dispersed two-phase flows.

Keywords: dust storms · electric field · intermittency

1 Introduction

Dust storms are an excellent natural laboratory for studying dispersed two-phase turbulent flows at friction Reynolds numbers up to $O(10^6)$. Due to the significant electrical charge carried by dust particles, electric fields as high as 100 kV m^{-1} generally exist in dust storms, constituting a unique electrohydrodynamic (EHD) turbulence regime [1–4]. Importantly, the intense particle-turbulence and particle-electrostatics interactions may lead to distinctive characteristics of such dust storm EHD turbulence compared to single-phase turbulence.

X. Zheng and S. Balachandar (Eds.): IUTAM 2023, IUTAM Bookseries 41, pp. 302–312, 2024.
https://doi.org/10.1007/978-3-031-47258-9_19

Turbulence exhibits spatio-temporal fluctuations across various scales. Measuring turbulence quantities at a fixed point in space can result in a time series with intermittent and localized intense peaks [5]. This phenomenon, known as turbulent intermittency, has been extensively studied in experiments and numerical simulations [6,7]. Small-scale intermittency focuses on increments or gradients of turbulent fields and is typically evaluated using the probability density function (p.d.f.) and higher-order moments of the field increments within the inertial and dissipative ranges. At large scales, velocity increment p.d.f.s tend to be near-Gaussian. As the scale decreases, the tails of these p.d.f.s become stretched and wider, deviating significantly from Gaussian distributions [8,9]. The Kolmogorov theory (K41) argues that, in the high Reynolds number regime, the structure function within the inertial range can be uniquely determined by the average energy dissipation rate, yielding the celebrated Kolmogorov linear (monofractal) scaling between the scaling exponent and order of the structure function. However, numerous experimental and numerical simulation studies have shown that the K41 theory holds only at lower orders. Instead, the scaling exponent exhibits a convex nonlinear relationship with order, deviating significantly from the Kolmogorov linear scaling. This deviation, referred to as anomalous (multifractal) scaling, along with the presence of non-Gaussian p.d.f.s of the field increments at small scales, characterizes small-scale intermittency [10].

In our previous study, the small-scale intermittency of the wind velocity, PM10 (particles less than $10\,\mu$m in diameter) dust concentration and electric field (hereafter referred to as multifield intermittency) in dust storm EHD turbulence was only investigated at a height of 5 m [4], in which the particle-turbulence and particle-electrostatics interactions are weak. It was found that, among the multiple physical fields, PM10 dust concentration exhibits the strongest intermittency, whereas the wind velocity shows the weakest intermittency. Moreover, the p.d.f.s of the streamwise wind velocity increments in dust storms are broader and fatter compared to those in clean-air conditions, indicating that the wind velocity at small scales becomes more intermittent in dust storms compared to clean-air conditions. However, these findings are limited to the weak (or negligible) multifield coupling regime and thus lack information on the relatively strong coupling regime.

In this study, we aim to evaluate the multifield intermittency in the near-surface layer during dust storms using datasets collected in an atmospheric turbulence observatory. In this region, multifield couplings are more pronounced compared to the region at and above 5 m in height, as investigated in our previous study [4]. The remainder of this paper is organized as follows. In Sect. 2, a detailed description of the used datasets and wavelet analysis is presented. Then, time-averaged and local multifield intermittency are quantified at different frequencies and in the time-frequency domain in Sects. 3.1 and 3.2, respectively. Finally, conclusions are provided in Sect. 4.

2 Methodology

2.1 Dataset

Field observations were carried out over the period from April 2021 to June 2021 at the atmospheric surface turbulence observatory-Qingtu lake observation array (QLOA)-which is located between the Badain Jaran Desert and Tengger Desert in China. The QLOA, designed by Professor Zheng [11], is considered a unique and high-tech scientific observation station. Zheng and co-workers have made outstanding contributions to the research of atmospheric surface layer turbulence and dust storm turbulence. The dry lake bed of Qingtu lake is a flat surface covering about 20 km^2, without roughness elements such as rocks, vegetation, sand dunes, etc. [2–4]. Such a large flat surface ensures that the atmospheric surface-layer flows are considered analogous to the flat-plate canonical turbulent boundary layers [12]. During March to June, dust events frequently occur at the QLOA site, owing to Mongolian cyclones that are accompanied by strong northwesterly surface winds [13].

The wind velocity, PM10 dust concentration, and electric field were synchronously measured at 0.9 m above the surface. The wind velocity was measured using a sonic anemometer (CSAT3B, Campbell Scientific) at a sampling frequency of 50 Hz, the PM10 dust concentration was measured using a DustTrak II Aerosol Monitor (Model 8530EP, TSI Incorporated) at a sampling frequency of 1 Hz, and the vertical electric field was measured using a vibrating-reed electric field mill (VREFM) developed by Lanzhou University at a sampling frequency of 1 Hz. The VREFM works based on the vibrating capacitor technique. As the vibrating electrode oscillates, an output DC voltage is proportional to the electric field to be measured and thus can be easily recorded. For further details on VREFM, the reader is referred to our previous studies [2,14].

The raw data collected from the observed dust storms were divided into multiple 1-h datasets, which were subsequently chosen based on stationary and near-neutral conditions to ensure the satisfaction of ergodicity and negligible thermal buoyancy requirements (for details, see refs. [3,12]). Furthermore, the synoptic scales that overlap with very large-scale motions (VLSMs) were removed by subtracting the large-scale trend across the entire measurement array from the raw signals [12]. In this study, three 1-h datasets selected from a clean-air condition and two dust storm events are used for the analysis. The main parameters of the three datasets are given in Table 1. The clean-air dataset was taken on May 26, 2021, at 09:00–10:00, the mild dust storm dataset was taken on April 28, 2021, at 14:00–15:00 UTC+8, and the severe dust storm dataset was taken on May 05, 2021, at 15:00–16:00 UTC+8. In Table 1, $\langle U \rangle$ is the mean convection velocity, $\langle \Theta \rangle$ is the mean ambient temperature, u_τ is the friction velocity, $\langle C \rangle$ is the mean PM10 dust concentration, and $\langle |E| \rangle$ is the mean magnitude of the vertical electric field. The friction Reynolds number Re_τ is estimated using a boundary layer height of $\delta = 150$ m, and the kinematic viscosity ν is calculated based on Sutherland's law [4].

Table 1. Main parameters of the three selected datasets.

Parameter	Clean-air	Mild dust storm	Severe dust storm
$\langle U\rangle$ (m s^{-1})	6.63	8.56	10.09
$\langle\Theta\rangle$ (K)	293.99	292.72	297.37
u_τ (m s^{-1})	0.35	0.47	0.57
$\langle C\rangle$ (mg m^{-3})	—	0.60	1.34
$\langle\lvert E\rvert\rangle$ (kV m^{-1})	0.32	9.97	44.27
ν (m^2 s^{-1})	1.77×10^{-5}	1.75×10^{-5}	1.80×10^{-5}
Re_τ	2.93×10^6	4.00×10^6	4.78×10^6

2.2 Wavelet Analysis

Fourier analysis represents data as a sum of trigonometric functions which extend to infinity, thus it is inefficient in dealing with local abrupt changes. In contrast to Fourier analysis, using a class of localized basis functions, termed wavelets, continuous wavelet transform allows us to unfold data into both space (or time) and scale (or frequency) domain and therefore can uncover local intermittent events effectively [15–18]. The continuous wavelet transform of a time series $\{X(n),\ n = 0, \cdots, N-1\}$ is defined as the convolution of $\{X(n)\}$ with a scaled and translated Morlet wavelet is defined as [16,18], which gives

$$W_{XX}(n,s) = \sum_{n'=0}^{N-1} \sqrt{\frac{\delta_t}{s}} X(n')\psi_0^* \left[\frac{(n'-n)\delta_t}{s}\right], \tag{1}$$

where n is the localized time index, s is the wavelet scale that is inversely proportional to frequency f (i.e. $1/s = 1.03f$, see ref. [18] for the details), Morlet wavelet $\psi_0(\eta) = \pi^{-1/4}e^{(i\omega_0\eta - \eta^2/2)}$ with $\omega_0 = 6$ being dimensionless frequency, $(\cdot)^*$ denotes the complex conjugate, and δ_t is the sampling interval of the time series $\{X(n)\}$. Here, the discrete scales $\{s(j)\}$ are selected as fractional powers of two:

$$s(j) = s_0 2^{j\delta_j}, \quad j = 0, 1, \cdots, J, \tag{2}$$

$$J = \delta_j^{-1} \log_2\left(\frac{N\delta_t}{s_0}\right), \tag{3}$$

where $s_0 = 2\delta_t$ is the smallest resolvable scale of the wavelet, J determines the largest scale, and δ_j is the spacing between discrete scales.

Since the square of a wavelet coefficient gives a fluctuating energy on a scale s at a time index n, we thus define the local wavelet power spectral density (PSD) of the time series $\{X(n)\}$ as [4,17,19,20]

$$\mathcal{P}_{XX}(n,s) = 2\delta_t \left|W_{XX}(n,s)\right|^2. \tag{4}$$

The global wavelet PSD is written as

$$\overline{\mathcal{P}_{XX}}(s) = \langle \mathcal{P}_{XX}(n,s) \rangle_n = \frac{1}{N} \sum_{n=0}^{N-1} \mathcal{P}_{XX}(n,s), \tag{5}$$

where the time average $\langle \cdot \rangle_n$ is performed over the whole period. It is demonstrated that the global wavelet PSD corresponds to the smoothed Fourier PSD and gives an unbiased and consistent estimation of the true global PSD [18,21].

The presence of intermittency can be evaluated by means of the non-Gaussian p.d.f. of the field difference between two points [4,22]. On the basis of wavelet transform, such deviations from the Gaussian PDF (with flatness factor equal to 3) is measured by the wavelet flatness factor defined as [4,19,23]

$$F_X(s) = \frac{\langle |W_{XX}(n,s)|^4 \rangle_n}{\langle |W_{XX}(n,s)|^2 \rangle_n^2}, \tag{6}$$

which quantifies the level of the time-averaged intermittency of time series $\{X(n)\}$ at the scale s.

In contrast to wavelet flatness factor, the so-called local intermittency measure (LIM) is defined as [17]

$$I_X(n,s) = \frac{|W_{XX}(n,s)|^2}{\langle |W_{XX}(n,s)|^2 \rangle_n}, \tag{7}$$

which is the ratio of local energy at the time index n and the scale s to the time-averaged energy and thus measures the level of local intermittency.

3 Results and Discussion

3.1 Time-Averaged Intermittency

Figure 1 shows an example of the raw time series of the mild dust storm dataset. It can be seen that the selected dataset is stationary and near-neutral. Clearly, the variations of wind velocity, PM10 dust concentration, and electric field are synchronized over time. However, compared to wind velocity, the latter two fields exhibit a higher probability of abrupt changes, suggesting the presence of significant differences in multifield intermittency. The intermittency of the electric field can also be observed from its wavelet PSD in the time-frequency domain. It is evident that the wavelet PSD shows changes and discontinuities both in time and frequency.

To quantitatively characterize the multifield intermittency in dust storms, we present the multifield wavelet flatness at different frequencies, as shown in the Fig. 2. Firstly, for mild and severe dust storm datasets, the multifield wavelet flatness increases with frequency. When the frequency exceeds a certain value, the multifield wavelet flatness becomes larger than three, indicating a heavy-tailed distribution of the multifield increments. Secondly, the flatness curves of PM10 dust concentration and electric field exhibit remarkable similarities,

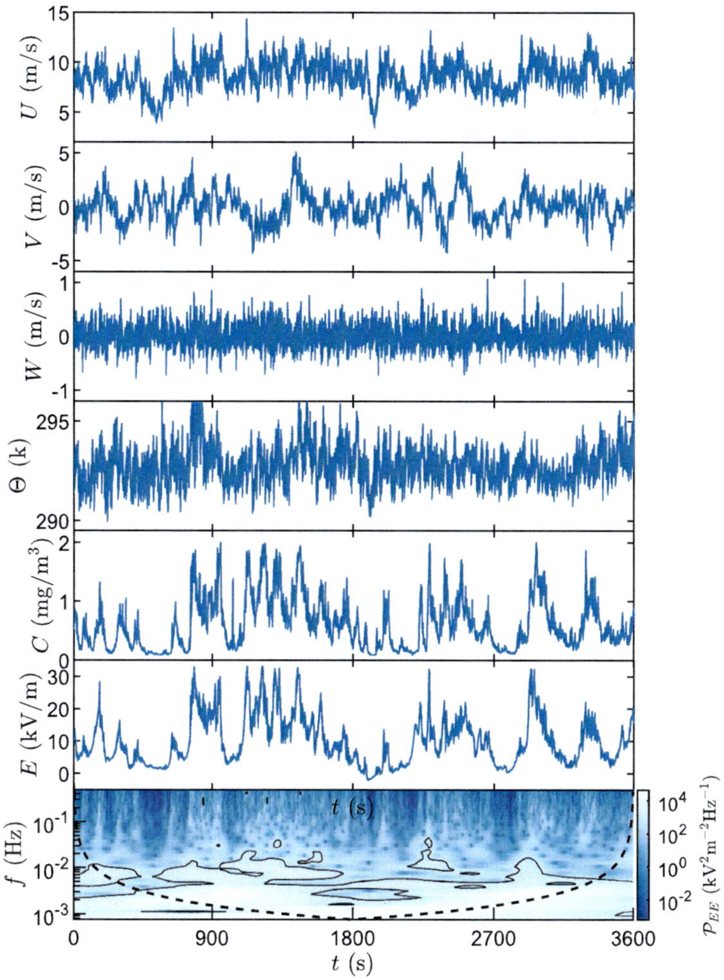

Fig. 1. Top-to-bottom: time series of streamwise (U), spanwise (V), and vertical (W) wind velocity, ambient temperature (Θ), PM10 dust concentration (C), vertical electric field (E), as well as wavelet PSD of vertical electric field ($\mathcal{P}_{EE}$). In the lowest panel, the thick contour (i.e. black solid lines) encloses regions of greater than 95% confidence for a red-noise process with a lag-1 coefficient of 0.72; the regions enclosed by dashed lines, t-, and f-axis represent the "cone of influence", where edge effects become important. Times are shown relative to April 28, 2021 at 14:00 coordinated universal time plus 8 h (UTC+8). All data are plotted with a frequency of 1 Hz for clarity.

with both being significantly higher than that of wind velocity. This suggests that the intermittency of PM10 dust concentration and electric field follows a similar pattern, but with a stronger intermittency than that of wind velocity. Furthermore, when compared to wind velocity under clean-air conditions, the wavelet flatness of wind velocity during dust storms is considerably amplified

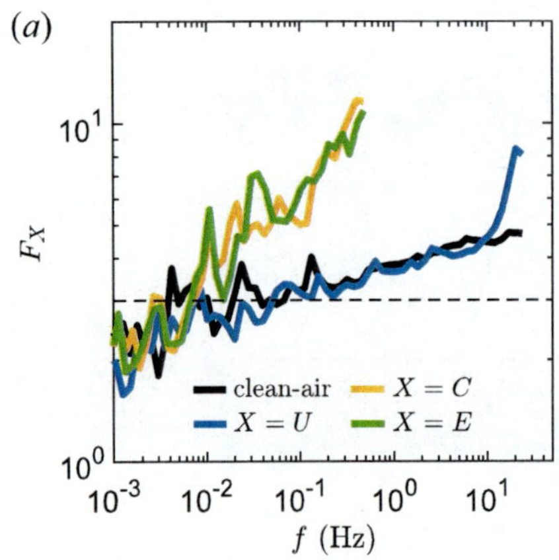

Fig. 2. (*a*) Wavelet flatness factors of the streamwise wind velocity for the clean-air dataset (black line) are compared with those of the streamwise wind velocity ($X = U$), PM10 dust concentration ($X = C$), and vertical electric field ($X = E$) for the mild dust storm dataset. (*b*) The same comparison is made, but for the severe dust storm dataset. The horizontal dashed lines represent $F_X = 3$, corresponding to the Gaussian PDF of the field difference.

beyond approximately 10 Hz. This amplification is particularly pronounced in the severe dust storm compared to mild one. This implies that the presence of dust particles strengthens the intermittency of wind velocity at small scales, and the intermittency becomes stronger with higher dust concentrations.

3.2 Local Intermittency

It is important to emphasize that wavelet flatness measures the time-averaged intermittency at different scales (or frequencies), but it lacks a quantitative characterization of how intermittency evolves over time. Therefore, Figs. 3 and 4 further illustrate the distribution of the multifield LIM during both mild and severe dust storm datasets in the time-frequency domain. It can be seen that the LIMs of PM10 dust concentration and electric field in the time-frequency domain exhibit remarkably similar patterns, but differ significantly from those of wind velocity, consistent with the results in Fig. 2. Particularly, when the frequency exceeds 0.01 Hz, it is evident that compared to the LIM of wind velocity, the LIM of PM10 dust concentration and electric field displays a similar distribution in frequency but a sparser distribution in time. This indicates that the relatively higher intermittency of PM10 dust concentration and electric field at small scales (or high frequencies) is primarily due to their higher temporal sparsity.

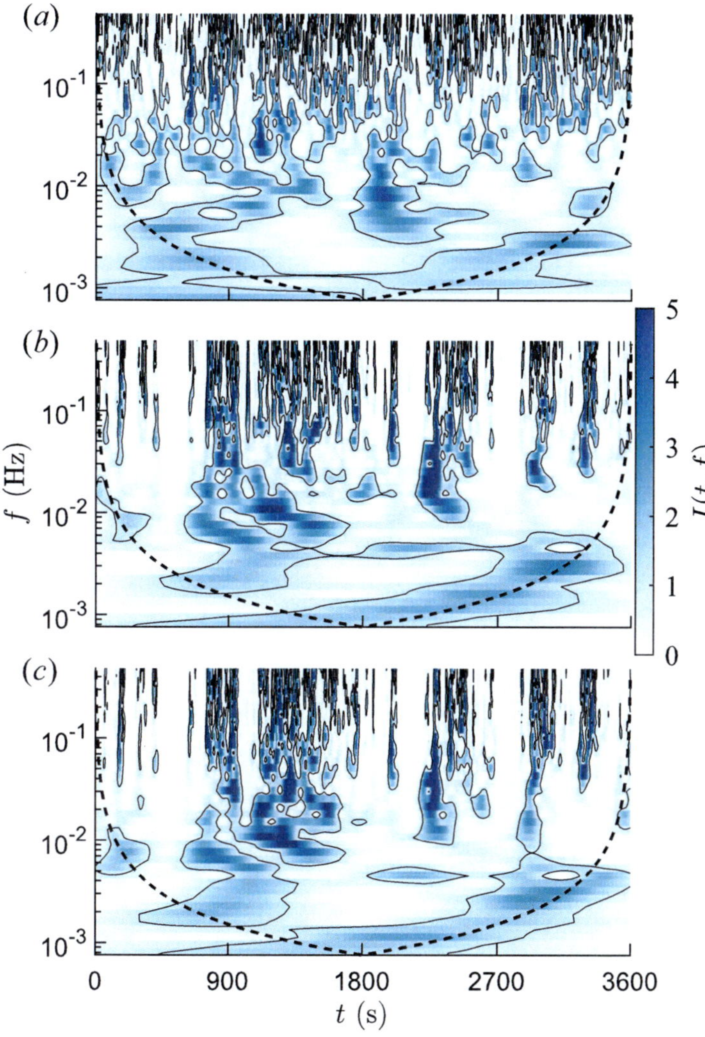

Fig. 3. (*a*) LIM of the streamwise wind velocity as a function of time and frequency for the mild dust storm dataset. (*b*) Same as (*a*) but for the LIM of the PM10 dust concentration. (*c*) Same as (*a*) but for the LIM of the vertical electric field. The thick contour (i.e. black solid lines) encloses regions of greater than 95% confidence for a red-noise process with a lag-1 coefficient of 0.72. Times are shown relative to April 28, 2021 at 14:00 coordinated universal time plus 8 h (UTC+8).

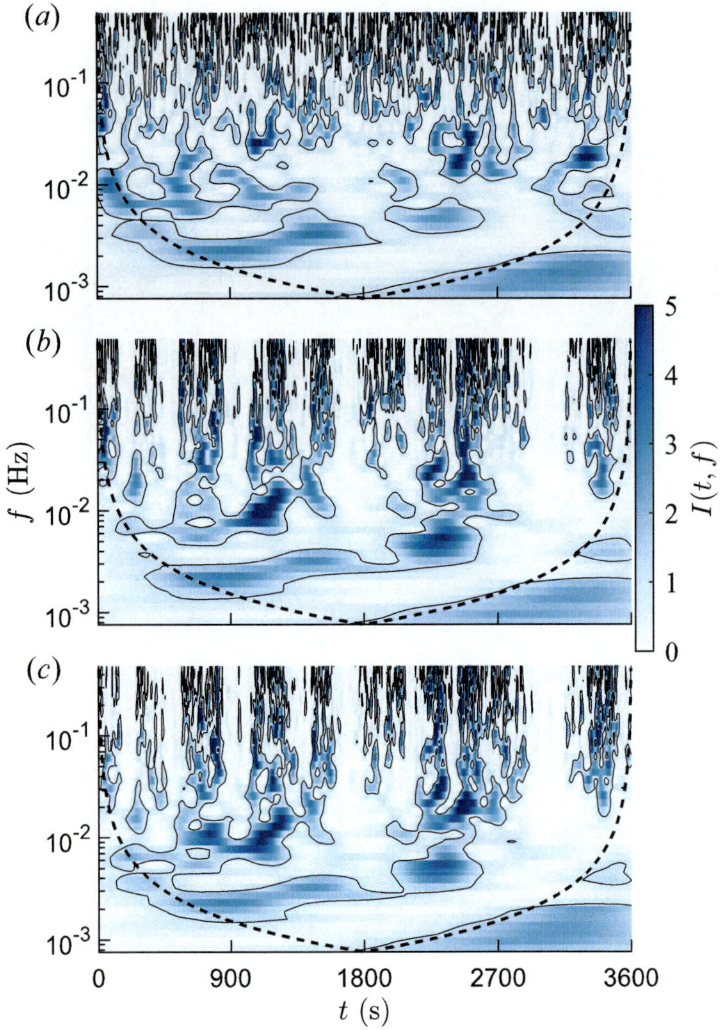

Fig. 4. (*a*) LIM of the streamwise wind velocity as a function of time and frequency for the severe dust storm dataset. (*b*) Same as (*a*) but for the LIM of the PM10 dust concentration. (*c*) Same as (*a*) but for the LIM of the vertical electric field. The thick contour (i.e. black solid lines) encloses regions of greater than 95% confidence for a red-noise process with a lag-1 coefficient of 0.72 Times are shown relative to May 5, 2021 at 15:00 coordinated universal time plus 8 h (UTC+8).

4 Conclusions

In this study, based on observed data of wind velocity, PM10 dust concentration, and electric field at a height of 0.9 m above the surface during dust storms, we use wavelet flatness factors and local intermittency measure to uncover the local intermittent characteristics of multiple physical fields in the time-frequency domain when the particle-turbulence and particle-electrostatics couplings are relatively strong. We found that dust particles lead to a substantial enhancement of intermittency of wind velocity in high frequencies (or small scales), which becomes more prominent with higher particle concentrations. Both PM10 dust concentration and electric field exhibit similar time-averaged intermittent behaviour at different frequencies (or scales), but with more pronounced intermittency than wind velocity. Furthermore, the PM10 dust concentration and electric field display similar local intermittent patterns in the time-frequency domain, with sparser intermittency compared to wind velocity in the time domain.

Acknowledgements. This study was supported by the National Natural Science Foundation of China (Grant No. 92052202) and the Fundamental Research Funds for the Central Universities (Grant No. lzujbky-2021-ey19).

References

1. Zheng, X.J., Huang, N., Zhou, Y.H.: Laboratory measurement of electrification of wind-blown sands and simulation of its effect on sand saltation movement. J. Geophys. Res. Atmos. **108**, 4322 (2003). https://doi.org/10.1029/2002JD002572
2. Zhang, H., Zhou, Y.H.: Reconstructing the electrical structure of dust storms from locally observed electric field data. Nat. Commun. **11**, 5072 (2020). https://doi.org/10.1038/s41467-020-18759-0
3. Zhang, H., Zhou, Y.H.: Unveiling the spectrum of electrohydrodynamic turbulence in dust storms. Nat. Commun. **14**, 408 (2023). https://doi.org/10.1038/s41467-023-36041-x
4. Zhang, H., Tan, X., Zheng, X.: Multifield intermittency of dust storm turbulence in the atmospheric surface layer. J. Fluid Mech. **963**, A15 (2023). https://doi.org/10.1017/jfm.2023.278
5. Batchelor, G.K., Townsend, A.A.: The nature of turbulent motion at large wave-numbers. Proc. R. Soc. Lond. Ser. A-Math. Phys. Eng. Sci. **199**, 238–255 (1949). https://doi.org/10.1098/rspa.1949.0136
6. Siggia, E.D.: Numerical study of small-scale intermittency in three-dimensional turbulence. J. Fluid Mech. **107**, 375–406 (1981). https://doi.org/10.1017/S002211208100181X
7. Faller, H., et al.: On the nature of intermittency in a turbulent von Kármán flow. J. Fluid Mech. **914**, A2 (2021). https://doi.org/10.1017/jfm.2020.908
8. Castaing, B., Gagne, Y., Hopfinger, E.J.: Velocity probability density functions of high Reynolds number turbulence. Physica D **46**, 177–200 (1990). https://doi.org/10.1016/0167-2789(90)90035-N

9. Shnapp, R.: On small-scale and large-scale intermittency of Lagrangian statistics in canopy flow. J. Fluid Mech. **913**, R2 (2021). https://doi.org/10.1017/jfm.2021.33
10. Kolmogorov, A.N.: A refinement of previous hypotheses concerning the local structure of turbulence in a viscous incompressible fluid at high Reynolds number. J. Fluid Mech. **13**, 82–85 (1965). https://doi.org/10.1017/S0022112062000518
11. Wang, G., Zheng, X.: Very large scale motions in the atmospheric surface layer: a field investigation. J. Fluid Mech. **802**, 464–489 (2016). https://doi.org/10.1017/jfm.2016.439
12. Hutchins, N., Chauhan, K., Marusic, I., Monty, J., Klewicki, J.: Towards reconciling the large-scale structure of turbulent boundary layers in the atmosphere and laboratory. Bound.-Layer Meteor. **145**, 273–306 (2012). https://doi.org/10.1007/s10546-012-9735-4
13. Zheng, X.: Mechanics of Wind-Blown Sand Movements. Springer, Berlin (2009). https://doi.org/10.1007/978-3-540-88254-1
14. Zheng, X.: Electrification of wind-blown sand: recent advances and key issues. Eur. Phys. J. E **36**, 138 (2013). https://doi.org/10.1140/epje/i2013-13138-4
15. Zhou, Y.H.: Wavelet Numerical Method and Its Applications in Nonlinear Problems. Springer, Berlin (2021). https://doi.org/10.1007/978-981-33-6643-5
16. Daubechies, I.: The wavelet transform, time-frequency localization and signal analysis. IEEE Trans. Inf. Theory **36**, 961–1005 (1990). https://doi.org/10.1109/18.57199
17. Farge, M.: Wavelet transforms and their applications to turbulence. Annu. Rev. Fluid Mech. **24**, 395–458 (1992). https://doi.org/10.1146/annurev.fl.24.010192.002143
18. Torrence, C., Compo, G.P.: A practical guide to wavelet analysis. Bull. Am. Meteorol. Soc. **79**, 61–78 (1998). https://doi.org/10.1175/1520-0477(1998)079⟨0061:APGTWA⟩2.0.CO;2
19. Alexandrova, O., Carbone, V., Veltri, P., Sorriso-Valvo, L.: Small-scale energy cascade of the solar wind turbulence. Astrophys. J. **674**, 1153–1157 (2008). https://doi.org/10.1086/524056
20. Ruppert-Felsot, J., Farge, M., Petitjeans, P.: Wavelet tools to study intermittency: application to vortex bursting. J. Fluid Mech. **636**, 427–453 (2009). https://doi.org/10.1017/S0022112009008003
21. Percival, D.P.: On estimation of the wavelet variance. Biometrika **82**, 619–631 (1995). https://doi.org/10.1093/biomet/82.3.619
22. Frisch, U.: Turbulence: The Legacy of A. N. Kolmogorov. Cambridge University Press, Cambridge (1995)
23. Meneveau, C.: Analysis of turbulence in the orthonormal wavelet representation. J. Fluid Mech. **232**, 469–520 (1991). https://doi.org/10.1017/S0022112091003786

Control the Migration of Self-propelling Particles in Thermal Turbulence via Reinforcement Learning Algorithm

Ao Xu[1,2(✉)], Hua-Lin Wu[1], and Heng-Dong Xi[1,2]

[1] School of Aeronautics, Northwestern Polytechnical University, Xi'an 710072, China
axu@nwpu.edu.cn

[2] Institute of Extreme Mechanics, Northwestern Polytechnical University, Xi'an 710072, China

Abstract. Planning energy-efficient trajectories can significantly benefit unmanned aerial vehicles (UAVs) by increasing their endurance. Inspired by birds that exploit warm rising atmospheric currents, flying vehicles can be controlled to utilize the updrafts to conserve energy. We adopt the reinforcement learning algorithm to train a self-propelling particle in a Rayleigh-Bénard (RB) convection cell with periodic vertical boundary conditions, where the large-scale circulation oscillates, and small-scale velocities fluctuate. The trained smart particle successfully learns to utilize the background flow structure to minimize energy consumption, enabling it to migrate with less energy consumption. Despite the complex flow structures that arise both in large-scale circulation and small-scale velocity fluctuations, an energy-efficient trajectory is identified by the particle through the strategy with the highest reward. This research has practical implications for UAVs patrolling in the convective layer of the atmosphere, where energy-efficient trajectories can enhance their endurance and cover a wider range.

Keywords: turbulent thermal convection · particle migration · reinforcement learning

1 Introduction

Soaring birds utilize warm rising atmospheric currents to remain in flight and increase their altitude [1]. However, staying airborne in the troposphere is not easy, as the winds of the air are turbulent. Meanwhile, gliders and unmanned aerial vehicles (UAVs) can also increase their endurance and save energy by utilizing thermals [2]. There are serval attempts at thermal soaring for UAVs. For example, in 2007, a research team from NASA built a 6.8 kg UAV with a 4.27 m wingspan [3]. The UAV flew 17 times and gained an average altitude of 173 m from 23 updrafts, and the strongest updraft allowed the UAV to ascend 844 m.

Modeling thermals plays a crucial role in thermal soaring. During the exploration and exploitation of the thermals, wind speed can only be measured at the current location. However, the rising thermal airflow exists over a larger area and for a certain duration. Lawrence et al. [4] investigated thermal soaring as a path-planning problem

X. Zheng and S. Balachandar (Eds.): IUTAM 2023, IUTAM Bookseries 41, pp. 313–325, 2024.
https://doi.org/10.1007/978-3-031-47258-9_20

in a simulation environment. Their study revealed that the solution is effective in the situation of partial or complete wind fields being detectable. Therefore, it is essential to model the thermals so that the flying vehicles can determine their states within the thermal and adopt an appropriate strategy. In 2006, Allen developed the chimney model [5] based on surface temperature and wind speeds collected at a station in the Nevada desert region. The model describes the size, spacing, height, and vertical velocity distribution of thermals, which extend from the ground to the highest level of the mixing layer. In 2009, Lawrance proposed the bubble model [6], which describes updrafts as a torus in a specific space, providing the size, height, and velocity distribution in three dimensions. The chimney and bubble thermal models depict idealized scenarios that assume neither the turbulent motion nor the modeled fluctuations. However, the wind field in the troposphere displays turbulent fluctuations that are considerably stronger. Turbulent fluctuations pose challenges in identifying effective thermal soaring strategies [7]. Turbulence affects all modes of flight and leaves an imprint on the trajectories of flying objects [8], as demonstrated by the similarity between the trajectories of golden eagles and those of particles carried by turbulent flows. The incorporation of turbulence is necessary for understanding the movement and behaviors of flying objects.

Atmospheric convection plays a vital role in shaping Earth's weather and climate. To gain a deep understanding of this complex natural phenomenon, we adopt the Rayleigh-Bénard (RB) convection system as a simplified model [9–11]. This model is useful in simulating wind patterns during strong convective weather. The RB system is controlled by several dimensionless parameters, including the Rayleigh number (Ra), Prandtl number (Pr), and cell aspect ratio (Γ). The Ra indicates the ratio of buoyancy forces to viscous forces resulting from temperature differences. The Pr reflects the thermophysical properties of the fluid. In the atmosphere, Ra typically ranges from 10^{18} to 10^{22} [10], and $Pr = 0.71$. The Γ reflects the geometric characteristics of the convection system, and it is approximately 100 for mesoscale convective systems [12]. In a fully developed turbulent thermal convection system, a thin temperature boundary layer is present near the upper and lower boundaries of the convection cell. Hot and cold plumes detach from the respective lower and upper walls and are driven by buoyancy forces. As these plumes move, they are self-organized into a large-scale circulation (LSC). When Γ increases, thermal plumes create a network of interconnected ridgelike structures comprising cold downwelling and hot upwelling fluids. This arrangement is commonly referred to as the superstructure of thermal turbulence [13, 14].

By adopting the RB turbulent environment, Reddy et al. successfully implemented the thermal soaring strategy of UAVs using a reinforcement learning algorithm in both simulations [15] and experimental environments [16]. The roll angle of the UAV was controlled to accomplish the gliding task. After 12 days of experiment and 3 days of evaluation, the UAV demonstrated excellent gliding performance, capturing the updraft's energy effectively. In our previous works [17, 18], we aimed to reduce energy consumption for a self-propelling particle by optimizing its trajectory in the RB turbulent convection. The optimized trajectories were designed to enable the agent to effectively utilize the kinetic energy of thermal turbulence. Compared to the straight-line trajectory, the optimized trajectory resulted in significantly less energy consumption, making it a much more efficient approach. It should be noted that in our previous works [17,

18], convection cells with no-slip vertical boundary conditions were adopted, where the large-scale circulation roll remains statistically steady. In the present work, we aim to extend our previous works by incorporating periodic vertical boundary conditions, where the large-scale circulation roll oscillates. This work will contribute to a more comprehensive and systematic understanding of the thermal soaring strategy in the turbulent RB environment.

2 Numerical Methods

2.1 The Kinematic Model for the Self-propelling Particle

Our approach to modeling self-propelling particles is based on the massless point-particle model. The dynamics of the self-propelling particles can be described as:

$$\mathbf{u}_{particle}(t) = \mathbf{u}_{fluid}(t) + \mathbf{u}_{propel}(t) \tag{1}$$

$$d\mathbf{x}_{particle}(t)/dt = \mathbf{u}_{particle}(t) \tag{2}$$

Here, $\mathbf{u}_{particle}$ and $\mathbf{x}_{particle}$ denote the velocity and position of the particle, $\mathbf{u}_{fluid}$ denotes the velocity of the carrier fluids, and $\mathbf{u}_{propel}$ denotes the velocity generated by the particle. Without control, the velocity of the particle is equivalent to that of the background flows because the particles are inertialess. However, with control, the velocity of the particle is the result of the background flow and the actively controlled propulsion velocity. We restricted the maximum propelling velocity of the particles to simulate the constraints of real-world scenarios. Specifically, it was limited to approximately one-third of the highest carrier flow velocity in the present work. Similar dynamics of the particles have been previously adopted by Krishna et al. [19, 20], Monthiller et al. [21] and Gunnarson et al. [22]. Alternatively, more complex kinematic model for the self-propelling particles, which incorporates inertial [23], rotational dynamics [24, 25], or multimodal locomotion [26], can be considered in future works.

2.2 Control Policy for the Self-propelling Particle

We employ reinforcement learning (RL) to optimize the control policy of the particle's trajectory to achieve energy-efficient migration. Compared with traditional control methods, reinforcement learning is advantageous in its agnosticism towards control and optimization tasks, and its robustness towards disturbances in chaotic systems [27–30]. In a typical RL framework, an agent observes the state of the external environment and subsequently selects an action to interact with the environment. Following the action, the particle receives a reward (or penalty), which serves to reinforce (or inhibit) the likelihood of the particle repeating (or forgoing) the action in future iterations. Overall, the goal of the particle is to maximize its cumulative reward [31].

The RL framework includes three critical components. The first component is the state, which describes the external environment. The second component is the action, which represents the decision of the particle. The set of all valid actions in a given environment is known as the action space. The third component is the reward, which evaluates the particle's action based on current state and next state. Followed by our previous works [17, 18], the state of the self-propelling particle includes the background flow velocity $\mathbf{u}_{fluid}$, the particle's spatial coordinate in the vertical direction y_{agent}, and the local temperature T. The action of the particle is the generating propelling velocity $\mathbf{u}_{propel}$. Here, we choose a local state and a laboratory frame for actions, while some other works reported both local states and local actions [32], or partially observed states [33].

The reward function is designed as:

$$r(t) = r_s(t) + r_e(t) + r_h(t) \tag{3}$$

where $r_s(t)$ denotes the reward related to the current state, $r_e(t)$ denotes the reward related to energy consumption, $r_h(t)$ denotes the reward related to time consumption. Specifically, they are given as:

$$r_s(t) = \begin{cases} -\phi, & \text{particle is out of the flow domain} \\ \varepsilon e_{basic}, & x^t_{particle} > x^{t-1}_{particle} \\ 0, & \text{otherwise} \end{cases} \tag{4}$$

$$r_e(t) = \begin{cases} \varepsilon[e_{basic} + (e_{max} - e)], & 0^\circ \leq \theta \leq 90^\circ \\ -2\varepsilon(e_{basic} + e), & 90^\circ \leq \theta \leq 180^\circ \end{cases} \tag{5}$$

$$r_h(t) = \begin{cases} (t_{max} - t)/\varepsilon, \; x^t_{particle} > L \\ 0, \qquad \text{otherwise} \end{cases} \tag{6}$$

Here, ϕ is the penalty factor, ε is a scale prefactor. e is the energy consumption of the particle, and $e_{\text{basic}} = e_{max} = 0.5\left(\left\|\mathbf{u}_{propel}\right\|\right)^2_{max}$. t is the simulation timestep and t_{max} is the time limit. In this work, we choose optimal hyperparameters of $\phi = 10$ and $\varepsilon = 10$ [18].

Model-free RL algorithms can be classified into two categories: policy optimization and Q-learning. Policy optimization involves the explicit representation and optimization of a policy, while Q-learning employs an approximator, typically in the form of neural networks, to learn the optimal policy. Several studies have explored the application of Q-learning in the control of self-propelling particles. For instance, Colabrese et al. [34] utilized Q-learning to demonstrate that gravitactic swimmers are capable of ascending higher in the Taylor-Green vortex flow. In another study, Muiños-Landin et al. [35] showed that artificial self-thermophoretic microswimmers can navigate effectively while under the influence of Brownian motion. Qiu et.al. [36] investigated the gyrotactic stabilities of microswimmers. Monderkamp et al. [37] trained active Brownian particles to traverse complex motility landscapes, while Gazzola et al. [38] and Verma et al. [39] developed optimal swimming strategies that minimize drag and energy consumption for the schooling fish. Tsang et al. [40] reported learning to swim at low Reynolds number flows, and Gustavsson et al. [41] reported swimming strategies in chaotic flows.

The Q-learning-based algorithm for self-propelling particles often adopts a discrete action space, where only a finite number of moves are available for the particle. Compared with the commonly used Q-learning method, the Soft Actor-Critic (SAC) algorithm allows continuous action spaces. Specifically, the SAC algorithm is an interpolation between policy optimization and Q-learning. The algorithm's objectives are to maximize the expected reward (i.e., successfully perform the task) and to maximize entropy (i.e., act as randomly as possible). In entropy-regularized reinforcement learning, the particle gets a bonus reward at each time step proportional to the entropy of the policy at that time step. This changes the RL problem to:

$$\pi^* = \arg \underset{\pi}{max}\, E_{\tau\sim\pi}\left[\sum\nolimits_{t=0}^{\infty} \gamma^t (R(s_t, a_t, s_{t+1}) + \alpha H(\pi(\cdot|s_t)))\right] \tag{7}$$

Here, π^* is the optimal policy. $R(s_t, a_t, s_{t+1})$ is the reward at timestep t concerning state s_t, action a_t, and the next state s_{t+1}. $H(\pi(\cdot|s_t))$ is the entropy. γ is the discount factor. α is the trade-off coefficient. The entropy H of τ is computed from its distribution π as $H(\pi(\cdot|s_t)) = E_{\tau\sim\pi}[-\log(\pi(\tau))]$. More details on the SAC algorithm can be found by Haarnoja et al. [42].

2.3 Simulation of the Turbulent Rayleigh–Bénard Convection

Based on Oberbeck–Boussinesq approximation, we simulate the incompressible turbulent environment in the RB convection cells. The temperature is considered as an active scalar, whose impact on the velocity field is conveyed through the buoyancy term. We neglected the effects of viscous heat dissipation and compression work while regarding all the transport coefficients as constants. Under these conditions, the governing equations for the RB thermal convection can be written as:

$$\nabla \cdot \mathbf{u} = 0 \tag{8}$$

$$\frac{\partial \mathbf{u}}{\partial t} + \mathbf{u} \cdot \nabla \mathbf{u} = -\frac{1}{\rho_0} \nabla P + \nu \nabla^2 \mathbf{u} + g\beta_T (T - T_0)\hat{\mathbf{y}} \tag{9}$$

$$\frac{\partial T}{\partial t} + \mathbf{u} \cdot \nabla T = \alpha_T \nabla^2 T \tag{10}$$

where $\mathbf{u} = (u, v)$, P and T are the velocity, pressure, and temperature of the fluid, respectively. ρ_0 and T_0 are reference density and temperature, respectively. $\hat{\mathbf{y}}$ is the unit parallel to gravity. With the scaling:

$$\begin{gathered} \mathbf{x}^* = \mathbf{x}/H, t^* = t/\sqrt{H/(\beta_T g \Delta_T)}, \mathbf{u}^* = \mathbf{u}/\sqrt{\beta_T g H \Delta_T}, \\ P^* \& = P/(\rho_0 g \beta_T \Delta_T H), T^* = (T - T_0)/\Delta_T \end{gathered} \tag{11}$$

Then, Eqs. (1)–(3) can be rewritten in dimensionless form as:

$$\nabla \cdot \mathbf{u}^* = 0 \tag{12}$$

$$\frac{\partial \mathbf{u}^*}{\partial t^*} + \mathbf{u}^* \cdot \nabla \mathbf{u}^* = -\nabla P^* + \sqrt{\frac{\mathrm{Pr}}{\mathrm{Ra}}} \nabla^2 \mathbf{u}^* + T^* \tilde{\mathbf{y}} \tag{13}$$

$$\frac{\partial T^*}{\partial t^*} + \mathbf{u}^* \cdot \nabla T^* = \sqrt{\frac{1}{\text{PrRa}}} \nabla^2 T^* \tag{14}$$

Here, H is the cell height and it is chosen as the characteristic length; $t_f = \sqrt{H/(\beta_T g \Delta_T)}$ is the free-fall time and it is chosen as the characteristic time. Δ_T is the temperature difference between heating and cooling walls. The two dimensionless parameters are the Ra and the Pr, which are defined as:

$$Ra = \frac{g\beta_T \Delta_T H^3}{\nu\alpha_T}, \; Pr = \frac{\nu}{\alpha_T} \tag{15}$$

In this study, we adopt the lattice Boltzmann (LB) method for simulating thermal convection. This method offers several advantages, such as easy implementation and high computational efficiency [43]. Further information about the numerical details of the LB method and the validation of our in-house solver can be found in our previous works [44–46]. We examine a two-dimensional Rayleigh-Bénard (RB) cell with a length of L and a height of H. The top and bottom horizontal walls of the cell are kept at constant cold and hot temperatures, respectively, and they are imposed with no-slip velocity boundary conditions throughout both the training and testing phases. During the training phase, the two vertical walls are also subject to no-slip velocity boundary conditions; while during the testing phase, periodic vertical boundary conditions are imposed. We fix the Prandtl number at $Pr = 0.71$, which corresponds to air, and the Rayleigh number at $Ra = 10^8$.

3 Results and Discussion

3.1 General Flow Patterns in the RB Convection

In Fig. 1, we show the typical snapshots of the temperature field and velocity field at $Ra = 10^8$, $Pr = 0.71$, and $\Gamma = 2$ for convection cells with vertical periodic boundary conditions. We can distinguish small-scale thermal plumes and LSC rolls. Thin thermal boundary layers appear near the bottom heating wall and the top cooling wall. Plumes that are released from the boundary layers penetrate upwards (or downwards) toward the opposite wall of lower (or higher) temperature, and intense mixing occurs in the central region.

Figure 2 illustrates the differences between the training and testing cells. In the training cell, the two vertical walls of the cell are subject to no-slip boundary conditions. The plumes emitted from the hot and cool boundary layers remain at fixed horizontal positions, approximately at the central and marginal regions of the cell respectively. Two LSC rolls are observed, which almost maintain a fixed position throughout the simulation, as shown in Figs. 2(a–c). In contrast, the testing cell has periodic vertical boundary conditions, and the horizontal position of the plumes varies over time, while the LSC undergoes horizontal oscillations with greater fluctuations. Figures 2(d–f) shows an example of the LSC oscillation over time. The different flow characteristics in these environments present a challenge to the testing of trained smart self-propelling particles.

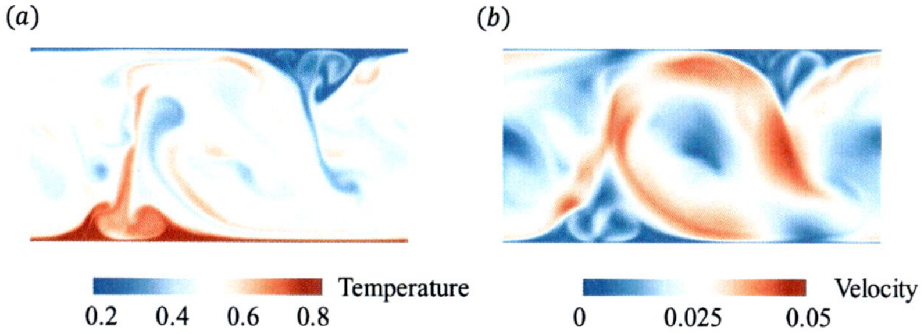

Fig. 1. Typical instantaneous (a) temperature field and (b) velocity field at $Ra = 10^8$, $Pr = 0.71$, and $\Gamma = 2$.

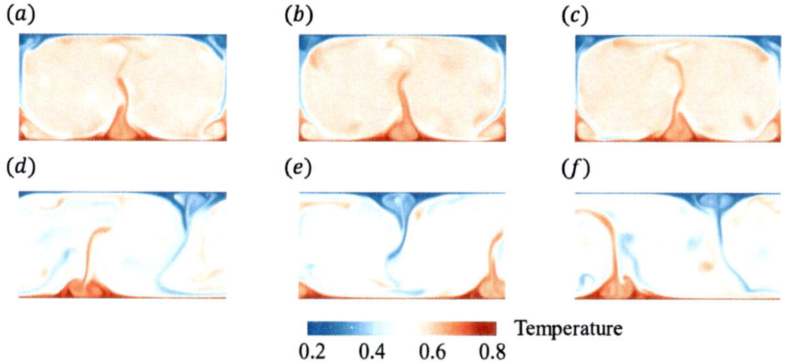

Fig. 2. Instantaneous temperature snapshot at (a) $t = 0$, (b) $t = 500$, (c) $t = 1000$ in the training cell and (d) $t = 0$, (e) $t = 500$, (f) $t = 1000$ in the testing cell.

3.2 Training and Testing of the Self-propelling Particle Migrating in the RB Cell

We train the particle to migrate through the turbulent RB cell with no-slip vertical boundary conditions. The particle is released at the bottom-left corner of the cell, namely, the initial position is at $(x, y) = (0, 0)$. Its goal is to reach the right-side boundary of the cell ($x = 2$) with limited time and minimal energy consumption. In the $\Gamma = 2$ cell, the large-scale coherent structure consists of a clockwise rotating primary roll and an anticlockwise rotating primary roll, which makes it a suitable environment for the learning particle to train in. For further information about the training results, we refer to our previous work [18].

Following the training phase, we then test whether the smart particle can migrate efficiently in convection cells with no-slip vertical boundary conditions. In the tests, the particles are released from a fixed point (0, 0) at different initial times. Figure 3(a) shows 100 example trajectories of these evaluations. Each trajectory starts by moving upward, driven by the clockwise corner roll located at the bottom-left of the cell. The particle then proceeds to travel along the counterclockwise LSC until it reaches the hot plumes, which correspond to the rising thermals. The particle utilizes these thermals to ascend to higher altitudes. Once the particle is close to the top layer of the cell, it propels towards the right direction and utilizes the horizontal currents to complete the migration task, until reaching the right-side boundary of the cell. Remarkably, all these trajectories overlap with each other, irrespective of the randomly assigned release times. This result indicates the training environment has fewer fluctuations and oscillations of LSC. The results also suggest the convergence and suitability of the trained policy in the training environment.

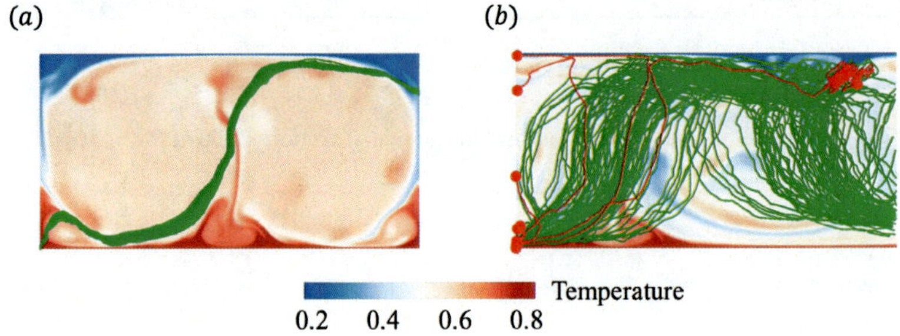

Fig. 3. Example trajectories in the (a) training cell with vertical no-slip boundary conditions and (b) testing cell with vertical periodic boundary conditions. Green lines represent the particle trajectory of successful attempts to complete the migration, while red lines represent unsuccessful attempts. Red dots indicate the termination point of unsuccessful attempts. The background contour shows the typical instantaneous temperature field.

We also plotted 100 example trajectories during testing phases in convection cells with periodic vertical boundary conditions, as shown in Fig. 3(b). The particles are released from the same start point at different initial times. For the successful attempts in each cell, the dispersed trajectories indicate the oscillation of the LSC over time. The failed attempts can be attributed to three reasons. The first reason is that particles exceed the left vertical boundary. This is due to the maximum propulsion velocity of the particles being limited to one-third of the maximum velocity in the training cell, and the maximum velocity occurring at the outer ring of the LSC rolls. When the particles enter the LSC rolls, the velocity of the background fluids dominates the total velocity of the particles. The circulation rolls are counterclockwise, particles sometimes fail to turn in the correct direction and are carried out by the rolls. The second reason is that particles exceed the upper or lower boundaries, mainly due to plumes carrying out the particles. The fluid velocity in the plume region near the LSC rolls is as large as the outer ring of the LSC rolls. Our experiments showed that this occurred only when hot plumes drift the particles from the upper bound. The third reason is that particles exceed the maximum simulation time. In this situation, the particles are captured by the LSC rolls or the corner rolls without being carried out. Some particles are also captured by the flow but then escape from the rolls and continue to finish the task within the time limit.

Next, we compare the energy consumed by smart particles and naive particles in the convection cell. We assume that the propelling velocity, $\mathbf{u}_{propel}$, is the only factor that affects the energy consumption of the particles. Therefore, we calculate the energy consumption as follows:

$$E_{propel}(t) = \int_t^0 \frac{1}{2} \left\| \mathbf{u}_{agent} \right\|^2 d\tau \tag{16}$$

In Fig. 4, we plot the average accumulative $E_{propel}(T)$ for the particles when it completes the migration task. Here, T is the time consumption of the particles. The green bar represents the smart policy, while the blue bar represents the naive policy. In this work, we define a naive particle that moves straight from the origin to the destination. We ensure that the naive particle took the same amount of time to complete the migration task as the smart particle and that its destination point is the same as the point where the smart particle left the right-side boundary. In the training environment, the smart policy achieved optimal performance, while in the testing environment, the average energy consumption of the smart policy was slightly higher than that in the training environment. Overall, we can observe that the smart policy has the potential to save energy.

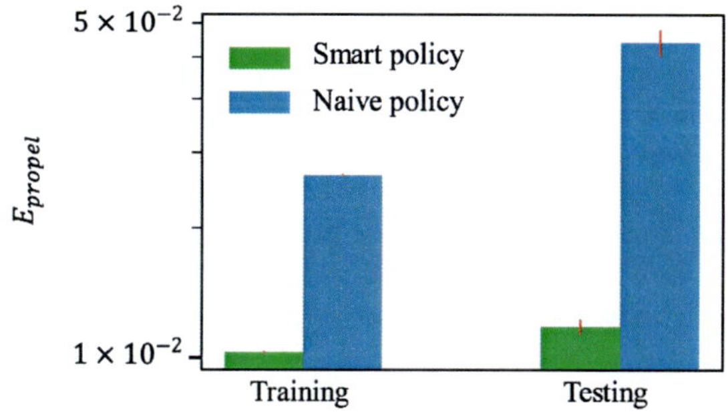

Fig. 4. The average propelling energy consumption of the smart particle (green bars) and the naive particle (blue bars) in the training cell and the testing cell, respectively. The error bars represent a 99% confidence interval over successful attempts.

To further test the robustness of the smart policy, we conducted more experiments in testing environments. In these experiments, particles were released from a random initial point with the restriction of $x < \Gamma/2$, which prevented the particles from being released too close to the outlet. The success rate of the training environment was 94%, with failures caused either by the center hot plume or being captured by the upper-left corner roll. However, in the test case, we observed the upper-left region of the cell contains more unsuccessful attempts. The particles released in this region are more likely to be carried out from the left boundary. This failure may have been caused by the training scheme, which rewarded the particles for avoiding this region, leading to the particles being inexperienced in this region and subsequently being carried out (Fig. 5).

To ensure a fair comparison of energy consumption across different random initial conditions, we adopt the energy consumption rate E_r, which is the energy consumed by particles divided by the horizontal displacement when the particles reach the right outlet, namely:

$$E_r = \frac{E_{propel}(T)}{\Gamma - x_{init}} \tag{17}$$

Here, T is the time consumption of the particle, $\mathbf{x}_{init}$ is the initial position of the x-coordinate of the particle. E_r describes the impact of varying initial conditions on energy

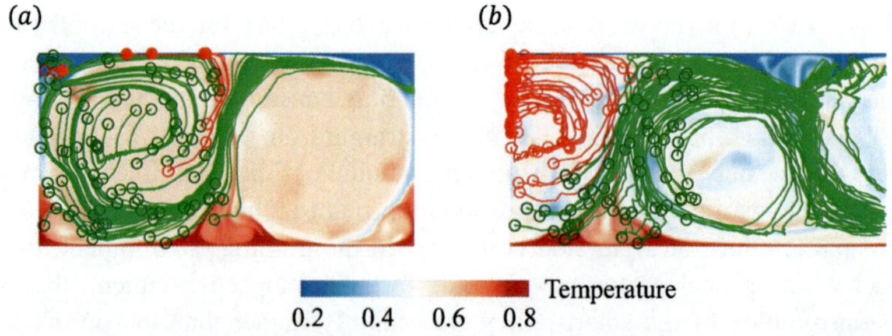

Fig. 5. Example trajectories with the random initial point in the (a) training cell with vertical no-slip boundary conditions and (b) testing cell with vertical periodic boundary conditions. Green lines represent the trajectory of successful attempts to complete the migration, while red lines represent unsuccessful attempts. Green and red rings indicate the initial points of each successful and unsuccessful attempt. Red dots indicate the terminal of unsuccessful attempts. The background contour shows the typical instantaneous temperature field.

consumption. From Fig. 6, we can see that in the training and testing environments, the average E_r is similar for both smart and naive policies. This suggests that random initial positions cause the particles to disregard the oscillation of the LSCs, thereby treating these environments almost equally. In contrast, the naive policy shows a significant difference in the average E_r in both the training and testing environments. This is because the smart policy can ignore the oscillation of the LSCs, whereas the naive policy fails to utilize the flow structures. The results indicate that in each training environment, the smart particles can save approximately two-thirds of the energy compared to the naive particles; while in the testing environment, the smart particles can save approximately 80% of the energy compared to the naive particles.

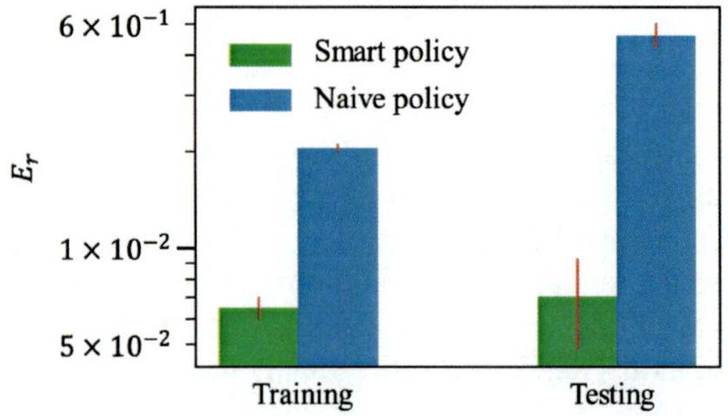

Fig. 6. The energy consumption rate E_r of the smart particle and the naive particle in the training cell and the testing cell, respectively. The error bars represent a 99% confidence interval over successful attempts.

4 Conclusion

In this work, we adopt the reinforcement learning algorithm to control a self-propelling particle migrating in a thermal turbulent environment. We choose the paradigmatic RB convection as the flow environment, which exhibits significant velocity and temperature fluctuations. The convection cell has no-slip vertical boundary conditions during the training phase and is changed to periodic vertical boundary conditions during the testing phase. We designed a reward function that incorporates the current state, energy consumption, and time consumption for the particles. The environment cues of the particle are chosen as the position as well as the velocity and temperature of the carrier flow. The training results show that the particles can learn to utilize the carrier flow currents to reduce energy consumption. We then apply the policy obtained from the training to the testing environments. These LSCs in the testing environments are suffering horizontal oscillation and larger fluctuations due to the horizontally periodic boundary conditions. Although these complex flow features bring challenges to identifying the trajectories for the smart particles, they can still obtain energy-efficient migrating trajectories. Compared to naive particles, our trained smart particles save propelling energy significantly.

Acknowledgments. This work was supported by the National Natural Science Foundation of China (NSFC) through Grants No. 12272311 and No. 12125204, and the 111 project of China (Project No. B17037). The authors acknowledge the Beijing Beilong Super Cloud Computing Co., Ltd for providing HPC resources that have contributed to the research results reported within this paper.

References

1. Williams, H.J., Shepard, E.L.C., Holton, M.D., Alarcón, P.A.E., Wilson, R.P., Lambertucci, S.A.: Physical limits of flight performance in the heaviest soaring bird. Proc. Natl. Acad. Sci. **117**(30), 17884–17890 (2020)
2. Ákos, Z., Nagy, M., Vicsek, T.: Comparing bird and human soaring strategies. Proc. Natl. Acad. Sci. **105**(11), 4139–4143 (2008)
3. Allen, M., Lin, V.: Guidance and control of an autonomous soaring vehicle with flight test results. In: 45th AIAA Aerospace Sciences Meeting and Exhibit, p. 867, January 2007
4. Lawrance, N.R.: Autonomous soaring flight for unmanned aerial vehicles (Doctoral dissertation) (2011)
5. Allen, M.: Updraft model for development of autonomous soaring uninhabited air vehicles. In: 44th AIAA Aerospace Sciences Meeting and Exhibit, p. 1510, January 2006
6. Lawrance, N., Sukkarieh, S.: Wind energy based path planning for a small gliding unmanned aerial vehicle. In: AIAA Guidance, Navigation, and Control Conference, p. 6112, August 2009
7. Ákos, Z., Nagy, M., Leven, S., Vicsek, T.: Thermal soaring flight of birds and unmanned aerial vehicles. Bioinspir. Biomim. **5**(4), 045003 (2010)
8. Laurent, K.M., et al.: Turbulence explains the accelerations of an eagle in natural flight. Proc. Natl. Acad. Sci. **118**(23), e2102588118 (2021)
9. Xia, K.Q.: Current trends and future directions in turbulent thermal convection. Theor. Appl. Mech. Lett. **3**(5), 052001 (2013)

10. Chillà, F., Schumacher, J.: New perspectives in turbulent Rayleigh-Bénard convection. Eur. Phys. J. E **35**, 1–25 (2012)
11. Xia, K.Q., Huang, S.D., Xie, Y.C., Zhang, L.: Tuning heat transport via coherent structure manipulation: recent advances in thermal turbulence. Natl. Sci. Rev. (2023)
12. Atkinson, B.W., Wu Zhang, J.: Mesoscale shallow convection in the atmosphere. Rev. Geophys. **34**(4), 403–431 (1996)
13. Stevens, R.J., Blass, A., Zhu, X., Verzicco, R., Lohse, D.: Turbulent thermal superstructures in Rayleigh-Bénard convection. Phys. Rev. Fluids **3**(4), 041501 (2018)
14. Pandey, A., Scheel, J.D., Schumacher, J.: Turbulent superstructures in Rayleigh-Bénard convection. Nat. Commun. **9**(1), 2118 (2018)
15. Reddy, G., Celani, A., Sejnowski, T.J., Vergassola, M.: Learning to soar in turbulent environments. Proc. Natl. Acad. Sci. **113**(33), E4877–E4884 (2016)
16. Reddy, G., Wong-Ng, J., Celani, A., Sejnowski, T.J., Vergassola, M.: Glider soaring via reinforcement learning in the field. Nature **562**(7726), 236–239 (2018)
17. Xu, A., Wu, H.L., Xi, H.D.: Migration of self-propelling particle in a turbulent environment with minimal energy consumption. Phys. Fluids **34**(3), 035117 (2022)
18. Xu, A., Wu, H.L., Xi, H.D.: Long-distance migration with minimal energy consumption in a thermal turbulent environment. Phys. Rev. Fluids **8**(2), 023502 (2023)
19. Krishna, K., Song, Z., Brunton, S.L.: Finite-horizon, energy-efficient trajectories in unsteady flows. Proc. Royal Soc. A **478**(2258), 20210255 (2022)
20. Krishna, K., Brunton, S.L., Song, Z.: Finite time Lyapunov exponent analysis of model predictive control and reinforcement learning. IEEE Access **11**, 118916–118930 (2023). https://doi.org/10.1109/ACCESS.2023.3326424
21. Monthiller, R., Loisy, A., Koehl, M.A., Favier, B., Eloy, C.: Surfing on turbulence: a strategy for planktonic navigation. Phys. Rev. Lett. **129**(6), 064502 (2022)
22. Gunnarson, P., Mandralis, I., Novati, G., Koumoutsakos, P., Dabiri, J.O.: Learning efficient navigation in vortical flow fields. Nat. Commun. **12**(1), 7143 (2021)
23. Colabrese, S., Gustavsson, K., Celani, A., Biferale, L.: Smart inertial particles. Phys. Rev. Fluids **3**(8), 084301 (2018)
24. Alageshan, J.K., Verma, A.K., Bec, J., Pandit, R.: Machine learning strategies for path-planning microswimmers in turbulent flows. Phys. Rev. E **101**(4), 043110 (2020)
25. Borra, F., Biferale, L., Cencini, M., Celani, A.: Reinforcement learning for pursuit and evasion of microswimmers at low Reynolds number. Phys. Rev. Fluids **7**(2), 023103 (2022)
26. Zou, Z., Liu, Y., Young, Y.N., Pak, O.S., Tsang, A.C.: Gait switching and targeted navigation of microswimmers via deep reinforcement learning. Commun. Phys. **5**(1), 158 (2022)
27. Biferale, L., Bonaccorso, F., Buzzicotti, M., Clark Di Leoni, P., Gustavsson, K.: Zermelo's problem: optimal point-to-point navigation in 2D turbulent flows using reinforcement learning. Chaos Interdisc. J. Nonlinear Sci. **29**(10), 103138 (2019)
28. Garnier, P., Viquerat, J., Rabault, J., Larcher, A., Kuhnle, A., Hachem, E.: A review on deep reinforcement learning for fluid mechanics. Comput. Fluids **225**, 104973 (2021)
29. Fang, W.Z., Xiong, T., Pak, O.S., Zhu, L.: Data-driven intelligent manipulation of particles in microfluidics. Adv. Sci. **10**(5), 2205382 (2023)
30. Brunton, S.L., Noack, B.R., Koumoutsakos, P.: Machine learning for fluid mechanics. Annu. Rev. Fluid Mech. **52**, 477–508 (2020)
31. Sutton, R.S., Barto, A.G.: Reinforcement Learning: An Introduction. MIT Press, Cambridge (2018)
32. Qiu, J., Mousavi, N., Gustavsson, K., Xu, C., Mehlig, B., Zhao, L.: Navigation of microswimmers in steady flow: the importance of symmetries. J. Fluid Mech. **932**, A10 (2022)
33. Kubo, A., Shimizu, M.: Efficient reinforcement learning with partial observables for fluid flow control. Phys. Rev. E **105**(6), 065101 (2022)

34. Colabrese, S., Gustavsson, K., Celani, A., Biferale, L.: Flow navigation by smart microswimmers via reinforcement learning. Phys. Rev. Lett. **118**(15), 158004 (2017)
35. Muiños-Landin, S., Fischer, A., Holubec, V., Cichos, F.: Reinforcement learning with artificial microswimmers. Sci. Robot. **6**(52), eabd9285 (2021)
36. Qiu, J., Mousavi, N., Zhao, L., Gustavsson, K.: Active gyrotactic stability of microswimmers using hydromechanical signals. Phys. Rev. Fluids **7**(1), 014311 (2022)
37. Monderkamp, P.A., Schwarzendahl, F.J., Klatt, M.A., Löwen, H.: Active particles using reinforcement learning to navigate in complex motility landscapes. Mach. Learn. Sci. Technol. **3**(4), 045024 (2022)
38. Gazzola, M., Tchieu, A.A., Alexeev, D., de Brauer, A., Koumoutsakos, P.: Learning to school in the presence of hydrodynamic interactions. J. Fluid Mech. **789**, 726–749 (2016)
39. Verma, S., Novati, G., Koumoutsakos, P.: Efficient collective swimming by harnessing vortices through deep reinforcement learning. Proc. Natl. Acad. Sci. **115**(23), 5849–5854 (2018)
40. Tsang, A.C.H., Tong, P.W., Nallan, S., Pak, O.S.: Self-learning how to swim at low Reynolds number. Phys. Rev. Fluids **5**(7), 074101 (2020)
41. Gustavsson, K., Biferale, L., Celani, A., Colabrese, S.: Finding efficient swimming strategies in a three-dimensional chaotic flow by reinforcement learning. Eur. Phys. J. E **40**, 1–6 (2017)
42. Haarnoja, T., Zhou, A., Abbeel, P., Levine, S.: Soft actor-critic: off-policy maximum entropy deep reinforcement learning with a stochastic actor. In: International Conference on Machine Learning, pp. 1861–1870. PMLR, July 2018
43. Xu, A., Shyy, W., Zhao, T.: Lattice Boltzmann modeling of transport phenomena in fuel cells and flow batteries. Acta. Mech. Sin. **33**, 555–574 (2017)
44. Xu, A., Shi, L., Zhao, T.S.: Accelerated lattice Boltzmann simulation using GPU and OpenACC with data management. Int. J. Heat Mass Transf. **109**, 577–588 (2017)
45. Xu, A., Shi, L., Xi, H.D.: Lattice Boltzmann simulations of three-dimensional thermal convective flows at high Rayleigh number. Int. J. Heat Mass Transf. **140**, 359–370 (2019)
46. Xu, A., Li, B.T.: Multi-GPU thermal lattice Boltzmann simulations using OpenACC and MPI. Int. J. Heat Mass Transf. **201**, 123649 (2023)

Volume-Fraction Distribution of Moderately Dense Particles in Re-Shocked Richtmyer-Meshkov Flow

Qi Wu[1], Baoqing Meng[3(✉)], Yousheng Zhang[2], Yipeng Shi[1(✉)], and Baolin Tian[2]

[1] Peking University, Beijing 100871, People's Republic of China
qiwu@stu.pku.edu.cn, syp@mech.pku.edu.cn

[2] Institute of Applied Physics and Computational Mathematics, Beijing 100094, People's Republic of China
{zhang_yousheng,tian_baolin}@iapcm.ac.cn

[3] Institute of Mechanics, Chinese Academy of Sciences, Beijing 100190, People's Republic of China
mengbaoqing92@foxmail.com

Abstract. Re-shocked Richtmyer-Meshkov flow laden with particles is extensively observed in supernova explosions and laser-driven inertial confinement fusion experiments. Studies on the dispersal of particles are fundamental in this topic, and the most intuitive quantity used to denote dispersal is the volume-fraction distribution. Previous studies on dilute particles have shown that the volume-fraction distribution deviates from the initially uniform distribution after re-shock. However, the dispersal of moderately dense particles remains an open question. To address this problem, we conducted three-dimensional simulations using the well-verified compressible multiphase particle-in-cell (CMP-PIC) method. We initially distributed the particles uniformly in a planar shock tube with an Air/SF_6 interface, setting different initial volume fractions for the particles, and compared the one-dimensional profile of particle velocity and volume fractions. Our findings reveal that the peak of the velocity profile is consistently located at the interface, and its magnitude decreases slightly as particle becomes denser. Moreover, the volume fraction distribution reveals an intriguing phenomenon in which particles tend to concentrate near the reflection wall after re-shock, with denser particles exhibiting a stronger tendency towards clustering. This study enhances our understanding of the following behaviors of particles under the coupled effect of interface instability and shock wave propagation.

Keywords: particle/fluid flow · shock waves

1 Introduction

The Richtmyer-Meshkov instability (RMI) is a complex multi-phase flow phenomenon that occurs when a shock wave interacts with a perturbed interface between two fluids of different densities. This instability leads to the growth of complex structures at the interface, resulting in the mixing of the two fluids. In

X. Zheng and S. Balachandar (Eds.): IUTAM 2023, IUTAM Bookseries 41, pp. 326–339, 2024.
https://doi.org/10.1007/978-3-031-47258-9_21

recent years, much attention has been focused on the study of the re-shocked RMI, which has been extensively studied due to its importance in various fields, including astrophysics, combustion, and inertial confinement fusion. Reshocked RMI occurs when the reflected shock wave from the end of the shock tube interacts with the already perturbed interface, leading to further amplification of the instability. This phenomenon can significantly enhance the growth rate of the RMI and produce highly turbulent mixing of the fluids.

In many practical situations, the presence of particles in the flow is inevitable. Understanding the behavior of particle-laden RM flows with re-shock is crucial in designing and optimizing various engineering systems that involve particle-fluid interactions. The movement, deposition, and concentration of particles are governed by their interactions with the flow field, as established by the following observations. A study by Ukai *et al.* [1] demonstrated that the motion of the interface induces movement of particles, indicating a strong relationship between interface motion and particle transport. In a separate study, Tian *et al.* [2] observed that moderately dense and dense particles follow the motion of the vortex structure rolled up at the RMI interface. These findings shed further light on the mechanisms underlying particle transport in RM flows.

Besides, the motion of particles in such flows is significantly influenced by the interaction with the shock waves generated during the RMI development. Numerous studies have been conducted to investigate the effects of shock waves on particles across a range of varying volume fractions, from dilute to dense. Theofanous *et al.* [3], for instance, conducted an experiment showing that the impact of shock waves on a dilute particle curtain causes its thickness to widen. Ouellet *et al.* [4] simulated the explosion experiment and found that the volume fraction distribution of moderately dense particles displayed a certain degree of non-uniformity following the explosion. DeMauro *et al.*'s experiment [5] focused on shock waves impacting dense particles with significant collision effects. They found that the deformation of the particle curtain was caused by a mismatch between upstream and downstream gas-phase velocities and pressures acting on the curtain. These findings provide valuable insights into the behavior of particles with varying volume fractions under shock wave impact in RM flows.

Additionally, the presence of particles also affects the properties of the flow, leading to changes in the dynamic behavior of particles due to the force exerted by the fluid on them. The theoretical work of Ukai *et al.* [1] has shown that dilute particles can affect the growth rate of the mixing width after the shock wave hits the interface. Building on this work, Meng *et al.* [6] have extended the theory to apply to the case of dense particles and found that particles with larger volume fractions have a strong inhibitory effect on the growth rate. Schulz *et al.* [7] have shown that dilute particles, with variations in Stokes numbers and initial volume fractions, have different effects on the growth rate following re-shock. Regarding the impact of particles on shock waves, Meng *et al.* [6] reported on the reflection of shock waves caused by dense particle layers. These findings highlight the significant impact that particles have on flow dynamics, particularly the growth of instability and configuration of shock waves.

Based on the studies mentioned above, it can be concluded that particles exhibit complex behaviors as a result of several factors, including the following behavior with the flow, the presence of shock waves, and changes in the flow configuration induced by particle effects. In recent studies, there has been a growing focus on the particle distribution in the re-shocked RM flows. For example, Schulz *et al.* [7] conducted pioneering work by examining the time evolution of the ratio of particle cloud volume fraction to initial volume fraction. However, the spatial distribution of moderately dense particles remains an area of limited understanding. As such, there is a pressing need for focused research in this area, in order to fully comprehend the behavior of particles in reshocked RMI flows and gain insights into their applications in various engineering fields.

However, simulating re-shocked RM flows poses a considerable challenge due to the intrinsic complexity of multi-phase flows, which are characterized by multiple scales of motion and the interaction between various physical processes, such as shock waves, interface instability, and turbulence. The situation is further complicated with the addition of particles, leading to extensive interactions between the particles and the fluid. To tackle these challenges, one effective approach is to use a Compressible Multiphase Particle-in-Cell (CMP-PIC) method [2], which can consider the compressibility of the fluid, including the presence of shock waves. CMP-PIC can simulate the movement of individual particles in a fluid by coupling Lagrangian particles with Eulerian fluid fields. This method is particularly useful for simulating all pattern flows of gas-particle systems, ranging from dilute to dense and granular flows, and has been adequately validated through numerous experimental, theoretical, and simulation studies.

The paper is organized as follows. The particle model, numerical method, simulation setup, grid resolution are presented in Sect. 2. In Sect. 3 the results of the velocity distribution are first shown to demonstrate the motion of particles. Following this, the results of the particle dispersal representing by volume-fraction profiles are presented In Sect. 4. Last, conclusions are presented in Sect. 5.

2 Methodology

The Lagrangian tracking method is used to solve the displacement $x_{p,i}$ and velocity $u_{p,i}$ of the particle parcel in the CMP-PIC method [2]. The parcels experience drag force, pressure gradient force, and inter-parcel collision force in the gas. The governing equations are given by

$$\frac{dx_{p,i}}{dt} = u_{p,i}, \tag{1}$$

$$\rho_p \frac{du_{p,i}}{dt} = \frac{\kappa_{p,i}}{V_{pc,i}} (u_{f,i} - u_{p,i}) - \nabla P_f - \rho_p A_{p,i}. \tag{2}$$

Here, the particle phase density ρ_p is set as $2520 kg/m^3$, referring to the material of soda lime which is commonly used in recent experiments [3,5]. The first drag force term, contains following parts: (1)The relative velocity between the ith parcel and gas at the specific position of the parcel. (2)Volume of the parcel $V_{pc,i}$

which is expressed as $V_{pc} = \alpha_p V_{cell}/N_p$. Here, α_p denotes the volume fraction of the particle phase, while V_{cell} represents the volume of a fluid cell, and N_p refers to the number of parcels present in the cell. (3)The coefficient $\kappa_{p,i}$, which is based on the drag coefficient C_d defined as follows

$$\kappa_{p,i} = \frac{1}{8} C_d \pi d_p^2 \rho_f |u_{f,i} - u_{p,i}|, \tag{3}$$

$$C_d = \frac{24}{Re_p} f_{drag}, Re_p = \frac{|u_{f,i} - u_{p,i}| d_p}{\nu_f}, \tag{4}$$

where the diameter d_p of each particle is assumed to be a constant value of $230\mu m$, while ρ_f is the density of gas. The particle Reynolds number Re_p, which is determined based on the relative velocity between the gas and parcels, the dynamic viscosity of the gas ν_f, and the particle diameter d_p. The Crowe's drag model [8] is employed to calculate the drag force term, which results in the expression of f_{drag}.

$$f_{drag} = \begin{cases} f_{base} \alpha_f^{-\zeta} & \alpha_p \leq 0.2 \\ 8.33 \frac{\alpha_p}{\alpha_f} + 0.0972 Re_p & \alpha_p > 0.2, \end{cases} \tag{5}$$

$$\zeta = 3.7 - 0.65 \exp(-0.5(1.5 - log_{10} Re_p)^2), \tag{6}$$

$$f_{base} = \begin{cases} 1 + \frac{1}{6} Re_p^{0.687} & Re_p < 1000 \\ 0.0183 Re_p & Re_p > 1000. \end{cases} \tag{7}$$

When a shock wave impacts the particles, the second term - pressure gradient force ∇P_f - has a considerable effect on the acceleration of each parcel. As for the third term, it represents the collision force between particle parcels. While this term can usually be ignored in dilute particle systems, neglecting it becomes impossible when the particles are dense. To address this issue, the coarse-grained discrete element method (DEM) introduced by Apte *et al.* [9] is utilized to solve collisions between particle parcels. The expression for this approach is as follows

$$A_{p,k} = \frac{1}{m_{pc}} \sum_l F_{kl,n}, \tag{8}$$

where $m_{pc} = \rho_p V_{pc}$ is the mass of the particle parcel. This approach models the interactions between parcels as a "spring-dashpot" system. The force of interaction in the normal direction between the kth and lth parcels is expressed as

$$F_{kl,n} = k_{n,pc} \delta_n - \gamma_{n,pc} v_{r,n}. \tag{9}$$

Here, δ_n is the overlap between the parcels defined based on the effective radius R_{eff},

$$\delta_n = max(0, R_{eff,k} + R_{eff,l} - \vec{r}_{kl}), \tag{10}$$

$$R_{eff} = \sqrt[3]{\frac{3V_{pc}}{4\pi}}, \tag{11}$$

where $\vec{r}_{kl}$ is the vector pointing from the centroid of parcel k to the centroid of parcel l. Moreover, $v_{r,n}$ is the relative velocity in the normal direction between the parcels. In Eq. 9, the stiffness and damping coefficients for parcels, $k_{n,pc}$ and $\gamma_{n,pc}$, respectively, are also used in the calculation. According to Crowe *et al.* [8], if the restitution coefficient e_{pc} is given, the two parameters can be related by the following equation

$$\gamma_{n,pc} = -\frac{2\ln e_{pc}}{\sqrt{\pi^2 + \ln e_{pc}}}\sqrt{m_{pc}k_{n,pc}}. \tag{12}$$

In the simulation, we take $k_{n,pc} = 2 \times 10^7$ and $e_{pc} = 0.1$.

The CMP-PIC model features a multi-component gas phase, for which a five-equation model is utilized for the control equations [2]. This model incorporates the gas volume fraction α_f, which satisfies the constraint $\alpha_p + \alpha_f = 1$, alongside the volume fraction of component k in the gas phase $\beta_k (k = 1, 2, ...)$, the density of the mixture gas ρ_f, and the density of component k, $\rho_{f,k}$ in the gas phase. It is assumed that the velocity and pressure at the interface between gas components are uniform, with values defined as u_f and P_f, respectively. The total energy of the system, E_f, is calculated as $E_f = 0.5u_f^2 + e_f$. The internal energy, e_f, is linked to the pressure and density of the gas via the ideal gas equation of state, i.e., $P_f = \rho_f e_f(\gamma - 1)$, where γ represents the specific heat ratio of the mixture gas. The thermodynamic parameters and viscosity of the mixture gas are obtained by weighting the mass fractions of each component, $Y_k = \beta_k \rho_{f,k}/\rho_f$, and the viscosity of a single component is calculated using the Sutherland formula [10]. The gas phase experiences reactive forces from the particle parcels, mainly including drag force and pressure gradient force. The detailed equations governing the gas phase are available in the reference by Tian *et al.* [2], which readers can refer to for more information.

We set up a planar shock tube as shown in Fig. 1 with dimensions of $L_x \times L_y \times L_z = 0.5\text{m} \times 0.1\text{m} \times 0.1\text{m}$. Reflective boundary conditions are set at the end of the shock tube, located at $x = 0.5m$, while outlet boundary conditions are set at the tube's left end ($x = 0$). Periodic boundary conditions are used in the y and z directions. A shock wave with $Ma = 1.5$ impacts the Air/SF_6 interface from the position $x = 0.05\text{m}$ to the right. We define the volume fraction of Air as β_f, and the volume fraction of SF_6 is then $1 - \beta_f$. The specific heat ratio and molecular weight of Air and SF_6 are $\gamma_{Air} = 1.4, \gamma_{\text{SF}_6} = 1.093$, and $M_{Air} = 29.04\,\text{kg}\,\text{kmol}^{-1}, \text{M}_{\text{SF}_6} = 146.07\,\text{kg}\,\text{kmol}^{-1}$, respectively. The initial pressure, temperature, and velocity of the unshocked gas are $P_{f0} = 23\,\text{kPa}$, $T_{f0} = 298\,\text{K}$, $u_{f0} = 0\,\text{m}\,\text{s}^{-1}$. The physical quantities behind the shock wave are determined by the Rankine-Hugoniot (RH) relations. The interface is located at $x = 0.1\,\text{m}$, and an initial multi-mode perturbation is added to the interface. The form of the perturbation is referred to Tritschler *et al.* [11]

$$\begin{aligned} \eta(y, z) &= a_0 \sin(k_0 y) \sin(k_0 z) \\ &\quad - 0.1a_0 \sum_{n=1}^{13} \sum_{m=3}^{15} a_{n,m} \sin(k_n y + \phi_n) \sin(k_m z + \chi_m), \end{aligned} \tag{13}$$

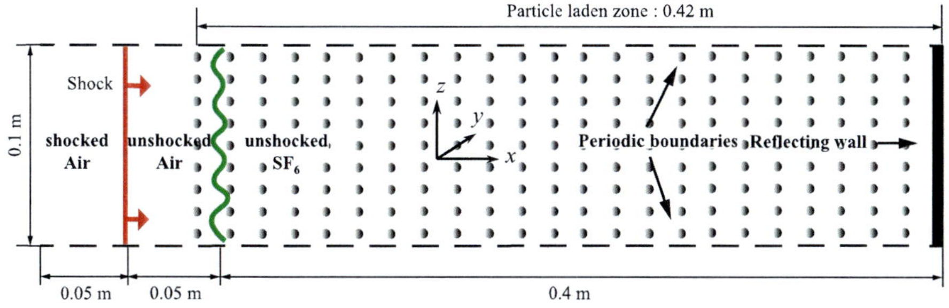

Fig. 1. Schematic of the simulation domain. For the multi-phase simulations, L_p is the length of the initial particle cloud.

where the constant amplitude is $a_0 = 0.0025\text{m}$, with wavenumbers $k_0 = 2\pi/\lambda$ and $\lambda = 0.02\text{m}$. For large wavenumbers, the amplitudes are calculated as $a_{n,m} = \sin(nm)/2$, while the phase shifts are given by $\phi_n = \tan(n)$ and $\chi_m = \tan(m)$. The pertubation is imposed on a finite thickness and then be prescribed in the gas volume fraction

$$\psi(x, y, z) = \frac{1}{2}\left(1 + \tanh\left(\frac{x-\eta(y,z)}{2\Delta_x}\right)\right), \tag{14}$$

$$\beta_f = \frac{1-\psi}{2}. \tag{15}$$

The particle parcels are initially uniformly distributed across the interface and occupy the entire length of the shock tube, which is $L_p = 0.42\text{m}$ in length as can be seen in Fig. 1. Each grid within the calculation domain is occupied by one particle parcel. We investigate the effect of five different initial particle volume fractions, namely $\alpha_{p0} = 0.2\%, 0.4\%, 0.6\%, 0.8\%$, and 1.0%; these values are displayed in Fig. 2.

A grid of 960×192^2 was utilized in combination with a fifth-order WENO scheme and Harten HLL-type solvers. The time step size was determined based on a CFL number of 0.25. To calculate inter-phase interaction forces, a cubic spline kernel function was applied to convert particle Lagrangian coordinates to gas Eulerian coordinates. For a more comprehensive understanding of the numerical methods utilized, reference can be made to the literature of Tian *et al.* [2]

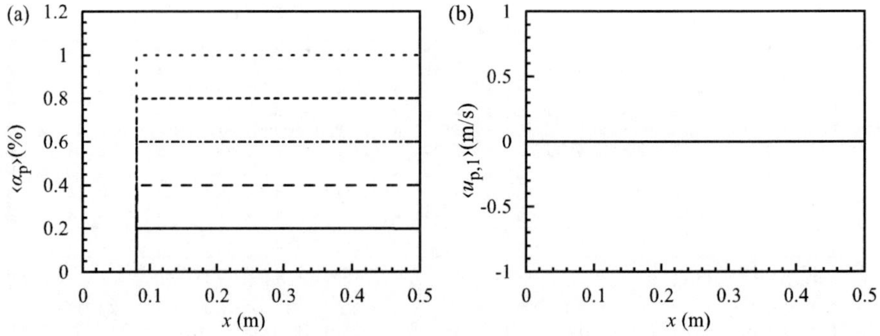

Fig. 2. (a)The initial volume fraction α_p, and (b) velocity $u_{p,1}$ distribution of particles along the flow direction, which are averaged in the $y-$ and $z-$ directions.

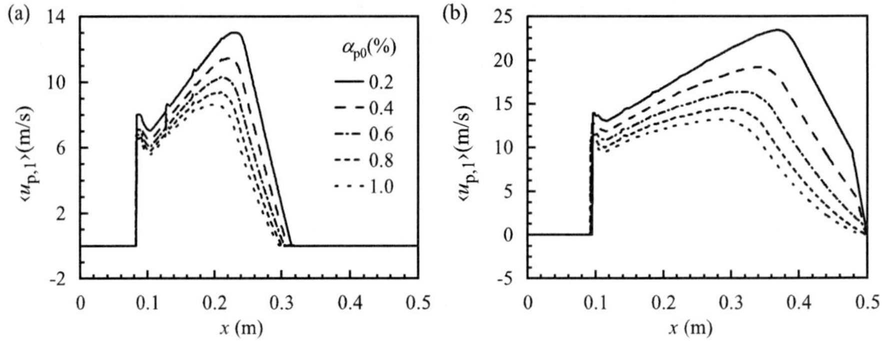

Fig. 3. The distribution of the particle velocity component $u_{p,1}$, along the flow direction measured at (a) $t = 1$ ms and (b) $t = 2.1$ ms before re-shock.

3 Velocity Distribution

Upon impact by a shockwave, the gas behind the wave gains velocity. Simultaneously, the initial small perturbations at the gas interface commence to develop and rapidly move into the nonlinear stage, resulting in a typical bubble-and-spike structure where light and heavy gases intertwine at the interface. Due to the multiple effects of shockwave impact, interface motion, and nonlinear structural evolution, the particles experience the force of the gas and also interact with each other through collision forces, causing initially stationary particles to acquire acceleration and move along the longitudinal direction. As particle motion mainly occurs in the x-direction, averaging particle velocity in the y- and z-directions will yield a one-dimensional distribution of particle velocity along the flow direction, denoted as $\langle u_{p,i}(x,t)\rangle$. As the particle velocity along the x-direction is dominant, our focus lies on comparing the spatial distribution of $\langle u_{p,1}\rangle 3$ at various moments before and after re-shock for different values of α_{p0}.

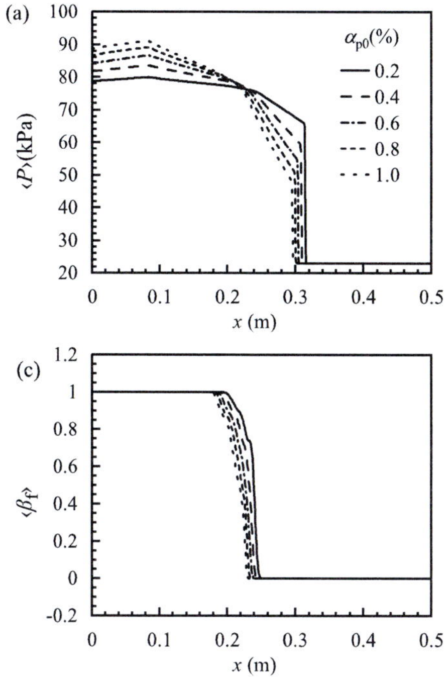

Fig. 4. The distribution of gas pressure P_f and air volume fraction β_f along the flow direction calculated by averaging the values in the $y-$ and $z-$ directions for (a) P_f at $t = 1.0$ ms, (b) P_f at $t = 2.1$ ms, (c) β_f at $t = 1.0$ ms, and (d) β_f at $t = 2.1$ ms before re-shock.

3.1 Before Re-Shock

A variation in $\langle u_{p,1} \rangle$ across the x direction, before re-shock, is presented in Fig. 3, with the velocity distribution at $t = 1.0$ ms depicted in Fig. 3(a). For all cases, the initial particle velocity experiences a gradual increase along the flow direction, followed by a rapid climb which peaks during the"rise-fall" process, eventually falling back to zero. An increase in α_{p0} results in an overall decrease in $\langle u_{p,1} \rangle$. The decline trend is more evident close to the peak position. The shift in peak value towards larger x values is an indication of delayed of particle motion with the increased values of α_{p0}. Notably, in the domain $x > 0.32$ m, $\langle u_{p,1} \rangle$ equals to zero because the shock wave has just arrived at this point and the particles are yet to be exposed to any force, causing them to remain stationary. The results presented in Fig. 3(b) correspond to a specific time instance, denoted by $t = 2.1$ ms. In comparison to Fig. 3(a), at this time, a majority of the particles spanning the L_p range have already initiated motion, except for a minor proportion of particles located near the reflected wall that remain stationary. Notably, the velocity of particle has increased overall in comparison with the earlier time instance at $t = 1.0$ ms, and the location of the peak $\langle u_{p,1} \rangle$ has shifted along the x direction.

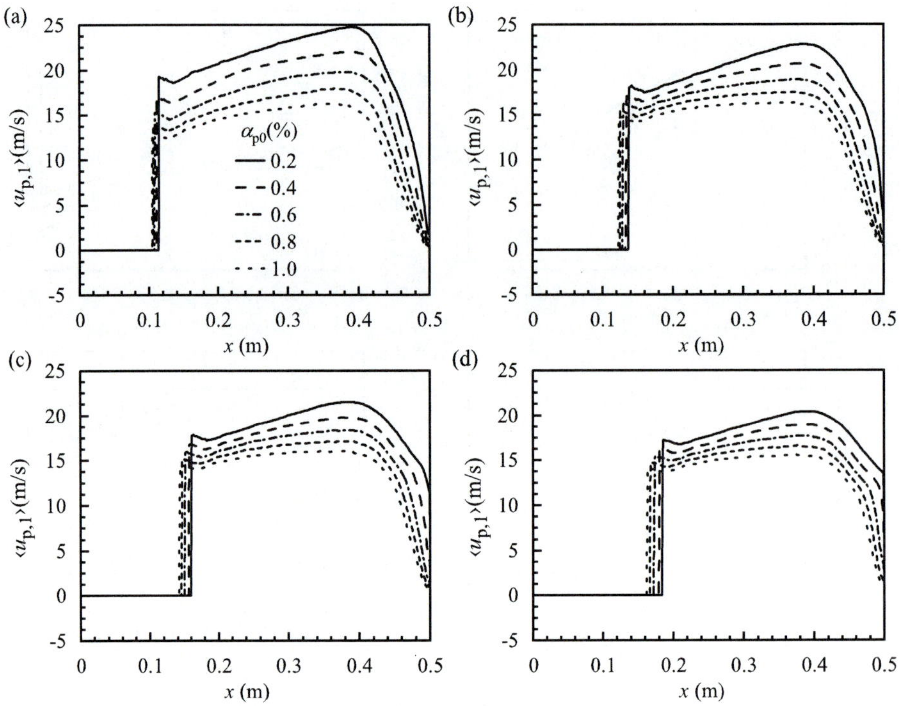

Fig. 5. The distribution of the particle velocity component $u_{p,1}$, along the flow direction measured after re-shock at (a) t = 3.2 ms,(b) t = 4.4 ms, (c) t = 5.8 ms, and (d) t = 7.2 ms after re-shock.

The particle velocity distribution before the re-shock highlights two important aspects: the range of particles that initiate motion and the location corresponding to the peak of the velocity distribution. To investigate the aforementioned issues, we compare the pressure P_f and air volum fraction β_f averaged in the y and z directions at two distinct time instants, as presented in Fig. 4. The pressure distribution along the flow direction prior to the re-shock is shown in Fig. 4(a) and (b). We observe that the locations of particles exhibiting non-zero velocities coincide with the locations of shock waves. This could be attributed to the higher velocity of shock wave propagation compared with that of the interface movement. Specifically, particles initially experience motion induced by the pressure gradient force, followed by gas drag force that accounts for their subsequent motion. Previous studies [3,6] investigated the effects of shock wave-particle interactions and showed that the intensity of transmitted shock waves attenuates when they impinge on particle layers. As demonstrated in Fig. 4(a), the region denoted by $x \in [0.24\,\text{m}, 0.3\,\text{m}]$ is located downstream of the shock wave. Figure 4(c) shows that this region is situated upstream of the particle interface. Herein, $\langle P_f \rangle$ decreased along the x direction.

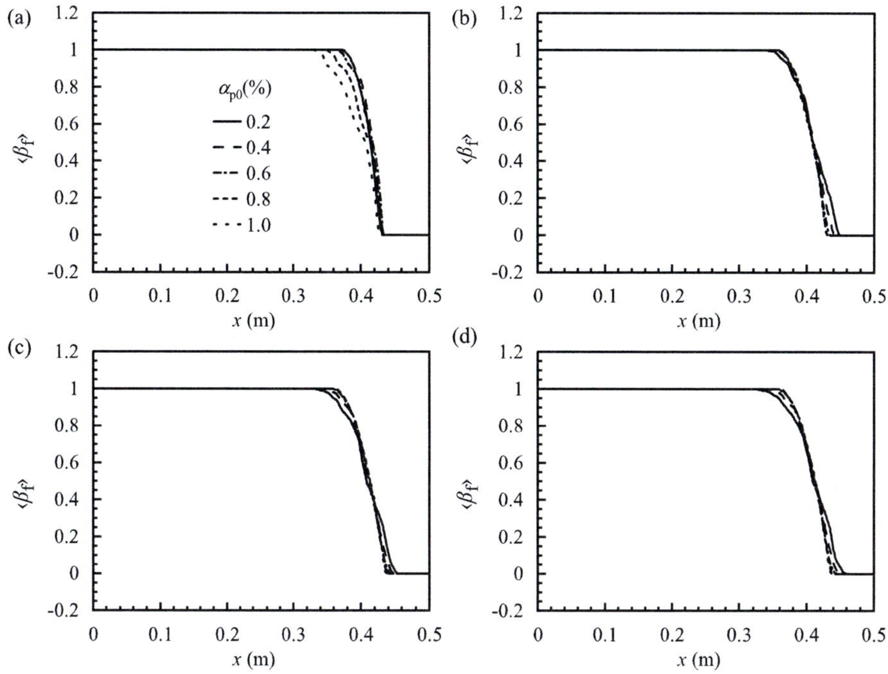

Fig. 6. The distributions of air volume fraction β_f along the flow direction, obtained by averaging their values in the $y-$ and $z-$ directions at (a) $t = 3.2$ ms,(b) $t = 4.4$ ms, (c) $t = 5.8$ ms, and (d) $t = 7.2$ ms after re-shock.

Consequently, the decrease in the force exerted by the pressure gradient on particles within this region led to reduced particle velocity along the x direction, as depicted in Fig. 3(a). Figure 4(c) and (d) exhibit the variation of air volume fraction along the flow direction, with the location of the interface marked by the descending region. The sustained motion of gas behind the shock wave would result in the movement of particles. Combined with Fig. 3, it was observed that the peak of particle velocity was approximately near the interface. Based on these results, we propose that the motion of the interface and the development of the nonlinear structural properties lead to the enhancement of the particle velocity.

3.2 After Re-Shock

The process in which the shock wave reflects from the end of the shock tube and the interface undergoes a second impingement is called re-shock RMI, during which the instability of the interface rapidly progresses to a turbulent mixing state. Figure 5 shows the particle velocity distribution at times $t = 3.2$ ms, 4.4 ms, 5.8 ms, 7.2 ms after the re-shock. The particles are gradually decelerated by the coupling effect of the reverse shock wave and interface movement. As

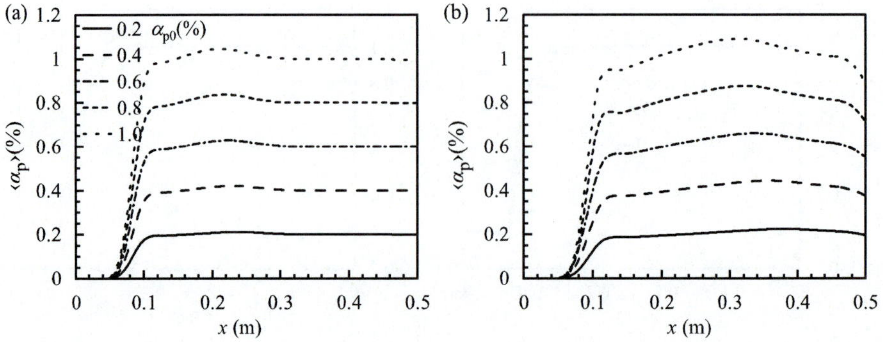

Fig. 7. The distribution of the particle volume fraction α_p along the flow direction measured at (a) $t = 1$ ms and (b) $t = 2.1$ ms before re-shock.

the time t increases, the entire velocity distribution curve shifts downwards. Although there is no significant change in the distribution law of $\langle u_{p,1}\rangle$ after the re-shock, the peak position becomes lower for increasing values of α_{p0}. One minor difference compared with prior-reshock is that the initial position of L_p continuously shifts to the right after the re-shock, which can be imagined as the movement of particles towards the end of the shock tube due to the persistent forward velocity along the x direction. Nevertheless, we did not observe any dependency of L_p on the pressure distribution after the re-shock (not displayed). We observe a dependence of the peak velocity on the interface position both prior to and after the re-shock. By comparing $\langle \beta_f\rangle$ at various time instants (Fig. 6) with the velocity distribution (Fig. 5), we find that the position of the interface does not change significantly after the re-shock and is roughly located at $x = 0.4m$ for presented cases, where the peak of $\langle u_{p,1}\rangle$ for each case also occurred. From the momentum equation, the pressure gradient force behind the shock wave is small, and particles mainly change their motion state due to gas drag. Although the velocity of the interface is decreasing, the mixing zone formed by the light and heavy gases may continue to develop, allowing particles to obtain a higher velocity than their surrounding gas. In addition, the compression of particles due to collisions may increase the collision frequency and keep the velocity at a higher level.

4 Volume-Fraction Distribution

The initial state of particles is characterized by a uniform distribution, which evolves into a non-uniform distribution as particles undergo acceleration and motion. To study the spatiotemporal evolution of particle distribution, we examine the particle volume fraction α_p.

4.1 Before Re-Shock

We obtained the one-dimensional distribution $\langle\alpha_p\rangle(x,t)$ of particle volume fraction by averaging α_p in the y and z directions. Figure 7 displays $\langle\alpha_p\rangle$ before the

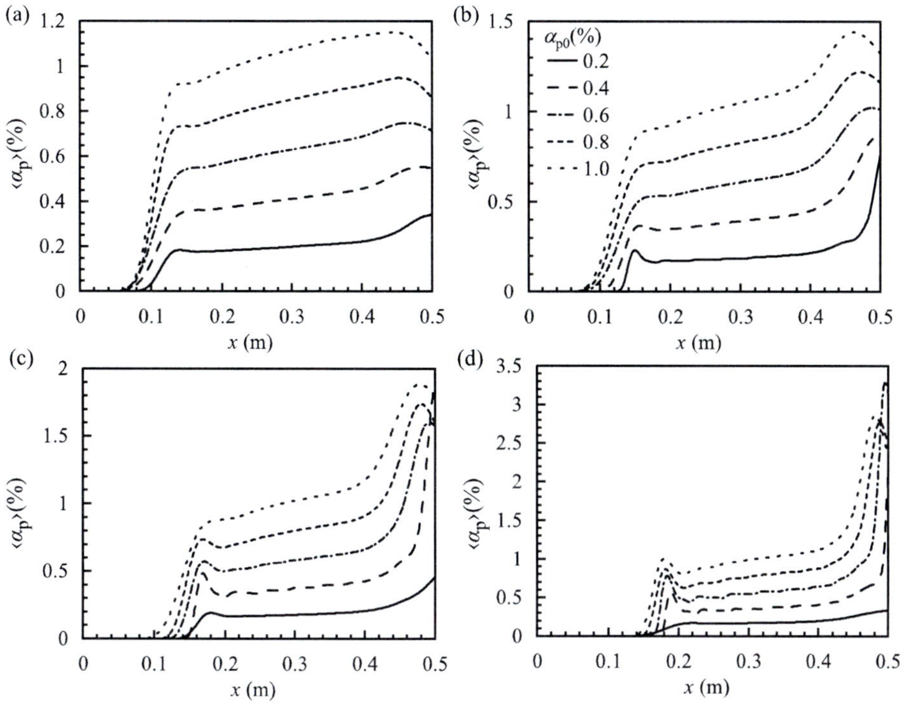

Fig. 8. The distribution of the particle volume fraction α_p, along the flow direction measured at (a) t = 3.2 ms,(b) t = 4.4 ms, (c) t = 5.8 ms, and (d) t = 7.2 ms after re-shock.

re-shock at times t = 1.0 ms, 2.1 ms, identical to the time selections for velocity distribution. Figure 7(a) displays a slight deviation of $\langle \alpha_p \rangle$ from the initial α_{p0}. The movement of particles towards the longitudinal direction causes the $\langle \alpha_p \rangle$ to be slightly lower than α_{p0} at the at the starting of L_p. Meanwhile, a minor rise is observed in the central region of L_p. Subsequently, the non-uniformity depicted in Fig. 7(b) becomes more prominent. $\langle \alpha_p \rangle$ displays low values at both ends of L_p and higher values in the central region. Additionally, a stronger non-uniformity in particle distribution is observed for cases yielding higher initial α_{p0}.

4.2 After Re-Shock

After the re-shock, particles are subjected to a complex wave impact, coupled with the non-linearity and even turbulence of the interface. Therefore, their distribution cannot be simply obtained based on experience. We observed the $\langle \alpha_p \rangle$ of particles after the re-shock and present the results in Fig. 8.

Figure 8(a) demonstrates that, at time t =3.2 ms, $\langle \alpha_p \rangle$ gradually increases along the flow direction and peaks where particles are most concentrated. Even cases with comparatively lower α_{p0}, such as 0.2%, show the continuous movement of particles in the positive direction of x, resulting in the peak value of $\langle \alpha_p \rangle$ being achieved at the end of the shock tube. Cases where $\alpha_{p0} \geq 0.6\%$ exhibit comparatively slower particle velocities, signifying that the peak has not yet reached the end of the shock tube. In Fig. 8(b), we observe that at time t =4.4 ms, two cases with initial particle volume fractions of 0.2% and 0.4%, respectively, display particle concentration near the reflective wall. In the case with $\alpha_{p0} =$ 0.2%, we observe an increase in $\langle \alpha_p \rangle$ near the wall surface, approximately to 0.66%, which is more than triple the initial value. We also observe, in Fig. 8(c), that high initial α_{p0} cases exhibit a gradual occurrence of particle accumulation near the wall surface by time t =5.8 ms. However, $\langle \alpha_p \rangle$ near the wall decreases for the case of $\alpha_{p0} = 0.2\%$, indicating lower accumulation. At time t =7.2 ms (as depicted in Fig. 8(d)), the volume fraction of particles for high α_{p0} cases exhibits a distinct peak at the reflection wall. The magnitude of the peak also increases with increasing α_{p0}. The concentration process is comparatively robust for high α_{p0} cases regardless of the re-shock.

5 Conclusions

In this study, we employed a planar shock tube to perform numerical simulations of re-shocked Richtmyer-Meshkov instability flow and analyze the velocity and volume fraction distributions of particles at different initial volume fractions. Our simulations focused on the range of moderate dense particle, and we utilized the CMP-PIC method owing to its versatility in handling multi-component gas mixtures and shock waves, while also being applicable across a wide range of particle volume fractions.

The distribution of particle velocity reveals several notable patterns. Particles with zero velocity are initially impacted by the shock wave and start moving. Specifically, particles' velocity exhibits a peak at the interface, with the peak value decreasing with an increase in α_{p0}. Based on our analysis, the appearance of the velocity peak before re-shock can be attributed to the particle response to the interface motion. Following re-shock, the interface experiences a decrease in speed, and the peak particle velocity value can be maintained in the vicinity of the interface. We attribute this observation to the enhanced particle motion resulting from the onset of turbulent mixing within the interface, as well as an increased collision rate caused by the continuous movement of particles in the flow direction. The crucial finding concerning the particle volume fraction distribution is that a higher value of α_{p0} leads to a more concentration effect, as particles tend to accumulate towards the end of the shock tube. Our study reveals that both shock wave impacts and instability developments tend to propel particles in the positive x direction, ultimately resulting in their compression against the reflected wall.

It should be noted that particle behavior in complex systems involving shock waves, multi-component gases, and perturbed interfaces is intricate. This study only analyzes the movement and aggregation behavior of heavy particles with a moderate volume fraction. Future studies involving statistical analyses of various types of initial conditions, as well as experimental guidance, will be essential.

Acknowledgements. The authors would like to thank the National Natural Science Foundation of China (NSFC) (nos. 11988102 and 12002063) and National Key Project (nos. GJXM92579) for financial support in carrying out the numerical simulations.

References

1. Ukai, S., Balakrishnan, K., Menon, S.: On Richtmyer-Meshkov instability in dilute gas-particle mixtures. Phys. Fluids **22**(10), 104103 (2010). https://doi.org/10.1063/1.3507318
2. Tian, B., Zeng, J., Meng, B., Chen, Q., Guo, X., Xue, K.: Compressible multiphase particle-in-cell method (CMP-PIC) for full pattern flows of gas-particle system. J. Comput. Phys. **418**, 109602 (2020). https://doi.org/10.1016/j.jcp.2020.109602
3. Theofanous, T.G., Mitkin, V., Chang, C.-H.: Shock dispersal of dilute particle clouds. J. Fluid Mech. **841**, 732–745 (2018). https://doi.org/10.1017/jfm.2018.110
4. Ouellet, F., Rollin, B., Koneru, R.B., Garno, J., Balachandar, S.: Effects of perturbing the particle volume fraction distribution in blast-driven multiphase instability. Shock Waves **31**, 337–360 (2021). https://doi.org/10.1007/s00193-021-01023-9
5. DeMauro, E.P., Wagner, J.L., DeChant, L.J., Beresh, S.J., Turpin, A.M.: Improved scaling laws for the shock-induced dispersal of a dense particle curtain. J. Fluid Mech. **876**, 881–895 (2019). https://doi.org/10.1017/jfm.2019.550
6. Meng, B., Zeng, J., Tian, B., Li, L., He, Z., Guo, X.: Modeling and verification of the Richtmyer-Meshkov instability linear growth rate of the dense gas-particle flow. Phys. Fluids **31**(7), 074102 (2019). https://doi.org/10.1063/1.5099996
7. Schulz, J.C., Gottiparthi, K.C., Menon, S.: Richtmyer-Meshkov instability in dilute gas-particle mixtures with re-shock. Phys. Fluids. **25**, 114105 (2013). https://doi.org/10.1063/1.4829761
8. Crowe, C.T., Schwarzkopf, J.D., Sommerfeld, M., Tsuji, Y.: Multiphase Flows with Droplets and Particles, 2nd edn. CRC Press, Boca Raton (2011). https://doi.org/10.1201/b11103
9. Apte, S. V., Mahesh, K., Lundgren, T.: A Eulerian-Lagrangian model to simulate two-phase/particulate flows. Center for Turbulence Research Annual Research Briefs 2003 (2003). http://refhub.elsevier.com/S0021-9991(20)30376-4/bib9A1158154DFA42CADDBD0694A4E9BDC8s1
10. Sutherland, W.: The viscosity of gases and molecular force. Lond. Edinb. Dubl. Phil. Mag. **36**(223), 507–531 (1893). https://doi.org/10.1080/14786449308620508
11. Tritschler, V.K., Olson, B.J., Lele, S.K., Hickel, S., Hu, X.Y., Adams, N.A.: On the Richtmyer-Meshkov instability evolving from a deterministic multimode planar interface. J. Fluid Mech. **755**, 429–462 (2014). https://doi.org/10.1017/jfm.2014.436

Author Index

X. Zheng and S. Balachandar (Eds.): IUTAM 2023, IUTAM Bookseries 41, p. 341, 2024.
https://doi.org/10.1007/978-3-031-47258-9